The Physics of Living Processes

The Physics of Living Processes

A Mesoscopic Approach

THOMAS ANDREW WAIGH

School of Physics and Astronomy and Photon Science Institute,
University of Manchester, UK

WILEY

This edition first published 2014
© 2014 John Wiley & Sons, Ltd.

Registered Office

John Wiley & Sons, Ltd, The Atrium, Southern Gate, Chichester, West Sussex, PO19 8SQ, United Kingdom

For details of our global editorial offices, for customer services and for information about how to apply for permission to reuse the copyright material in this book please see our website at www.wiley.com.

Library of Congress Cataloging-in-Publication data applied for.

A catalogue record for this book is available from the British Library.

ISBN: 9781118449943

Set in 10/12pt Times by SPi Publisher Services, Pondicherry, India

Printed and bound in Singapore by C.O.S. Printers Pte Ltd

1   2014

Contents

Preface

This book is based around a two-semester course on biological physics that has been taught over the past few years at the University of Manchester. Students on the course are predominantly physics undergraduates who have good mathematics and physics foundations, but often lack any advanced biology or chemistry background. Therefore in a bid to make the course self-contained we teach the students the necessary biology, chemistry and physical chemistry as the course goes along. It is therefore hoped that anyone with a reasonably good high/secondary school background in both physics and mathematics can follow the material in this introductory course on the physics of living processes.

The book is divided into five principal sections: *building blocks*, *soft matter*, *experimental techniques*, *systems biology* and *spikes, brains and the senses*. The first section describes the basic building blocks used to construct living organisms. The next four sections introduce a series of useful tools for biological physicists to solve problems in the life sciences. This list of tools is not exhaustive, but it is hoped that they will introduce the reader to some of the possibilities.

The subheading of the book: 'A Mesoscopic Approach' is there to emphasize the range of length scales considered in the solution of biological problems. Here the mesoscale is taken to include the length scales from molecules to the microscale. Thus, for the most part the underlying quantum mechanical details are ignored, i.e. the approach is coarse grained and on the nano/micrometre scale. The quantised details of the molecules in biological problems are neglected in order to make them tractable, i.e. to provide an approximate solution in real time. Again this is a pragmatic approach due to time constraints and the unwieldy nature of current state-of-the-art *ab initio* quantum mechanics simulations. Some fascinating biologically relevant processes certainly do depend sensitively on quantum mechanical details for a satisfactory explanation. Photosynthesis and photodetection are classic examples, but for brevity these areas will be predominantly ignored and readers are directed to additional literature for self-study.

In the first three sections of the book the field of molecular biophysics will be introduced. The presentation will focus on the simple underlying concepts and demonstrate them using a series of up-to-date applications. It is hoped that the approach will appeal to physical scientists who are confronted with biological questions for the first time due to the current biotechnological revolution.

The fields of biochemistry and cellular physiology are vast and it is not the aim of the current textbook to encompass the whole area. The first three sections of the book functions on a reductionist, nuts and bolts approach to the subject matter. They aim to explain the constructions and machinery of biological molecules very much as a civil engineer would examine the construction of a building or a mechanical engineer examine the dynamics of a turbine. Little recourse is taken to the specific chemical details of the subject, since these important areas are better treated in other dedicated biochemistry courses. Instead, modern physical ideas are introduced to explain aspects of the phenomena that are confronted. These ideas provide an alternative complementary set of tools to solve biophysical problems. It is thus hoped that the book will equip the reader with these new tools to approach the subject of biological physics.

The reductionist approach to molecular biology has historically been very successful. Organisms, drugs and foods can all be studied in terms of their constituent molecules. However, knowing all the musicians in an orchestra still leaves us none the wiser with respect to the music they are playing as an ensemble. The concerted motion and interaction of many thousands of different types of molecules gives rise to life. The pulsating quivering motion of amoeba cells under a microscope, jet lag in bumble bees when they are moved between continents, the kaleidoscope of colours detected in the ape retina when we view a Van Gogh painting, the searing pain when we burn ourselves, and the miraculous healing processes when we have been cut; all these phenomena and many more require explanation. Often such studies have been labelled physiology and in the modern era predominantly have been studied by medics and vets, who often function in an engineering, the-organism-is-broken-how-do-we-fix-it, -type approach. Pharmacologists also study the effect of drugs on the metabolism of creatures, generally concentrating their interest in the development of new medicines, but again practically this takes a pragmatic engineering approach. What are the large-scale effects of the addition of a single chemical on the millions of biochemical processes upon which it could possibly have an impact? People do the experiments and hope they will discover a well-targeted drug (a magic bullet if you will), which will modify only a single faulty mechanism associated with a particular disease with limited side effects. Necessity (time constraints) requires that they ignore most of the holistic effects required for a complete understanding of a disease.

However, increasingly a quantitative understanding of the phenomena involved in living processes is required, returning physiology to the domain of the physical sciences[*]. This would allow us to make a rigorous connection between the structure and dynamics of biological molecules on the nanoscale and their concerted behaviour in living processes over time at larger lengths scales (at the scale of organs and organisms). At first sight this seems an impossible task due to the complexity of the phenomena involved (a huge intractable many-body problem), but a range of new tools are available in the twenty-first century to explore physiology that give us some hope; high-resolution noninvasive *experimental probes* now exist that will not disrupt or burn the specimens (magnetic resonance imaging, optical coherence tomography, ultrasound, fluorescence micros-copy, positron emission tomography, and X-ray tomography, to name just a few); *postgenomic technology* (the human DNA genome has been sequenced, how do we make sense out of all this information?) holds a vast amount of information relevant to living processes that still needs to be properly mined; *network theory* provides mathematical tools to describe the geometry and connectivity of interacting components be they neurons, metabolic processes or individual biochemical products; *soft condensed-matter physics* demonstrates how the tools of conventional physics can be applied to the unusual behaviour of biological soft matter (e.g. statistical mechanics, fluid mechanics, elasticity theory, and novel model biomimetic materials); *systems biology* explores the robustness and diversity of biochemical processes in terms of the circuit diagrams of individual biochemical reactions; and *synthetic biology* allows cells to be completely reprogrammed to test the fundamental requirements for life. All these methods offer new approaches to solve physiological problems.

The discussion in this book is extended on from that previously presented in the textbook 'Applied Biophysics', also written by the current author, and it incorporates some of the same material. Applied Biophysics considered the application of soft-matter physics in molecular biology. The approach is now extended to the study of living processes with additional emphasis on modern deterministic themes concerning the behaviour of cells, action potentials and networks. These ideas and tools are applied to a series of problems in agriculture, medicine and pharmaceutical science. Many of these areas are currently being revolutionised and it is hoped that the text will provide a flavour of the fields that are being developed with relation to their biological physics and provide a bridge towards the relevant research literature.

The connections between the different themes discussed in the book need to be stressed, since they join together many of the different chapters e.g. signalling from **Chapter 23** and motility from **Chapter 7**. This integration of themes lends strength to quantitative descriptions of physiology in the last two sections of the book and highlights many important facets of a physiological question, e.g. how is insulin metabolised;

---

[*] The word *physiology* derives from the Greek word *Physis* (nature, origin) and *logos* (knowledge), so we see we have come full circle with the near tautology that is the physics of physiology.

what are the biochemical circuits? How is the motion of a mouse tracked by an owl; what is the activity of its neural networks? And how do the contractions of the heart give rise to the fluid mechanics of a pulse; what is the electrophysiology of the heart?

Much can be learnt from cells for would-be nanobiotechnologists. For example, currently we can make synthetic nanomotors, but switching them on or off when required is an ongoing challenge. Nature achieves this task with great speed and efficiency using virtuoso performances of ion channels and action potentials inside nerve cells. Physiology thus provides a treasure trove of processes that could be borrowed for synthetic biological designs.

Ethical questions are a concern when considering experiments with live organisms. These questions are worthy of careful thought and their solution requires a consensus amongst a broad community. The ethical consideration of such complex subjects as stem-cell research, cloning, genetic engineering and live animal studies should be considered hand in hand with the clear benefits they have for society [1]. Specifically it is a shock to an experimental physicist when samples arrive for experiments that are associated with an actual person: a sputum sample from individual X or a blood sample from individual Y. Thus, for ethical reasons it should be emphasised that, although experimental medical biological research can be fascinating, it needs real support from health-care professionals and should not be attempted independently (not to mention the legal implications)!

A few rudimentary aspects of medical molecular biophysics will be considered in **Parts I** and **II**. In terms of the statistics of the cause of death, heart disease, cancer and Alzheimer's disease are some of the biggest issues that confront modern society. An introduction will thus be made to the action of striated muscle (heart disease), DNA delivery for gene therapy (cancer is a genetic disease) and amyloid diseases (Alzheimer's). These diseases are major areas of medical research, and combined with food (agrochemical) and pharmaceutics provide the major industrial motivation encouraging the development of molecular biophysics. Physiological aspects of some of these industrial biotechnological challenges are also briefly examined towards the end of the text.

Please try to read some of the books highlighted at the end of each section, they will prove invaluable to bridge the gap between undergraduate studies and active areas of research science.

Further Reading

Reiss, M.J. & Straughan, R. (2001) *Improving Nature: The Science and Ethics of Genetic Engineering*, Cambridge University Press.

 Readers can access PowerPoint slides of all figures at http://booksupport.wiley.com

Acknowledgements

I would like to thank the scientists and colleagues who have been an inspiration to me: Dame Athene Donald, Sir Sam Edwards, Eugene Terentjev, Wilson Poon, Peter Pusey, Mike Evans, Jian Lu, Henggui Zhang, David Quere, Claudine Williams, Pierre Giles de Gennes, Ian Ward, Ralph Colby, Rama Bansil, Andy Murray, Mark Dickinson, Tom Mullin, Helen Gleeson, Ingo Dierking, Emanuela di Cola, Ben Cowsil, Paul Coffey, Aris Papagiannopoulos, Stive Pregent, Sanjay Kharche, Mike Gidley, Pei-Hua, Tanniemola Liverpool, Peter Higgs, Andrew Turberfield, Stuart Clarke, Adrian Rennie, Richard Jones, Pantelis Georgiadis, Manlio Tassieri, Alex Malm, James Sanders, Dave Thornton, Viki Allan, Phil Woodman, Tim Hardingham, Alex Horsley, Shaden Jaradat, Matt Harvey, Gleb Yakubov, Paul Pudney, Andrew Harrison, David Kenwright, Peter Winlove, Mark Leake, John Trinick, Amalia Aggeli, and Neville Boden. Prof. Wouter den Otter should be thanked for corrections to 'Applied Biophysics' that were carried through to the current text.

I would also like to thank my family Sally, Roger, Cathy, Paul, Bronwyn, Oliver, Christina and last, but not least, my daughter Emily. The majority of this book was written in the physics department of the University of Manchester. The students and staff members who have weathered the ongoing teaching experiments should be commended. I am indebted to the staff at the University of Edinburgh, the University of Cambridge, the College de France and the University of Leeds for helping to educate me concerning the behaviour of soft condensed-matter and biological physics. The Institute of Physics should also be thanked for their support of the biological physics community in the UK through the creation of a dedicated biological physics group. The final stages of the preparation of the book's manuscript were completed at the Kavli Institute for Theoretical Physics in Santa Barbara, USA. The Institute's organizers should be thanked for making the author's sabbatical visit such a pleasant experience.

Part I

Building Blocks

Every life form discovered so far on the planet Earth follows the same blueprint. The genetic information is carried by nucleic acids that code for a small number of families of other carbon-based molecules. The actual number of possible molecules within these families is huge (in principle infinite due to polymerisation), but many of the general features of the molecules can be deduced by comparison with other similar molecules contained within the same family e.g. the classifications of *proteins*, *lipids*, *nucleic acids* or *carbohydrates*. Furthermore living organisms are constructed from cells that conform to two basic blue prints; *prokaryotes* and *eukaryotes*. The basic nanomachinery is held in common between all members of either of these two cellular forms, although specialisations are possible (human cells come in over 200 varieties). Thus, if you are presented with a new unknown organism on the planet Earth you can rapidly make some educated guesses about how it works, how its cells function and what molecules it is made of. Provided, that is, you have been given a little appropriate training.

A study of the building blocks in cells is also very satisfying with respect to the nanomachinery that life has on offer. Thousands of tiny molecular machines are in action in every cell, elegantly optimised by evolution. There is thus much to be learnt for people building new devices on the nanoscale through the inspection of naturally occurring cellular machinery.

It is hoped that a strong case has been made that time spent learning about cells and biomolecules is extremely well spent. Furthermore, much of the most useful generic information can be acquired relatively quickly, although complete mastery of even a small area of biochemistry can require a lifetime's work. This section of the course is normally greeted with a little scepticism by physics undergraduates, since it is more descriptive in style than they are familiar with, and it is the most biologically and chemically demanding section of the course. However it is surprising the mileage that results from learning a small number of key facts, and it is of course impossible to discuss any biological problems whatsoever without naming the key players; *molecules* and *cells*.

Suggested Reading

Goodsell, D.S. (2010) *The Machinery of Life*, Springer. A simple discursive introduction to biochemistry with some attractive illustrations. The book includes a gallery of pen and ink drawings of different biomolecules. I particularly liked the perspective of the congested environment inside cells.

Goodsell, D.S. (2004) *Bionanotechnology: Lessons from Nature*, Wiley-Blackwell. Another excellent book from Goodsell, similar in style to *The Machinery of Life*.

1

Molecules

It is impossible to pack a complete biochemistry course into a single introductory chapter. Some of the basic properties of the structure of simple biological macromolecules will therefore be covered. The aim of this chapter is to give the reader a basic grounding in the rich variety of molecules that life presents and some respect for the extreme complexity of the chemistry of biological molecules in a wide range of cellular processes.

Cells are predominantly composed of water that is structured and organised by inorganic ions and carbon containing (organic) molecules. The extracellular matrix in organisms can, in addition, contain solid crystalline composites such as calcium carbonate and silicates that form bones and exoskeletons. However most of the processes vital for life occur in aqueous solutions, although they are typically highly congested, with a huge variety of competing molecular nanoparticles present. How robust well-regulated living processes occur in such congested environments is still a matter of ongoing research.

There are four main classes of organic macromolecules inside cells: these are the *lipids*, *proteins*, *carbohydrates* and *nucleic acids*. Also, mixtures are possible such as *glycolipids* (carbohydrates fused to lipids) and *glycoproteins* (carbohydrates fused to proteins).

The subject of the molecular structure of materials will be first approached and how this arises from the underlying quantum mechanics. Then, the concept of chirality will be introduced for molecules, cells and organisms. Finally, a rapid tour will be made of the main classes of biological molecules that occur in living cells.

1.1 Chemical Bonds and Molecular Interactions

Ernest Rutherford described the atom as a minute nucleus surrounded by a huge expansive cloud of electrons; 'a mosquito in a cathedral'. Niels Bohr extended this picture, since only certain energy states are permitted by the quantum principle of Max Planck (1900) and de Broglie's wave/particle duality. Electrons are confined to discrete shells as they orbit the nucleus (**Figure 1.1**). Subsequently, Erwin Schrödinger showed how to calculate the energies and the spatial distributions of these electronic orbitals.

The Physics of Living Processes: A Mesoscopic Approach, First Edition. Thomas Andrew Waigh.
© 2014 John Wiley & Sons, Ltd. Published 2014 by John Wiley & Sons, Ltd.

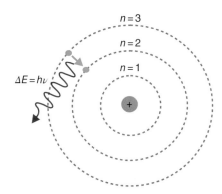

Figure 1.1   *In the Bohr model of an atom electrons move around the heavy nucleus, at a characteristic distance ~0.1 nm. The nucleus acts as a 'mosquito' in the huge empty volume of the electronic cathedral. Transitions between different energy levels (principal quantum number, n) can be detected by the emission or absorption of quantised photons (energy $\Delta E = h\nu$, h is Planck's constant and ν is the frequency.)*

Exact calculations of electron wave functions that use the Schrödinger equation will be left to more specialised quantum mechanics courses, as too will be more accurate quantum electrodynamics calculations, which are currently the most accurate theories to simulate the behaviour of atoms and molecules. Both approaches with complex biomolecules tend to be horrendously difficult and require extensive computational power. However, a few more details of quantum mechanics are useful for the development of an intuitive picture of molecular structure and dynamics, and will thus be introduced here.

Max Born postulated a radical reinterpretation of the quantum theory; different quantum states of electron distributions in the vicinity of an atomic nucleus are characterised by different probability distributions. These probability distributions are the fundamental quantities predicted by theory, and intrinsic uncertainties in their values are written in to the laws of nature (*Heisenberg's uncertainty principle*). The Schrödinger equation predicts the probability that an electron is found at a certain position. The full information about a bound electron inside an atom is now thought to be described by four quantum numbers; n (the principle quantum number related to the energy), l (the total angular momentum), m (the z component of the angular momentum) and s (the angular spin momentum). Based on these four quantum numbers, combined with the symmetry of the electrons' wave functions, Pauli deduced his exclusion principle, which is that 'no two electrons can be associated with the same nucleus and have precisely the same values of all four of the quantum numbers'. This principle stops matter collapsing in on itself, since electrons repel each other to form shells such that each electron has a different permutation of the four quantum numbers. This stability is reflected in the repulsive hard-core interactions experienced by all atoms at short distances, i.e. the excluded volume potential.

The quantum theory also explains the formation of molecules. Neighbouring atoms share or transfer electrons to create more energetically stable quantised electronic orbital structures. The geometries of atomic orbitals are classified as s (spherical), p (double lobed), d (double lobe threading a doughnut) and f (double lobe threading two doughnuts), **Figure 1.2**. Molecular orbitals tend to be even more complicated, since they require hybridisation of neighbouring atomic orbitals.

The prediction of chemical bonding patterns tends to be a job for the intuition of a good synthetic chemist with many years of experience, or an extremely hard *ab initio* quantum mechanical calculation for a computer. Chemical bonding can be classified under the broad (and often overlapping) headings of *ionic*, *covalent*, *metallic* and *hydrogen*.

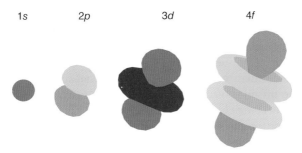

Figure 1.2   *Schematic diagram of some varieties of probability distribution geometries found in the electronic orbitals of atoms. 1s spherical, 2p double lobed, 3d double lobed threading a doughnut, and 4f double lobed threading two doughnuts.*

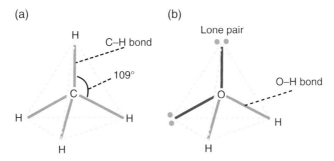

Figure 1.3   *Schematic diagram that shows a comparison of the molecular bonding geometries. (a) Methane (CH_4) is symmetrical with a constant 109° angle between C–H bonds. (b) Water is asymmetrical with a 104.5° between the H–O–H bonds.*

Ionic bonding occurs when atoms ionise to form electrolytes with the classic example being table salt, NaCl. An ionisation energy is associated with the movement of an electron from one atom to another, so their electronic structures are more energetically favourable.

When the difference in electronegativity is small between two atoms, they may form molecules through *covalent bonding*, e.g. H_2, HF, H_2O or *metallic bonding*, e.g. Na, K, Fe. Covalent bonding requires electrons to be shared and they are closely localised between the two bonded atoms. Metallic bonding involves delocalised electronic wave functions that allow rapid mobility of electrons through the crystalline lattices.

The development of a good intuition for the relationship between molecular structures and chemical formulae requires years of experience, but it is instructive to ask a simple question on tetrahedral bonding to start to build some understanding. It is useful to ask why the bonding pattern in methane (CH_4) is symmetrical while water (H_2O) is not (**Figure 1.3**). The answer is found in the quantum mechanics, because carbon has two fewer electrons than oxygen. In water $2s$ and $2p$ orbitals of oxygen hybridise (four orbitals in total); two electrons bond with two Hs, and the other two electrons become lone pairs. The lone pairs are negatively charged and attract the H atoms from neighbouring water molecules (*hydrogen bonding*). Water thus has a distorted tetrahedral structure due to the different interactions between the lone pairs and the hybridised orbitals. In contrast methane's carbon atom forms covalent bonds with four hydrogen atoms in a perfectly symmetrical tetrahedral structure.

Molecules assemble together to form different phases of matter (gas, liquid, solid, etc.), but retain their individual identities at the atomic scale. The exact phase adopted is determined by intermolecular forces between the molecules and they are weaker than the intramolecular forces already considered (ionic bonding, covalent bonding, etc.).

The most common intermolecular force is that of van der Waals. This can be thought of as a default interaction that occurs between all molecules and dominates the interactions if all the other forces are switched off, e.g. with liquid helium at low temperatures. In fact van der Waals forces correspond to a family of three or more types of force. *Debye, Keesom*, and *van der Waals* interactions are the principle subclassifications, but higher-order multipolar interactions also occur (**Chapter 4**). In general, van der Waals forces arise from instantaneous stochastic dipole moments associated with the motions of individual electrons. The instantaneous distribution of electrons can influence those of surrounding atoms, and the net outcome can be a weakly positive attraction. In biology, this force can explain the miraculous manner in which geckos can crawl up windows and flies can stick to ceilings. It is, however, not a universal mechanism of adhesion, e.g. tropical frogs use capillary forces to hold themselves onto surfaces, which in turn depends on the humidity.

Unlike *covalent bonding*, the van der Waals force does not have a unique direction, it cannot be saturated (the number of atoms or molecules involved is based on geometrical conditions, not on the electronic structure of the orbitals), it depends on the sample geometry and requires quantum electrodynamics (QED) calculations for a quantitative treatment. Although the derivations are complicated, analytic QED solutions do exist for standard geometries, e.g. see Adrian Parsegian's book, 'van der Waals forces in biology', for more details (also **Chapter 4**).

The importance of *hydrogen bonding* in biology cannot be more emphasised (**Figure 1.4, Section 1.7**). All living organisms contain a large amount of water, and water has its unique properties due to hydrogen bonding. For example, water expands when it freezes, which helped life's origins in the sea immeasurably. Water is also important for the self-assembly of many biomolecules. Indeed from a certain perspective, humans are predominantly structured self-assembled water. There is a nice analogy with a wobbly children's party jelly that is 98% water and 2% protein, but appears (at the low frequencies people are familiar with) to be a solid. The water component in human cells is slightly lower at ~70% w/w, but the jelly analogy helps the development

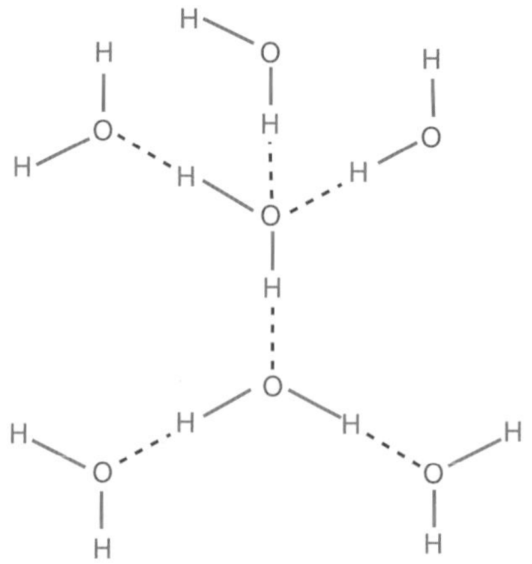

Figure 1.4 *Schematic diagram of the network structure formed by water molecules in liquid water. Dashed red lines indicate hydrogen bonds. Chains of hydrogen-bonded water molecules occur over a wide range of angles for liquid water, but become more restricted in ice crystals.*

of an intuition of the cell's varied dynamic properties at different time scales (its viscoelasticity). At long time scales cells appear to be solid, whereas at short time scales they are liquid-like.

Hydrogen bonding in water has a number of subtle secondary interactions that are important in the determination of its effects. One phenomenon is the hydrophobic effect that is primarily due to entropy. Consider a fat chain surrounded by water molecules. The normal hydrogen-bonded network in liquid water is disrupted and the water molecules become structured due to the loss of mobility as they orient themselves away from the fat chains. There is consequently a penalty in the free-energy term due to the loss of entropy ($-TS$ in $F = U - TS$, the Helmholtz free energy, **Section 3.4**). Entropy will be considered in more detail in **Chapter 3**. In a hand-waving manner, entropy is a measure of the randomness of a system, or equivalently the number of accessible states to a system. The hydrophobic effect is the molecular origin of why oil and water do not mix on the macroscale.

In general, all the forces between different biomolecules, subcellular compartments, cells, organisms and nanoscaled particles are of interest to biological physics. These interactions lead to relatively weak forces compared to chemical bonds, but help explain important physical and biological phenomena such as phase transitions and self-assembly. Such coarse-grained interactions over nanometer length scales are called mesoscopic forces, and will be considered in more detail in **Chapter 4**.

1.2   Chirality

Chirality is a symmetry operation in which molecules do not superpose with their mirror images (**Figure 1.5**), e.g. right-handed B DNA molecules (that exist naturally) do not superpose with left-handed B DNA molecules (an artificial construct). The occurrence of chirality can be determined for many organic molecules directly by inspection of their molecular structure. A carbon atom with four different groups attached to it can immediately be deduced to be chiral (**Figure 1.5c**) and this chirality could in turn affect the macroscopic behaviour of the material. All amino acids have chiral centres, except glycine, and so too do the nucleic acids, DNA and RNA, many carbohydrates and many lipids. This implies a fundamental molecular cause for the macroscopic chirality observed in biological materials, although spontaneous chiral symmetry breaking is also possible (in this case both chiralities occur on average in equal amounts).

Chiral interactions between molecules often dramatically perturb the phase of matter formed, e.g. the crystalline or liquid-crystalline phase adopted will become twisted with characteristic orientational defect structures (**Chapter 6**). Furthermore, chiral phases of matter often have unusual optical properties when they interact with photons (used commercially in liquid-crystalline displays with synthetic molecules). Cilial chirality in human embryos is thought to lead to the partial left/right asymmetry of the body parts in humans, e.g. the heart is on the left in most individuals.

1.3   Proteins

Polymers consist of a large number of identical subunits (monomers) connected together with covalent bonds. A protein is a special type of polymer; in a protein there are up to twenty different amino acids (**Figure 1.6**) that can function as monomers and all the monomers are connected together with identical peptide linkages (C–N bonds, **Figure 1.7**). Only twenty amino acids occur in nature that are used to create proteins in eukaryotic cells, although three additional protein forming amino acids occur in bacteria and over 140 nonprotein-forming amino acids are known. A particularly persuasive evolutionary argument of why only twenty amino acids exist in proteins in eukaryotic cells is still lacking. The best partial explanation is that twenty is enough, and reflects the common evolutionary origins of all life on the planet Earth. Synthetic chemists have made new amino acids

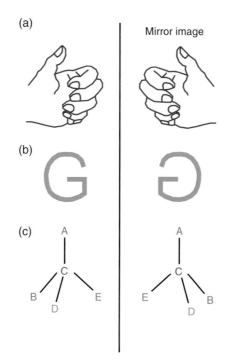

Figure 1.5 Chirality ('handedness') occurs when an object cannot be superposed with its mirror image. (a) A human hand and (c) a carbon molecule with four different substituents are chiral objects. (b) A flat letter 'G' is achiral in three dimensions, since it can be lifted up and overlayed with its mirror image. (c) Such chiral carbon atoms are found in many organic molecules and give rise to their macroscopic chiral phases, e.g. amino acids in the α helices of proteins can form twisted cholesteric liquid-crystalline phases. The letters A, B, D, and E denote arbitrary distinct chemical substituents, whereas C is a carbon atom.

and connected them together to form novel synthetic proteins, so there is in principle no clear chemical barrier that restricts the number of possibilities (which is explored in the field of synthetic biology), but natural life only uses twenty (or twenty three if bacterial life is included in the list).

The twenty eukaryotic protein forming amino acids can be placed in different families dependent on the chemistry of their different side groups. Five of the amino acids form a group with lipophilic (fat-liking) side chains: glycine, alanine, valine, leucine, and isoleucine. Proline is a unique circular amino acid that is given its own separate classification. There are three amino acids with aromatic side chains: phenylalanine, tryptophan, and tyrosine. Sulfur is in the side chains of two amino acids: cysteine and methionine. Two amino acids have hydroxyl (neutral) groups making them water loving: serine and threonine. Three amino acids all have very polar positively charged side chains: lysine, arginine and histidine. Two amino acids form a family with acidic negatively charged side groups and they are joined by two corresponding neutral counterparts that have a similar chemistry: aspartate, glutamate, asparagine, and glutamine. More generally, the protein forming amino acids can be separated into three principle families; *hydrophobic* (they hate water), *polar* (they like water) and *charged* (they like water and are charged when they are incorporated into proteins).

The linkages between amino acids all have the same chemistry and basic geometry (**Figure 1.7**), which greatly simplifies their classification on the atomic scale. The peptide linkage that connects all amino acids together consists of a carbon atom attached to a nitrogen atom through a single covalent bond. The condensation of two α-amino acids to form a dipeptide is shown in **Figure 1.8**. There is only one way that

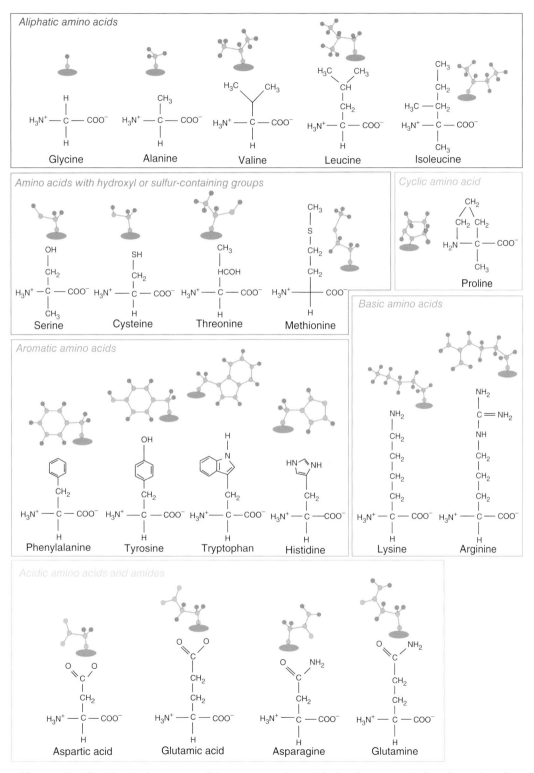

Figure 1.6 *The chemical structure of the twenty amino acids that form proteins in eukaryotic cells.*

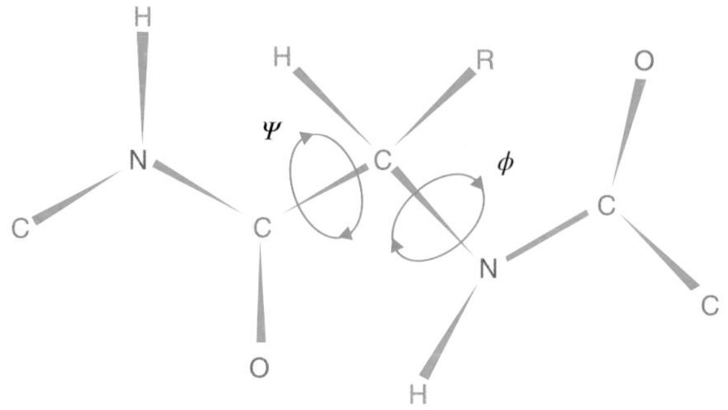

Figure 1.7   *All of the amino acids have the same primitive structure and are connected with the same peptide linkage through C–N bonds (O, N, C, and H indicate oxygen, nitrogen, carbon and hydrogen respectively). R is a pendant side group that provides the amino acid with its identity, i.e. proline, glycine, etc (**Figure 1.6**). Defining the rotational angles (Ψ, φ) for each amino acid gives a reasonably compact description of the peptide backbone conformation (it leads to Ramachrandran maps of protein conformation).*

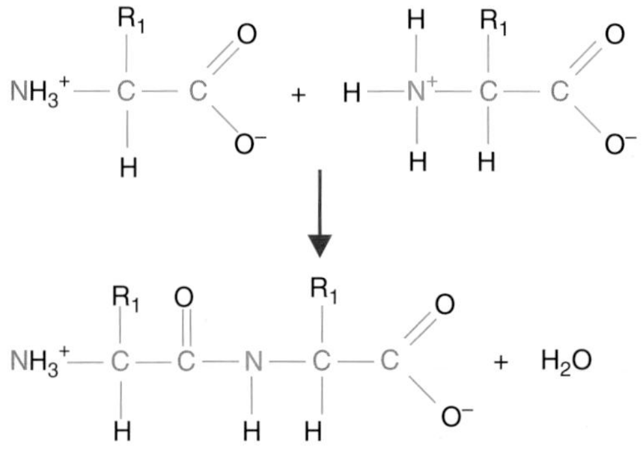

Figure 1.8   *A condensation reaction between two peptides can create a peptide linkage and a water molecule.*

two amino acids can be connected in proteins, the *peptide linkage*, and the chemical formula for four amino acids connected in line is

$$-CR_1HCONH-CR_2HCONH-CR_3HCONH-CR_4HCONH-$$

where the Rs are the groups that differentiate the 20 different amino acids and the hyphens are the peptide linkages. The peptide linkage has a directionality and in general -A-B- is usually different from -B-A-. Peptides are thus conventionally written with the N-ending peptide (the unbound amine terminus) on the left and the C-ending peptide (the unbound carboxyl terminus) on the right.

Although the chemistry of peptide linkages is fairly simple, to relate the primary sequence of amino acids (the combination of amino acids along the chain) to the resultant three-dimensional structure of the proteins is very complicated and predominantly remains an unsolved problem. To describe protein structure in more detail it is useful to describe motifs of secondary structure that occur in their morphology. The motifs include *alpha helices*, *beta sheets* and *beta barrels* (**Figure 1.9**). The first beta sheet and alpha helical structures were suggested by William Astbury during the 1930s. He also proposed that adjacent proteins were held together in hair and wool by hydrogen bonds. The first refined models of protein structure required more quantitative

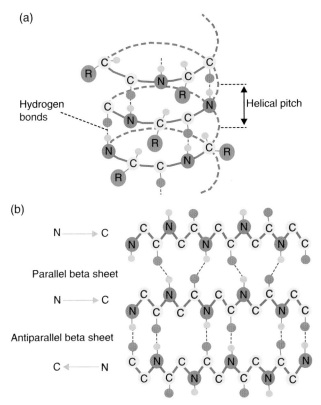

Figure 1.9   *Simplified secondary structures of (a) an α-helix and (b) a β-sheet that commonly occur in proteins (hydrogen bonds are indicated by dotted red lines). Beta sheets can be parallel or antiparallel. Adapted with permission from F. Pang, PhD Thesis, University of Manchester.*

analysis of X-ray diffraction patterns and the exact geometries of peptide bonds were calculated by Linus Pauling and Robert Corey in 1951.

Standard *alpha helices* in proteins are right-handed (amino acid carbon atoms are often chiral, since they are attached to four different groups, **Figure 1.5c**) and the pitch (repeat distance) is 3.6 amino acid groups (**Figure 1.9a**). The helical chirality can also be observed on the next larger length scale in a coiled coil structure, a helix of helices, although the pitch is less well defined on the helical superstructure that exists on the largest length scale. A standard molecular protein alpha helix is 0.54 nm in width. Each amino acid is displaced at ~100° to its predecessor along the chain and there is a small 0.15 nm displacement along the axis of the helix. The alpha helix structure is stabilised by hydrogen bonds between neighbouring amino acids (N–H groups interact with C=O groups). Typically an amino acid has hydrogen bonds that interact most strongly with its neighbour four steps along the chain.

Peptide chains can fold back on themselves with one strand adjacent to the next, in the form of *beta sheets*. In a similar manner to alpha helices, the structures are stabilised by hydrogen bonds between neighbouring amino acids, principally the N–H and C=O groups. The pleating of chains causes the distance between peptides to be ~0.6 nm along a single chain. The distance between adjacent chains is ~0.5 nm. The carbon atoms in beta sheet peptides are chiral, so the sheets have a chiral twist to them. Peptide chains have a directionality (N terminus versus the C terminus), which is indicated by an arrow on a diagram (**Figure 1.9b**). Adjacent beta sheet strands

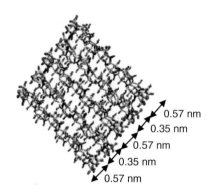

0.57 nm
0.35 nm
0.57 nm
0.35 nm
0.57 nm

Figure 1.10 *Packing of antiparallel beta sheets found in silk protein. Distances between the adjacent beta sheets are shown.*

can be arranged antiparallel, parallel or in combinations of the two. The antiparallel arrangement (alternating N terminus directions) is slightly more energetically stable than the other possibilities.

The full three-dimensional tertiary structure of a protein usually takes the form of compact globular morphologies (the globular proteins) or long extended conformations (fibrous proteins, **Figure 1.10**). Globular morphologies typically consist of a number of secondary motifs combined with more disordered regions of peptide. Quaternary globular structures are also possible where two or more whole protein chains are assembled side by side into a superstructure, e.g. haemoglobin is an assembly of four individual globular chains.

Charge interactions are very important for the determination of the conformation of biological polymers. The degree of charge on a polyacid or polybase (protein, nucleic acid, etc.) is determined by the pH of a solution, i.e. the concentration of hydrogen ions. Water has an ability to dissociate into oppositely charged ions, and this process depends on temperature,

$$H_2O \underset{\leftarrow}{\overset{\rightarrow}{}} H^+ + OH^- \tag{1.1}$$

The product of the hydrogen and hydroxide ion concentrations formed from the dissociation of water is a constant (K_w) at room temperature,

$$c_{H^+} c_{OH^-} = 1 \times 10^{-14} M^2 = K_w \tag{1.2}$$

where c_{H^+} and c_{OH^-} are the concentrations of hydrogen and hydroxide ions, respectively. Addition of acids and bases perturbs the equilibrium dissociation process of water and the acid/base equilibrium phenomena involved are a cornerstone of the physical chemistry of aqueous solutions. Due to the vast range of possible hydrogen ion (H^+) concentrations typically encountered in aqueous solutions, it is normal to use a logarithmic scale (pH) to quantify them. The pH is defined as the negative logarithm (base 10!) of the hydrogen ion concentration,

$$pH = -\log c_{H^+} \tag{1.3}$$

Typical values of pH range from 6.5–8 in physiological intracellular conditions. pHs in the range 1–2 are strongly acidic and pHs in the range 12–13 are strongly basic.

When an acid (HA) dissociates in solution it is possible to define an equilibrium constant (K_a) for the dissociation of its hydrogen ions (H^+),

$$HA \underset{\leftarrow}{\rightarrow} H^+ + A^- \qquad K_a = \frac{c_{H^+} c_{A^-}}{c_{HA}} \qquad (1.4)$$

where c_{H^+}, c_{A^-} and c_{HA} are the concentrations of hydrogen ion, acid ion and acid respectively. Since the hydrogen ion concentration follows a logarithmic scale it is natural to define the dissociation constant on a logarithmic scale (pK_a) as well,

$$pK_a = - \log K_a \qquad (1.5)$$

The logarithm of both sides of equation (1.4) can be taken to give a relationship between the pH and the pK_a value,

$$pH = pK_a + \log \left\{ \frac{c_{A^-}}{c_{HA}} \right\} \qquad (1.6)$$

where c_{A^-} and c_{HA} are the concentrations of the conjugate base (e.g. A^-) and acid (e.g. HA), respectively. This equation enables the degree of dissociation of an acid (or base) to be calculated and it is named after *Henderson and Hasselbalch*. Thus, from a knowledge of the pH of a solution and the pK_a value of an acid or basic group the charge fraction on a protein can be found to a first approximation. The propensity of the amino acids to dissociate in water is described in **Table 1.1**. In contradiction to what their name might imply, only amino acids with acidic or basic side groups are actually charged when incorporated into proteins. These charged amino acids are arginine, aspartic acid, cysteine, glutamic acid, histidine, lysine and tyrosine.

Another important interaction between amino acids is determined by the degree to which they are able to form hydrogen bonds with the surrounding water molecules. This amino acid hydrophobicity (the amount they

Table 1.1 *Fundamental physical properties of the twenty amino acids that form proteins in eukaryotic cells. [Ref. Data adapted from C.K. Matthews, K.E. van Holde, K.G. Ahem, Biochemistry, 3rd edition, Prentice Hall, 1999.]*

Name	pK$_a$ value of side chain	Mass of residue (Da)	Occurrence in natural proteins (% mol)
Alanine	Neutral	71	9.0
Arginine	12.5	156	4.7
Asparagine	Neutral	114	4.4
Aspartic acid	3.9	115	5.5
Cysteine	8.3	103	2.8
Glutamine	Neutral	128	3.9
Glutamic acid	4.2	129	6.2
Glycine	Neutral	57	7.5
Histidine	6.0	137	2.1
Isoleucine	Neutral	113	4.6
Leucine	Neutral	113	7.5
Lysine	10.0	128	7.0
Methionine	Neutral	131	1.7
Phenylalanine	Neutral	147	3.5
Proline	Neutral	97	4.6
Serine	Neutral	87	7.1
Threonine	Neutral	101	6.0
Tryptophan	Neutral	186	1.1
Tyrosine	10.1	163	3.5
Valine	Neutral	99	6.9

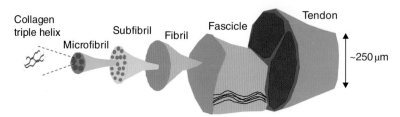

Figure 1.11 *Hierarchical structure for the collagen triple helices in found in mammalian tendons. Collagen triple helices are combined into microfibrils, then into subfibrils, fibrils, fasicles and finally into tendons.*

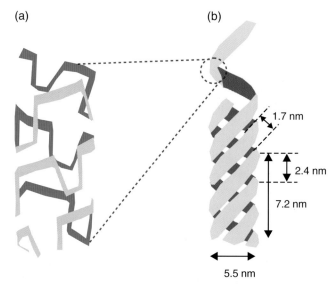

Figure 1.12 *The β turns in elastin form (a) a secondary elastic helix that is subsequently assembled into (b) a superhelical fibrous structure.*

dislike water) is an important driving force for the conformation of proteins. Crucially it leads to the compact conformation of globular proteins, as the hydrophobic groups are buried in the centre of the globule to avoid contact with the surrounding water.

Covalent interactions are possible between adjacent amino acids and can produce solid protein aggregates (**Figures 1.10** and **1.11**). For example, disulfide linkages are possible in proteins that contain cysteine and these disulfide bonds form the strong interprotein linkages that are found in fibrous proteins, e.g. keratins in hair.

The internal secondary structures of protein chains (α helices and β sheets) are stabilised by hydrogen bonds between adjacent atoms in peptide groups along the main chain. The important structural proteins such as keratins, collagens (**Figure 1.11**), silks (**Figure 1.10**), arthropod cuticle matrices, elastins (**Figure 1.12**), resilin and abductin are formed from a combination of both intermolecular disulfide and hydrogen bonds.

Some examples of the globular structures adopted by proteins are shown in **Figure 1.13**. Globular proteins can be denatured in a folding/unfolding transition. Typically the complete denaturation transition is a first-order thermodynamic phase change with an associated latent heat (thermal energy is absorbed during the transition). The unfolding process involves an extremely complex sequence of molecular origami transitions. There are a vast number of possible molecular configurations (2^{N-1} for an N residue protein, with

(a)

(b)

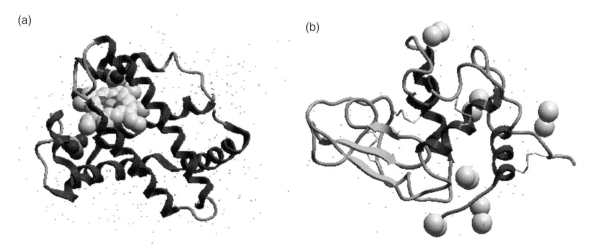

Figure 1.13 *Two structures of globular proteins calculated using X-ray crystallography data. (a) Myoglobin (an oxygen carrier in muscle), (b) lysozyme (an antibacterial enzyme found in tears).*

the very restrictive assumption that each peptide linkage only has two possible conformations; clearly a lower limit on the calculation) during the reverse process of protein folding, when the globular protein is constructed from its primary sequence by the cell, and thus frustration can be important. At first sight it appears a certainty that protein molecules will become trapped in an intermediate state and never reach their correctly folded form, since the process appears to require astronomically long time scales. This is called *Levinthal's paradox*, the process by which natural globular proteins manage to find their native state among the billions of possibilities in a finite time. The current explanation of protein folding is that there is a funnel of energy states that guides the kinetics of folding across the complex energy landscape that has a huge number of degrees of freedom (**Figure 1.14**).

There are two main types of interchain interaction between different proteins in solution; those in which the native state remains largely unperturbed in processes such as protein crystallisation and the formation of filaments in sheets and tapes, and those interactions that lead to a loss of conformation, e.g. heat set gels (e.g. table jelly, and boiled eggs) and amyloid fibres (e.g. Alzheimer's disease, and Bovine Spongiform Encephalopathy).

A wide range of protein structures with atomic resolution are known from X-ray crystallography experiments. This data and the associated three-dimensional visualisation software are freely available on the internet and are well worth a quick investigation (search online for the 'protein data bank').

1.4 Lipids

Cells are divided into a series of subsections and compartments by membranes that are formed predominantly from lipids (**Chapter 2**). The other main roles for lipids are as energy-storage compounds and for cell signalling in messenger molecules such as steroid hormones (oestrogen and testosterone).

Lipids are amphiphilic; the head groups like water (and hate fat) and the tails like fat (and hate water). This amphiphilicity drives the spontaneous self-assembly of the molecules into membranous morphologies.

There are four principle families of lipids, fatty acids with one or two tails (including carboxylic acids of the form RCOOH, where R is a long hydrocarbon chain), and steroids and phospholipids, where two fatty acids are

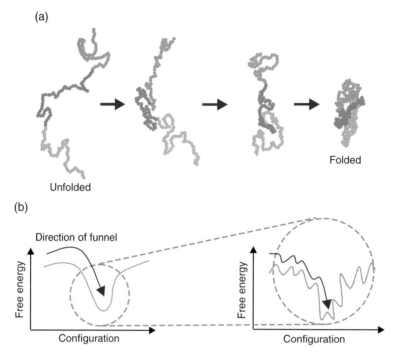

Figure 1.14 (a) Schematic diagram of a molecular dynamics simulation of the folding of a short polymer chain into a well-defined globular structure. Three different regions of the chain are colour coded to improve visibility. (b) Schematic diagram indicating the funnel that guides the process of protein folding through the complex configurational space that contains many local minima. The funnel avoids the frustrated misfolded protein structures described in Levinthal's paradox.

linked to a glycerol backbone (**Figure 1.15**). The type of polar head group differentiates the particular species of naturally occurring lipid. Cholesterol is a member of the steroid family and these compounds are often found in membrane structures. Glycolipids also occur in membranes and in these molecules the phosphate group on a phospholipid is replaced by a sugar residue. Glycolipids have important roles in cell signalling and the immune system. For example, these molecules are an important factor for the determination of blood cell compatibility during a blood transfusion.

Fatty acids are some of the simplest lipids and often consist of 16 to 18 carbon atoms arranged in a chain, e.g. oleic acid (a free fatty acid, found in olive oil). This chain can contain single or double covalent bonds and has nonpolar C–H bonds except for at the end group e.g. $CH_3(CH_2)_7CH=CH(CH_2)_7COOH$. Fatty acids are stored long term in the form of triacylglycerols.

Phospholipids are the main components of eukaryotic cell membranes. They consist of two fatty acids chains joined together with a polar head group and are thus amphiliphic. Examples include phosphatidylcholine (a pulmonary surfactant) and phosphatidylethanolamine (a constituent of the membranes of nervous tissue).

1.5 Nucleic Acids

The '*central dogma of biochemistry*' according to F.C. Crick is illustrated in **Figure 1.16a**. DNA contains the basic blueprint that guides the construction of the vast majority of living organisms. To implement this

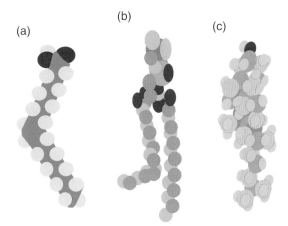

Figure 1.15 *Space-filling models of some lipid molecules typically encountered in biology (a) fatty acid with one tail (oleate), (b) phospholipid and (c) cholesterol.*

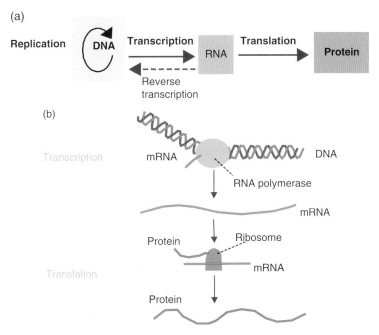

Figure 1.16 *(a) The* central dogma of molecular biology *considers the transcription and translation of DNA. DNA is* transcribed *to form a messenger RNA (mRNA) chain and this information is* translated *into a protein sequence. The dotted line indicates* reverse transcription *that can happen through the action of retroviruses, e.g. HIV or leukaemia virus. DNA is also duplicated from a DNA template (*replication*). (b) RNA polymerase is required for transcription, whereas a ribosome is used for translation.*

blueprint, cells need to *transcribe* the DNA to RNA and this structural information is subsequently *translated* into proteins that use specialised protein factories (the ribosomes). The resultant proteins can then be used to catalyse specific chemical reactions (enzymes) or be used as building materials to construct cells.

This simple biochemical scheme for the transfer of information has powerful implications. DNA can now be altered systematically using *recombinant DNA technology* and then placed inside a living cell, to hijack the cell's mechanisms for translation. The proteins that are subsequently formed can be tailor-made by the genetic engineer to fulfil a specific function, e.g. bacteria can be used to form fibrous proteins that are used to create biodegradable plastics.

Modern biochemistry experiments show that some minor corrections are required to the central dogma. Some viruses (retroviruses) are able to reverse the direction of information transfer in the transcription step (**Figure 1.16a**). RNA is translated into DNA that resides long term in the infected organism's genome. This is a reason for the high infection rates for lentiviruses such as HIV and leukemia virus. It does however imply an efficient mechanism for the genetic modification of organisms can be obtained using genetically modified lentiviruses, and this is now a standard technique in molecular biology.

The monomers of DNA are made of a sugar, an organic base, and a phosphate group (**Figure 1.17**). There are only four organic bases that naturally occur in DNA and these are *thymine, cytosine, adenine,* and *guanine* (*T, C, A,* and *G*). The sequence of bases in each strand in a double helix contains the genetic code. The base

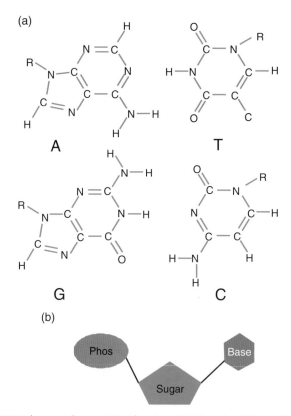

Figure 1.17 (a) The four DNA bases Adenine (A), Thymine (T), Guanine (G) and Cytosine (C) are constructed from carbon, nitrogen, oxygen and hydrogen atoms. (b) The generic chemical structure of the base of a nucleic acid consists of a phosphate group, a sugar and a base.

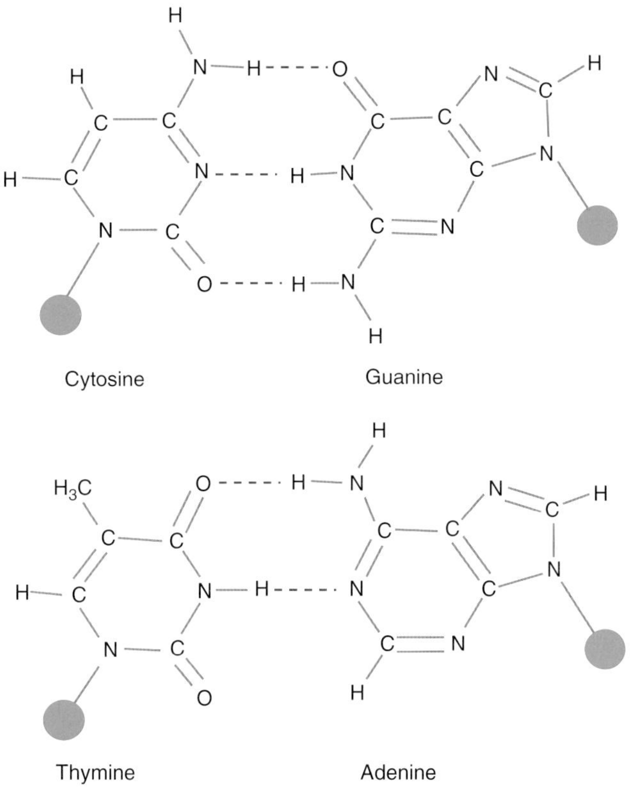

Figure 1.18 *Bases are carefully matched due to their geometry and the number of hydrogen bonds formed. Base pairing drives the formation of DNA helices and provides two copies of the same molecular information. Cytosine and guanine form three hydrogen bonds, whereas thymine and adenine form two hydrogen bonds.*

pairs in each strand of double-helical DNA are complementary, *A* has an affinity for *T* (they form three hydrogen bonds) and *G* for *C* (they form two hydrogen bonds) (**Figure 1.18**). The interaction between the base pairs is driven by the geometry of the hydrogen-bonding sites. Thus, each strand of the DNA helix contains an identical copy of the genetic information to its complementary strand and replication can occur by separation and resynthesise of two additional chains on each of the two original double-helical strands. Methylation of cytosine and adenine in DNA is common after replication and is thought to be an important factor in gene regulation.

There is a major groove and a minor groove on the biologically active A and B forms of the DNA double helix. The individual polynucleotide chains have a sense of direction, in addition to their individuality (a complex nucleotide sequence). DNA replication *in vivo* is conducted by a combination of the DNA polymerases (I, II and III).

The formation of helical secondary structures in DNA drastically increases the rigidity of each separate chain and is called a *helix–coil transition*. DNA in its double helical form can store torsional energy, since the monomers are not free to rotate (like a telephone cable). The ends of a DNA molecule can be joined together in a circle to form a compact supercoiled structure, that often occurs in bacteria, and their behaviour presents a series of fascinating questions in both statistical mechanics and topological analysis (**Chapter 10**).

Table 1.2 *Averaged structural parameters of polynucleotide helices calculated from X-ray fibre diffraction patterns.*

Property	A form	B form	Z form
Direction of helix rotation	Right	Right	Left
Number of residues per turn	11	10	12
Rotation per residue	33°	36°	30°
Rise in helix per residue (nm)	0.26	0.34	0.37
Pitch of helix (nm)	2.8	3.4	4.5

DNA has a wide variety of structural possibilities (**Table 1.2**, **Figure 1.19**). There are three standard types of double helix labelled *A*, *B*, and *Z* that occur in solid fibres and are deduced from fibre diffraction experiments that average over a huge number of base pair conformations. Typically, DNA in solution has a structure intermediate between *A* and *B*, dependent on the chain sequence and the exact aqueous environment. An increase in the level of hydration tends to increase the number of *B*-type base pairs in an aqueous double helix. Z-type DNA is favoured in some extreme nonphysiological conditions and has little importance for naturally occurring cells.

There are a number of local structural modifications to the helical structure that are dependent on the specific chemistry of the individual strands and occur in addition to the globally averaged *A*, *B* and *Z* classifications. The *kink* is a sudden bend in the axis of the double helix which is important for complexation in the nucleosome. The *loop* contains a rupture of hydrogen bonds over several base pairs and the separation of two nucleotide chains produces loops of various sizes. During the process of DNA transcription RNA polymerase is bound to DNA to form a loop. In the *breathing* of a double helix a temporary break in the hydrogen bonds is caused by a rapid partial rotation of one base pair, which makes the hydrogen atoms in the NH groups accessible and enables them to be exchanged with neighbouring protons in the presence of a catalyst. The *Cruciform structure* is formed in the presence of self-complementary palindromic sequences separated by several base pairs. Hydrophobic molecules (e.g. DNA active drugs) can be *intercalated* in DNA, i.e. slipped between two base pairs. Helices that contain three or four nucleic acid strands are also possible with DNA, but do not occur *in vivo*. Real double helices experience thermal fluctuations on the nano scale, so there are breathing modes of the base pairs as well as flexural modes of the chains' backbone. Furthermore the local ordering depends on the base sequence to some extent and phonons (quantised lattice vibrations) travel through the structure (not to mention the possibility of solitons due to nonlinear interactions). The DNA chain is highly charged and interacts Coulombically with itself, neighbouring chains, and a fluctuating ion cloud that contains a large number of associated counterions. There is clearly a great deal of molecular diversity in the structure of DNA chains and it represents some knotty structural problems.

DNA has interesting features with respect to its polymer physics. The persistence length (l_p, a measure of the chains' flexibility) of DNA is on the order of 60 nm for E-coli (which depends on ionic strength), it can have millions of monomers in its sequence and a correspondingly gigantic contour length (L) (for humans L is 95 mm!). The large size of DNA has a number of important consequences; fluorescently labelled DNA is visible with an optical microscope and the cell has to solve a tricky packaging problem *in vivo* (it uses chromosomes) to fit the DNA inside the nucleus of a cell, which is at most a few micrometers in diameter (**Chapter 18**).

DNA and RNA are both made from nucleic acids and their chemistry is closely related. Ribonucleic acids have the same base pairs except thymine is replaced by uracil (U). DNA can form double helices, whereas RNA predominantly does not. RNA is chemically less stable, which implies an improved evolutionary fitness for long-term data storage in DNA, rather than in RNA.

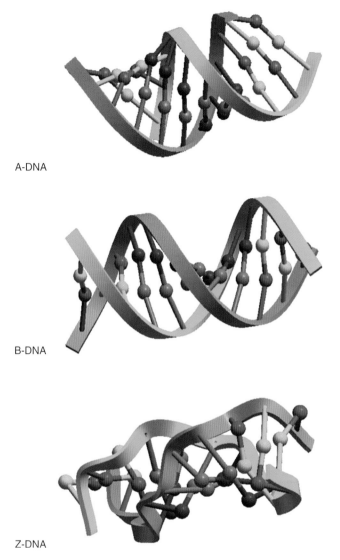

A-DNA

B-DNA

Z-DNA

Figure 1.19 *Molecular models of A-, B- and Z-type double helical structures of DNA. A- and B-type helical structures typically occur in biological systems. Z-DNA helical structures crystallise under extreme nonphysiological conditions.*

The range of applications for DNA analysis and manipulation in molecular biology is huge, with possibilities such as transgenic organisms, genetic analysis of disease, and a molecular understanding of evolution (**Chapter 2**).

1.6   Carbohydrates

Historically, advances in carbohydrate research have been overshadowed by developments in protein and nucleic acid science. This has in part been due to the difficulties in the analysis of the structure of carbohydrates and the extremely large number of chemical structures that exist naturally.

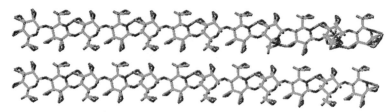

Figure 1.20   *Sheet-like structures formed in cellulosic materials. The β(1 → 4) linkages between glucose molecules induce extended structures and the chains are linked together with hydrogen bonds.*

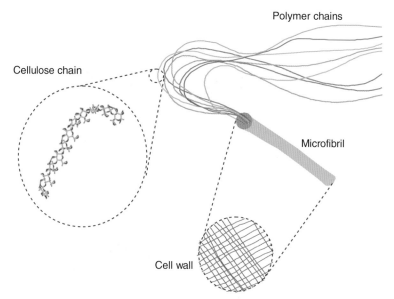

Figure 1.21   *The hierarchical structure of cellulose found in plant cell walls. Polymeric cellulose chains are combined into microfibrils that form the walls of plant cells.*

There are two important glucose polymers that occur in plants (the first and second most important biopolymers in terms of biomass on the Earth) that are differentiated by the linkage between the monomers; cellulose and amylopectin. *Cellulose* is a very rigid polymer and has both nematic liquid-crystalline and semicrystalline phases. It is used widely by plants as a structural material. The straight chain formed by the β(1–4) linkage between glucose molecules is optimal for the construction of fibres, since it gives them a high tensile strength (**Figures 1.20** and **1.21**) in the chain direction and reasonable strength perpendicular to the chain due to the substantial intrachain hydrogen bonding in sheet-like structures. *Amylose* and its branched form, *amylopectin* (starch), are used in plants to store energy and often this material adopts smectic liquid-crystalline phases (**Figure 1.22**). Starches form the principle component of mankind's food sources. In amylose the glucose molecules are connected together with α(1–4) linkages. α-Linkages between glucose molecules are well suited to the formation of an accessible sugar store, as they are flexible and can be easily degraded by enzymes. Amylopectins are formed from amyloses with additional α(1–6) flexible branched linkages between glucose molecules (**Figure 1.23**) and contain both double-helical and amorphous structures. Amylopectin is stored in a hierarchical liquid-crystalline architecture. Glycogen is an analogous amorphous hyperbranched glucose polymer similar to amylopectin that is used in animals as an energy store.

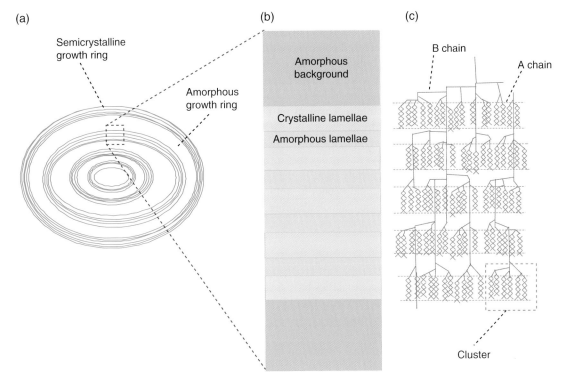

Figure 1.22 *Four length scales are important in the hierarchical structure of starch, a glucose storage material found in plants; (a) the whole granule morphology (~μm), and the growth rings (~100 nm), (b) the crystalline and amorphous lamellae (~9 nm), and (c) the molecular structure of the amylopectin (~Å). [Reproduced with permission from T.A. Waigh, PhD Thesis, University of Cambridge, 1996.]*

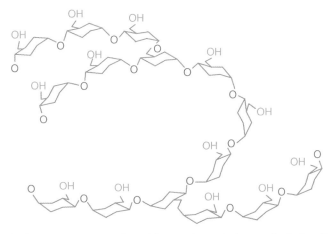

Figure 1.23 *The branched primary structure found for amylopectin in starch. Both α(1–4) (linear regions) and α(1–6) (branched regions) flexible linkages occur between glucose monomers.*

Chitin is another structural polysaccharide that forms the exoskeleton of crustaceans and insects. It is similar in its functionality to cellulose, it is a very rigid polymer and has a cholesteric liquid-crystalline phase.

It must be emphasised that the increased complexity of linkages between sugar molecules, compared with nucleic acids or proteins, provides a high density mechanism for encoding information. A sugar molecule can be polymerised in a large number of ways e.g. the six carbons of a glucose molecule could each be polymerised that provides an additional 6^{N-1} arrangements for a carbohydrate compared with a protein of equivalent length (N) in which there is only one possible mechanism for the connection of amino acids, the peptide linkage. These additional possibilities for information storage with carbohydrates are used naturally in a range of immune response mechanisms and signalling pathways.

Pectins are extracellular plant polysaccharides that form gums (used in jams), and similarly *algins* can be extracted from sea weed. Both are widely used in the food industry. *Hyaluronic* acid is a long negatively charged semiflexible polyelectrolyte and occurs in a number of roles in animals. For example, it is found as a component of cartilage (a biological shock absorber) and as a lubricant in synovial joints.

1.7   Water

Water is a unique polar solvent and its properties have a vast impact on the behaviour of biological molecules (**Figure 1.24**). Water has a high dipole moment of 6.11×10^{-30} C m, a quadrupole moment of 1.87×10^{-39} C m^2 and a mean volume polarisability of 1.44×10^{-30} m^3.

Water exists in a series of crystalline states at subzero temperature or elevated pressures. The structure of ice formed in ambient conditions has unusual cavities in its structure due to the directional nature of hydrogen bonds and it is consequently less dense than liquid water at its freezing point. The polarity of the OH bonds formed in water drives the association into dimers, trimers, etc. (**Figure 1.14**) and causes a complex many-body problem for the description of water in both liquid and solid condensed phases.

Antifreeze proteins have been designed through evolution to have an alpha-helical dipole moment that disrupts the ability of the surrounding water to crystallise through the disruption of the hydrogen-bonded network structure. These antifreeze molecules are often found in organisms that exist in subzero temperatures, e.g. arctic fish and plants. Genetically engineered fish antifreeze proteins are added to ice cream to increase its shelf life and improve its consistency.

Images of biological processes can be created in vivo using the technique of nuclear magnetic resonance that depends on the mobility of water to create the images (**Section 19.9**). This powerful noninvasive method allows water to be viewed in a range of biological processes, e.g. in the cerebral activity of living organisms.

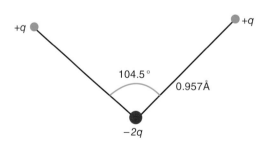

Figure 1.24   *The geometry of a single water molecule that becomes tetrahedral once hydrogen bonded to other water molecules in ice crystals (**Figure 1.4**). The oxygen molecule is shown in red and the hydrogen molecule is shown in green.*

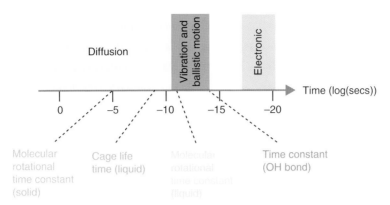

Figure 1.25 *The range of time scales that determine the physical properties of liquid water shown on a logarithmic scale. [Reproduced with permission from The Structure and Properties of Water by D. Eisenberg & W. Kauzmann (2005). By permission of Oxford University Press.]*

Even at very low volume fractions water can act as a plasticiser, to change solid biopolymers between glassy and nonglassy states. This ingress of water can act as a switch, to trigger cellular activity in plant seeds, and such dehydrated cellular organisms can remain dormant for many thousands of years e.g. seeds found entombed in Egyptian pyramids can still be made to successfully germinate.

A wide range of time scales (10^{-18}–10^3 s) of water are important to understand its biological function (**Figure 1.25**). The range of time scales includes such features as the elastic collisions of water at ultrafast times (~10^{-15} s) to the macroscopic hydrodynamic processes observed in blood flow at much slower times (~s) (**Chapter 14**).

The physical interaction between water and other molecules is of key importance in the determination of the properties of biological cells. Water is a polar molecule (the hydrogen–oxygen covalent bonds are dipolar, since they are readily polarised). The hydrogen atoms have a slight positive charge and the oxygen atoms have a slight negative charge. Water molecules can form hydrogen bonds due to the interaction energy of the dipoles and they also interact with other charged molecules, e.g. charged ions.

As a result of water's high dielectric permittivity ions and polar molecules are readily soluble, i.e. the attractive electrostatic interactions are readily screened by the charged water molecules. Furthermore, uncharged molecules that can form hydrogen bonds often dissolve in water. Together, both sets of molecules are called hydrophilic. Nonpolar molecules tend to avoid contact with water in solution and tend to closely associate with one another (hydrophobic). It is possible for a single molecule to have both hydrophobic and hydrophilic regions, and these schizophrenic molecules play a crucial role in the determination of many biological structures, e.g. the shape of globular proteins (hydrophobic sections are tucked away in the centres) and the structure of membranes (hydrophobic regions are placed back to back in a bilayer) are determined by hydrophobicity.

1.8 Proteoglycans and Glycoproteins

Proteoglycans (long carbohydrate molecules attached to short proteins) and *glycoproteins* (short carbohydrate molecules attached to relatively long proteins) are constructed from a mixture of protein and carbohydrate molecules (glycosoaminoglycans). In common with carbohydrates, proteoglycans and glycoproteins exhibit extreme structural and chemical heterogeneity. Furthermore, the challenges presented to crystallography by their noncrystallinity, means that a full picture of their biological function is still being developed. There still

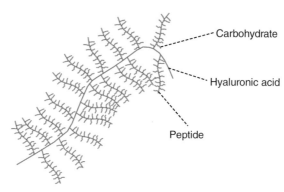

Figure 1.26 *A schematic diagram of the aggrecan aggregate found in cartilage. The aggrecan monomers (side brushes) consist of a core protein (green) with highly charged side chains (red). The bottle brushes are physically bound to the linear hyaluronic acid backbone chain (blue) forming a bottle-brush of bottle-brushes super structure. The structure self-assembles as a function of pH. [Reprinted with permission from A. Papagiannopoulos, T.A. Waigh, T. Hardingham, M. Heinrich, Biomacromolecules, 2006, 7, 2162–217. Copyright © 2006, American Chemical Society.]*

remains a 'here lies beasties' sign on the glycoprotein area of the map of molecular biology. This is both metaphorically and technically correct, since a vast range of micro-organisms spend their lives swimming through mucins inside human stomachs and intestines.

Many proteoglycans and glycoproteins used in the extracellular matrix (regions outside the cells that are often carefully regulated) have bottle-brush morphologies (**Figures 1.26** and **1.27**). An example of a sophisticated proteoglycan architecture is aggrecan and it consists of a bottle-brush of bottle-brushes (**Figure 1.26**). These materials have a very large viscosity and are used to dissipate energy in collagenous cartilage composites and to reduce friction in synovial joints. An example of an extracellular glycoprotein is the mucin (MUC5AC) found in the stomach of mammals. These molecules experience telechelic (either end) associations to form thick viscoelastic gels that protect the stomach lining from autodigestion (**Figure 1.27**). In addition to the secreted gelling mucins, many genes in the mammalian genome express mucins that exist tethered to the surfaces of epithelial cells. These provide steric protection for the cellular membranes (polymeric brushes) and their malfunction is implicated in some respiratory diseases, e.g. chronic obstructive pulmonary disorder or cystic fibrosis.

Other examples of glycoproteins occur in enzymes (Ribonuclease B), storage proteins (egg white), blood clots (fibrin) and antibodies (Human IgG).

1.9   Viruses

Viruses are intracellular parasites; biological entities that multiply by the invasion of cells. Often viruses are simple molecular constructions that consist of a single nucleic acid chain covered with a repeated pattern of proteins that form a coat. In addition to aspects related to their biological role in disease, viruses have attracted a great deal of attention from biophysicists for their physical properties. Viruses often self-assemble into well-defined monodisperse geometrical shapes (rods and polyhedra) (**Figure 1.28**), from their constituent components. Such materials have proven ideal model systems for the examination of the phase behaviour of charged colloids and lyotropic liquid crystals (**Chapter 6**), and in terms of the self-assembly of their native structure in solution (**Chapter 8**).

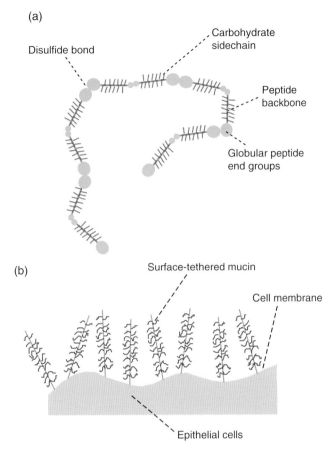

(a)

Disulfide bond

Carbohydrate sidechain

Peptide backbone

Globular peptide end groups

(b)

Surface-tethered mucin

Cell membrane

Epithelial cells

Figure 1.27 *(a) Mammalian stomach mucin contains a series of sticky telechelic bottle-brushes. The dumbbell monomers assemble end-on-end to form symmetric dimers that subsequently polymerise to form a network structure. Crosslinking/branching of fibres occurs at low pH and creates thick viscoelastic gels. Peptide regions are shown in blue and carbohydrates are shown in red. (b) Many mucin genes code for epithelial mucins that form membrane-tethered coats on epithelial cells (surface-tethered polymeric brushes). Peptide regions are shown in green and carbohydrates are shown in red.*

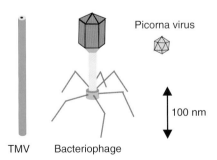

Picorna virus

100 nm

TMV   Bacteriophage

Figure 1.28 *Schematic diagram of a range of virus structures; rod-like (TMV), asymmetric (bacteriophage), and icosohedral (picorna).*

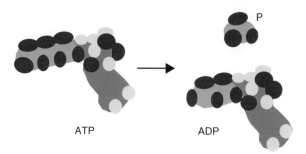

Figure 1.29 *ATP is the standard currency of energy in the cell. The high-energy state (ATP) is converted into a low-energy state (ADP) with the loss of a phosphate group (P) and the liberation of ~20 kT of free energy (the exact amount depends on the specific details of the reaction's environment.)*

Spherical (icosahedral) type viruses have well-defined symmetries and the rules for their tessellation (the recipes they use to fill geometrical space) are now mathematically well described. Many of these viruses are important for human health, e.g. polio virus.

Tobacco mosaic virus (TMV) is a much-studied fibrous virus that infects tobacco plants. TMV forms long extended structures, which can be self-assembled *in vitro* by the correct choice of physical conditions (pH, temperature, etc.) and the viruses so formed can then go on to infect tobacco plants (**Section 8.2**).

1.10   Other Molecules

The inorganic ions of the cell include sodium (Na^+), potassium (K^+), magnesium (Mg^{2+}), calcium (Ca^{2+}), chloride (Cl^-), hydrogen carbonate (HCO_3^-), and phosphate (HPO_4^{2-}). These ions only constitute a small fraction of the cell's weight with less than 1% of the cell's mass, but they play a critical role in cell function and a huge amount of the human body's energy consumption is spent pumping them into and out of cells with membrane proteins. Inorganic ions are particularly important in electrically active cells (e.g. neurons), since they give rise to action potentials used in signalling.

Adenosine diphosphate (ADP) and adenosine triphosphate (ATP) are the 'currency of energy' in many biochemical processes. One molecule can release around $20 \, kT$ (8.2×10^{-20} J) of biochemically useful energy. The energy is stored by the addition of an extra phosphate in the ATP structure and can be released when it is metabolised into ADP (**Figure 1.29**). Phosphate groups are highly charged and electrostatic energy is liberated once the end phosphate group of the row of three in ATP is chopped off. Guanosine triphosphate (GTP), hydrogen ions and sodium ions are also used as fuels in cells, but they occur less frequently.

There is a vast range of other biomolecules (polyphenols, vitamins, etc.) that have not been covered in this short introductory section and the reader should refer to a specialised biochemistry textbook for details, e.g. Berg, Tymoczko and Stryer.

Suggested Reading

If you can only read one book, then try:

Alberts, B. *et al.* (2010) *Essential Cell Biology*, 3rd edition, Garland. Good up-to-date introductory text on cellular biochemistry that requires little biological background.

Alberts, B., Johnson, A., Lewis, J., Ruff, M., Roberts, K., & Walter, P. (2002) *The Molecular Biology of the Cell*, Garland Science. A very detailed account of cellular biochemistry, useful once the contents of Stryer and the introductory Alberts book have been digested.

Berg, J.M., Tymoczko, J.L. & Stryer, L. (2011) *Biochemistry*, 7th edition, W.H.Freeman. Classic in depth introduction to biochemical processes.

Branden, C.J. & Tooze, J. (1998) *Introduction to Protein Structure*, 2nd edition, Garland Science. Clear introductory text on protein structure from a biochemistry perspective.

Caladine, C.R., Drew, H., Luisi, B. & Travers, A. (2004) *Understanding DNA: The Molecule and how it Works*, 3rd edition, Academic Press. Good introductory book on DNA properties from an engineering perspective.

Parsegian, A. (2006) *van der Waals Forces*, Cambridge University Press. A catalogue of analytic results for van der Waals forces in different geometries.

Tutorial Questions 1

1.1 Make a list of ten biological molecules and give a disease associated with the absence/malfunction of each, e.g. insulin and diabetes.

1.2 Metals occur in a range of biological processes and form a key component of the structure in a number of biological molecules. Make a list of some of the biological molecules in which metal atoms occur.

1.3 Explain how peptide chains find their unique folded structure. Calculate the number of permutations for a chain of 200 amino acids if it is assumed there are two possible conformational states for each amino acid monomer.

1.4 Describe the practical barriers that impede our understanding of the structure of carbohydrates and their complexes (glycoproteins, glycolipids, etc.).

1.5 The primary sequence of proteins can be predicted from the genome. The human genome project was completed in 2003. Explain why there still are many unsolved problems in the molecular biology of human cells.

1.6 Cancers are genetic diseases that affect DNA. Name some different carcinogenic materials. Mechanistically explain the difference between a cancerous and noncancerous genetic mutation.

1.7 A DNA chain has a molecular weight of 4×10^8 and the average monomer molecular weight of a nucleic acid is 660 Da. For an A-type helix there are 11 residues per helical pitch, and the translation per residue is 2.6 Å. For a B-type helix there are 10 residues per helical pitch, and the translation per residue 3.4 Å. For a Z-type helix there are 12 residues per helical pitch and the translation per residue is 3.7 Å. Calculate the length in cm of a duplex DNA chain if it is in the A, B and Z helical forms. Find the average size of the nucleus in a mammalian cell, e.g. from a Google search. Explain how the cell manages to accommodate the DNA chain in its nucleus.

1.8 Suppose that a micelle is isolated that contains a single protein that normally exists as a transmembrane molecule. Describe how the lipid and protein would be expected to be arranged on the surface of the micelle.

1.9 Calculate the pH of a 0.2 M solution of the amino acid arginine if its pK_a value is 12.5.

2

Cells

The Big Bang is thought to have occurred around 13.7 billion years ago, which was followed by the creation of the Earth 4 billion years ago (probably through the aggregation of interstellar dust). Very simple life forms first evolved 3.5 billion years ago and cellular life has therefore been on the planet Earth for the majority of its history. This chapter will begin with a discussion of the possible origins of the first cells. This provides an illuminating perspective for the requirements of more advanced life forms. Then, a quick discussion of energy metabolism is made; how cells obtained the necessary energy for living processes to move between ecological niches and invade the whole surface of planet Earth. From thermodynamic considerations, energy metabolism is seen as a prerequisite for successful life, since energy is needed to drive the nonequilibrium cellular environment (equilibrium cells are dead/dormant cells) and produce well ordered living low-entropy structures.

The discussion then returns to the central dogma of biology and the molecular events involved are considered in more detail. Evolution is treated, since it provides a wider overarching framework to consider interrelationships between molecules, cells and organisms. Mutations and cancer are also introduced, because cancers play an important role in human disease and provide biological physics with a wide range of sophisticated unsolved problems. The two standard modern types of cells are then discussed combined with their associated nanomachinery (the organelles), which are the *prokaryotes* and *eukaryotes*. Mechanisms for DNA storage are considered with chromosomes that occur in human cells compared with supercoiling in bacterial cells. Next, the cell cycle and the genetic code are covered. More modern ideas on the interaction of genes in networks are also approached. Advances in the human genome project are described, i.e. the challenge to sequence the entire human genome. Then, two modern applications of DNA biotechnology are considered in genetic fingerprinting and genetic engineering. The chapter ends with some specific examples of human cell types, standard experimental cell models and stem cells.

Cells are packed full of molecular machinery. Dilute solution studies often provide incomplete information on processes inside live cells that naturally occur in a congested environment. Thus, an ongoing challenge for biological physics is to understand how a robust well-regulated environment results from the myriad of competing interactions between molecules and organelles inside live cells.

The Physics of Living Processes: A Mesoscopic Approach, First Edition. Thomas Andrew Waigh.
© 2014 John Wiley & Sons, Ltd. Published 2014 by John Wiley & Sons, Ltd.

2.1 The First Cell

The nature of the first cell is a hard question to answer unequivocally, since no fossils are left due to the fragility of the materials from which the cells were constructed. However, a reasonably credible possibility can be motivated based on chemical constraints, and the need for self-replicating information transfer.

Life is thought to have begun spontaneously on the planet Earth 3.5 billion years ago. During the 1950s Stanley Miller demonstrated that electrical sparks discharged into mixtures of hydrogen, methane, ammonia and water (a model prebiotic soup) resulted in the formation of amino acids and other simple organic molecules such as lipids. Indeed, the spectroscopic signature of amino acids has been found in interstellar space and on comets in astrophysics experiments, so conditions for their creation seem to be fairly common in the universe. Macromolecules can be formed when mixtures of amino acids are heated together to create polypeptides (**Figure 1.8**). However, it is critical for the evolution of life that the macromolecules are able to self-replicate. Only a macromolecule that can synthesise more copies of itself would be a suitable candidate for the origin of life. Polypeptides (proteins) can store information and have a wide range of roles due to their folding patterns, but do not self-replicate.

In fact, all organic macromolecules can store information (carbohydrates, lipids, proteins and nucleic acids), but only nucleic acids are also able to self-replicate. In the 1980s RNA was demonstrated to catalyse more copies of itself in a test tube. Thus, RNA is the foremost candidate molecule for the origin of life on Earth. Some viruses exist today whose genome (the information for replication that also controls its life cycle) is stored in RNA, not in DNA that is used in most present day organisms. DNA is thought to have displaced RNA during evolution for the majority of nonviral organisms, since it is slightly more chemically stable and thus a better candidate for long-term high-fidelity data storage. The interaction between primordial proteins and RNA is thought to have given rise to the more sophisticated, self-replicating genetic codes observed in modern eukaryotic cellular organisms (**Section 2.10**).

The first living cell is deduced to have been a self-replicating RNA chain surrounded by a self-assembled (driven by amphiphilicity) bag of phospholipids (**Figure 2.1**). The membrane allowed the cell to enrich the concentrations of molecules favourable for its life, e.g. food, ATP, and ions. Furthermore, the membrane helped protect the cell against fluctuations in the concentration of its surroundings, which would disrupt its

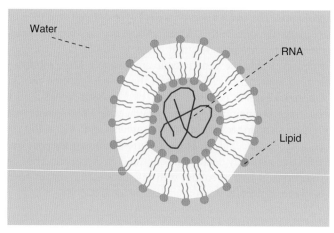

Figure 2.1 *Primordial cells may have consisted of an RNA chain encapsulated by a self-assembled lipid bilayer arranged as a spherical vesicle.*

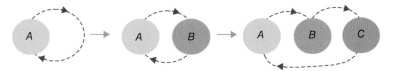

Figure 2.2 *Steps in the evolution of biochemical reactions. Evolution proceeds from left to right as the more complex reactions outperform the simpler ones, e.g. ABC outperforms, A and AB. All the reactions schemes need to be autocatalytic and RNA is the main candidate for the initial biological molecule A.*

life cycle. RNA-directed protein synthesis would have led to increases in the specialisation and diversification of protein nanomachinery inside the cell as it evolved.

To provide increased molecular complexity during replication, chemical reaction loops are thought to have formed (due to the interaction of RNA molecules). These loops are required to form the next generation of molecules in a process of autocatalysis. The autocatalytic reactions could become sequentially more complex as evolution proceeds (**Figure 2.2**). Each new generation could include additional reaction steps, if they had an evolutionary advantage and if they were chemically permissible.

An alternative theory is that autocatalytic sets self-organised from a soup of biological molecules in abrupt steps once the soup became sufficiently complex (*self-organising autocatalytic sets*), i.e. a sequential increase in the level of complexity was not required due to large-scale fluctuations. Emergent order from random events (order from chaos) is observed in a range of nonbiological physical phenomena (e.g. avalanches in granular materials) and is an interesting alternative for early evolution, i.e. molecular evolution may have occurred in bursts.

Modern molecular biology techniques are able to probe the basic requisites for cellular life and there are many current research projects in the area of *synthetic biology* that do so. Examples include the reduction of the number of genes expressed by an organism to construct a minimal genome for a live cell, and nuclear replacement of the DNA in one organism with that of a closely related organism so that the cell is reprogrammed by the new DNA to become a cell type of a different species.

2.2 Metabolism

The metabolism of an organism provides the energy to drive all of its internal processes. The evolution of metabolism thus gives another interesting perspective for the requirements of living organisms. Once the information storage and compartmentalisation of the cell had been achieved by evolution, the next main biochemical element to evolve may have been a portable energy source.

All cells now use adenosine $5'$-triphosphate (ATP) as their source of metabolic energy to drive the synthesis of cell constituents and fulfill activities that have energy requirements such as movement (muscle contraction). Three key chemical processes are necessary for the simplest model of energy metabolism; glycolysis, photosynthesis and oxidative metabolism. Together through evolution these processes have created ATP as the primary currency of energy. Photosynthesis is the chemical process by which photonic energy from the Sun is stored in sugar molecules,

$$\text{Photosynthesis} \quad 6CO_2 + 6H_2O \xrightarrow{\text{Photons}} C_6H_{12}O_6 + 6O_2$$

Photosynthesis provides a long-term store of the Sun's energy in glucose molecules, whereas glycolysis and oxidative metabolism provide two methods to release the stored energy, e.g. for the creation of ATP,

$$\text{Glycolysis} \quad C_6H_{12}O_{6\,(glucose)} \rightarrow 2C_3H_6O_3 \text{ (lactic acid)}$$

$$\text{Oxidative metabolism} \quad C_6H_{12}O_6 + 6O_2 \rightarrow 6CO_2 + 6H_2O$$

The development of an on-demand energy source provides a clear evolutionary advantage for a simple cellular organism, which will greatly improve its robustness and ability to explore a wider variety of environments. Modern quantitative treatments of metabolism require ideas from systems biology that will be introduced in Part IV.

2.3   Central Dogma of Biology

Evolution eventually determined that the vast majority of higher organisms depend on the central dogma of biology for their hereditable differences (**Figure 1.16a**). Very simply this follows from the scheme

$$\text{DNA} \rightarrow \text{RNA} \rightarrow \text{Protein} \rightarrow \text{Vast range of other molecules}$$

The dogma holds up very well to scientific scrutiny. However, modern advances indicate there are a number of subtleties in its interpretation. For example, the expression of sections of DNA can be switched on and off by environmental factors. Such issues are considered in the emerging field of *epigenetics*. A good illustration of research in this area is given by studies on identical twins. Identical twins are not necessarily identical in terms of their cellular biochemistry. There are numerous examples of environmental factors that lead to different expression of twins' DNA. For example, if one twin smokes and the other does not, the smoker's cells will be more aged and so too will be their DNA. Furthermore, there are other examples of large weight differences in identical twins, and the occurrence of cancer in one twin and not the other. Recent studies also indicate the possibility to inherit epigenetic traits from generation to generation, e.g. the level of expression of a certain gene can be modified by environmental conditions and transferred to an organism's off spring.

20 687 protein-coding genes are currently known in the human genome, which constitutes about 3% of the total amount of DNA. 76% of the other bases in the genome are transcribed into RNA, which implies that the majority of bases do not code for proteins, but have a regulatory function (the genome is predominantly not 'junk' as previously thought).

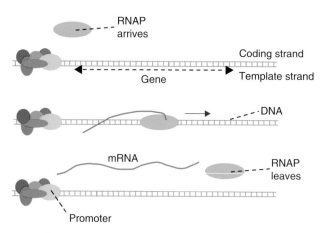

Figure 2.3   *Messenger RNA can be transcribed from DNA by a RNA polymerase (the RNAP complex). Stretches of DNA that code for proteins are called genes.*

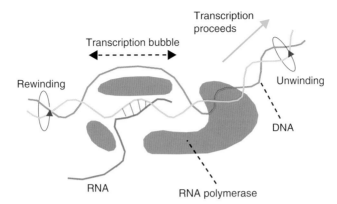

Figure 2.4 *RNA polymerase is required during the transcription of mRNA from a section of a DNA chain. RNAP creates a bubble of DNA as it transcribes the messenger RNA chain.*

There are three main types of RNA that are used in the translation of genes; mRNA (messenger RNA), rRNA (ribosomal RNA) and tRNA (transfer RNA). Messenger RNA carries information about a protein sequence, ribosomal RNA is a catalyst used in ribosomes and transfer RNA is an adaptor for protein synthesis.

Transcription is the first stage of gene expression, which involves the synthesis of mRNA from a DNA template by a RNA polymerase (**Figure 2.3**). mRNA is synthesised from the template DNA strand and has the same sequence as the coding strand (**Figure 2.4**).

Translation is a process by which cells synthesise proteins (**Figure 2.5**). During translation information encoded in mRNA is used to specify the amino acid sequence of a protein. Transfer RNA molecules play a key role in this process through delivery of amino acids in the correct order specified by the mRNA sequence.

Replication is the process by which a cell copies its DNA prior to division. Replication is necessary so that the genetic information present in cells can be passed on to daughter cells. DNA is copied by enzymes called DNA polymerases (**Table 2.1**).

There are *three phases of transcription* from a DNA chain; initiation, elongation, and termination. With the *initiation step* RNA polymerase (RNAP) binds to a *promoter*, which is a segment of DNA upstream from the coding sequence of a gene. RNAP forms a closed promoter complex (which includes transcription factors) with the entwined DNA strands. An open complex then forms as the DNA strands separate to form a transcription bubble. Next, during *elongation* (**Figure 2.4**) the RNA polymerase moves away from the bound promoter along the DNA chain and expresses a long molecule of mRNA as it goes; it adds nucleotides to the $3'$ end as the RNA molecule grows and the order of addition is determined by that of the template strand. Finally, there is the *termination* step that is caused by two alternative signals. Rho-independent termination occurs when the newly synthesised RNA molecule forms a G–C-rich hairpin loop followed by a run of Us. The mechanical stress breaks the weak rU–dA bonds that fill the DNA–RNA hybrid. This effectively terminates the process of transcription. Alternatively rho-dependent termination can occur where a protein factor called 'Rho' (ρ) destabilises the interaction between the template and the mRNA. This releases the synthesised mRNA from the elongation complex. Transcription requires the unwinding of about 13 base pairs of DNA, which generates a torsional stress on the DNA strands that needs to be relieved by topoisomerase enzymes.

Gene regulation is important in cells so that genes are expressed in the correct order and to the correct level (**Figure 2.6**). *Transcription initiation* is required for RNA synthesis to begin from a DNA chain. A combination of activator and repressor proteins, and other transcription factors, enable the RNA polymerase (an enzyme) to bind to a segment of DNA upstream of the coding sequence of the gene (the promoter). The TATA binding

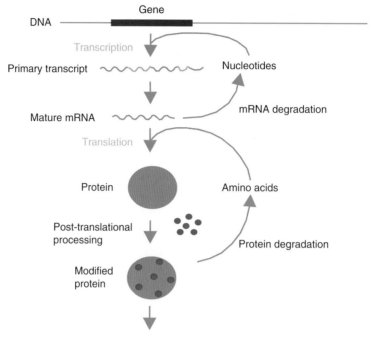

Figure 2.5 *The genes on a DNA chain are transcribed into mRNA chains and then translated into protein structures. Subsequent post-translational modifications are then possible, e.g. phosphorylation or glycosylation.*

Table 2.1 *Comparison of the processes of replication and transcription of DNA chains. Replication involves the creation of new DNA chains, whereas during transcription short mRNA chains are produced from sections of DNA.*

Replication	Transcription
The entire chromosome is copied.	Only selected fragments of DNA are copied.
It requires primers.	No DNA primers are required.
Both DNA strands are copied (predominantly).	Only one DNA strand is copied.
DNA Polymerase has proofreading ability.	RNA Polymerase has no proofreading ability.

protein is complementary to the DNA sequence present in most promoter regions of genes. The TATA sequence is thus a simple signal to begin the construction of the transcription complex and initiate gene expression.

Post-transcriptional processing of DNA stored information is also now known to be important to understand the actual expression of genes. *Splicing* is a modification of an mRNA chain after transcription, in which introns (noncoding gene regions) are removed and exons (coding regions) are joined together. Typically this is needed to create a correct piece of mRNA before it can be used to produce the correct protein during the process of translation. For many eukaryotic introns, splicing is done in a series of reactions that are catalysed by complexes of small nuclear ribonucleoproteins (**Figure 2.7**).

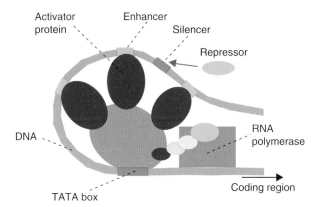

Figure 2.6 *Activators, repressors, coactivators and basal transcription factors work together to regulate the expression of genes along a DNA chain. They primarily do this by modulating the activity of RNA polymerase.*

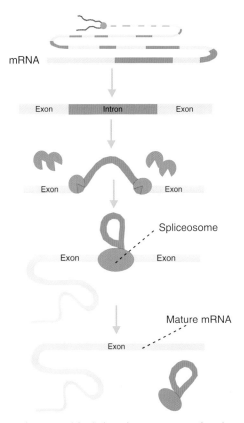

Figure 2.7 *Messenger RNA is often modified by the process of splicing before it is translated into protein structures. Sections of mRNA called* introns *are removed from the chain and only* exon *sections are translated.*

2.4 Darwin's Theory of Natural Selection

Darwin's theory of natural selection is the process by which genetic mutations that enhance reproduction become and remain more common in successive generations of a population. The theory follows from three simple facts: heritable variations exist within a population of organisms, organisms produce more offspring than can survive, and these offspring vary in their ability to survive and reproduce (they have different evolutionary fitnesses).

Darwin's theory is an important qualitative argument and there is a huge amount of evidence that proves these general principles guide the evolution of life. It is possible to create an evolutionary tree that contains all the organisms on planet Earth based on Darwin's theory (**Figure 2.8**). Recently, the theory has been made much more quantitative due to advances in genetic techniques, detailed studies of organisms that rapidly divide and stochastic mathematical modelling.

Genetic data now points to the origin of all humans on planet Earth in Africa, and that they dispersed from the continent around 100 000 years ago. Two separate measurements that support this picture of evolutionary archaeology are possible from genetic studies based on mitochondrial and nuclear DNA. Furthermore, the existence of mitochondrial DNA indicates the origin of these organelles as ancient symbiotic bacteria that became trapped inside eukaryotic cells during evolution many hundreds of millions of years ago.

Developmental strategies in animals are ancient and highly conserved. In essence, a mammal, a worm and a fly are put together with the same basic genetic building blocks and regulatory devices. For example, the morphology of a human embryo a few days after fertilisation is hard to tell from a fish embryo, which implies some common evolutionary ancestors for all vertebrate life. In Part IV quantitative models for this process of morphogenesis will be investigated. Morphogens can diffuse across sheets of cells and induce pattern formation (e.g. why most humans have five fingers and not four or six) which causes the bilateral symmetry seen in most organisms.

Before DNA-based genetic studies there was still strong evidence for evolutionary trees for the relationships between organisms. The arrangement of bones in the wing of a bat, the wing of a bird and the hand of a human, show a very well conserved pattern (**Figure 2.9**). Even fossilised dinosaurs have similar bone blueprints to modern vertebrates and this demonstrates a common lineage for all the organisms many millions of years ago.

A wide range of factors affect speciesation; the processes that control the development of a new species of organism (**Table 2.2**); the detailed mechanism that drives the process of natural selection. *Directional selection* rejects most mutations that do not increase the fitness in the environmental niche, *balancing* causes

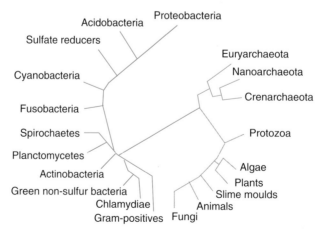

Figure 2.8 *Simple schematic evolutionary tree that shows the main families of organisms on planet Earth.*

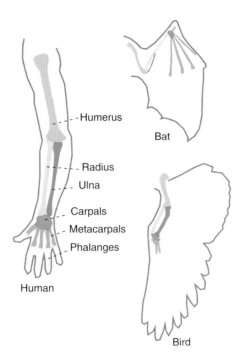

Figure 2.9 *Evolutionary connections can be deduced by comparison of the bone structure in different organisms. The similarity of bones in the arm of humans, wings of bats and wings of birds indicates that they share common evolutionary ancestors.*

Table 2.2 *Some of the qualitative driving factors through which evolution can increase (+) or decrease (−) variations within and between populations.*

Driving factor	Variations within populations	Variations between populations
Inbreeding or genetic drift	−	+
Mutation	+	−
Migration	+	−
Directional selection	−	+/−
Balancing	+	−
Incompatible	−	+

extreme mutations to rapidly increase fitness, and *incompatibility* mechanisms stop interbreeding so only self-fertilisation is possible. Similar evolutionary factors give rise to common biological solutions in the process of *evolutionary convergence*. Thus, swallows and swifts appear very similar, although they only have very distant evolutionary ancestors, due to a common environmental niche.

The evidence for evolution is huge, although the time scales for changes are so slow it is difficult to intuitively appreciate them (millions of years are often required). However, evolution can be observed in real time with micro-organisms that have short life spans and consequently need to reproduce quickly. Hundreds of generations can be observed per day, e.g. E.coli bacteria can divide every 20 min. This phenomenon of real-time evolution has recently been put to important effect in studies of the evolutionary relationship between cancer cells in a single tumour. Furthermore, another example of real time evolution is antibiotic resistance in bacteria. Bacteria can swap sections of DNA between themselves (horizontal gene transfer) that provide increased resistance to a particular antibiotic. MRSA (named after the antibiotic to which the bacteria are resistant) infections picked up in hospitals have made headlines in national newspapers as the bacteria evolve to become antibiotic resistant, frequently with fatal consequences.

Yet more molecular evidence for evolution is seen by the comparison of the amino acid sequence from similar proteins in different organisms. The rate of neutral evolution for an amino acid sequence of a protein depends on the sensitivity of the proteins' function to amino acid changes. Changes during *neutral mutations* commonly have a linear dependence on time, but in general the rate of amino acid variations with time depends on the specific type of protein considered. Two examples of linear mutation rates are observed with Cytochrome C (which controls the oxidation reaction in mitochondria) and haemoglobin (an oxygen-carrying molecule in blood). The mutation rates can be used as clocks to measure evolutionary significant time scales.

There is no regular relation between how much DNA change takes place in evolution and how much functional change results. Under identical conditions of natural selection, two populations may arrive at two different genetic compositions as a direct result of natural selection (a stochastic effect), and the process is called *evolutionary divergence*.

Evolution is the conversion of heritable variation between individuals within populations into heritable differences between populations in time and space by population genetic mechanisms. By definition, different *races* of a species can exchange genes and different *species* cannot.

The complexity of the proteins expressed by an organism tends to scale exponentially with the number of base pairs, since each base pair can have a number of different roles. Humans are not the most complicated organisms in terms of their length of DNA, e.g. amoebas and ferns have two orders of magnitude more base pairs in their genome than humans. Thus, many organisms express a much wider range of proteins than humans and are thus more complicated at the molecular level. Interactions between genes in an organism can greatly increase their functional complexity. Many of the regulatory interactions between genes are still unknown (For more information on how systems biologists aim to solve this problem, read Part IV).

Sophisticated mathematical ideas from *game theory* (such as Nash equilibria) and stochastic differential calculus can provide much more quantitative predictions for evolutionary processes, e.g. models for the development of cancers.

2.5   Mutations and Cancer

The possibility of discrete mutations in an organism's genome was known before the molecular nature of DNA was explained. Gregor Mendel founded the field of genetics during the mid-1800s by crossbreeding peas through careful observation of their different phenotypes, i.e. the characteristic resultant morphologies of the organisms (specifically the shapes of pea seeds). Lots of examples were also known at the time for human genetics, such as haemophilia in Europe's Royal families, sickle cell anaemia in Africans, colour blindness in men, and Huntingdon's disease. The list of well-characterised mutations in humans is now huge, not to mention those observed in cultivated plants, livestock and pets.

Mutations are changes in the DNA sequence of a cell's genome and are caused by radiation, viruses, transposons (jumping genes) and mutagenic chemicals, as well as errors that occur during meiosis or DNA replication. Mutations can also be induced by the organism itself by cellular processes such as hypermutation. Several different types of change in DNA sequence can result in mutations. They can have no effect, alter the product of a gene, or prevent a gene from functioning.

Cancer is not a single disease, but a grouping for over 200 separate diseases in humans. All cancers have a genetic origin. Some are due to spontaneous mutations and some are viral in origin. Cancers result from self-amplifying mutations, due to the failure of an organism's natural DNA error correction and prevention mechanisms. For example, mutations in the error-correcting machinery of a cell might cause the cell and its children to accumulate errors more rapidly. A mutation in signalling machinery of the cell can send erroneous messages to nearby cells. A mutation could cause cells to become neoplastic such that they migrate and disrupt more healthy cells. A mutation may cause a cell to become immortal, which causes it to disrupt healthy cells for the whole lifespan of the organism. The resultant immortal cancerous cells will never self-destruct no matter how harmful they are to their neighbours and they may be fatal for the organism.

Cancer is of huge importance to human medicine. It is the second biggest killer after cardiovascular disease. Furthermore, it is very common to perform cellular biology experiments with cancer cells (e.g. HeLa cells from Henrietta Lack, an African American who died from cancer in the 1950s, are a standard cell line), not only for their interest in cancer treatment, but because immortal cell lines are much easier to culture than noncancerous cell lines.

Game theory is starting to play a role in the understanding of evolution in cancerous cell lines. Experiments are now able to explore the genetic relationship between hundreds of cells in a single tumour, which allows the evolutionary family tree to be probed and quantitative theories to be developed. The steady accumulation of genetic errors in cancer cells in each new generation can thus be explored.

2.6 Prokaryotic Cells

There are two main types of cell in modern organisms; *prokaryotic* and *eukaryotic* cells. Prokaryotic cells are smaller and simpler than eukaryotic cells, but they share the same general molecular mechanisms that governs their lives, and that indicates their common ancestry. The main differences between the two cell types are that prokaryotes do not have a nucleus and rarely have membrane-bound organelles. Both types of cell have DNA as the genetic material, exterior membranes, ribosomes, accomplish similar functions, and are very structurally diverse.

Bacteria are small morphologically simple cellular organisms and are prokaryotic. Only a minority of bacterial species have developed the ability to cause disease in humans. Most bacteria take the form of spheres, rods and spirals with diameters of 1–10 μm. They will be encountered again in terms of their mechanisms of molecular motility in **Chapter 7** and **Chapter 16**.

Archaebacteria are commonly classified as the second main variety of prokaryote after bacteria, but modern evidence points to them being on a separate evolutionary tree from other prokaryotes and thus a completely different class of cell. Examples of archaebacteria include the extremophiles, but they have also been found living as oceanic plankton and can form methane in human intestines.

Escherichia coli (E.coli) is a common inhabitant of the human intestinal tract and is a standard model bacterium that is used for research (**Figure 2.10**). E.coli is rod-shaped, about 1 μm in diameter and 2 μm long. The bacterium is surrounded by a rigid wall that is composed of polysaccharides and peptides. Within the cell wall is the plasma membrane that consists of a phospholipid bilayer and associated proteins. The double-helical DNA chain forms a single circular molecule in the nucleoid and it is not surrounded by a membrane that separates it from the cytoplasm. The cytoplasm contains some 30 000 ribosomes, the sites for protein synthesis.

2.7 Eukaryotic Cells

Eukaryotic cells are surrounded by plasma membranes (**Figure 2.11**). In addition, they contain a *nucleus* (~5 μm diameter), the site where DNA replication and RNA synthesis occurs; *ribosomes* where the translation of RNA

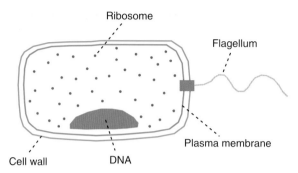

Figure 2.10   *Schematic diagram of a prokaryotic cell, e.g. an E.coli bacterial cell. The DNA is super-coiled and there are no chromosomes and no nucleus. Pili, fimbrae, gas vacuoles, carboxysomes, cytoskeletons, magnetosomes and endospores can also occur (not shown).*

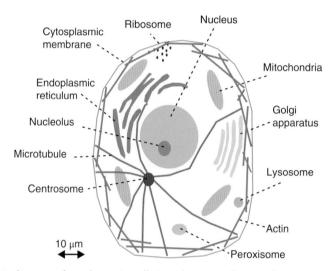

Figure 2.11   *Schematic diagram of a eukaryotic cell. A wide range of organelles exist inside the cell. The DNA is housed inside a specialised nucleus.*

into proteins takes place (molecular protein factories); *mitochondria* that play a role in energy metabolism; *chloroplasts* that are the sites of photosynthesis and are found only in plant cells and green algae; and *lysosomes* that provide specialised metabolic compartments for digestion of macromolecules and oxidative reactions.

The *nucleus* is the most obvious organelle in any eukaryotic cell under an optical microscope (**Figure 2.12**). It is a membrane-bound organelle and is surrounded by a double membrane. It communicates with the surrounding cytosol via numerous nuclear pores in the double membrane. The DNA held inside the nucleus is responsible for determination of the cell's unique characteristics. The DNA is identical in every cell of an organism, but its pattern of expression depends on the specific cell type. Some genes may be turned on or off, which is why a liver cell is different from a muscle cell, and a muscle cell is different from a fat cell. When a cell divides, the DNA and surrounding proteins condense into chromosomes that are visible by microscopy. The prominent structure inside the nucleus near to the DNA is the *nucleolus*. The nucleolus produces ribosomes, which move out of the nucleus to positions on the endoplasmic reticulum where they allow proteins to be synthesised.

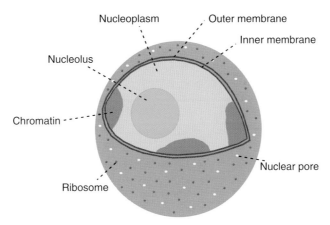

Figure 2.12 *Schematic diagram of a nucleus in a eukaryotic cell. Chromatin (dark green) contains the DNA combined with histones. Ribosomes are created in the nucleolus (blue) and pass through the nuclear pores into the cytoplasm.*

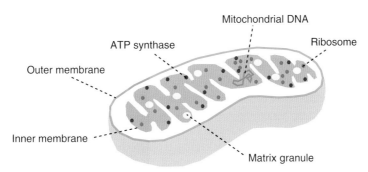

Figure 2.13 *Schematic diagram of the structure of mitochondria from a eukaryotic cell. ATP, the main currency of cellular energy, is synthesised in these organelles.*

Mitochondria provide the energy a cell needs to move, divide, and produce secretory products. In short, they are the power centers of the cell (**Figure 2.13**). They are about the size of bacteria, but may have different shapes that depend on the cell type. Mitochondria are membrane-bound organelles, and like the nucleus, have a double membrane. The outer membrane is fairly smooth, but the inner membrane is highly convoluted, and forms folds called cristae. The cristae greatly increase the inner membrane's surface area. It is on these cristae that food (sugar) is combined with oxygen to produce ATP, the primary energy source for the cell.

Transport, sorting and processing are important factors in the control of the extremely complex intracellular environment. The *endoplasmic reticulum* functions not only in the processing and transport of proteins, but also in the synthesis of lipids. From the endoplasmic reticulum, proteins are transported to the *Golgi apparatus*, where they are further processed and sorted for transport to their final destinations. Carbohydrates are synthesised in the Golgi apparatus and it also serves as a site of additional lipid synthesis.

The *cytoskeleton* is a network of protein filaments that extend throughout the cytoplasm. It determines the structural framework of the cell, the cell shape and it also can be responsible for the movement of entire cells (**Chapter 16**).

Multicellular organisms evolved from unicellular eukaryotes. Some unicellular eukaryotes continue to form temporary multicellular aggregates that appear to represent an evolutionary transition from single-cellular to multicellular organisms, e.g. slime molds. Slime molds can live as both single cells and as multicellular aggregates dependent on the stage of their life cycle. Continued cell specialisation in multicellular aggregates has led to the observed multicellular complexity of present-day plants and animals. Plants are composed of fewer cell types than animals, although their genomes are often much larger.

2.8 Chromosomes

Human DNA when unravelled is a piece of string one meter in length (3×10^9 base pairs $\times 3.4$ Å for B-type DNA), but only 1 nm in diameter (if all 46 chromosomal DNA strings are place end-on-end). This huge length of DNA needs to fit in a nucleus that is only ~5 μm in size. The cell accomplishes this feat by the creation of chromosomes. Negatively charged DNA is wound around positively charged histones, and then combined with scaffolding proteins, much like pieces of cotton wound around a sequence of bobbins. The electrostatic phenomena involved in chromosomal complexation is described in more detail in **Chapter 18**.

Down's syndrome, Edwards syndrome and other genetic diseases occur when additional copies of chromosomes are created during cell fertilisation. This can now be screened for, before birth, during early stages of pregnancy, e.g. by counting the number of chromosomes in the cells of an embryo or by the study of characteristic biochemical metabolites.

In humans, each cell normally contains 23 pairs of chromosomes, which makes a total of 46 chromosomes (**Figure 2.14**). Twenty-two of these pairs, called autosomes, look the same in both males and females. The 23rd pair, the sex chromosomes, differ between males and females. Females have two copies of the X chromosome, while males have one X and one Y chromosome.

Bacterial DNA is not stored in chromosomes inside their cells to reduce the DNA's size as with eukaryotic cells. Instead, the bacterium has solved the packaging problem by using supercoils (it forms packages of tightly twisted DNA called plectonomes). Double-helical DNA chains cannot rotate independently about each base pair, so the circular chains can store twist energy. Specialised topoisomerase enzymes twist the DNA chains in bacteria, so that they can form plectonomic structures (compact tightly wound morphologies).

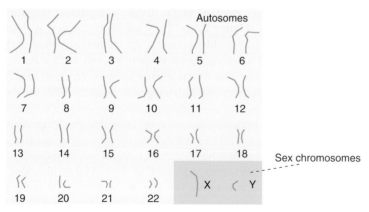

Figure 2.14 *Relative sizes of the 23 pairs of chromosomes in human cells. The 23rd pair determines the sex of the individual; XX is female and XY is male.*

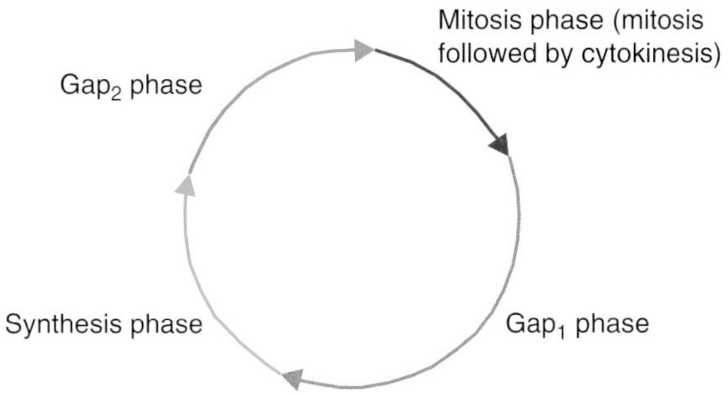

Figure 2.15   *There are four phases in the life cycle of a eukaryotic cell; gap1, synthesis, gap2 and mitosis. Gap1 and gap2 allow time for cells to grow in size and create additional organelles. In the synthesis phase the nucleus duplicates its DNA. Mitosis is when the nucleus splits in two and cytokinesis is when the whole cell subsequently divides in two.*

2.9   Cell Cycle

Cells run through cycles of expansion and division, with multiple checkpoints for correct replication. Clearly a cell must double the amount of DNA it contains before division if the information is to be transferred without degradation to both its daughter cells. Furthermore, in eukaryotic cells the chromosomes must be accurately distributed to each of the daughter cells. There are four principal steps in the cycle of a eukaryotic cell (**Figure 2.15**). Two Gap phases (*Gap*1 and *Gap*2) allow the cell to increase its size and create new organelles. A *synthesis* phase allows new DNA to be made and a *mitosis* phase occurs when the cell actually divides. The mitosis phase is further divided into the prophase (two sets of chromosomes condense and two centrosomes assemble with their mitotic spindles), prometphase (the nuclear envelope breaks down), metaphase (chromosomes are aligned along the equator of the mitotic spindle), anaphase (two sets of chromosomes separate), telophase (chromosomes arrive at the spindle poles and the new nuclear envelopes assemble) and cytokinesis (the contractile ring pinches the cell in two, with the creation of two daughter cells).

2.10   Genetic Code

The genetic code for amino acids for protein-forming amino acids in eukaryotic cells (**Figure 2.16**) describes how stretches of DNA (parts of genes) that code for amino acids are created in eukaryotic cells by transcription and translation. The amino acids are then strung together to form proteins. The code involves triplets of base pairs. The triplets describe the 20 eukaryotic amino acids found in nature and the STOP codon that instructs the ribosome to stop the production of proteins. Each amino acid has a number of corresponding nucleic acid codes. The groups of three base pairs thus have 4^3 (64) possible arrangements. There is thus some redundancy in coding for the twenty amino acids and the STOP codon ($64-21 = 43$).

2.11   Genetic Networks

The expression of different genes in an organism's genome are closely interrelated. Indeed it is possible to map the connection between different genes to create a genetic network diagram for a cell. Two simple genetic networks in which the output of the first gene represses or activates the second are shown in **Figure 2.17**.

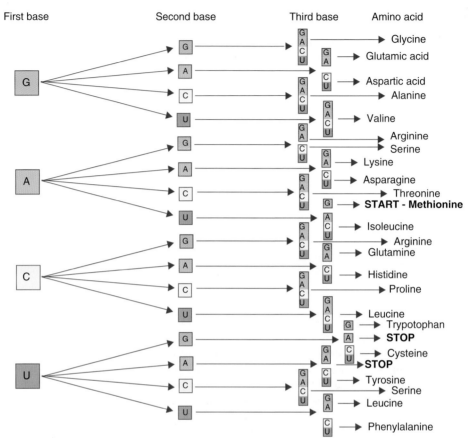

Figure 2.16 *The genetic code of eukaryotic DNA for proteins consists of groups of three nucleic acids. In eukaryotic cells the nucleic acid triplets code for 20 amino acids and the STOP codon. There is thus some redundancy, i.e. there are multiple codes that describe the same amino acid. The code is slightly different in prokaryotes, since three additional amino acids are included in their proteins.*

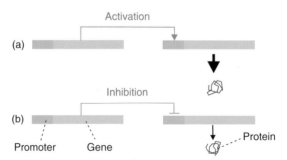

Figure 2.17 *Genes interact with one another in genetic networks. (a) They can activate one another (increase in gene expression) or (b) they can inhibit one another (reduction in gene expression).*

More generally genes interact with one another in complex chemical circuits and can perform useful tasks, such as the logical operations (AND, OR, NOR, etc.) or analogue calculations (integration, division, subtraction, etc.). They are described in more detail in **Chapter 22** on systems biology.

The action of gene network circuits has been measured in detail in some simple micro-organisms, e.g. bacteria. Bacterial studies are aided by the particularly short, ~2.5 hour, life cycle of the organisms. Gene circuits are arranged to switch genes on in the correct order and allow them to perform crucial cellular programs. The phenomena demonstrated by genetic circuitry are also described in much more detail in **Chapter 22**.

2.12   Human Genome Project

The *human genome project* was completed in 2003. Also, the genome of chimpanzees, the majority of Neanderthal man and more than thirty other species have to date been sequenced. The number of organisms' genomes now known in their entirety is rapidly increasing due to the rapidly decreasing cost of DNA sequencing. In the future everyone's individual DNA sequences may be determined directly after birth. This will facilitate personalised medicine based on the sequence information, but will also create a series of ethical dilemmas, e.g. premiums on health insurance policies would be harder to regulate if people knew accurately at what age they would die from natural causes.

The total amount of DNA in the chromosomes of a single human cell contains ~3.3 Giga (3.3×10^9) base pairs, so it posed a serious technological challenge to sequence this huge number of base pairs. A brief tour of DNA gel electrophoresis techniques is needed to appreciate how the full genome was actually sequenced. Electrophoresis can be used to separate different lengths of DNA chain (**Figure 2.18**), and has played a central role in DNA sequencing. The method is fairly simple to set up experimentally, it just requires an electric field, a hydrogel and a source of DNA. The detailed physical analysis of gel electrophoresis can be subtle (a process

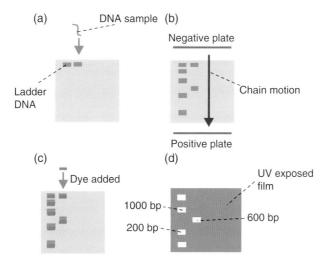

Figure 2.18   *DNA electrophoresis measures the mobility of DNA chains in an electric field across a hydrogel. Sizes are calibrated by comparison with a standard ladder sample that runs in parallel. (a) Ladder DNA and sample DNA are loaded on a gel. (b) Electrophoresis begins when the electric field is applied. (c) A DNA binding dye is added so that the chains can be located. (d) An image of the gel is taken by exposure of a photographic film (or a CCD camera) to UV light to locate the position of the DNA chains.*

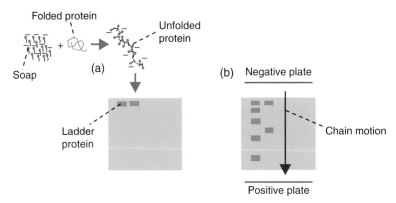

Figure 2.19 *SDS electrophoresis can be used to separate different lengths of protein. (a) A negatively charged soap (SDS) is added to unwind the proteins. (b) Application of an electric field causes the chains to separate in size, and their sizes can be calculated using a ladder sample.*

of biased reptation, **Chapter 10** and **Section 19.14**), but to obtain quantitative size information is relatively simple, it just requires comparison of a sample with that of a precalibrated DNA ladder mixture. The current trend is to make smaller and smaller electrophoresis apparatus using microfluidics, which provides increased sensitivity to a wider range of lengths with reduced amounts of DNA. The eventual optimal solution may be to provide real-time sequencing using the motion of DNA through a single nanopore and this has been experimentally demonstrated. Changes in the conductance of a protein membrane channel are used to read off base-pair sequences as the DNA chain passes through the pore. Modern genome sequencing techniques also need to be massively parallelised to increase the rate of throughput.

Electrophoresis is possible with proteins if SDS (a strong denaturant) is added to linearise their morphologies (**Figure 2.19**). The *Western blot* is a related analytical technique used to detect specific proteins. In this method gel electrophoresis separates native or denatured proteins by the length of the polypeptide chain. Proteins are then transferred to a membrane where they are mixed with antibodies specific to a target protein that allows them to be identified. The *Southern blot* is used to detect specific DNA sequences. Electrophoresis-separated DNA is placed on a membrane and fragments are detected by hybridisation; a coil–double-helix transition with a complementary DNA strand. The *Northern blot* is used to study gene expression by the detection of RNA. Electrophoresis is used to separate RNA by size and another hybridisation step is then used to probe the occurrence of a specific sequence.

Standard electrophoresis techniques can only discriminate between different lengths of DNA for fairly small number of base pairs. Thus, the genome in the human genome project was first broken into smaller sections for sequencing, approximately 150k base pairs in length. The smaller DNA sections were then spliced into bacterial chromosomes that had been genetically engineered. The bacteria replicated and the genome material was copied. Each 150k piece was then sequenced separately after this amplification procedure was complete.

The *hierarchical shot gun approach* was used to sequence the 150k sections of DNA in the human genome project. Shotgun sequencing occurs when DNA is broken up randomly into small segments that are sequenced using the chain termination method to obtain 'reads'. Multiple overlapping reads for the target DNA are obtained by several rounds of fragmentation and sequencing. Computer programs can use overlapping ends of different reads to assemble a continuous sequence. The *chain-termination method* was created by the English chemist Sanger, and it led to the award of one of his two Nobel Prizes. The method requires a single-stranded DNA template, a DNA primer strand (needed to help the DNA polymerase), a DNA

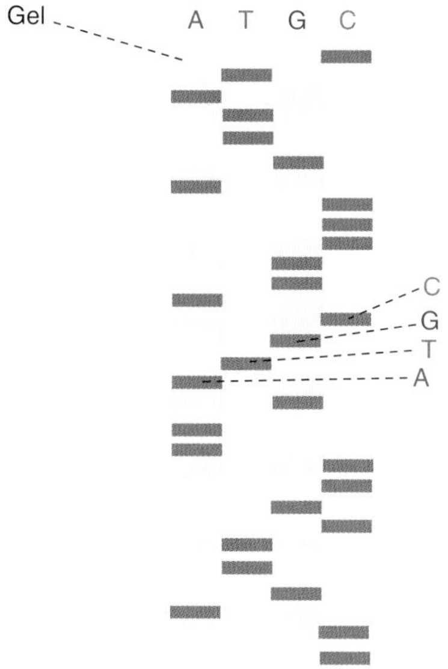

Figure 2.20 *ATGC bases can be read off an electrophoresis gel using the Saenger chain termination method. Each of the four lanes corresponds to a different termination reaction. A sequence of four base pairs, CGTA are highlighted.*

polymerase (which copies pieces of DNA), fluorescently labelled nucleotides, and nucleotides that terminate DNA strand elongation. There are four separate sequencing reaction mixtures. Each sequencing reaction has a different nucleotide added to terminate the reaction. A dark band on a gel results from a chain-termination reaction. The relative positions of different bands among the four lanes can be used to read off the DNA sequence (**Figure 2.20**).

2.13 Genetic Fingerprinting

The *Polymerase chain reaction* (PCR) is a technique that can amplify a few copies of a DNA sequence to generate thousands of millions of copies. The technique uses thermal cycling; cycles of repeated heating and cooling induce successive DNA melting transitions and are followed by enzymatic DNA replication reactions. As PCR progresses the DNA generated by the first reaction is itself used as a template for further replication, and sets in motion a chain reaction in which the quantity of the DNA template is exponentially amplified. To allow the DNA melting to occur without destroying enzymatic activity, PCR requires a heat stable DNA polymerase and such polymerases have been isolated from heat resistant bacteria, e.g. those found in geysers in Yellowstone National Park.

In *genetic fingerprinting* a reference sample is taken from an organism and then compared to another to test for a match. The method is based on PCR to amplify minute quantities of reference sample and then short tandem repeats (STR) are compared in the DNA chains, i.e. highly polymorphic regions that have short repeated sequences of DNA, e.g. four bases repeated. The STRs are targeted with sequence specific primers and then amplified using PCR. Each individual STR polymorphism is shared by 5–20% of the population. The study of thirteen loci thus results in an accidental match probability of $(1/20)^{13} \sim 1/10^{18}$, which gives an extremely high level of discrimination without the need to sequence the complete genome.

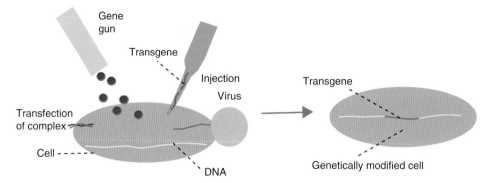

Figure 2.21   *Possible methods to insert a gene into an organism, to create a genetically modified organism include injection, viral vectors, gene guns and transfection complexes.*

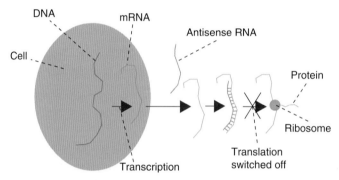

Figure 2.22   *Antisense RNA technology can allow genes to be temporarily switched off. The antisense RNA chain complexes with a complementary mRNA chain and interrupts the process of translation.*

2.14   Genetic Engineering

The field of genetic engineering considers the introduction of an altered gene into an organism. Genes can be transferred in a number of ways. One method is to highjack the life cycle of lentiviruses and use them to make permanent useful changes to an organism's DNA (rather than just make more copies of themselves, which is their usual strategy). Alternatively, gene guns can propel DNA chains into the cell's nucleus that can then integrate with the genome, or the new DNA can be injected or transferred with chemical vectors such as cationic polyelectrolytes or liposomes (**Figure 2.21**). There are a wide range of commercial applications for gene transfer, such as the creation of goats that secrete medicinally useful antibiotics in their milk or plants that have an arctic fish antifreeze gene introduced in their genome to make them frost resistant. Other possibilities include curing diseases with single-point mutations such as cystic fibrosis, to more whimsical goals such as the production of red grass for lawns or purple tomatoes for greengrocers.

The gene transferred during genetic engineering is called a *transgene* and the engineered product is called a *transgenic* organism. The moral/ethical issues and controls for such processes are still being widely debated.

Another powerful technique in genetic engineering is to use complementary strands of mRNA to reduce the expression of DNA (**Figure 2.22**), since they can prevent the translation step. This is a temporary measure, because it is not inherited by any daughter cells and mRNA is broken down fairly quickly. Genes in an organism can thus be switched off and the resultant changes to the cell provide valuable information on their roles in cell metabolism.

2.15 Tissues

Cells act co-operatively in multicellular organisms and are hierarchically arranged into tissues, organs and organ systems. Tissues contain both cells and other materials such as the extracellular matrix. The human body is composed of more than two hundred different kinds of cells that are distributed among five main types of tissue; *nerve*, *blood*, *fibroblast*, *muscle* and *connective tissue*.

There are four distinct forms of mammalian *muscle cells*; skeletal and cardiac (which both form striated muscular tissues), smooth muscle (found in blood vessels and intestines) and myoepithlial cells (again present in intestines). Muscle cells are responsible for the production of force and movement.

Nerve cells are used in signalling (**Part E**). They are highly branched and their morphology allows them to receive up to 10^5 inputs from other cells. The electrochemistry of nerve cells is a fascinating area; the efficiency and time response of these electrical circuits has been carefully optimised by evolution. Nervous tissue is composed of nerve cells (neurons), and sensory cells (ear cells, retinal cells, etc.).

Blood cells have a squashed doughnut shape (**Figure 2.23**), which is related to the elasticity of their cytoskeleton. Red blood cells carry carbon dioxide and oxygen, towards and away from the lungs respectively. White blood cells play a role in the fight against infections and inflammatory reactions (granulocytes, monocytes and macrophages). Lymphocytes contribute to immune responses.

Fibroblast cells are largely responsible for the secretion and regulation of the extracellular matrix, e.g. the production of molecules such as collagen or aggrecan. *Epithelial cells* control the passage of material across the boundary of organs, e.g. in the interior of the intestinal tract or the skin of the organism.

Connective tissues include bones and cartilage. They play a central role in the organisation of the whole organism anatomy.

The relative arrangement of cell types into organs is described by anatomical studies and reference should be made to the extensive literature if specific maps of the relative distribution of cells in a particular organ of an organism are required.

2.16 Cells as Experimental Models

Cells have evolved from a common ancestor, so they have conserved mechanisms. The diversity of present-day cells thus means that a certain biophysics experiment may often be more readily done with one type of cell than with another, e.g. experiments with yeast cells are often easier to perform than those with human cells.

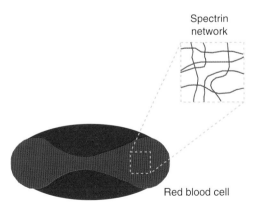

Figure 2.23 *Cross section through a human red blood cell, which naturally adopts the squashed doughnut shape. The spectrin network (formed from a fibrous protein) in the cell wall is a dominant factor for the determination of the morphology of the cell.*

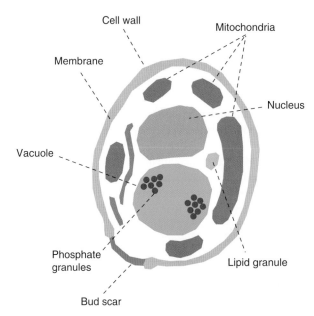

Figure 2.24 *A schematic diagram of the organelles inside a yeast cell. The cell is a relatively simple example of a eukaryote and the DNA is stored in a nucleus.*

Prokaryotic cells (e.g. bacteria such as E.coli) are ideal models for the study of many fundamental processes in biochemistry and molecular biology that include molecular genetics, DNA replication, genetic coding, gene expression, and protein synthesis. This is facilitated by their small genome size, rapid growth rate (they replicate every 20–40 min) and simple nutritional requirements. E.coli are often used to engineer proteins and to amplify DNA for genetic analysis.

Yeast is the simplest eukaryotic cellular organism, and is thus a model for fundamental eukaryotic cell biology. The word yeast is derived from the Old English word for beer. *Saccharomyces cerevisiae* (from the latin for beer) is the most frequently studied species of yeast (**Figure 2.24**). It consists of three times more DNA than the genome of E. coli, but can be readily grown in the laboratory and divides every two hours. Intensive research studies for the creation of a complete circuit diagram for the gene interactions in a eukaryotic cell have thus focused on yeast cells as an important first step.

Human and *animal cells* are the most difficult to study, although there is a large medical and veterinarian impetus which motivates the study. The human body contains more than two hundred different types of cells and a large genome. The cells are delicate, easily damaged and lots of expertise is needed to keep them alive and replicating. However, isolated human cells and other mammalian cells can be grown in culture. This has allowed DNA replication, gene expression, protein synthesis and cell division to be examined in detail. Mice are often used as a model mammalian system when human studies are impossible.

2.17 Stem Cells

Stem cells are found in most multicellular organisms. These cells are able to renew themselves by mitotic cell division and differentiate into a diverse range of cell types. Two broad types of stem cells are found in mammals. *Embryonic* stem cells are derived from the initial ball of cells, which result from division of the initial

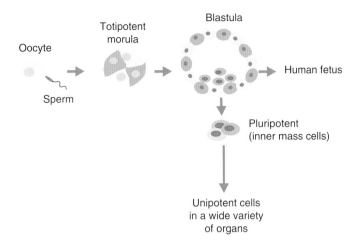

Figure 2.25 *Stem cells are categorised in terms of the number of cell types into which they can differentiate. Pluripotent cells can differentiate into nearly any cell type.*

fertilised egg cell (**Figure 2.25**). *Adult* stem cells are found in a range of adult tissues and are important for renewal of tissue (many cell types have a limited number of divisions before they self-destruct).

Stem cells vary in their degree of potency. *Totipotent cells* can differentiate into embryonic cells and other cell types. *Pluripotent* cells can differentiate into nearly all cell types. *Multipotent* cells can differentiate into a closely related family of cells. *Oligopotent* cells can differentiate into only a few cell types. *Unipotent* cells can produce only more of the same type of cell, but they are still stem cells because they can renew themselves indefinitely.

An important current research challenge is to reverse engineer the developmental clock on cells to return them to the pluripotent state. Ethical issues are then reduced as the need for foetuses or umbilical cord storage is removed, which were the previous sources for pluripotent stem cells. The scope of the possible benefits of stem-cell research should not be oversold, but they currently appear to be huge. In principle, pluripotent stem cells can be plugged into injured regions of the body to treat: strokes, brain injury, learning defects, Alzheimer's, Parkinson's, baldness, blindness, deafness, missing teeth, bone marrow, spinal cord injury, myocardial infarction, muscular dystrophy, osteoarthritis, rheumatoid arthritis, Crohn's disease, muscular dystrophy, diabetes, and cancers.

Suggested Reading

If you can only read one book, then try:

Alberts, B. *et al.* (2010) *Essential Cell Biology*, 3rd edition, Garland. This is a good introductory textbook for people with no cellular biology background.

Alberts, B., Johnson, A., Lewis, J., Ruff, M., Roberts, K. & Walter, P. (2008) *Molecular Biology of the Cell*, 5th edition, Garland Science. More advanced coverage of cellular processes, once the contents of the introductory book have been digested.

Church, G. & Regis, E. (2012) *Regenesis*, Basic Books. Popular and very enthusiastic introduction to synthetic biology which describes some of the key experiments performed.

Nadeau, J. (2012) *Introduction to Experimental Biophysics*, CRC Press. Good introduction to useful experimental biochemistry procedures for physicists.

Nowak, M.A. (2006) *Evolutionary Dynamics*, HUP. Mathematical introduction to evolutionary theory in biology.

Sheehan, P. (2009) *Physical Biochemistry*, Wiley. Good emphasis on modern biochemical characterisation techniques.

Weinberg, R.A. (2013) *The Biology of Cancer*, 2nd edition, Garland Science. Comprehensive coverage of the subject of cancer.

Tutorial Questions 2

2.1 The creation of complex nanostructures (rotary motors in bacteria that require the concerted action of over twenty separate proteins) or the creation of a complex organ, such as the eye, are often presented as inexplicable in terms of the theory of evolution. Choose an example and demonstrate that closer inspection shows more than ample evidence exists for evolutionary intermediates.

2.2 Two cells are stuck together in a piece of tissue. List the range of forces that could contribute to this process of adhesion.

2.3 Do a Google search on Max Perutz's work on protein structure; how oxygen is transported by myoglobin and haemoglobin. Explain how this was an extension of the X-ray diffraction techniques developed by other physicists (Bragg, von Laue) to biological systems.

2.4 Describe the molecular features of oil mixing with water.

2.5 Give five more examples of the uses of genetic engineering in addition to those given in the text.

2.6 Use your imagination to speculate how a cell communicates with a neighboring cell. Describe a possible method to monitor this process.

Part II
Soft Condensed-Matter Techniques in Biology

Classically, condensed matter was considered under the archetypal classifications of gases, liquids, and solids. A list that was sometimes extended to include plasmas. These idealisations are very useful for the introduction of reference models, but unfortunately there are a huge range of everyday materials that just do not fit within this classification scheme: foams, granular matter, membranes, polymers, liquid crystals, colloids, materials confined to surfaces, etc. The full list of these other alternative phases of matter is huge.

The term 'soft matter' was coined by the Nobel Prize winning physicist Pierre Gilles de Gennes. Initially, the classification was restricted to materials whose geometries experience large fluctuations due to thermal energies, but rapidly it became a banner used for a wide range of materials, often found ubiquitously (household items), that are neither gases, liquids nor solids. The community of soft-matter physicists has been actively exploring synthetic materials for many years now, and there are consequently lots of experimental/theoretical tools that have been developed in soft condensed matter for synthetic materials that can be directly applied to biology. For fundamental insights into soft matter, synthetic materials have the advantage of improved simplicity and increased robustness to experimentation, when compared to biological systems (many cells and biological molecules are extremely complicated and fragile).

In this Part a selection of results from soft-matter physics will be introduced that are deemed useful for biology. The discussion begins with some general statistical tools, which is followed by the study of mesoscopic forces and phase transitions; then three classes of biological molecules that have clear synthetic analogues are covered: polymers, liquid crystals and membranes; dynamic phenomena are then studied that includes cellular and intracellular motility; self-assembly is examined, the unusual manner in which many biological assemblies automatically structure themselves; surface phenomena are highlighted, since a reduction in dimensionality has a dramatic impact on the forces experienced by materials; in addition the areas of charged polymers, continuum mechanics, fluid mechanics, rheology (viscoelasticity), motor proteins, biomaterials and DNA complexation will be approached. This is the largest Part in the book, which emphasises the rich palette of ideas developed in condensed-matter physics that can subsequently be applied to biology.

The Physics of Living Processes: A Mesoscopic Approach, First Edition. Thomas Andrew Waigh.
© 2014 John Wiley & Sons, Ltd. Published 2014 by John Wiley & Sons, Ltd.

Suggested Reading

Piazza, R. (2011) *Soft Matter: The Stuff that Dreams are Made of*, Springer. Well written, amusing, popular account of soft matter physics.

Hamley, I.W. (2007) *Introduction to Soft Matter: Synthetic and Biological Self-Assembling Materials*, Wiley-Blackwell. Compact textbook that introduces the physical chemistry of soft matter.

Mitov, M. & Weiss, G. (2012) *Sensitive Matter: Foams, Gels, Liquid Crystals and other Miracles*, Harvard University Press. Another good short popular introduction to soft condensed-matter physics.

de Gennes, P.G., Badoz, J. & Reisinger, A. (1996) *Fragile Objects: Soft Matter, Hard Science and the Thrill of Discovery*, Springer. Interesting examples of research into soft-matter physics from one of the founders of the field.

3

Introduction to Statistics in Biology

Statistics and statistical mechanics are now standard analytical tools in biology. Secondary school educations introduce statistics as a way to check for the consistency between theory and experiment, not as a predictive theory in its own right. The field of *statistical mechanics* gives exactly such a theory for soft matter, where accurate predictions can be made on the behaviour of soft systems based on the enumeration of the number of possible states of the systems. Fundamental research into statistical mechanics is a large field of study, so only a few principle results will be introduced that are required as tools in subsequent sections. Reference should be made to the suggested reading list for readers who are new to these ideas and want to see a systematic development of the key results. Other more specialist areas of biological statistics, for example data mining from the human genome project, undoubtedly have a long rosy future, but have been omitted due to space constraints.

Often, a key contribution of physicists in collaborations with biologists involves the statistical analysis of large data sets, and biophysicists should thus be familiar with as many statistical tools as possible.

3.1 Statistics

A very rapid review of some statistical tools will first be made. Mathematics, analogous to a musical instrument, needs to be practised for pleasing results. Thus, reference should be made to the problem sets in the specialised statics books described at the end of the chapter, and this should be a priority for the reader if any of the ideas in this section appear to be novel.

It is useful to define characteristic values of a probability distribution to develop a quantitative understanding. The simplest characteristic value is the *average* or *mean* value. For a discrete distribution the mean ($<x>$) is simply calculated if the underlying probability distribution ($P(x_i)$) is known,

$$\langle x \rangle = \sum_i x_i P(x_i) \tag{3.1}$$

The Physics of Living Processes: A Mesoscopic Approach, First Edition. Thomas Andrew Waigh.
© 2014 John Wiley & Sons, Ltd. Published 2014 by John Wiley & Sons, Ltd.

where x_i are the discrete values of x that occur in the distribution. For a continuous probability distribution ($P(x)$) the discrete summation becomes an integral and an expression for the mean value is

$$\langle x \rangle = \int_{-\infty}^{\infty} xP(x)dx \tag{3.2}$$

The *variance* quantifies the spread or dispersion of the data and is often the second most useful parameterisation of a probability distribution after the mean. The variance (σ^2) can be defined in terms of the variation from the mean value,

$$\sigma^2 = \frac{1}{N}\sum_i (x_i - \langle x \rangle)^2 = \langle x^2 \rangle - \langle x \rangle^2 \tag{3.3}$$

where N is the number of data points in the distribution. This is the variance of a discrete distribution, but the extension to continuous distributions is simply performed using equation (3.2) and the second definition in equation (3.3) in terms of expectation values.

The *skew* parameter (γ) is also a useful measure, since it helps describe the asymmetry of a probability distribution (it is based on an odd moment),

$$\gamma = \frac{1}{N\sigma^3}\sum_i (x_i - \langle x \rangle)^3 \tag{3.4}$$

The *kurtosis* (k) is another parameter that is frequently used and is defined as

$$k = \frac{1}{\sigma^4}\left\langle (x - \langle x \rangle)^4 \right\rangle - 3 \tag{3.5}$$

The kurtosis is useful to quantify how much a distribution varies from a Gaussian, since the kurtosis is zero for a Gaussian probability distribution. Gaussian probability distributions are the most commonly observed variety in physics and biology, so deviations of the kurtosis from zero can have important implications, e.g. they can indicate nonergodicity and glassy phenomena. In principle, infinitely many moments of a probability distribution can be calculated and mathematical techniques exist especially for the job of moment creation, i.e. *characteristic functions*.

Often in biology people are interested in the correlation between two variables. A useful method to quantify the degree of correlation between variables x and y is given by the *covariance*,

$$\text{cov}(x,y) = \frac{1}{N}\sum_i (x_i - \langle x \rangle)(y_i - \langle y \rangle) \tag{3.6}$$

An example of the use of covariance is provided by fluorescence microscopy experiments in which colocalisation is probed. Two different coloured fluorophores can be used to label different biomolecules that interact with one another in a cell. The covariance can be used as a measure of the cooperative transport of the two molecules. A high value of the covariance implies a high degree of cooperativity.

The *Gaussian distribution* is the most common probability distribution that occurs in physics and biology (**Figure 3.1**). The functional form of a Gaussian probability distribution is

$$P(x;\langle x \rangle,\sigma) = \frac{1}{\sigma\sqrt{2\pi}}e^{-(x-\langle x \rangle)^2/2\sigma^2} \tag{3.7}$$

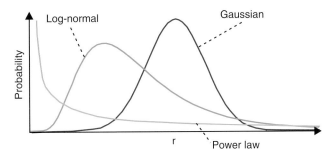

Figure 3.1 *Schematic diagram of Gaussian, log-normal and power law (fat-tailed, Zipf) probability distributions. The probability is shown as a function of the size of the event (r).*

The reason for the preponderance of Gaussian distributions in nature is due to the *central limit theorem*. This theorem states that when the sum X ($X = \sum_i x_i$) of N independent variables (x_i) is taken from a distribution of mean μ_i and variance σ_i^2, then the following three statements are true:

a. X has a mean value of

$$\langle X \rangle = \sum_i \mu_i \tag{3.8}$$

b. X has a variance of

$$V(X) = \sum_i \sigma_i^2 \tag{3.9}$$

where σ_i^2 is the variance of variable x_i.

c. X follows a Gaussian probability distribution in the limit of large sample sizes (N), as $N \rightarrow \infty$.

Thus any quantity calculated from the cumulative effect of many independent factors will be Gaussian to a good approximation, almost independently of the probability distributions of the underlying variables. There are a few exceptions to this rule, such as Levy flights and some other fat-tailed probability distributions, but these are relatively rare. Such awkward probability distributions will, however, be discussed further in the section on random walks, since they can be invoked to explain subdiffusive motion, which is a dominant mode of transport at short times in live cells.

A second common distribution (which is approximated by a Gaussian for large sample sizes) is the *Poisson distribution*. The probability of the observation of r independent events with mean number μ (where μ is also the constant rate at which events occur) is

$$P(r;\mu) = \frac{e^{-\mu}\mu^r}{r!} \tag{3.10}$$

The *binomial distribution* gives the probability of the observation of r events out of n tries, if an event in each try has a constant probability p of success. The functional form is

$$P(r;p,n) = p^r (1-p)^{n-r} \frac{n!}{r!(n-r)!} \tag{3.11}$$

The distribution can also be approximated by a Gaussian for large sample sizes.

The *log-normal distribution* provides a useful analytic form for skew non-Gaussian distributions. The distribution is given by

$$P_r(r;\mu,\sigma) = \frac{1}{r\sigma\sqrt{2\pi}}e^{-\frac{(\ln r - \mu)^2}{2\sigma^2}} \tag{3.12}$$

Some unusual probability distributions have *fat tails*, i.e. there is a small, but significant, probability that large events (*r*) can occur. An example of a fat-tailed probability distribution is the Levy flight,

$$p(r;q,w) = \sqrt{\frac{w}{2\pi}}\frac{e^{-\frac{w}{2(r-q)}}}{(r-q)^{3/2}} \tag{3.13}$$

where *q* is a location parameter and *w* is a scale parameter. The Levy flight probability distribution provides an alternative model for random walks (**Chapter 7**) and has been used widely for models of the motion of biological organisms, e.g. the motility patterns of animals (sharks, albatross, etc.). Another example of a fat-tailed distribution is the *Zipf* power law probability distribution

$$p(r) = ar^{-(s+1)} \tag{3.14}$$

where *a* and *s* are positive constants.

Such simple probability distributions can be used to construct Monte Carlo simulations, which are a useful tool if analytical expressions are too hard to calculate for a model of a particular phenomenon, e.g. in the construction of a stochastic model of motor protein stepping with many internal degrees of freedom. Monte Carlo simulations require a cumulative probability density function for efficient algorithmic implementation, although some badly behaved probability distributions can be simulated (albeit slowly) if inverse sampling methods are used.

3.2   Entropy

When an isolated system is left alone for a sufficiently long time, it evolves into a state of thermal equilibrium. This equilibrium state consists of a range of microstates populated so that their probability distribution has the greatest disorder that is allowed by the physical constraints on the system. A good measure of the amount of disorder in a system is its entropy (*S*), which was introduced by Boltzmann,

$$S \equiv k_B \ln\Omega \tag{3.15}$$

where k_B is Boltzmann's constant, and Ω is the number of microstates allowed by the macroscopic constraints.

For example, an experiment is performed where a single coin is flipped and it has two possible microstates, i.e. heads and tails. The entropy of the experiment is highest if the coin is unbiased, i.e. the probability of a head is equal to the probability of a tail (0.5). Each flip of the coin corresponds to a bit of information, a concept that will be returned to in **Section 3.3**. The entropy of the experiment is zero if the outcome of the experiment is known, e.g. a double-headed coin.

When a whole series of coins (*N* in number) are flipped, the number of microstates of the system, where each coin can have one of two states (e.g. head or tail), can be calculated from the number of permutations,

$$\Omega = \frac{N!}{L!(N-L)!} \tag{3.16}$$

(a)

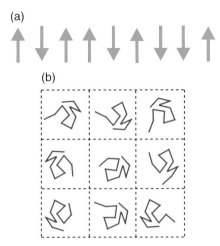

(b)

Figure 3.2   *Calculation of a system's entropy is a powerful tool in statistical mechanics. For example, the entropy can be calculated by counting the number of possible microstates in (a) ferromagnets (the green arrows are spins) or (b) macromolecules (the red lines are distinct protein conformations).*

where L is the number of heads, and $N–L$ is the number of tails. For an unbiased coin the case where the number of heads equals the number of tails is the most probable outcome, since it has the largest number of microstates associated with it. Pictured from another perspective the most likely state is that in which the system's entropy has been maximised. For very large systems in equilibrium the state with maximum entropy is very sharply defined and allows the equilibrium state to be clearly identified. This principle of counting the number of microstates to calculate the entropy can be applied very generally, e.g. the calculation of the spin states of a ferromagnet (**Figure 3.2a**), the possible conformations of a protein chain (**Figure 3.2b**) or the open/closed states of an ion channel.

It is possible to define the temperature of a system in terms of its entropy and energy. The temperature is the quantity that is equalised in two subsystems that exchange energy, when they come to equilibrium. Heat flows to maximise the total disorder in the two systems.

3.3   Information

Another measure of the entropy (S) of a system equivalent to equation (3.15) is

$$S = -k_B \sum_{i=1}^{M} p_i \ln p_i \tag{3.17}$$

where p_i is the probability of state i and M is the number of states in the system. This expression is useful for the consideration of the process of information transfer (or more specifically the Shannon version of it, equation (23.73), where the Boltzmann constant is neglected and $\log_2$ is used instead of ln), particularly in neurons, and is also used in **Chapter 5** for the construction of the free energy during phase separation of liquid mixtures.

3.4 Free Energy

The cost of the creation of order (the reduction of entropy) is that some organised energy must be degraded into thermal form. The description of this process is helped by the introduction of the concept of free energy. The *Gibbs free energy* (G) is defined as

$$G = U + pV - TS \tag{3.18}$$

where U is the internal energy, p is the pressure, V is the volume, T is the temperature and S is the entropy.

In contrast, the *Helmholtz free energy* (H) is defined as

$$F = U - TS \tag{3.19}$$

The *chemical potential* is the Gibbs free energy per particle. Formally, the Helmholtz free energy can be used when the temperature, volume and particle number are fixed in a system and the energy fluctuates. Similarly, the Gibbs free energy can be used when the temperature, volume and chemical potential are held fixed and both energy and particle number can fluctuate. Under some conditions the Helmholtz's and Gibb's free energies can be equal, e.g. if the temperature and volume are held constant.

An example of the minimisation of free energy that allows the combination of entropic randomness with energy to be illustrated is the self-assembly of amphiphilic molecules in solution (**Figure 3.3**). Under the correct physical conditions (pH, temperature, salt concentration, etc.) the micelles can self-assemble into low-entropy well-organised micellar structures, if there is a corresponding increase in the internal energy to offset it (associated with the hydrophobicity of the chain). Thus, the free energy is minimal.

In some cases the change in entropy can give rise to a force. An entropic force (f) can be exerted by a system if its constraints are varied, and the force is given by derivatives of the free energy (F). Consider the entropic force on a system whose length is its constraint, e.g. the size of a box of a gas or the length of a polymer chain (useful for rubber elasticity with polymer chains, **Chapter 10**). The entropic force is given by

$$f = -\frac{dF}{dL} \tag{3.20}$$

where L is the length of the constraint. The cost of upgrading energy from thermal to mechanical form is that the system's order is reduced. Free-energy transduction is least efficient when it proceeds by the uncontrolled release of a big constraint. It is most efficient when it proceeds by the incremental controlled release of many small constraints.

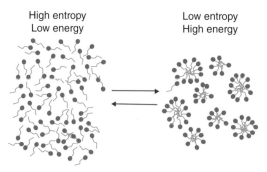

High entropy
Low energy

Low entropy
High energy

Figure 3.3 *Micellar self-assembly can be rationalised through a balance of a system's need to increase the energy of interactions with its need to increase the entropy. Single lipids in solution have high entropy and low energy, whereas micelles have low entropy and high energy. A change in the environmental conditions (e.g. pressure, temperature etc) can shift the equilibrium towards self-assembly or disassembly.*

Enthalpy (*H*) is a further useful thermodynamic quantity to define, since it is often measured in chemical reactions and during changes in phase with calorimetry experiments,

$$H = U + pV \tag{3.21}$$

where *U* is the internal energy, *p* is the pressure and *V* is the volume.

3.5 Partition Function

The probability distribution ($p(E)$) for a small classical system to be in a microstate with energy *E* is found to be equal to a normalisation constant (*a*) multiplied by an exponential factor,

$$p(E) = ae^{-E/kT} \tag{3.22}$$

where *T* is the absolute temperature of the surroundings and *k* is the Boltzmann constant. Equation (3.22) is called the *Boltzmann distribution*. There is a kinetic motivation for the Boltzmann distribution. Consider a population of molecules that can flip between two states *A* and *B*,

$$A \underset{k_-}{\overset{k_+}{\rightleftharpoons}} B \tag{3.23}$$

where k_+ is the rate constant for the forward reaction and k_- is the rate constant for the reverse reaction. The ratio of the number of molecules in the second state to the first is equal to the Boltzmann factor,

$$\frac{B}{A} = e^{-\Delta E/kT} \tag{3.24}$$

where ΔE is the difference in energies between the two states. In **Section 3.4** the Helmholtz free energy of a molecular system was introduced as equation (3.19). It can be shown that the minimal value of the free energy is just

$$F = -kT \ln Z \tag{3.25}$$

where the partition function (*Z*) is defined as

$$Z = \sum_j e^{-E_j/kT} \tag{3.26}$$

and *j* is the number of energy levels in a system and E_j is the value of each energy level. Systems in equilibrium move to minimise their free energy and this can in principle be used as a tool to model the behaviour of systems.

To find the probability $p(E_i)$ that an energy state E_i is occupied the partition function can again be used,

$$p(E_i) = \frac{e^{-E_i/kT}}{Z} \tag{3.27}$$

Here, the partition function (*Z*) forms the normalisation factor ($a = 1/Z$) required in equation (3.22). Average quantities can then be constructed from such probabilities in the usual manner for a discrete distribution, e.g. the average energy of the system is

$$\langle E_i \rangle = \sum_i p(E_i) E_i = \frac{\sum_i E_i e^{-E_i/kT}}{Z} \tag{3.28}$$

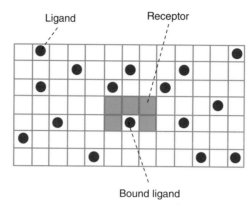

Figure 3.4 *Schematic diagram of a microstate during the association of a ligand with a receptor on a square lattice, e.g. nicotine adsorbing to nicotine receptors in the brain. A Langmuir adsorption isotherm is observed for the adsorption kinetics of the ligand to the receptor.*

As an example of the power of the mathematical machinery of statistical mechanics, consider the calculation of ligand binding to a substrate, e.g. binding of an enzyme to an ion. A lattice model is used to model the binding behaviour of the ions (**Figure 3.4** shows a microstate of the arrangements of the enzyme and the ions on the lattice). The number of microstates is again given by equation (3.16), but here N is the total number of available lattice sites and L is the number of ions.

The partition function (Z) can be written as a sum of two components; one where all the ions are unbound and the other where a single ion is bound

$$Z(L,\Omega) = \underbrace{\sum e^{-\beta L \varepsilon_s}}_{\text{unbound}} + e^{-\beta \varepsilon_b} \underbrace{\sum e^{-\beta(L-1)\varepsilon_s}}_{\text{bound}} \tag{3.29}$$

where ε_s is the energy of an ion in solutions, ε_b is the energy of an ion bound to the enzyme and $\beta = 1/kT$ is defined to make the equations slightly more compact. The summations can be evaluated by the consideration of the number of microstates,

$$\sum_{\text{unbound}} e^{-\beta L \varepsilon_s} = e^{-\beta L \varepsilon_s} \frac{\Omega!}{L!(\Omega-L)!} \tag{3.30}$$

and

$$\sum_{\text{bound}} e^{-\beta(L-1)\varepsilon_s} = \frac{\Omega!}{(L-1)!(\Omega-(L-1))!} e^{-\beta(L-1)\varepsilon_s} \tag{3.31}$$

The partition function, equation (3.29), can therefore be written as equation (3.29),

$$Z(L,\Omega) = e^{-\beta L \varepsilon_s} \frac{\Omega!}{L!(\Omega-L)!} + e^{-\beta \varepsilon_b} e^{-\beta(L-1)\varepsilon_s} \frac{\Omega!}{(L-1)!(\Omega-(L-1))!} \tag{3.32}$$

For large numbers of ions,

$$\frac{\Omega!}{(\Omega-L)!} \approx \Omega^L \tag{3.33}$$

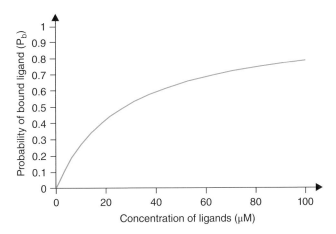

Figure 3.5 *Langmuir adsorption isotherm for the probability that a ligand binds to a receptor as a function of ligand concentration.*

Equation (3.32) can be simplified to give

$$p_{\text{bound}} = \frac{e^{-\beta\varepsilon_b}\dfrac{\Omega^{L-1}}{(L-1)!}e^{-\beta(L-1)\varepsilon_s}}{\dfrac{\Omega^L}{L!}e^{-\beta L\varepsilon_s} + e^{-\beta\varepsilon_b}\dfrac{\Omega^{L-1}}{(L-1)!}e^{-\beta(L-1)\varepsilon_s}} \tag{3.34}$$

This expression can be further simplified algebraically to give

$$p_{\text{bound}} = \frac{(L/\Omega)e^{-\beta\Delta\varepsilon}}{1 + (L/\Omega)e^{-\beta\Delta\varepsilon}} \tag{3.35}$$

where $\Delta\varepsilon = \varepsilon_b - \varepsilon_s$. Equation (3.35) is equivalent to

$$p_{\text{bound}} = \frac{(c/c_0)e^{-\beta\Delta\varepsilon}}{1 + (c/c_0)e^{-\beta\Delta\varepsilon}} \tag{3.36}$$

where c is the ligand concentration and c_0 is a reference concentration. Equation (3.36) is the *Langmuir adsorption isotherm*, a classic model that describes the adsorption of molecules to surfaces, and can be used to describe the lock and key model of ligand binding, e.g. with protein receptors (**Figure 3.5**). A more specific example would be as a model of the adsorption of nicotine to nicotine receptors in the brain. The average occupancy of the receptor can be calculated as a function of the number of nicotine molecules using the Langmuir model.

3.6 Conditional Probability

In general terms, the conditional probabilities of two events A and B are related by *Bayes' formula*,

$$p(A/B) = \frac{p(B/A)p(A)}{p(B)} \tag{3.37}$$

where $p(A/B)$ is the probability of A if B is known and vice versa for $p(B/A)$. $p(A)$ is the probability of A and $p(B)$ is the probability of B. This simple looking equation has some important consequences and is used as the cornerstone for the scientific paradigm of *Bayesian inference*. Probability is defined as a degree of belief, rather than just a measure of the proportion of possible outcomes.

Bayesian inference can provide some important practical tools. For example, consider optical tracking of a particle using a CCD camera and a microscope. Each new frame in a movie could be considered as a separate measurement of the particles coordinates x and y. However, this estimate can be improved upon, because the positions are not totally independent. With a reasonable estimate of a function that describes the motion of the particle (to predict where it next will be from its previous position), Bayes formula allows a more accurate calculation of the particle's position to be made. This reasoning forms part of the *Kalman filter*, which is used widely in satellite global positioning systems (GPS), and has found wider applications in biological data analysis, e.g. the interpretation of electrophysiological recordings.

3.7 Networks

A network is defined as a collection of nodes that are connected by links. The total number of connections to a node is called the *degree* of the node. The highest degree of a node in **Figure 3.6** is thus 7.

Networks occur in biology in a number of different guises, e.g. the interactions of neurons in the brain, interaction of organisms in an ecosystem, the interactions of individuals in a society, the interactions of genes in transcription networks, etc. Network theory (graph theory) aims to describe the emergent properties of networks in terms of their statistics. Useful characteristic quantities include the *shortest path length* between two nodes, the *degree distribution* of the nodes, measures of *clustering* of nodes and the distribution of *loops*.

Random networks are constructed if the probability that a link joins any two nodes is a global constant, independent of the position of any of the other links. However, many of the networks observed in biology are not random. A classic example is the *small-world network* (**Figure 3.6**). In these networks nodes are densely connected to their nearest neighbours, but there are also a small fraction of long-distance connections (the probability of long-distance connections follows a fat-tailed distribution). The long-distance connections have dramatic consequences for the network's behaviours. The shortest path length between two nodes

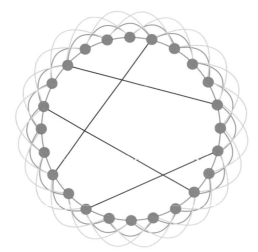

Figure 3.6 *A schematic diagram of a small-world network. The long-range (red) links play a crucial role in the determination of the networks properties. They cause a dramatic decrease in the average shortest path length between two nodes. Green circles indicate network nodes and blue links are short-range connections.*

becomes much smaller in small-world networks and leads to the classic experimental observation that everyone in the world knows everyone else through at most 6 people (the 6 degrees of separation experiments were performed by Stanley Milgram in 1967).

Rigorous mathematical work into small-world networks is a surprisingly recent event, given that these networks underpin the basic properties of a vast number of examples, e.g. the interconnectivity of neurons, the world wide web and telecommunication networks.

Specific network motifs have become important in systems biology, where researchers are trying to reverse engineer gene transcription networks in terms of the interactions of their chemical circuits. Network motifs are considered in more detail in **Chapter 22**.

Suggested Reading

If you can only read one book, then try:

Top recommended book: Dill, K. & Bromberg, S. (2011) *Molecular Driving Forces*, 2nd edition, Garland Science. Excellent, clearly written, undergraduate textbook that considers entropy and thermodynamics.

Allon, U. (2007) *An Introduction to Systems Biology: Design Principles of Biological Circuits*, CRC. Classic introductory text in systems biology that introduces some tools from network theory.
Barrat, A., Barthelemy, M. & Vespignani, A. (2008) *Dynamical Processes on Complex Networks*, Cambridge University Press. Useful account of the statistical physics of complex networks.
Bialeck, W. (2012) *Biophysics: Searching for Principles*, Princeton. Excellent advanced discussion of information and entropy in biological systems, although at a postgraduate level.
Denny, M. & Gaines, S. (2000) *Chance in Biology: Using Probability to Explore Nature*, Princeton. Simple discursive introduction to useful ideas from statistics.
Dorogovtsev, S.N. (2009) *Lectures on Complex Networks*, Oxford University Press. Good compact discussion of physical processes associated with complex networks.
Krapivsky, P.L., Redner, S. & Ben-Naim, E.A (2010) *A Kinetic View of Statistical Physics*, Cambridge University Press. Very advanced text, but considers some modern results in statistical physics that are ripe for investigation in biology.
Nelson, P. (2007) *Biological Physics*, W.H.Freeman. Demonstrates that biology can be used as a vehicle to learn statistical physics. Has a wonderful range of biological problems.
Newman, M.E.J. & Barkema, G.T. (1999) *Monte Carlo Methods in Statistical Physics*, Clarendon. Clearly explained account of computational techniques in statistical physics that are equally useful for the simulation of biological phenomena.
Newman, M.E.J. (2010) *Networks: An Introduction*, Oxford University Press. Another useful overview of the physics of networks.
Phillips, R., Kondev, J., Theriot, J. & Garcia, H.G. (2013) *Physical Biology of the Cell*, Garland. Expansive account of biological physics with a good introduction to statistical physics.
Sethna, J.P. (2006) *Statistical Mechanics: Entropy, Order Parameters and Complexity*, Oxford University Press. Useful modern undergraduate textbook on statistical mechanics.

Tutorial Questions 3

3.1 The concentration profile for particles in solution in a vessel as a function of height (z) depends on their density (through their net mass m_{net}). The particles have a Boltzmann distribution for their energies

$$c(z) = A_1 e^{-m_{net}gz/kT}$$

where g is gravity and kT is the thermal energy (k is the Boltzmann constant and T is the temperature). Calculate a value for A_1.

4

Mesoscopic Forces

The reader will be familiar with some simple manifestations of the fundamental forces that drive the interactions between matter such as electrostatics, gravity and magnetism. However, nature has used a subtle blend of these forces in combination with geometrical and dynamical effects to determine the interactions of biological molecules. Forces that act on the micro/nanoscale are called mesoscopic. These mesoscopic forces are not fundamental as with electromagnetism or gravity, but the separation into the different contributions of the elementary components would be very time consuming, if indeed it is practically possible, with molecular dynamic simulations. Therefore, in this chapter a whole series of simple models for mesoscopic forces will be studied to help with their quantification and some generic methods to experimentally measure the forces will be reviewed. There are a rich variety of mesoscopic forces that have been identified at the nanoscale. These include *van der Waals force*, *hydrogen bonding*, *electrostatic interactions*, *steric forces*, *fluctuation forces*, *depletion forces*, and *hydrodynamic interactions*.

4.1   Cohesive Forces

The predominant force of cohesion between matter is the *van der Waals interaction*. Regions of objects made of the same material always attract each other due to spontaneously induced dipoles. The strength of van der Waals bonds is relatively weak $\sim 1 \text{ kJ mol}^{-1}$, but the forces act between all types of atom and molecule (even neutral ones).

  A fundamental definition of the van der Waals interactions is a force of quantum mechanical origin that operates between any two molecules, and arises from the interaction between spontaneously oscillating electric dipoles. Although it is not wanted to overburden the description with the detailed quantum mechanics of the calculations, the potential ($V_{12}(r)$) which gives rise to the van der Waals force between molecules 1 and 2 can be defined as

The Physics of Living Processes: A Mesoscopic Approach, First Edition. Thomas Andrew Waigh.
© 2014 John Wiley & Sons, Ltd. Published 2014 by John Wiley & Sons, Ltd.

$$V_{12}(r) = -\frac{1}{24(\pi\varepsilon_0)^2} \frac{1}{r^6} \sum_{n,k} \frac{\left|\langle n|\vec{m}|0\rangle_1\right|^2 \left|\langle k|\vec{m}|0\rangle_2\right|^2}{(E_1^n - E_1^0) + (E_2^k - E_2^0)} \equiv -\frac{C_{12}}{r^6} \qquad (4.1)$$

where $\langle n|\vec{m}|0\rangle$ is the transition dipole moment from the quantised state n to 0 for molecule 1 and similarly $\langle k|\vec{m}|0\rangle$ is that for molecule 2 from quantised state k to 0, r is the distance between the two molecules, and C_{12} is a constant. E_1^n and E_1^k are the energies of quantum states n and k for molecules 1 and 2, respectively. Thus, there is a characteristic $1/r^6$ dependence of the potential between point-like molecules, with a single characteristic constant of proportionality (C_{12}, the Hamaker constant), which depends on the variety of molecules considered. Van der Waals forces are sometimes called a dispersion interaction, because the same quantities determine both the optical properties of the molecules (the dispersion of light) and the forces between them. It is therefore possible to observe the effects of van der Waals forces optically with micrometre-sized colloidal particles in solution.

The manipulation of van der Waals forces tends to be more complicated in practice than many of the fundamental interactions that are encountered by students in foundation physics courses. Van der Waals forces are *long range* and can be effective from large distances (>10 nm) down to interatomic spacings (<0.1 nm). The forces may be *repulsive* or *attractive*, and crucially, in general they do not follow a single simple power law of the separation distance as is illustrated in **Figure 4.1** for four separate possible geometries ($V(r)$ is the potential). Indeed even these power laws are just the first-order terms of a much more complicated power-series expansions. Van der Waals forces tend to both bring molecules together and mutually align or orient them. Unlike gravitational and Coulomb forces, van der Waals forces are *not generally additive*, e.g. the sum of the force of material A with material B (F_{AB}) and material A with material C (F_{AC}) is not necessarily equal to the force of material A with both material B and material C at the same time ($F_{ABC} \neq F_{AB} + F_{AC}$). At larger separations (>10 nm) the effect of the finite speed of propagation (the speed of light, c) of the interaction also becomes important. This is the *retardation effect* and is observed experimentally in a r^{-7} dependence on separation (r) for the interaction potential of point objects rather than r^{-6} at close distances, and for semi-infinite

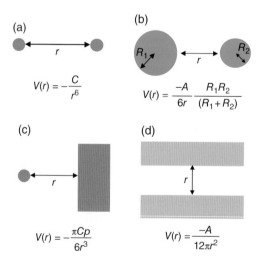

Figure 4.1 *The leading term in the potential (V) of the van der Waals interactions between surfaces as a function of separation (r) depends on the geometry. Four geometries are shown (a) two point atoms, (b) two spheres, (c) a point atom and a plane, and (d) two plane surfaces.*

sheets it is r^{-3} rather than r^{-2} for close distances. The interaction potentials (and thus forces laws from $F = -\dfrac{dV}{dr}$) illustrated in **Figure 4.1** can be proved by careful summation of the contributions in equation (4.1) over an extended body.

Question: Explore the strength of the van der Waals forces on a fly stuck to the ceiling. There are 3000 hairs per foot and six feet per fly. Calculate the maximum mass (m) of the fly that maintains contact with the ceiling. Model the interaction experienced by each hair with the ceiling as that due to a sphere with a planar surface. The Hamaker constant (A) for the interaction is 10^{-19} J, the radius of curvature (d) of each of the fly's hairs is 200 nm and the separation distance (r) between the hair and the surface is 1 nm. The van der Waals force is given by (the corresponding potential is not shown on **Figure 4.1**)

$$F_{vw} = -\frac{Ad}{6r^2} \tag{4.2}$$

Answer: In equilibrium the fly's weight (mg) is balanced by the adhesive van der Waals force (F_{vw}),

$$mg = 6 \times 3000 \times 3.33 \times 10^{-9} \, \text{N} \tag{4.3}$$

Therefore, the maximum mass of the fly is 6.12 μg.

In molecular dynamics simulations the interactions between biomolecules are often captured using the Lennard Jones 6–12 potential,

$$V(r) = \varepsilon \left[\left(\frac{r_0}{r} \right)^{12} - 2 \left(\frac{r_0}{r} \right)^{6} \right] \tag{4.4}$$

where r_0 is the equilibrium separation between the particles and ε is a characteristic energy constant. The attractive (negative) term corresponds to the van der Waals force for a point particle and the repulsive force is the hard-sphere term (it originates from the Pauli exclusion principle; there is a large energy penalty for the overlap of filled electronic orbitals when atoms approach closely, **Section 1.1**).

4.2 Hydrogen Bonding

Water exhibits an unusually strong interaction between molecules, which persists into the solid state (**Figure 4.2**, **Section 1.7**). This unusual interaction is given a special name, *hydrogen bonding*, and is an important effect in a wide range of hydrogenated polar molecules with different molecular geometries (**Figure 4.3**). Hydrogen has special polar properties because its nucleus only contains a single proton and thus its charge density has the highest possible value for a stable nucleus.

Hydrogen bonds are typically stronger than van der Waals in the range 10–40 kJ mol^{-1}, but are still weaker than ionic or covalent interactions by an order of magnitude. Hydrogen bonding plays a central role in molecular self-assembly processes such as virus formation, the construction of biological membranes and the determination of protein conformation. Hydrogen bonds (D–A) occur between a proton donor group D, which is a strongly polar group in such molecules as FH, OH, NH, SH and a proton acceptor atom A which is slightly electronegative such as F, O, N and S.

Hydrogen bonding also has important consequences for apolar biomolecules in aqueous solutions. *Clathrate structures* form in the arrangement of water molecules that surround hydrophobic compounds, e.g. the hydrophobic tails of lipids (**Figure 4.4**). The clathrates are labile (the water molecules can exchange position with their neighbours), but the water molecules are more ordered and often assume pentagonal cages. Thus, for apolar biomolecules (e.g. hydrocarbons) the free energy of transfer into an aqueous environment is

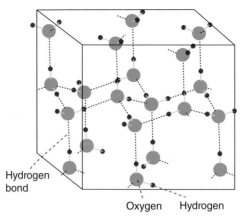

Figure 4.2 *The molecular structure of crystalline water (ice I). The hydrogen bonds are indicated by dotted black lines and the covalent bonds by continuous blue lines. Oxygen atoms are green and hydrogen atoms are red.*

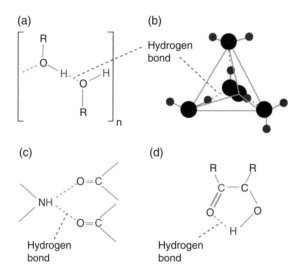

Figure 4.3 *Examples of the range of geometries of hydrogen bonds encountered in organic molecules, (a) chain structure, (b) three-dimensional structure (also seen in **Figure 4.2**), (c) bifurcated structure, and (d) intramolecular bond.*

proportional to the surface area of the molecules, since the entropy change is proportional to the area of the clathrates. This hydrophobic interaction is sometimes given the status of a separate mesoscopic force, since the reduction in free energy causes hydrophobic molecules to be driven together in aqueous solutions. The entropy of the associated water molecules plays a critical role in this case. However, to model such interactions in aqueous solutions care must be taken not to double count the effects under both the 'hydrogen bonding' and 'hydrophobic' banners. Surface force apparatus evidence has been provided on the long-range nature of the hydrophobic effect. It is still an active area of study, but the intersurface repulsive potential ($V(r)$) is thought to have the basic form

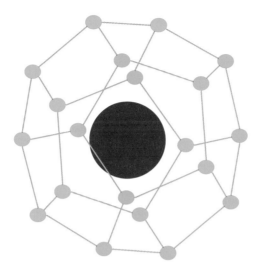

Figure 4.4 *Schematic diagram of clathrate water molecules (blue) around a hydrophobic compound (red). The pentagonal structure is a common theme among clathrates, although there are many different specific realisations.*

$$V(r) = V_0 e^{-r/\lambda} \tag{4.5}$$

where λ is the decay length, typically on the order of nanometers, V_0 is a constant and r is the distance between the surfaces.

Ab initio computational methods to quantify the strength of hydrogen bonds are still at the rudimentary level. One stumbling block is the ability of hydrogen bonds to bifurcate (e.g. a single oxygen atom can interact with two hydrogen molecules simultaneously) that leaves a would-be modeller with a tricky multibody problem. Another challenge is the wide spectrum of dynamic phenomena possible in hydrogen-bonded solutions and care must be taken in the determination of the critical time window for the phenomenon that needs to be modelled (**Section 1.5**). On the experimental side, a series of important advances in the dynamics of hydrogen bonds have been made with the arrival of pulsed femtosecond lasers. The lifetime of water molecules around solution state ions has been directly measured to be of the order of 10 ps. It is hoped that such detailed experiments will allow the refinement of tractable potentials that accurately describe both the structure and dynamics of hydrogen bonds.

4.3 Electrostatics

4.3.1 Unscreened Electrostatic Interactions

In principle, the electrostatic interaction between biomolecules can be calculated explicitly in a molecular dynamics simulation by the addition of all the different long-range contributions (although rapidly fluctuating quantities are always a challenge). Ion–ion, ion–dipole and dipole–dipole interactions need to be treated, not to mention higher multipole terms (**Figure 4.5**). The directionality of the interaction is paramount in the calculation of dipolar interactions, but also becomes important with the lowest order ion–ion interactions with extended objects, e.g. the parallel alignment of charged rods.

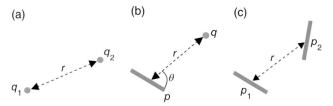

Figure 4.5 *Geometry for the calculation of the strength of interaction between electrostatically charged ions and dipoles, (a) ion–ion interactions (q_1, q_2), (b) ion–dipole interactions (q, p) and (c) dipole–dipole interactions (p_1, p_2).*

Coulomb's law for the interaction energy (E_c) between two point charges is given by

$$E_c = \frac{q_1 q_2}{4\pi\varepsilon\varepsilon_0 r} \tag{4.6}$$

where ε is the relative dielectric permittivity, ε_0 is the permittivity of free space, q_1 and q_2 are the magnitude of the two charges and r is the distance between the charges. The next most important electrostatic interaction is between ions and dipoles. The energy of interaction (E_p) between a dipole (p) and a point charge (q) is given by

$$E_p = -\frac{p^2 q^2}{6(4\pi\varepsilon\varepsilon_0)^2 kT r^4} \tag{4.7}$$

where kT is the thermal energy. Similarly, there is an interaction energy between two separate dipoles (E_{pp}) that is given by

$$E_{pp} = \frac{p_1^2 p_2^2}{3(4\pi\varepsilon\varepsilon_0)^2 kT r^6} \tag{4.8}$$

where p_1 and p_2 are the strengths of dipoles 1 and 2, respectively.

4.3.2 Screened Electrostatic Interactions

Ionic bonds between molecules due to electrostatic interactions can be very strong with strengths on the order of ~500 kJ mol^{-1}. Thus, for a large range of biological molecules, electrostatics is vitally important for their correct function (**Figure 4.6**) and provides the dominant long-range interaction.

An *electric double layer* forms around charged groups in aqueous solution (**Figure 4.7**) and the process by which it screens the Coulombic interaction is important for the determination of electrostatic forces. The concept of screening allows a complex many-body problem of two strongly charged objects immersed in an electrolyte that contains many billions of simple ions to be approximated by a simple two-body problem with a modified potential between the two strongly interacting objects. The addition of charge onto a solid surface in a liquid can happen in two ways; by the dissociation of surface groups or by the adsorption of ions onto the surface. For example, surface carboxylic groups can be charged by dissociation -COOH → -COO$^-$ + H$^+$, which leaves behind a negatively charged surface. The adsorption of an ion from solution onto a previously uncharged or oppositely charged surface (e.g. binding Ca^{2+} onto a negatively charged protein) could charge the surface positively.

The chemical potential (μ, the free energy per molecule, **Section 3.4**) for ions in the electric double layer that surrounds a charged aqueous system is the sum of two terms,

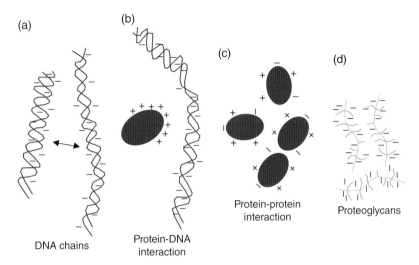

Figure 4.6 *Schematic diagram of some molecular systems in which the electrostatic interaction dominates the intermolecular forces (a) nucleic acids, (b) nucleic acids interacting with charged proteins, (c) the aggregation of charged proteins, and (d) proteoglycans.*

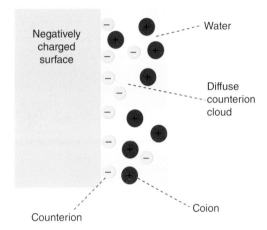

Figure 4.7 *The distribution of counterions in an electric double layer around a negatively charged solid surface. The diffuse counterion cloud is formed from salt ions in the solution combined with ions that dissociate from the solid surface.*

$$\mu = zeV + kT \ln \rho \qquad (4.9)$$

where V is the electric potential, ρ is the number density of counterions, kT is the thermal energy, z is the valence of the charged groups, and e is the electronic charge. The first term on the right-hand side of equation (4.9) is due to the direct contribution of electrostatics, whereas the second term is from the entropy of the counterions. The form of the chemical potential (equation (4.9)) is consistent with the *Boltzmann distribution* for the density of the counterions (equation (3.22)) and can be re-expressed as

$$\rho = \rho_0 e^{-zeV/kT} \tag{4.10}$$

where ρ_0 is related to the chemical potential

$$\rho_0 = e^{\mu/kT} \tag{4.11}$$

A fundamental formula from the theory of electromagnetism is the *Poisson equation* for electrostatics. It relates the potential (V) to the free ion concentration ($\rho_{freeion}$) immersed in a dielectric external to the surface,

$$\frac{d^2V}{dx^2} = -\frac{ze\rho_{freeion}}{\varepsilon\varepsilon_0} \tag{4.12}$$

where ε_0 is the permittivity of free space and ε is the relative permittivity of the dielectric (e.g. water) in which the ions are embedded. The one-dimensional version of the Poisson equation, dependent only on the perpendicular distance from the surface (x), is quoted for simplicity. The Poisson equation can be combined with the Boltzmann distribution for the thermal distribution of ion energies to give the *Poisson–Boltzmann (PB) equation*,

$$\frac{d^2V}{dx^2} = -\frac{ze\rho_0}{\varepsilon\varepsilon_0}e^{-zeV/kT} \tag{4.13}$$

It is a nonlinear differential equation, which makes analytic solution a challenge (**Section 11.1**), but accurate numerical solutions can be made with computers. The PB equation can be solved to give the potential (V), the electric field ($E = \partial V/\partial x$) and counterion density (ρ) at any point in the gap between two planar surfaces. The density of counterions and coions from a planar surface can therefore be calculated as shown schematically in **Figure 4.8**.

There are some limitations on the validity of the PB equation at short separations that include: ion correlation effects (electronic orbitals become correlated), finite ion effects (ions are not point like), image forces (sharp boundaries between dielectrics affect the solutions of the electromagnetism equations), discreteness of surface charges (the surface charge is not smeared out smoothly) and solvation forces (interaction of water molecules with the charges). These important questions will be analysed in more detail in **Chapter 11**.

The pressure (P) between two charged surfaces in water can often be calculated using the *contact value theorem*. It relates the force between two surfaces to the density of contacts (or ions in this case) at the midpoint ($\rho(r)$),

$$P(r) = kT[\rho(r) - \rho(\infty)] \tag{4.14}$$

where $\rho(\infty)$ is the ion concentration at infinity (e.g. the bulk salt concentration), kT is the thermal energy and r is the separation between the surfaces. The force between charged surfaces is discussed in more detail with

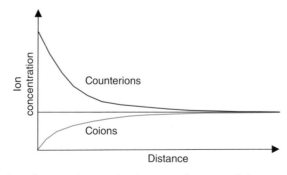

Figure 4.8   *The concentration of counterions and coions as a function of distance from a charged planar solid surface in water, e.g. that is shown in **Figure 4.7**.*

respect to the physics of cartilage in **Chapter 17**. Equation (4.14) can be used with a series of mesoscopic forces. Specifically, with aqueous electrostatics the intersurface pressure is given by the increase in the ion concentration at the surfaces as they approach each other. The theorem is valid as long as there is no specific interaction between the counterions and the surfaces. The contact value theorem also functions well in other calculations such as those of double-layer interactions, solvation interactions, polymer-associated steric and depletion interactions, undulation and protrusion forces.

Example: The solution to the PB equation at a distance x from a charged membrane surface is

$$V = \left(\frac{kT}{ze}\right) \ln\left(\cos^2 \kappa x\right) \tag{4.15}$$

and the characteristic length scale (κ^{-1}) that defines the screening by the ions (**Section 11.2**) is given by

$$\kappa^2 = \frac{(ze)^2 \rho_0}{2\varepsilon\varepsilon_0 kT} \tag{4.16}$$

where ρ_0 is the charge density on the surfaces, z is the valence of the counterions, kT is the thermal energy, ε is the relative permittivity and ε_0 is the permittivity of free space.

Two surfaces with charge density (σ) of 0.4 C m^{-2} are placed at a separation (D) of 2 nm and the inverse Debye screening length (κ) is 1.34×10^9 m^{-1}. Calculate the repulsive pressure (P) between the two surfaces.

Answer: From the contact value theorem and equation (4.16) the pressure can be calculated directly,

$$P = kT\rho_0 = 2\varepsilon\varepsilon_0 \left(\frac{kT}{ze}\right)^2 \kappa^2 = 1.68 \times 10^6 \, \text{Nm}^{-2} \tag{4.17}$$

where ρ_0 is the ion concentration, P is the pressure, κ is the Debye screening length and kT is the thermal energy. This is thus a large pressure between the surfaces equivalent to 16.6 atmospheres.

4.3.3 The Force Between Charged Aqueous Spheres

A surprisingly successful theory for the forces between colloidal particles in aqueous solution is that due to Derjaguin, Landau, Verwey and Overbeek (DLVO). It is surprisingly successful, since it involves two simple approximations for the contributions of both van der Waals forces and electrostatic interactions, and it assumes they are simply additive. The DLVO expression has received confirmation from a wide range of experimental techniques such as optical tweezers, light scattering, neutron/X-ray scattering, coagulation studies and surface-force apparatus.

The competition between attractive van der Waals and repulsive double layer forces is thought to determine the stability or instability of many colloidal systems. The DLVO potential includes both these terms (**Figure 4.9**). Algebraically the potential ($V(r)$ as a function of separation distance (r) between two charged plane surfaces) is

$$\frac{V(r)}{\text{area}} = -\frac{H}{12\pi r^2} + \frac{64kTc\Gamma^2}{\kappa} e^{-\kappa r} \tag{4.18}$$

where H is the Hamaker constant for the van der Waals force, κ^{-1} is the Debye screening length, c is the bulk salt concentration, Γ is defined as tan ($ze\psi/4kT$), and ψ is the electrostatic surface potential (a measurable constant). The first component on the right-hand side of equation (4.18) is due to the van der Waals interaction and the second is from the screened electrostatic potential. The agreement between the DLVO model and experiment is often excellent for charged surfaces.

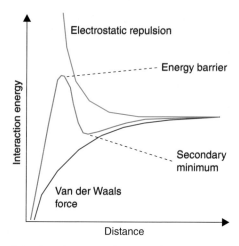

Figure 4.9 *Schematic diagram of the DLVO potential (green) between two colloidal particles. The interaction energy is shown as a function of the distance of separation. The secondary minimum is due to the interplay between electrostatic (blue) and van der Waals (red) forces.*

For a number of colloidal materials it is found that the critical coagulation concentration (ρ_∞) varies as the inverse sixth power of the valency (z) of the counterions in the salt that surrounds the colloids, i.e. $\rho_\infty \propto 1/z^6$. The total DLVO interaction potential between two spherical particles that interact at constant surface potential (using **Figure 4.1b** for the van der Waals interaction) is

$$V(r) = \left(\frac{64\pi kTR\rho\gamma^2}{\kappa^2}\right)e^{-\kappa r} - \frac{AR}{12r} \tag{4.19}$$

where r is the interparticle distance, A is the Hamaker constant, ρ is the density of the colloids, kT is the thermal energy, κ^{-1} is the Debye screening length, γ is the surface potential and R is the radius of the colloids.

The critical coagulation concentration occurs when both the potential ($V = 0$) and the force ($dV/dr = 0$) are equal to zero. The condition of zero potential ($V = 0$) upon substitution in equation (4.19) leads to

$$\frac{\kappa^2}{\rho} = 768\pi kTr\gamma^2 \frac{e^{-\kappa r}}{eA} \tag{4.20}$$

The second condition on dV/dr leads to the condition that $\kappa r = 1$, which shows that the potential maximum occurs at $r = \kappa^{-1}$. Insertion of this expression into equation (4.20) provides the relationship

$$\frac{\kappa^3}{\rho} = \frac{768\pi kT\gamma^2}{eA} \tag{4.21}$$

Now, from the definition of the Debye screening length, equation (4.16), it is known that

$$\kappa^2 \propto \frac{\rho z^2}{\varepsilon T} \tag{4.22}$$

The surface potential (γ) is known to be constant at high surface potentials ($=1$). Equation (4.21) can be squared and substituted in equation (4.22), which provides an expression for the dependence of the critical concentration on the counterion valence,

$$z^6 \rho \propto \varepsilon^3 T^5 \frac{\gamma^4}{A^2} \tag{4.23}$$

Therefore, the critical coagulation concentration follows the expression $\rho \propto 1/z^6$, which proves the relationship.

4.4 Steric and Fluctuation Forces

The packing constraints on solvents in confined geometries produce oscillatory force curves with a period that is determined by the solvent size, e.g. they can be measured using a surface-force apparatus. These steric forces are most readily measured experimentally between two smooth hard surfaces (**Figure 4.10**). For example the *packing force* (F) due to solvent molecules confined between two hard planar surfaces can be approximated by

$$F(r) = B \cos\left(\frac{2\pi r}{\lambda}\right) e^{-r/\lambda} \tag{4.24}$$

where r is the separation of the surfaces, B is a constant and λ is the diameter of the molecules.

Polymers at surfaces can also give rise to *steric entropic forces*, which are induced by the entropy of the chains. Realisations of this phenomenon of entropic stabilisation include proteins on the surface of interacting membranes, friction-reducing hyaluronic acid chains in synovial joints, and DNA chains absorbed onto histones. For a polymer attached to a colloid to cause effective colloidal stabilisation with steric forces the polymer must adopt an extended conformation, e.g. good solvent conditions apply to its configurational statistics (**Section 10.2**). The range of interaction of steric forces is governed by the distance from the surface that the polymer chains can extend (**Figure 4.11**). Typically in biophysical examples the polymer chains that provide the steric stabilisation for a surface are attached by adsorption from solution or are grafted on to a surface by specialist enzymes.

For *polymerically stabilised systems* the repulsive energy per unit area ($W(r)$) between the surfaces is roughly exponential,

$$W(r) \approx 36kT e^{-r/R} \tag{4.25}$$

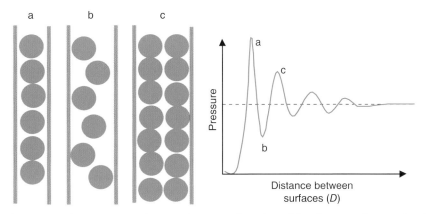

Figure 4.10 *The pressure between two planar surfaces as a function of the distance of separation (D) combined with schematic diagrams of the confined molecules at three different separations (a, b and c). The force is mediated by the excluded volume of the spherical molecules trapped between the surfaces and is called a* depletion potential.

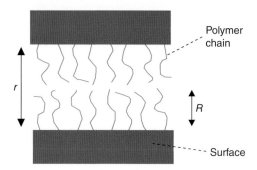

Figure 4.11 *The steric forces between the two surfaces are produced by the entropic contribution of the grafted flexible polymer chains to the free energy. R is the size of the chains and r is the distance of separation between the two surfaces.*

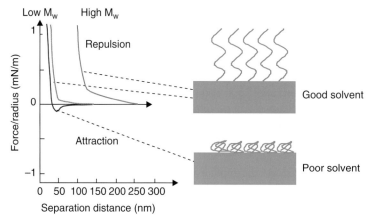

Figure 4.12 *The force between two surfaces with surface adsorbed polymers in a good or poor solvent as a function of the separation distance. The length of the interaction increases with the molecular weight (M_w) of the adsorbed polymer. This experimental data is from SFA measurements. [Reprinted with permission from G. Hadzioannu, G. Patel, S. Granick, M. Tirrell, 1986, J. Am. Chem. Soc, 108, 2869–2876. Copyright © 1986, American Chemical Society.]*

where R is the unperturbed radius of the polymer chains and r is the separation between the surfaces. The steric force between two mica surfaces with surface grafted polymer chains in a good solvent is shown in **Figure 4.12**.

Interacting membranes experience a steric force due to the fluctuations of membrane structures called *membrane forces* (**Figure 4.13**), without the need to attach polymer chains to the surfaces. The entropic force per unit area due to collisions between the two surfaces is given by the contact value theorem (similar to equation (4.14))

$$P(r) = kT[\rho(r) - \rho(\infty)] \tag{4.26}$$

where $P(r)$ is the pressure, $\rho(r)$ is now the volume density of molecular contacts at a separation distance r, kT is the thermal energy and $\rho(\infty)$ is the number of molecular contacts at infinity (**Figure 4.14**). A simple scaling calculation can be made for the magnitude of the *undulation forces* between two membranes. The density of contacts at a certain height (D) is equal to the inverse volume of a single flexural mode of the membrane,

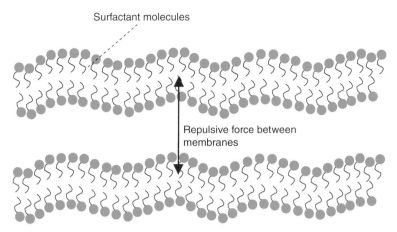

Figure 4.13 *Repulsive undulation forces occur between flexible membranes composed of surfactant molecules (or lipids) due to the thermally driven collisions.*

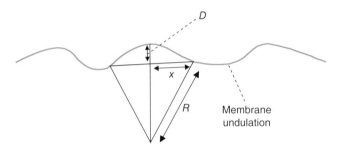

Figure 4.14 *Geometry for the calculation of the magnitude of undulation forces experienced by a membrane. R is the membrane curvature of the bending mode, D is the height of the bending mode and x is the radius of the contacts (quarter of the wavelength of the bending mode).*

$$\rho(r) = 1/(\text{volume of mode}) = 1/\pi x^2 r \tag{4.27}$$

where x is equal to the radius of the contacts ($r = D$). The density of the contacts at infinite separation ($\rho(\infty)$) of the membranes must be zero,

$$\rho(\infty) = 0 \tag{4.28}$$

From a simple continuum elasticity model the energy of a bending mode (E_b per unit area, **Chapter 12**) is

$$E_b = \frac{2k_b}{R^2} \tag{4.29}$$

where k_b is the membrane bending rigidity and R is the membrane curvature of the bending mode. Each bending mode occupies an area that depends on its wavelength (πx^2) and the bending energy can be equated with the thermal energy (kT) to give

$$kT \approx \frac{2\pi x^2 k_b}{R^2} \tag{4.30}$$

The 'Chord theorem' for the geometry of the membrane (**Figure 4.14**) relates x to R and r,

$$x^2 \approx 2Rr \tag{4.31}$$

This allows equation (4.30) to be re-expressed as

$$kT \approx \frac{4\pi r k_b}{R} \tag{4.32}$$

The entropic force per unit area ($P(r)$) between the membranes can be constructed from the contact value theorem (equation (4.26) and (4.27)) and it is

$$P(r) = \frac{kT}{\pi x^2 r} \approx \frac{kT}{2\pi R r^2} = \frac{(kT)^2}{k_b r^3} \tag{4.33}$$

This r^3 dependence of the pressure between membranes has been verified experimentally.

Membranes also experience *peristaltic* (from hydrodynamic effects that cause the membranes to be pulled together at close separations; a similar phenomenon can occur at the macroscopic level between the hulls of closely approaching boats) and *protrusion forces* (the detailed nature of the excluded volume of the membranes is important for very close molecular overlap). These intermembrane forces are weaker than the undulation forces, but can still provide a significant contribution to the interactions.

4.5 Depletion Forces

Depletion forces are another mesoscopic interaction that is formed by a subtle range of more fundamental forces. An illustrative example is when colloidal spheres are mixed with polymers in aqueous solution. The colloids can experience an effective attractive interaction when the polymers are excluded from the volume between spheres when they approach one another (**Figure 4.15**). Such phenomena were originally verified through macroscopic measurements on the phase separation of polymer/colloid mixtures. Recent experiments with dual-trap optical tweezers have provided direct evidence for the interaction potential between two spherical colloidal probes in DNA solutions (**Figure 4.16**).

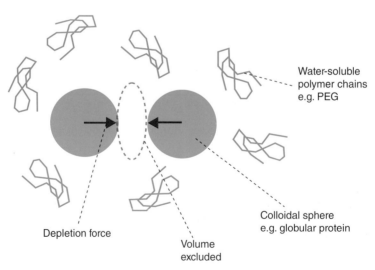

Figure 4.15 *Depletion forces between two colloids in a solution of water-soluble polymer chains (e.g. polyethylene glycol) due to the volume exclusion effects.*

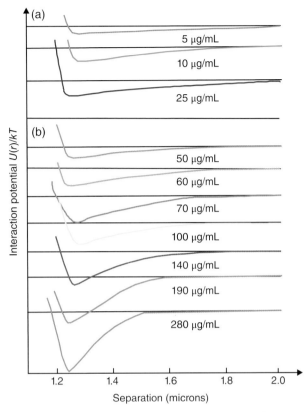

Figure 4.16 *Depletion potential between two 1.25-μm silica spheres as a function of the separation distance for DNA concentrations in (a) dilute and (b) semidilute solutions measured with dual trap optical tweezer experiments. [Reprinted with permission from Ref. R. Verma, J.C. Crocker, T.C. Lubensky, A.G. Yodh, Macromolecules, 2000, 33, 177–186. Copyright © 2000, American Chemical Society.]*

The depletion force can be understood from an analysis of the thermodynamics of the polymer/colloid mixture. The addition of the polymer lowers the solvent's chemical potential, and creates a depletion force that drives the colloidal surfaces together. This is an effect often used to promote protein crystallisation for crystallographic structural studies (**Section 5.4**).

For a dilute concentration of polymers in a colloidal solution the osmotic pressure (π_0) is proportional to the number density of polymer chains (N/V), and each chain contributes kT to the osmotic pressure,

$$\pi_0 = \frac{N}{V}kT \tag{4.34}$$

This pressure is analogous to that of an ideal gas, that gives rise to the van der Waals equation ($P = NkT/V$, where P is the pressure, N is the number of gas molecules, and V is the volume).

The depletion energy (W_{dep}) of the polymer/colloid mixture is approximately the product of the osmotic pressure with the volume from which the chain is depleted,

$$W_{dep} = -\pi_0 V_{dep} = -\pi_0 \frac{4}{3}\pi R^3 \tag{4.35}$$

where R is the radius of the depleted polymer molecules. Thus, a high molecular weight polymer (a large R) and a high polymer concentration (a large π) are required for a strong depletion force. The quantitative predictions of equation (4.35) were first verified experimentally by the measurement of the force between two interacting bilayer surfaces in a concentrated dextran solution and has subsequently achieved a number of other quantitative successes e.g. **Figure 4.16** with optical tweezers.

Naturally occurring intracellular environments are extremely crowded and there is a wide range of colloidal particles. Thus, a wide hierarchical selection of excluded volume depletion interactions can occur.

4.6 Hydrodynamic Interactions

Each of the mesoscopic forces discussed in the previous section have a time scale associated with their interaction, since Einstein's theory of special relativity states that they cannot occur instantaneously. Therefore, the dynamics of the components of each system (e.g. solvents, counterion clouds and tethered polymers) needs to be understood to realistically gauge the strength of the interaction potentials. More advanced treatments of mesoscopic forces therefore often need to consider how to evaluate the dependence of the forces on time and in aqueous solution this requires an understanding of the fluid's hydrodynamics (**Chapter 14**).

4.7 Bell's Equation

The measurement of mesoscopic forces as a function of the distance of separation are sensitive to the adhesion of specimens to surfaces, and there are a number of subtleties involved, particularly when the single-molecule limit of sensitivity is approached.

As seen previously with the brief discussion of hydrodynamic interactions, to fully describe the detachment of a molecule requires the specification of the time scale of the measurement. If the average time of the molecular vibration or collision of a molecule with a surface is t_0, then the mean lifetime of the molecule on the surface is found to be given by *Bell's equation* (it is derived from Arrhenius kinetics, **Section 20.3**),

$$t = t_0 e^{-w_0/kT} \tag{4.36}$$

where w_0 is the adhesion (bond) energy and kT is the thermal energy. The reciprocal of the mean lifetime given by equation (4.36) gives a detachment rate (v),

$$v = v_0 e^{w_0/kT} \tag{4.37}$$

where $v_0 = 1/\tau_0$ is the collision frequency. If w_0 is large $> kT$ and negative, the mean lifetime at the surface is greater than the vibration time ($\tau > \tau_0$), and the molecule remains bound or becomes free, dependent on the waiting time. Even at zero pulling force Bell's equation predicts that the molecule will eventually detach from the surface. A molecule that desorbs for the surface will have received kinetic and potential energy that is in excess of the mean energy of the molecules on the surface lattice. Thus, the temperature of the surface will fall as a consequence of the energy loss, as required by the conservation of energy. No pair potential can fully describe the interaction, which is really a multitude of molecular interactions that follow each other in space and time. It is better to think of the interactions that result in molecular detachment as a process.

The effective adhesion force for small numbers of molecules is not simply given by the maximum in the derivative of the bond potential dV/dr, but by the Bell equation (4.36). Substitution of $w_0 = Fr_0$ in the Bell equation, i.e. energy = force × bond length, gives

$$t = t_0 e^{-Fr_0/kT} \tag{4.38}$$

where t_0 is the constant natural lifetime of the bond under zero external force.

For example, consider a ligand of molecular weight 2 kDa bound noncovalently to a receptor via a lock and key type mechanism with a bond of length (r_0) 1 nm and an energy (w_0) of $-35\ kT$ at 37 °C. A challenge is to calculate the force that will be needed to detach the ligand within ~1 s. The mean velocity of the ligand can be estimated, since the kinetic energy is equal to the thermal energy,

$$\frac{1}{2}mv^2 = \frac{1}{2}kT \tag{4.39}$$

For a ligand of molecular weight 2000 Da its mass is 3.3×10^{-24} kg. At 37 °C equation (4.39) gives a velocity of 36 m s^{-1}, and a mean collision time of $t_0 = r_0/v = 2.8 \times 10^{-11}$ s. The mean natural bond lifetime is therefore given by Bell's equation (4.36) and is approximately 12 h (4.4×10^4 s). To reduce the bond lifetime requires a force given by the rearranged Bell equation with $w_0 = rF_0$ (equation (4.38)),

$$F = \left(\frac{kT}{r_0}\right) \ln\left(\frac{t_0}{t}\right) \tag{4.40}$$

Numerical values can be substituted into the expression and give F to be 46 pN when t is 1 s. Thus, a force of 46 pN is required to decrease the bond lifetime to 1 s.

When multiple detachments and attachments occur repeatedly, n times, the average of the exponential of the work done (W) is a constant and equal to the exponential of the thermodynamic work w_0. However, the *Jarzynski equation* (another important result from nonequilibrium thermodynamics) states

$$\left\langle e^{W/kT} \right\rangle \xrightarrow{n \to \infty} e^{-w_0/kT} \tag{4.41}$$

The Jarzynski equation can be combined with the Bell equation to give

$$t_0 = \frac{1}{\langle 1/t \rangle} \tag{4.42}$$

The strange practical implications of this equation are best illustrated with an example. Consider a single pull-off measurement for the detachment of a macromolecule from a membrane. When subjected to a given pulling force the detachment time (t) is found to be 10 μs. Three more repeat measurements give a distribution of detachment times 1, 10 and 100 μs. The average pull-off time for four measurements is

$$\langle t \rangle = (1 + 10 + 10 + 100)/4 = 30\ ms \tag{4.43}$$

However, a better estimate of the molecular oscillation time (t_0) is given from the Jarzynski relationship equation (4.42),

$$t_0 = \frac{1}{1 + \dfrac{1}{10} + \dfrac{1}{10} + \dfrac{1}{100}} = 3.3\ s \tag{4.44}$$

Consider two *bonds in series* ($N = 2$) subjected to a pulling force F, where the mean lifetime t_i for each bond is $t_{i=1} = 2$ s and $t_{i=2} = 4$ s as given by the Bell equation. The probability that each will break during a 1s interval is

$$p_1 \approx \frac{1}{t_1} = 0.5, \quad p_2 \approx \frac{1}{t_2} = 0.25 \tag{4.45}$$

If the two bonds are linked together in series the probability that they are both unbroken is $(1 - p_1)(1 - p_2) = 0.375$. The probability that one of the two bonds is broken is $1 - 0.375 = 0.625$. This is bigger than the probability that a single bond breaks (0.5). The mean lifetime of the two bond molecule is $1/0.625 = 1.6$ s, which is shorter than either of the 2 constituent molecules. As $N \to \infty$ (e.g. a long macromolecule) the probability of rupture tends to 1.0 and the lifetime of the junction that contains the bonds in series goes to zero.

Bell's equation shows that the force to rupture a bond depends on the time to rupture. Thus, high pulling forces generally lead to short rupture times; $t \to t_0$ as $F \to \frac{w_0}{r_0}$. At the other extreme of weak forces, the rupture time approaches the natural lifetime of the bond; $t \to t_0 e^{-w_0/kT} = t_0$ as $F \to 0$. Thus, when a junction held by a number of bonds in series is subjected to a pulling force (F) the bond that actually breaks first may be different, dependent on the magnitude of F.

The rupture of a chain that consists of N bonds in series occurs when the first link is broken; the rupture of an adhesive junction that consists of N *bonds in parallel* occurs when the last link is broken. Consider the two-bonds example again, this time in parallel. The probability that a junction that consists of two bonds in parallel will open in 1 s is $0.5 \times 0.25 = 0.125$, and this is less than the rupture probability of the strongest bond, as expected.

For N independent bonds in parallel the probability of rupture is just a product of their individual (independent) probabilities $p_1 p_2 p_3 .... p_N$, which tends to zero for large N. Such parallel junctions will therefore be long lived.

Thus, time is seen to be important with molecular forces. A weak force will open a junction eventually, but a much larger force will do so quicker, although not immediately.

4.8 Direct Experimental Measurements

There are a large number of experimental probes for intermolecular forces, which often operate using only a small range of physical principles (**Figure 4.17**). Some of the most important methods will be quickly reviewed (also see **Chapter 19**).

The *thermodynamic properties* of gases, liquids and solids (*PVT* data, boiling points, latent heats of vaporisation, and lattice energies) provide important information on short-range interparticle forces. Similarly, adsorption isotherms provide information on the interactions of molecules with surfaces.

A range of *direct physical techniques* on gases, liquids and solids (e.g. molecular beam scattering, viscosity, diffusion, compressibility, NMR, X-ray, and neutron scattering experiments) can provide information on short-range interactions of molecules with particular emphasis on their repulsive forces. A sensitive method for the characterisation of hydrated molecules is to measure the separation of the molecules (using X-ray scattering) in parallel with osmotic pressure measurements. This provides a noninvasive piconewton measurement of intermolecular forces.

Thermodynamic data on solutions (phase diagrams, solubility, partitioning, miscibility, and osmotic pressure) provides information on short-range solute solvent and solute–solute interactions. With colloidal dispersions coagulation studies as a function of the salt concentration, pH or temperature yield useful information on interparticle forces.

Adhesion experiments provide information on particle adhesion forces and the adhesion energies of solid surfaces in contact.

Thermodynamic properties
(PVT phase diagram)

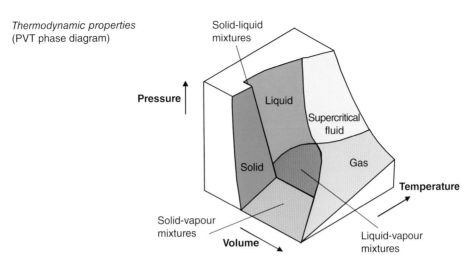

Direct physical data

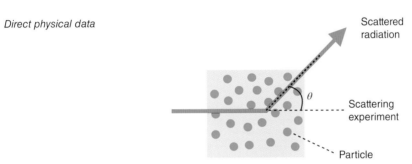

Thermodynamic data on solutions
Phase stability

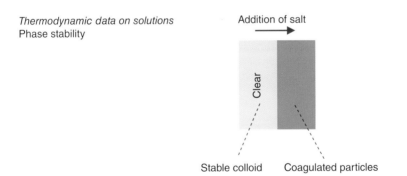

Adhesion experiments

Figure 4.17   *(continued)*

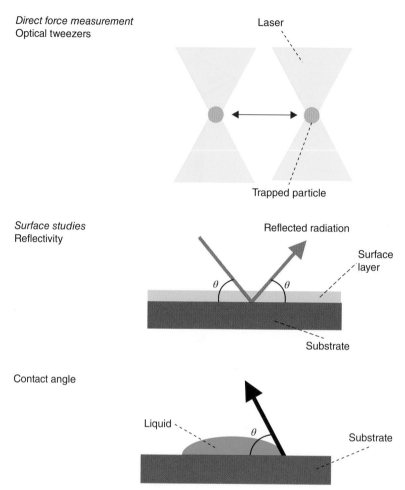

Figure 4.17 *The range of techniques for the measurement of intermolecular forces include thermodynamic phase diagrams, scattering experiments, thermodynamic phase stability, adhesion experiments, optical tweezers, surface reflectivity and contact angles.*

Direct force measurement between two macroscopic/microscopic surfaces as a function of surface separation can provide the full force law of interaction, e.g. surface-force apparatus (SFA), optical/magnetic tweezers and atomic force microscopes (AFM).

Surface studies such as surface tension and contact-angle measurement can give information on liquid–liquid and solid–liquid adhesion energies. Similarly, reflectivity of particles (neutrons) and waves (X-rays and light) can provide invaluable information on surface energies. With *film balances,* the thicknesses of free soap films and liquid films adsorbed on surfaces can be measured as a function of salt concentration or vapour pressure, again this provides a direct measurement of intersurface potentials.

Hydrodynamic studies of liquids can be made using nuclear magnetic resonance spectroscopy (NMR), fluorescence microscopy, bright-field microscopy and the elastic/inelastic scattering of light, X-rays and neutrons. These methods are particularly useful for the measurement of hydrodynamic effects that lead to time-dependent mesoscopic forces.

Suggested Reading

If you can only read one book, then try:

Top recommended book: Israelichvilli, J. (2011) *Intermolecular and Surface Forces*, 3rd edition, Academic Press. Classic authoritative text on mesoscopic forces. The 3rd edition contains a good discussion of Bell's equation.

Barrat J.L. & Hansen, J.P. (2003) *Basic Concepts for Simple and Complex Fluids*, Cambridge University Press. Clear mathematical approach to soft matter.
Dill, K. & Bromberg, S. (2010) *Molecular Driving Forces: Statistical Thermodynamics in Biology, Chemistry, Physics and Nanoscience*, Garland Science. Excellent introduction to statistical models of colloidal forces.
Evans, S.F. & Wennerstrom, H. (1994) *The Colloidal Domain*, Wiley. Useful account of forces in colloidal systems.
Parsegian, V.A. (2005) *Van der Waals Forces: A Handbook for Biologists, Chemists, Engineers and Physicists*, Cambridge University Press. Extensive collection of algebraic expressions for van der Waals forces in different geometries.
Tabor, D. (1991) *Solids, Liquids and Gases*, Cambridge University Press. Simple refined presentation of the basic phenomena in condensed matter.

Tutorial Questions 4

4.1 Two adjacent atoms experience a Lennard-Jones potential that determines their atomic spacing equation (4.4),

$$V(r) = \varepsilon \left\{ \left(\frac{r_0}{r}\right)^{12} - 2\left(\frac{r_0}{r}\right)^{6} \right\}$$

The energy constant for the interaction (ε) is 0.8×10^{-18} J and the equilibrium distance (r_0) is 0.33 nm. Calculate the force the atoms would experience if they are compressed to half of their equilibrium distance.

4.2 Calculate the Debye screening length for solutions of sodium chloride at concentrations of 0, 0.001, 0.01, 0.1 and 1 M. Water has an intrinsic dissociation constant of around 10^{-14} M^2 (equation (1.2)). Also calculate the screening length for a divalent salt solution (e.g. $MgCl_2$) at the same concentration. Finally calculate the Debye screening length in standard physiological conditions (0.1 M salt).

4.3 A charged polymer can adopt both globular and extended linear conformations. It is assumed that the charge is conserved during the change in conformation, and there is a negligible amount of salt in the solution and no charge screening. Explain in which geometry the potential decreases most rapidly with the distance from the chain. The charge is smeared out on a planar surface. Compare the potential with that of the linear and globular morphologies.

4.4 Charged spherical viruses have polymer chains attached to their surfaces, e.g. they are 'PEGylated' in the language of a synthetic chemist. Calculate the distance of separation at which the entropic force due the chains become significant compared to that due to the intervirus electrostatic repulsion.

4.5 Explain the significance of Bell's equation for single molecule measurements.

5

Phase Transitions

Everyone has experience of a range of phase transitions in everyday life. These could be boiling a kettle (a *liquid–gas* phase transition) or melting a wax candle (a *solid–liquid* phase transition). In terms of their molecular arrangement, phase transitions almost always involve a change between a more ordered state and a less well ordered state. Materials can be in stable equilibrium or nonequilibrium if their form evolves with time. A huge range of phase transitions occur in biological systems and there is a comprehensive theoretical framework, originally developed to describe simple synthetic materials, that also describes their behaviour.

5.1 The Basics

A material can adopt a number of phases simultaneously at equilibrium. Thus, *phase diagrams* are required to describe the relative amounts of the coexisting phases. The standard everyday states of matter are the crystalline (a regular incompressible lattice), the liquid (an irregular incompressible lattice) and the gas (an irregular compressible lattice) phases. However, there are also other less conventional phases such as amorphous solids, rubbers and glasses that occur in Nature and our intuition concerning their mechanical behaviour and microstructure needs careful consideration. Liquid crystals and gels also commonly occur in biology and present additional possibilities for the thermodynamic state of a material. Other more exotic examples of phase changes studied in this book include wetting (e.g. water drops on lotus leaves), and complexation (e.g. DNA compaction in chromosomes); the range of possible thermodynamic phases of a biological molecule is vast.

Some of the basic phenomena that concern phase transitions will be recapped before they are applied to a range of biophysical problems. Hopefully readers will have encountered many of these concepts previously in an introductory thermodynamics course. The two states between which a *first-order* phase transition occurs are distinct, and occur at separate regions of the thermodynamic configuration space. *First-order* phase transformations experience a discontinuous change in all the dependent thermodynamic variables, except

The Physics of Living Processes: A Mesoscopic Approach, First Edition. Thomas Andrew Waigh.
© 2014 John Wiley & Sons, Ltd. Published 2014 by John Wiley & Sons, Ltd.

the free energy, during the phase transition. In contrast, the states between which a *second-order* phase transition occurs are contiguous states in the thermodynamic configuration space. In these continuous phase transitions (2nd, 3rd, 4th, etc.), the dependent variables such as the heat capacity, compressibility and surface tension, diverge or vanish as the independent variables approach a critical value, e.g. a critical temperature (T_c). This divergence occurs in a characteristic manner around the critical point, example if the heat capacity (C_v, the ability of the material to store thermal energy) diverges as a function of the temperature near to the critical temperature for a phase transition (T_c) its functional form is $C_v \sim (T-T_c)^{-\alpha}$. The exponents α are found to be universal, dependent on the class and symmetry of the continuous phase transition, but not on the exact details of the material's molecular components.

The differences between a first- and a second-order phase transition, can be illustrated by the consideration of the heat capacity (C_p) and the enthalpy (H) as a function of temperature (**Figure 5.1**). The discontinuous nature of the first-order phase transition is clearly evident in this example.

The *Gibbs phase rule* gives the number of parameters that can be varied independently in a system that is in phase equilibrium. Consider an ideal gas whose properties depend on the volume (V), number of particles (n), pressure (P) and temperature (T). The equation of state (χ), is a unique function dependent on the four variables describing the system, and is given mathematically by

$$\chi(V,n,P,T)=0 \tag{5.1}$$

With an ideal gas it is well known that the equation of state takes the form of the ideal gas law,

$$\frac{PV}{nRT}-1=0 \tag{5.2}$$

where R is the ideal gas constant. Only three independent variables are thus needed to define the ideal gas system, the fourth variable is dependent. Generally, if the number of degrees of freedom in a system is f,

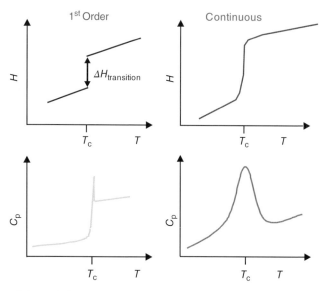

Figure 5.1   *Comparison of the enthalpy (H) and heat capacity (C$_p$) as a function of temperature (T) for first-order and continuous phase transitions (2nd, etc.). T$_c$ is the critical temperature for the phase transition.*

(the number of independent intensive variables that remain after all the possible constraints have been taken into account), the number of simultaneously existing phases is p and the number of components is c, the Gibbs phase rule states they are related,

$$f + p = c + 2 \tag{5.3}$$

This rule is very useful to predict the phase behaviour for multicomponent colloidal systems and can be proved from general thermodynamic principles for systems in thermal equilibrium.

To describe a phase transition in more detail it is useful to define an *order parameter*. This order parameter takes a zero value in the disordered phase and a finite value in the ordered phase. How the order parameter varies with temperature (or other independent variable such as pressure, volume, etc.) describes the nature of the transition. Acceptable order parameters for biological phase transitions include the density of a sample, its volume and the degree of molecular orientation. The choice of order parameter is guided by the nature of the phase transition that requires description, e.g. orientational order parameters are crucially important for an understanding of liquid-crystalline phases and will be examined in **Chapter 6**.

The behaviour of *continuous phase transitions* was extensively developed during the second half of the twentieth century. Universal behaviour was found for continuous phase transitions near to the critical point, dependent on only the symmetry of the constituent phases. Continuous phase transitions are thus characterised by the values of critical exponents needed to describe the divergence of the order parameters at the phase-transition temperature.

For a continuous phase transition a limited number of thermodynamic variables diverge or become zero at the critical point. For example the heat capacity (C_p, at constant pressure) near the critical temperature (T_c, **Figure 5.1**) is given by

$$C_p = \begin{cases} A(T - T_c)^{-\alpha} & T > T_c \\ A(T_c - T)^{-\alpha} & T < T_c \end{cases} \tag{5.4}$$

where A is a constant, T is the temperature and α is the critical exponent. The same critical exponent (α) applies either side of the transition point.

For a gas–liquid phase transition that approaches the critical temperature (T_c), the density (ρ) also diverges with a characteristic exponent (β),

$$\rho_{\text{liquid}} - \rho_{\text{gas}} = B(T_c - T)^{\beta} \tag{5.5}$$

where B is a constant and T is the temperature. This can be pictured experimentally with a transparent kettle full of a subcritical boiling liquid close to a second-order gas–liquid transition. All the dissolved gas is assumed to be removed (which obscures the effect), and the liquid will become milky at the transition temperature. Large fluctuations in the density (the order parameter) are found at the critical point (the boiling point of the fluid). The density fluctuations can be quantified and are related to the compressibility of the fluid (how much the volume (V) of a material changes in response to a change in pressure (P), $\kappa = -\dfrac{1}{V}\dfrac{\partial V}{\partial P}$), which diverges at the critical point,

$$\kappa = C(T - T_c)^{-\nu} \tag{5.6}$$

where C is a constant and ν is the critical exponent. The expression for the compressibility is valid for temperatures above the critical temperature ($T > T_c$). As explained previously these large density fluctuations can be viewed experimentally in the form of *critical opalescence*; liquids become cloudy due to large fluctuations of density, and incident light is scattered as they approach a phase transition. The correlation length for the physical size of the density fluctuations also diverges at the critical point.

Another example of a continuous phase transition relates to the thermodynamics of surfaces. The surface tension (γ) between two liquid phases approaches zero at the critical point, although the width of the interface diverges,

$$\gamma = \gamma_0 \left(1 - \frac{T}{T_c}\right)^{\mu} \tag{5.7}$$

where μ is the characteristic critical exponent and γ_0 is the average surface tension. Thus, phase transitions are not limited to bulk three-dimensional systems.

Other sources of continuous phase transitions include: systems in which finite-size effects play a dominant role (formation of small lipid micelles that leads to a broad critical micelle concentration), systems that contain impurities or inhomogeneities that broaden the phase transition, and systems in which equilibrium times are long compared to observation times, e.g. the glass transition of polymers (**Section 12.3**). In this chapter a range of phase transitions will be considered in detail that are important for molecular biophysics; the helix–coil transition, the globule–coil transition, crystallisation and liquid–liquid phase separation.

5.2   Helix–Coil Transition

Helix–coil transitions occur in a wide variety of biological situations with an immense range of biological molecules, e.g. carbohydrates, proteins and nucleic acids. Reversible thermodynamic double and single helix–coil transitions both commonly occur in Nature (**Figure 5.2**). The chains can be transferred from the helix to the coil state by a change in a range of environmental factors such as the temperature, the quality of the solvent or the pH of the solution.

The *Zipper model* is the simplest method for a description of the thermodynamics of the helix–coil transition (**Figure 5.3**). Let s be defined as the equilibrium constant for the creation of a new helical unit at the end of a helical sequence. This can be written as

$$s = \frac{...cchh\underline{h}hcc...}{...cchh\underline{c}cc...} \tag{5.8}$$

where h is the helical state of a polymer chain, c is the coil state of a polymer chain and the underlining corresponds to the chain link considered. The equilibrium constant (s) is called the *propagation* step and provides the statistical weight for the growth of a helical section of a polymer chain provided the nucleation step has occurred. σ is defined as the equilibrium constant for the *nucleation* step; the formation of a helical unit on a coil from the flexible chain state,

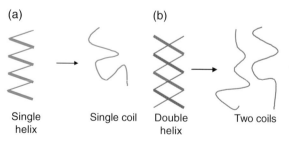

Figure 5.2   *Schematic diagram of the helix–coil transition in (a) single helix (e.g. polypeptides) and (b) a double helix (e.g. DNA).*

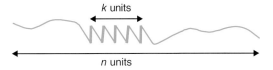

Figure 5.3 *The Zipper model for an alpha helix consists of a stretch of* k *helical units arranged along a polymeric chain of* n *units.* (n − k) *units are in the coil state.*

$$\sigma = \frac{(...cc\underline{h}cc...)}{(...cc\underline{c}cc...)} \tag{5.9}$$

Normally this nucleation step has a very small value, i.e. $\sigma \ll 1$. For a chain of n units the partition function (Z) for the helix–coil transition can be constructed (**Chapter 3**),

$$Z = 1 + \sum_{k=1}^{n} \Omega_k \sigma s^k \tag{5.10}$$

where Ω_k is the number of ways that k helical units can be placed on a chain of n monomeric units. The number of distinct permutations for Ω_k can be calculated and is equal to

$$\Omega_k = (n - k + 1) \tag{5.11}$$

The bulk fractional helicity (θ) of a sample is the order parameter that can be most easily measured using polarimetry, X-ray diffraction or NMR. The helicity (θ) is simply defined as the number of monomers in the helical state (k) divided by the total number of available monomers (n),

$$\theta = \frac{k}{n} \tag{5.12}$$

The helicity can be calculated from the partition function defined in equation (5.10) using

$$\theta = \frac{1}{n}\frac{\partial \ln Z}{\partial \ln s} \tag{5.13}$$

This zipper model is found to be in good agreement with experimental data for the behaviour of short α-helical chains (**Figure 5.4**). However, the zipper model breaks down for long chains, since thermodynamic fluctuations can lead to sections of helix interspersed with parts of random coil. The possibility of such fluctuations is not included in the zipper partition function and is thus badly described by the model (it is a mean-field model).

The *Zimm–Bragg* (Ising) model provides a more sophisticated description of the helix–coil transition that allows for fluctuations in the helicity along the chain. As before, due to the cooperative nature of the helical conformations, links are assumed to exist in two clearly differing discrete energy states; helical and coil-like. The junction between helical and coil sections carries a large positive free energy (Δf) that encourages long lengths of helix. There are two Zimm–Bragg parameters introduced for the model, the statistical weights of the states s and σ,

$$s \equiv \exp(-\Delta f / kT), \quad \sigma \equiv \exp(-2\Delta f_s / kT) \tag{5.14}$$

where Δf is the free energy change due to the addition of an extra helical section, Δf_s is the free energy change for nucleation, k is the Boltzmann constant and T is the temperature. For naturally occurring biopolymers the statistical weight for nucleation (σ) is typically very small on the order of $\sim 10^{-3}$–10^{-4}. The model constitutes a

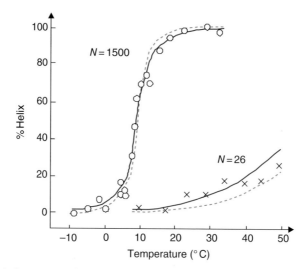

Figure 5.4 *Percentage helicity as a function of temperature for α helices. The agreement of the Zipper model (dashed blue line) with data (red line) for the helicity of a polypeptide of two separate lengths N = 1500, and N = 26 is shown. The polypeptide forms an alpha helical structure at high temperatures and is a coil at lower temperatures. This is an opposite trend to that observed with amylose and DNA double helices, that melt at high temperatures. [Reprinted with permission from B.H. Zimm, J.K. Bragg, J. Chem. Phys, 1959, 31, 526. Copyright 1959, AIP Publishing LLC.]*

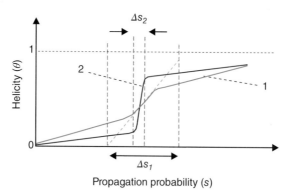

Figure 5.5 *The helicity (θ) plotted as a function of the strength of the propagation step (s) for the Zimm–Bragg model of the helix–coil transition (Δs₁ and Δs₂ are the width of the transitions for low and high cooperativities, respectively).*

simple method for coarse graining the complex network of hydrogen bonds required for helix formation with their well-defined positions and angles, to simply describe the phase transition.

The partition function for the Zimm–Bragg model becomes more complicated than with the zipper model and only numerical solutions to the behaviour will be examined (**Figures 5.5** and **5.6**). It is found that the helix–coil transition in a single-stranded homopolymer occurs over a very narrow temperature interval, which becomes narrower as the cooperativity parameter (σ) decreases, i.e. strong cooperativity. The mean lengths of the helical and the coil sections are finite and independent of the total chain length even as the polymer length (N) tends to infinity.

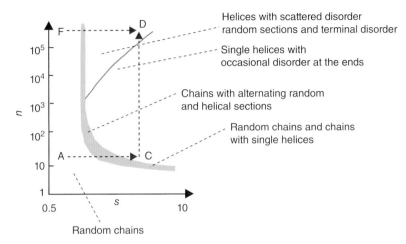

Figure 5.6 *The phase diagram for the states of a helical Zimm–Bragg chain with $\sigma = 10^{-4}$. n is the length of the polymeric chain and s is the propagation step. [Reprinted with permission from B.H. Zimm, J.K. Bragg, J. Chem. Phys, 1959, 31, 526. Copyright 1959, AIP Publishing LLC.]*

The helicity (θ) defined in equation (5.12) for the Zimm–Bragg model as a function of the degree of co-operativity ($\sigma = 1$, no cooperativity and $\sigma <<1$, strong cooperativity) is shown in **Figure 5.5**. Curve 1 indicates a small degree of cooperativity and curve 2 has strong cooperativity. Biological helices normally demonstrate a high degree of cooperativity.

Figure 5.6 shows the complete phase diagram for the Zimm–Bragg model of a helical chain with regions of disorder. A range of distinct phases are possible that include random chains, chains with alternating random and helical sections, random chains and chains with single helices, and single helices with occasional disorder at the ends. The phase diagram is very rich, even for such a simple idealised system. Single-molecule experiments are able to investigate each of these scenarios on a molecule-by-molecule basis (**Chapter 19**) and are in reasonable agreement with the model.

Although frequently regarded as a melting of helices (the process is accompanied by a differential scanning calorimetry endotherm; a differential scanning calorimeter (DSC) is a sensitive device for the measurement of the amount of heat required to change a samples temperature, **Section 19.2.1**), the single helix–coil transformation should not be considered a true thermodynamic phase transition. In a one-dimensional system the equilibrium coexistence of macroscopic phases is prohibited by a theorem due to Landau. The *Landau theorem* is deduced from the exceedingly small energy associated with phase separation for one-dimensional systems, it is impossible for a true phase transition to occur in one dimension. For a short chain the width of the helix–coil transition is thus anomalously large due to the dominance of end effects.

A helix–coil transition in charged polymers can be initiated by a change in the pH of the medium, e.g. with nucleic acids. In this case, the transition is accompanied by a sharp change in the average charge of the molecule, which provides another experimental method for the study of the phase behaviour using titration. However, there are a number of subtleties for the quantification of counterion condensation with charged polymers that will be returned to in **Chapter 11** and need to be considered in detail to describe the helix–coil phenomena involved.

The analysis of phase transitions with double helices is slightly different to the case of single helices. With the *double helix–coil transition*, internal coil sections of the double chain are loops. The strong entropic disadvantage of long loops results in increased cooperativity of the phase transition when compared to that for single helices. The loop factor leads to an abrupt sharpening of the helix–coil transition, and calculations show that it can be considered a true phase transition in this case.

Other rearrangements of secondary structure can be described using a modified Zimm–Bragg-type model. The formation of beta sheets can be satisfactory described with such a model, but again charge effects need to be carefully considered in real beta sheet–coil phase transitions.

In heteropolymers the helicity constants of the monomers for each type of chemistry are different and the character of the helix–coil transition therefore depends on the primary structure. Such a model is thus more realistic for naturally occurring nucleic acids (there are four varieties of monomer for the DNA heteropolymer) and proteins (twenty amino acids could be involved). In a real heteropolymer, the helix–coil transition proceeds by consecutive melting of definite helical sections, whose primary structures possess a sufficiently high concentration of low melting temperature links, although thermodynamic fluctuations still need to be incorporated for an accurate analysis. More sophisticated Zimm–Bragg models can describe this behaviour and again there is good agreement with experiment. The statistical descriptions of the helix–coil and beta sheet–coil transitions are thus a success story of molecular biophysics.

5.3   Globule–Coil Transition

There is a wide range of experimental evidence for the globule–coil transition with polymeric chain molecules. For example, the temperature can be reduced for an extended polymer in a good solvent that changes the quality of the solvent for the chain and causes it to shrink into a dense spherical globule (**Figure 5.7**). Some of the clearest data on the globule–coil transition has been measured using fluorescence microscopy with DNA molecules. Under an optical microscope, fluorescently tagged DNA molecules can be clearly seen to contract in an abrupt transition when an attractive interaction is introduced between the monomers, e.g. through the introduction of positive multivalent counterions (indeed there are naturally occurring molecules that perform this task such as spermidine). Furthermore, neutron, X-ray and light scattering, dynamic light scattering, differential scanning calorimetry, electron microscopy and atomic force microscopy, all clearly point to the existence of a globularisation phase transition with macromolecules. The globule–coil transition is closely associated with the phenomena of folding in proteins and the compaction of nuclear DNA, which makes it of large biological significance.

A swelling coefficient (α) is normally used as the order parameter for the globule–coil transition and is defined as the ratio of the end-to-end radius (R) of the chain to the end-to-end radius (R_0) of the coil $\alpha = R/R_0$. The end-to-end radius will be discussed in detail in **Chapter 10**, and it is a measure of chain size. When the molecule shrinks α is less than one ($\alpha < 1$, a bad solvent) and when it swells α is greater than one ($\alpha > 1$, a good solvent). The globule–coil transition is found to be a first-order phase transition; there is a sharp change in α as a function of temperature, solvent quality or pressure.

The entropic contribution to the free energy (F_{ent}) when a polymer coil is stretched by a factor α is given by

$$F_{ent} = -TS(\alpha) \tag{5.15}$$

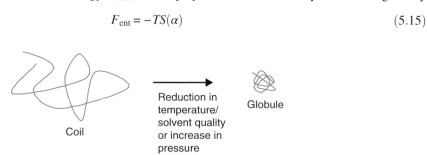

Coil   Reduction in
              temperature/
              solvent quality
              or increase in
              pressure   Globule

Figure 5.7   *Schematic diagram of the coil–globule transition for a single chain induced by a reduction in temperature/solvent quality or an increase in pressure.*

where T is the temperature, and S is the entropy, which depends on the degree of stretching. Remember that the total free energy (F, **Chapter 3**) for a system is related to the internal energy (U), temperature (T) and entropy (S) by $F = U–TS$. The energy of self-interactions as a function of the chain expansion can be defined as $U(\alpha)$. The standard form of the free energy of a globule–coil transition is thus the sum of the entropy (F_{ent}) and the self-interaction terms (U),

$$F(\alpha) = F_{ent}(\alpha) + U(\alpha) \tag{5.16}$$

The internal energy can be expanded in powers of the monomer density n (similar to the approximation for a van der Waals gas that leads to the ideal gas equation $PV = nRT$),

$$U = VkT\left[n^2 B + n^3 C + ..\right] \tag{5.17}$$

where V is the volume of the coil and kT is the thermal energy. B and C are second (two-body collision) and third (three-body collision) virial coefficients, which describe the strength of the intersegment attraction. Negative B implies an attractive intrachain potential, positive B implies a repulsive intrachain potential and a negligible B ($B \sim 0$ in a theta solvent) causes C to dominate the behaviour. The monomer density (n) is of the order of the degree of polymerisation of a chain (N) divided by the chain size (R), N/R^3, so equation (5.17) can be written

$$U \sim R^3 kT\left[B\left(\frac{N}{R^3}\right)^2 + C\left(\frac{N}{R^3}\right)^3\right] \tag{5.18}$$

By definition $\alpha = R/R_0$ and the unperturbed radius is $R_0 = aN^{1/2}$ (**Section 10.1**), where a is the size of a monomer, can be used. Therefore, equation (5.18) can be rewritten as

$$U(\alpha) = kT\left[\frac{BN^{1/2}}{\alpha^3 a^3} + \frac{C}{\alpha^6 a^6}\right] \tag{5.19}$$

where N is the number of monomers in the chains, B and C are the virial constants, and kT is the thermal energy.

The entropy contribution to the free energy can also be calculated using a simple statistical model and is given by

$$F_{ent}(\alpha) \sim kT\left(\alpha^2 + \alpha^{-2}\right) \tag{5.20}$$

Through minimisation of the total free energy (equation (5.16)) of the globule–coil transition using equations (5.20) and (5.19), for the entropic and internal free energies, respectively, it is possible to construct a phase diagram for a single polymer chain, **Figure 5.8**. The coil condenses onto itself when the second virial coefficient is sufficiently negative. There is a jump in the molecular size (parameterised by α) as the polymer chain condenses onto itself and the polymer chain experiences a first-order phase transition.

Quasielastic light scattering studies show clear noninvasive evidence for coil–globule transitions in biopolymers in solution as their temperature is reduced. However, such experiments with large numbers of polymer chains are very sensitive to the total monomer concentration. Low concentrations of biopolymer are necessary to avoid aggregation of the globules in poor solvent conditions due to the strong chain-chain attraction that can easily confuse the results. The use of highly charged biopolymers can circumvent these experimental problems, but additional terms must be added to the free energy (equation (5.17)) to describe the Coulombic repulsion between monomers and the entropy of the counterions associated with the chains, which complicates the theoretical analysis.

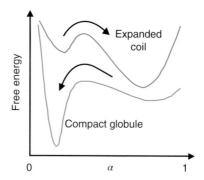

Figure 5.8 *Schematic diagram of the free energy for a flexible polymeric globule as a function of the degree of expansion (α). Minima for the expanded coil and compact globule are shown. At low temperatures the chain adopts a globular conformation, whereas at higher temperatures it expands up into a flexible extended chain conformation.*

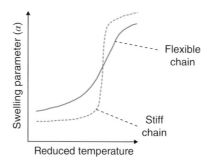

Figure 5.9 *The dependence of the swelling parameter (α) on the reduced temperature for flexible (green) and rigid (blue) polymers that experience a globule–coil transition. The chains form globules at low temperatures and the transition is very sharp for the stiffer chains.*

Detailed theoretical studies show that large neutral globules formed from a single large neutral polymer chain consist of a dense homogeneous nucleus and a relatively thin surface layer, a fringe of the less-dense material. In equilibrium, the size of the globule adjusts itself so that the osmotic pressure of the polymer in the globules nucleus equals zero. As the θ point (where B the second virial coefficient is zero) of a polymer/solvent mixture is approached from poor solvent conditions, a globule gradually swells and its size becomes closer to that of a coil. The transition becomes continuous, a second-order phase transition, as the second virial coefficient approaches zero. The width of the globule–coil transition in a chain with N monomers is proportional to $N^{-1/2}$ and becomes infinitely sharp as N tends to infinity.

The character of the globule–coil transition also sensitively depends on the stiffness of the chains. For stiff chains (e.g. DNA, helical proteins and highly charged biomacromolecules), the transition is very sharp and is close to a first-order phase transition. For flexible chains it is much smoother and is a second-order phase transition (**Figure 5.9**).

The globule–coil transition of a macromolecule of moderate length and significant stiffness can result in the creation of distinctive small globule morphologies. When the size of a small globule is comparable with the persistence length of the chain, the structure of the globule sensitively depends on the flexibility mechanism.

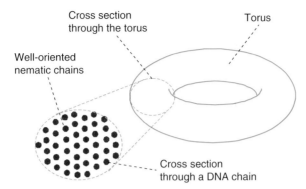

Figure 5.10   *Schematic diagram of a DNA doughnut (torus) formed by a globularisation phase transition that shows the liquid-crystalline internal structure. The structure results from the rigidity of the DNA chains to bending (they are semiflexible).*

Experimentally for a semiflexible chain (e.g. DNA), a small globule is found to take the shape of a doughnut (a torus, **Figure 5.10**). Furthermore, these toroidal shapes often have a liquid-crystalline internal structure. DNA chains are rigid and charged, but compaction can be induced by the addition of a flexible polymer to change the water chemical potential or the addition of multivalent counterions, e.g. Mg^{2+} or spermidine to induce an attractive electrostatic intrachain force.

Giant DNA molecules inside cells naturally exist in a very complex compacted state, often combined with proteins. The structure and phase transitions of chromosomes will be covered in more detail in **Chapter 18**.

The native structure of a *globular protein* is characterised by a precise series of amino acids that possess the property of self-assembly. The compactness of a protein globule is maintained primarily by the hydrophobic effect. Hydrophobic groups are mainly located inside the globule and hydrophilic ones on the surface. The protein globule is a system of rigid blocks of secondary structure and its surface bristles with the side groups of amino acids. Van der Waals interactions between side groups of neighbouring blocks also contribute to the details of the tertiary structure. Electrostatic interactions play an essential role in globular protein folding (when the globule is far from the isoelectric point) and charged links are often only located on the surface of the globule, which provide electrostatic stabilisation of the whole globule against aggregation through DLVO forces, equation (4.17). Together with coil and native globular states, the phase diagram of a protein molecule also includes a molten globular state and it is this state that is qualitatively explained by simple globule–coil theories such as equations (5.16)–(5.20).

The *self-assembly* of the tertiary structure of a globular protein is found to proceed in two stages: a rapid globule–coil transition driven by hydrophobicity and electrostatics (the molten globule), followed by the slow formation of the native structure inside the globule. This complicated molecular origami is exceedingly subtle and reference should be made to the literature of a specific protein for exact details. Often additional proteins (chaperones) are required along the folding pathway, to produce the biologically active native structure.

5.4   Crystallisation

Protein crystallisation in three dimensions is of little relevance to the function of living organisms (except for a few unusual examples such as some seed storage proteins and extracellular proteins), but is of central importance to structural biology. Only through the production of large high-quality defect-free crystals can the structure of proteins be obtained using diffraction techniques (**Figure 5.11**). A billion-pound question

Figure 5.11 *Optical photograph of a single crystal of a globular protein prepared for crystallographic studies. The preparation of large single crystals facilitates the process of structural elucidation.*

is thus posed by the biotechnology industry; how to form high-quality protein crystals for structure determination to understand structure/function relationships with Ångstrom level resolution.

The *liquid–solid transition* is much more complicated than the liquid–liquid transition (**Section 5.5**), since an infinite number of order parameters are in principle required to completely describe the resultant crystalline structure; typically the Fourier components of the density are chosen for this role. The liquid-solid transition is invariably a first-order phase transition.

The process of crystallisation occurs in many materials when the temperature is reduced. At the melting temperature (T_m), a liquid material never freezes, because it costs energy to form an interface. Without impurities the sample must be undercooled ($T < T_m$) to induce crystallisation (homogeneous nucleation). The free energy change ($\Delta G(r)$) upon crystallisation can be constructed as the sum of the energy to form the crystalline nuclei (the first term on the right-hand side) and the energy to form the surfaces (the second term on the right-hand side),

$$\Delta G(r) = \frac{4}{3}\pi r^3 \Delta G_b + 4\pi r^2 \gamma_{sl} \tag{5.21}$$

where r is the crystallite radius, ΔG_b is the bulk energy and γ_{sl} is the surface free energy of the crystal/liquid interface. The free energy is plotted as a function of temperature in **Figure 5.12**.

The change in entropy (ΔS_m) in a liquid–solid transition at constant pressure (P) can be related to the latent heat released (ΔH) using a standard thermodynamic expression,

$$\Delta S_m = \left(\frac{\partial G_s}{\partial T}\right)_p - \left(\frac{\partial G_1}{\partial T}\right)_p = \frac{\Delta H}{T_m} \tag{5.22}$$

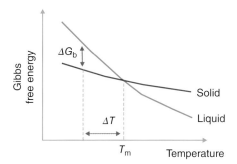

Figure 5.12 *Schematic diagram of the Gibbs free energy for the solid (orange) and liquid (green) phases of a material as a function of the temperature (T_m is the melting temperature, ΔG_b is the difference in free energy between the solid and liquid phases). At low temperatures a solid phase is adopted, whereas at high temperatures the materials adopt a liquid phase.*

where G_s and G_l are the free energy of the solid and liquid phases, respectively. The subscript p corresponds to the differential at constant pressure. The bulk contribution to the Gibbs energy associated with a temperature change (ΔT) is related to the change in enthalpy upon melting (ΔH_m) by an approximate relationship,

$$\Delta G_b \approx -\frac{\Delta H_m}{T_m}\Delta T \tag{5.23}$$

Thus, the total energy from equation (5.21) for the formation of a crystal is

$$\Delta G(r) = -\frac{4}{3}\pi r^3 \frac{\Delta H}{T_m}\Delta T + 4\pi r^2 \gamma_{sl} \tag{5.24}$$

$G(r)$ can be differentiated with respect to the crystal radius, and shows that the free energy has a maximum at a critical radius (r^*) given by

$$r^* = \frac{2\gamma_{sl}T_m}{\Delta H_m \Delta T} \tag{5.25}$$

Substitution in equation (5.24) shows that the free energy barrier is associated with the formation of stable nuclei due to the surface energy is given by

$$\Delta G^* = \frac{16\pi}{3}\gamma_{sl}^3 \left(\frac{T_m}{\Delta H_m}\right)^2 \frac{1}{\Delta T^2} \tag{5.26}$$

The functional form of the free energy with respect to the crystallite radius is shown in **Figure 5.13**. Thus, crystallites must spontaneously nucleate with sizes above this critical size for crystallisation to take place.

Arrhenius dynamics are often used to model the kinetics of crystallisation with an activation energy given by ΔG^* (**Section 20.3**) and the probability of a crystal being nucleated is found to be proportional to

$$\exp(-\Delta G^*/kT) \tag{5.27}$$

From the combination of equations (5.26) and (5.27) it is observed that the surface energy (γ_{sl}) governs the growth of the crystals. For colloidal crystals the rate of nucleation (Γ) can be written as a product of two factors,

$$\Gamma = e^{-\Delta G^*/kT} v \tag{5.28}$$

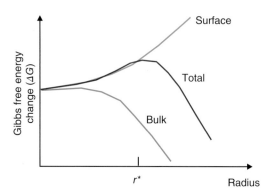

Figure 5.13 *The Gibbs free energy for the crystallisation of a solid as a function of the crystallite radius. The crystals have a critical radius (r*) which depends on the interplay between the bulk and surface energies for the creation of stable crystallites. Only crystallites with radii > r* are stable, i.e. there is a nucleation barrier for the first-order transition.*

where ν is a measure of the rate at which critical nuclei, once formed, transform into larger crystallites. An ongoing challenge is to relate the phase behaviour (e.g. through ΔG and ν) to the interprotein potential and this is still a hot area of research. Advances have been made that map the behaviour of globular protein crystallisation to that expected for spherical colloids that contain a number of adhesive patches on their surfaces, in addition to the standard electrostatic and van der Waals forces.

Once small protein crystals are formed, the surface free energy continues to play a role in the development of crystalline morphology as described by equation (5.21). Small crystals are absorbed by large crystals as their surface free energy is minimised in a process called *Ostwald ripening*. It is found to be an important effect in the production of ice cream, as anyone who has eaten melted and subsequently refrozen ice cream will testify, i.e. large ice crystallites are not very palatable.

Naturally occurring solid proteins often adopt fibrous semicrystalline morphologies, and the kinetics and morphologies of these materials is much more complicated than the case of perfectly crystalline globular proteins. Furthermore, many of these materials adopt intermediate liquid-crystalline mesophases due to their extended molecular structures, and this behaviour will be examined in **Chapter 6**.

5.5 Liquid–Liquid Demixing (Phase Separation)

Another common process in biological systems is liquid–liquid phase separation (**Figure 5.14**). Examples include the production of food gels, aggregation of ocular proteins in the eye and the partitioning of intracellular ionic species. Liquid–liquid phase separation can also happen in lower dimensional systems, e.g. protein/lipid mixtures on the surface of cell membranes can experience a process of two dimensional liquid–liquid phase separation which creates a raft-like heterogeneous morphology.

A useful simple reference system, before more complicated biological molecules are considered, is the free energy of mixing of simple molecular liquids. It represents an important reference model that needs to be understood, before the more sophisticated phenomena involved in the phase separation of colloids, surfactants and polymers can be considered.

The change in free energy upon mixing (F_{mix}) of the two fluids A and B is the difference in free energies before (F_{A+B}) and after ($F_A + F_B$) phase separation,

$$F_{\mathrm{mix}} = F_{A+B} - (F_A + F_B) \tag{5.29}$$

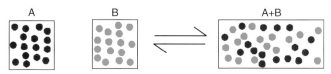

Figure 5.14   *Two phase-separated systems A and B are involved in a reversible mixing liquid-liquid phase transition.*

The Boltzmann formula for the entropy of mixing (S, equation (3.17)) is given by

$$S = -k_B \sum_i p_i \ln p_i \qquad (5.30)$$

where p_i is the probability of occupation of a lattice site i by a liquid molecule of type A and k_B is the Boltzmann constant. The internal degrees of freedom of the two fluids are neglected in this model.

The molecules (As and Bs) are assumed to interact in a pairwise additive manner on a lattice, i.e. the energies of nearest neighbours are added. ϕ_A and ϕ_B are the volume fractions of the A and B molecules, respectively. The mean-field assumption (i.e. concentration fluctuations are neglected) is that a given site has $z\phi_A$ A neighbours and $z\phi_B$ B neighbours. The interaction energy per site is therefore

$$\frac{z}{2}\left(\phi_A^2 \varepsilon_{AA} + \phi_B^2 \varepsilon_{BB} + 2\phi_A \phi_B \varepsilon_{AB}\right) \qquad (5.31)$$

where ε_{AA}, ε_{BB}, and ε_{AB} are the binary interaction energies between AA, BB and AB molecules, respectively.

The energy of the unmixed state is

$$\frac{z}{2}\left(\phi_A \varepsilon_{AA} + \phi_B \varepsilon_{BB}\right) \qquad (5.32)$$

The difference in the two interaction energies associated with the process of mixing is therefore

$$U_{mix} = \frac{z}{2}\left[\left(\phi_A^2 - \phi_A\right)\varepsilon_{AA} + \left(\phi_B^2 - \phi_B\right)\varepsilon_{BB} + 2\phi_A \phi_B \varepsilon_{AB}\right] \qquad (5.33)$$

If every site is occupied by either A or B there is a further condition on the sum of the two volume fractions,

$$\phi_A + \phi_B = 1 \qquad (5.34)$$

The mathematics is simplified by the definition of an interaction parameter (χ),

$$\chi = \frac{z}{2kT}\left(2\varepsilon_{AB} - \varepsilon_{AA} - \varepsilon_{BB}\right) \qquad (5.35)$$

The mixing energy equation (5.33) can therefore be expressed as

$$U_{mix} = \chi \phi_A \phi_B \qquad (5.36)$$

The equation for the free energy is $F = U - TS$ (**Chapter 3**) and the total free energy of phase separation in the mixture can then be constructed (**Figure 5.15**),

$$\frac{F_{mix}}{kT} = \phi_A \ln \phi_A + \phi_B \ln \phi_B + \chi \phi_A \phi_B \qquad (5.37)$$

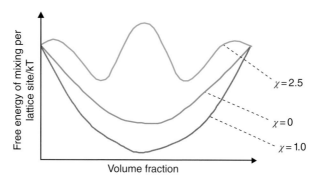

Figure 5.15 *The free energy of mixing of liquids A and B as a function of the volume fraction. χ is the interaction parameter. For χ = 0 and χ = 1 the phases are mixed, whereas for χ = 2.5 phase separation occurs.*

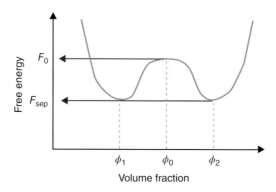

Figure 5.16 *Graphical construction for the free energy of separation of liquids A and B. The free energy for the phase-separated fluids (F$_{sep}$) is lower than the mixed free energy (F$_0$) that causes the mixture to separate into two volume fractions ϕ$_1$ and ϕ$_2$.*

If the mixture phase separates into two distinct coexisting phases, the total free energy of the separated mixture is the arithmetic average of the two mixture free energies that contribute (**Figure 5.16**),

$$F_{sep} = \frac{\phi_0 - \phi_2}{\phi_1 - \phi_2} F_{mix}(\phi_1) + \frac{\phi_1 - \phi_0}{\phi_1 - \phi_2} F_{mix}(\phi_2) \tag{5.38}$$

Coexisting compositions ϕ_1 and ϕ_2 are formed, since the separated free energy is smaller than the homogeneous mixture ($F_{sep} < F_0$) and in equilibrium the system seeks to minimise its free energy. Whether the system is stable to fluctuations is important in the determination of the phase behaviour. The stability of the system depends on the second derivative of the free energy ($d^2 F/d\phi^2$) (**Figure 5.17**). It is interesting that purely repulsive interactions can promote the formation of an ordered phase in this model and this behaviour has been demonstrated experimentally with hard-sphere colloidal systems that can phase separate.

A good example of phase separation in biocolloids is demonstrated by proteins from the eye, e.g. the gamma crystallin–water system. The compressibility and correlation length for the phase separation of gamma crystallins are shown in **Figures 5.18**. At the point of phase separation the compressibility and correlation length measured with light scattering diverges with a power law and a characteristic exponent is measured as described in **Section 5.1**. Such aggregation phenomena are thought to be associated with the formation of cataracts.

Chemically different neutral polymers mix very poorly in solution and the slight repulsion between the monomeric links is often sufficient to separate the mixture into two virtually pure phases. The degree of

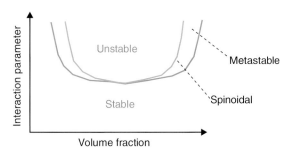

Figure 5.17 *The interaction parameter (χ) of a binary mixture of liquids as a function of the volume fraction. Both the spinoidal line and the metastable region are shown.*

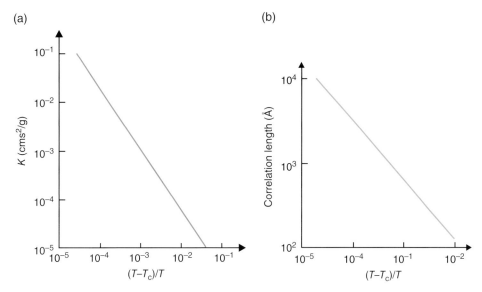

Figure 5.18 *(a) The compressibility and (b) the correlation length measured with light scattering as a function of the rescaled temperature. Both parameters diverge for ocular proteins as the critical temperature for phase separation (T$_c$) is approached. The line of best fit is from linear regression. [Reprinted with permission from P. Schurtenberger, R.A. Chamberlin, G.M. Thurston, J.A. Thomson, G.B. Benedek, PRL, 1989, 63, 19, 2064–2067. Copyright 1989 by the American Physical Society.]*

monomeric repulsion is magnified by a high degree of polymerisation. The separation of the two polymers is a phase transition and it can be realised by either mechanisms of spinoidal decomposition or nucleation and growth (as with the model of liquid–liquid transition discussed previously). Mixtures of polymers occur in countless biological systems, so it is an important effect to study for its impact *in vivo*, e.g. the phase separation of aggrecan/collagen mixtures in cartilage. Furthermore, the morphologies formed upon phase separation are sensitively dependent on the dynamics of the constituent molecules. Large degrees of dynamic asymmetry (e.g. long slow-moving polymers mixed with fast moving colloidal particles) lead to a range of novel time dependent phase separated morphologies. **Figure 5.19** shows the phase-separated morphology that can occur in biopolymer food gels during their preparation and simulations indicate it is a common physical phenomenon, e.g. it is even thought to occur during the creation of the aggregate structure of the early universe.

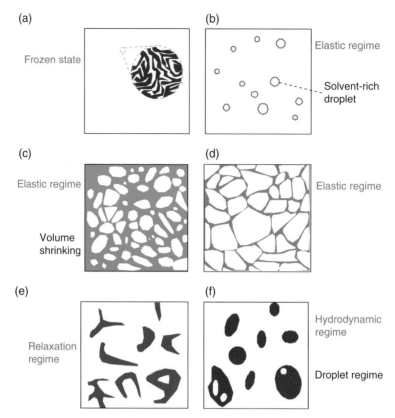

Figure 5.19   *The viscoelastic phase separation of food biopolymers has a complex series of evolutionary steps. A frozen state (a) is followed by an elastic regime (b, c and d), followed by a relaxation regime (e), followed by a hydrodynamic regime (f). The unusual morphologies are driven by the dynamic asymmetry of the phase separating components and are observed in a wide range of systems, e.g. astrophysical models of the early universe. [Reproduced from H. Tanaka, J. Phys: Condensed Matter, 2000, 12, R207–264. © IOP Publishing. Reproduced with permission. All rights reserved.]*

The compatibility of two polymers improves substantially after weakly charging of one of the components. This is due to the contribution of the entropy of the counterions to the free energy.

During the early stages of tissue differentiation and morphogenesis cells undergo a sorting process that resembles liquid–liquid phase separation. Here, the Flory interaction parameter (χ) is equivalent to the energy of adhesion between the cells. Such phenomena are vitally important to life in multicellular organisms and a coherent picture of this extremely complicated process is only slowly emerging.

Suggested Reading

If you can only read one book, then try:

Top recommended book: Dill, K. & Bromberg, S. (2011) *Molecular Driving Forces*, 2nd edition, Garland Science. Simple introduction to statistical models of phase transitions.

Barrat, J.L. & Hansen, J.P. (2003) *Basic Concepts for Simple and Complex Fluids*, Cambridge University Press. Clear compact account of theoretical models of complex fluids.

Callen, H.B. (1985) *Thermodynamics and an Introduction to Thermostatics*, Wiley. Classic text on thermodynamics.

Grosberg, A.Y. & Khoklov, A.R. (1994) *Statistical Physics of Macromolecules*, AIP Press. A good pedagogic account of polymer physics.

Jones, R.A.L. (2002) *Soft Condensed Matter*, Oxford University Press. Reasonably simple introductory treatment of soft condensed matter physics.

Kleman, M. & Lavrentovich, O.P. (2003) *Soft Matter Physics*, Springer. Similar in level and scope to Chaikin and Lubensky.

Lubensky, T. & Chaikin, P. (1995) *Principles of Condensed Matter Physics*, Cambridge University Press. Mathematically sophisticated coverage of soft-matter physics.

Rubinstein, M. & Colby, R.H. (2004) *Polymer Physics*, Oxford University Press. Similar in level to Grosberg's book with a great range of tutorial exercises.

Tutorial Questions 5

5.1 The DSC endotherm for a helix–coil transition for an α helical polypeptide becomes narrower as the length of the peptide is increased. Explain the phenomenon in terms of the thermodynamics of the transition.

5.2 Peptides in low-water conditions are only partially plasticised and can demonstrate glassy (nonergodic) behaviour at room temperature. Explain how this might alter the behaviour of the globule–coil transition of a long peptide chain.

5.3 Calculate the free-energy barrier for the nucleation of a lysozyme crystal if the critical crystallite size is 50 nm, the surface free energy is 1.2 mJ m^{-2}, the melting temperature is 50 °C and the temperature is reduced by 1 °C below the melting temperature.

6

Liquid Crystallinity

Rod-like molecules can spontaneously align themselves in solution to form anisotropic fluids of reduced viscosity, if the concentration of the molecules is increased beyond a critical value (*lyotropic liquid crystals*) or if the correct temperature range is chosen (*thermotropic liquid crystals*). Mankind has recently developed synthetic examples of such materials in a wide range of roles such as the displays of television screens, bullet-proof jackets and soap powders, but Nature has already been using the rich variety of phenomena associated with liquid crystals in a range of biological processes for many millions of years.

Research on synthetic liquid crystals has been predominantly driven by electro-optical applications (display screens), although there exist other niche markets for these materials, e.g. Kevlar bullet-proof jackets and puncture-resistant bicycle tyres. Processing intermediates of many synthetic lyotropic colloids often have liquid-crystalline intermediates. Such studies with synthetic materials have motivated the development of good quantitative theories for liquid crystals that can also be applied to biology.

6.1 The Basics

Liquid crystals are an intermediate state of matter (a mesophase) between a liquid and a solid. They are characterised by orientational ordering of the molecules (solid-like behaviour), while they maintain an ability to flow (liquid-like behaviour).

There are a wide range of biological molecules that form liquid-crystalline phases. These include lipids, nucleic acids, proteins, carbohydrates and glycoproteins. There is therefore a correspondingly wide range of liquid-crystalline phenomena that are biologically important. For example, cell membranes are maintained in liquid-crystalline phases that are used to compartmentalise the cell and still permit the transfer of important molecules (**Figure 6.1a**), slugs move on nematic trails of glycoprotein molecules whose viscoelasticity is intimately connected with their chosen form of locomotion (**Figure 6.1b**), starch assembles into smectic structures as a high density energy store in plant storage organs (**Question 6.1**), spider silk has a low viscosity liquid-crystalline phase as it is extruded from the spiders spinneret to form the supertough materials that are used to make its web (**Figure 17.14**), protocollagen forms nematic phases during the construction of

The Physics of Living Processes: A Mesoscopic Approach, First Edition. Thomas Andrew Waigh.
© 2014 John Wiley & Sons, Ltd. Published 2014 by John Wiley & Sons, Ltd.

(a)   (b)   (c)

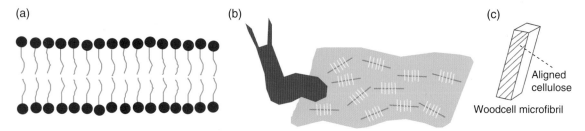

Aligned
cellulose

Woodcell microfibril

Figure 6.1   *Schematic diagrams of naturally occurring examples of liquid-crystalline materials that include, (a) cell membranes, (b) slug slime, and (c) cellulose microfibrils.*

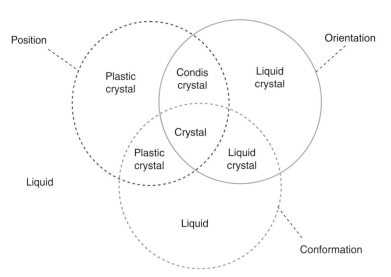

Figure 6.2   *Types of phase found in condensed matter that result from a mixture of conformation, position and orientation. [Reproduced from C. Viney in Protein Based Materials, Ed. K. McGrath, D. Kaplan, Birkhauser, 1997, 281–311. with kind permission from Springer Science + Business Media B.V.]*

the tough viscoelastic collagen networks in mammalian skin and cellulose microfibrils form chiral nematic phases in plant cell walls that help provide trees with their strength and resilience (**Figure 6.1c**).

The principle structural *phases* formed by soft condensed matter are due to an interplay between *positional*, *orientational* and *conformational disorder* (**Figure 6.2**). There are thus a wide variety of different mesophases that can be further refined with in the broad category of liquid-crystalline materials, i.e. materials with orientational and conformational ordering (**Table 6.1, Figure 6.3**). In addition to liquid-crystalline phases the possibility for internal conformational ordering of molecules combined with positional ordering (with a related lattice) leads to the phases of condis crystals and plastic crystals. A further mesophase subclassification is possible upon the inclusion of *molecular chirality*, i.e. the molecules have a well-defined handedness. Many biological molecules are chiral (e.g. DNA is normally right handed) and their mesophase structure reflects the chiral interaction between subunits. The principal chiral mesophases are the *cholesterics* (chiral nematics) and *tilted smectics* (chiral smectics). Chirality also has a large impact on the defect textures that liquid-crystalline molecules adopt and thus their macroscopic properties.

Table 6.1 *The range of mesophases commonly encountered with biological molecules are primarily determined by a combination of the positional and orientational order parameters.*

Phase	Positional Order	Orientational Order	Mesophase Order Parameter
Liquid	None	None	None
Nematic	None	Yes	Legendre polynomials (e.g. P_2) of orientation
Smectic	One dimensional	Yes	Fourier components of displacement in 1D (e.g. ψ)
Columnar	Two dimensional	Yes	Fourier components of displacement in 2D
Crystalline	Three dimensional	Yes	Infinite number of Fourier components in 3D

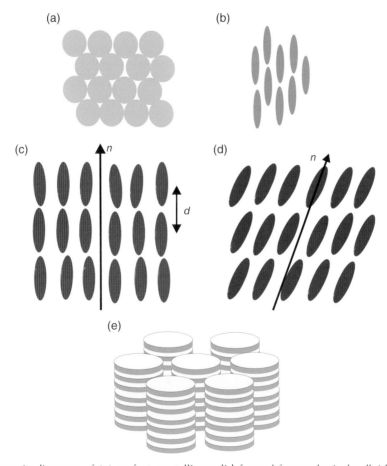

Figure 6.3 *Schematic diagrams of (a)* perfect crystalline solid *formed from spherical colloidal particles, (b) a* nematic *phase formed from rod-like mesogens, (c)* smectic A *(perpendicular) phase formed from rod-like mesogens, (d)* smectic C *(tilted) phases formed from chiral rod-like mesogens, and (e)* hexatic *phases are a rarer liquid-crystalline phase in biology formed from disk-like mesogens. Hexatic phases have been observed with nucleosome particles.*

To detect *liquid-crystalline phase transitions* a wide range of experimental techniques can be used. These include differential scanning calorimetry to study the latent heat absorbed, polarising microscopy to view the strength and variety of defect textures, X-ray and neutron scattering to measure orientational and lattice order parameters, and atomic force microscopy and ellipsometry to measure the thickness of the terraces on surfaces.

In practice liquid crystals are often detected through their optical textures using crossed polarisers under an optical microscope, since it is a cheap readily available technique. These *defect textures* can enable the exact liquid-crystalline phase to be assigned. The quantitative evaluation of these defect textures in biological liquid crystals will be considered in **Section 6.2**.

Differential scanning calorimetry reveals the presence of liquid-crystalline phase transitions in a material through the detection of the associated enthalpy changes. The isotropic–nematic phase transition is clearly demonstrated to be a true thermodynamic event with an associated endotherm (**Figure 6.4**). Typically, a liquid-crystalline material can experience a sequence of thermodynamic phase transitions as a function of temperature, e.g. a crystal transforms into a smectic liquid crystal, then into a nematic liquid crystal and finally it is converted into an isotropic liquid as a function of increasing temperature.

Following the discussion of phase changes in **Chapter 5** it is useful to consider the relevant *order parameters* for a liquid-crystalline phase transition. The three simplest phases of liquid crystals are *nematics* (a single direction of preferred orientation), *cholesterics* (nematics with the orientational direction twisting along a helix) and *smectics* (with long-range layered ordering in one dimension).

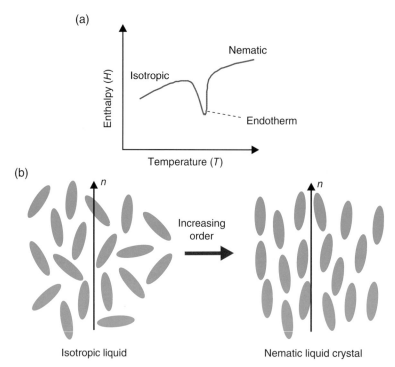

Figure 6.4 *(a) Differential scanning calorimetry can measure the enthalpy as a function of temperature for a liquid-crystalline material and the endotherm due to an isotropic–nematic phase change is shown. (b) Schematic diagram that indicates the increased orientational ordering upon an isotropic–nematic phase transition for small rigid molecules that results in the endotherm.*

With *nematic liquid crystals* it is conventional to use the second-order Legendre polynomial (P_2), which is based on a spherical coordinate system (θ, ϕ, r), to quantify the degree of orientational alignment of the molecules,

$$S = \langle P_2(\cos\theta)\rangle = \left\langle \frac{3}{2}\cos^2\theta - \frac{1}{2}\right\rangle \tag{6.1}$$

where θ is the angle the long axes of the molecules make with the nematic director (**Figure 6.5**). The director is a unit vector that indicates the average direction of alignment of the molecules. The calculation for the nematic order parameter can be pictured through the calculation of a $\frac{3}{2}\cos^2\theta - \frac{1}{2}$ term for each molecule in the solution that is then averaged over all the molecules. This nematic order parameter (S) is equal to 1 for perfect alignment parallel to the director, −1/2 for a perfect perpendicular alignment, and zero for an isotropic liquid. A typical plot for the variation of S as a function of temperature during the isotropic/nematic phase change is shown in **Figure 6.6**. The nematic order parameter (S) is seen to decrease with an increase in

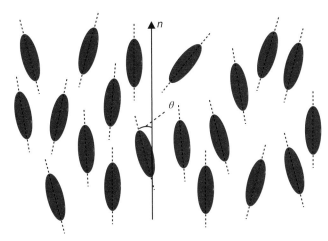

Figure 6.5 *The long axes of mesogens in a nematic phase of biological molecules become ordered along an axis. n is the direction of the average value of the director field. θ is the angle the rods make with the director, n.*

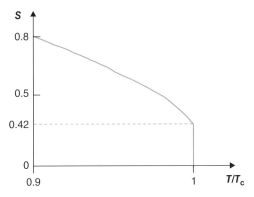

Figure 6.6 *The nematic order parameter (S) for a nematic liquid crystal as a function of temperature (T). Above the critical temperature (T_c) the material forms an isotropic liquid phase, whereas below T_c it is a nematic liquid crystal.*

temperature and is lost abruptly at T_c, the temperature of the phase transition from a nematic liquid crystal to an isotropic liquid.

An additional order parameter in combination with that for nematicity is required to describe the lamellar ordering found with smectic liquid crystals, i.e. the one-dimensional lattice structure needs to be described. A *lamellar order parameter* is defined through an expansion of the electron density of the periodic smectic stack as a Fourier series of the displacement in the direction perpendicular to the stack (z). The first cosine term of the infinite series is then kept as a good first approximation for the stack density. The order parameter (ψ) is defined as the amplitude of this cosine function,

$$\rho(z) = \rho_0 \left(1 + \psi \cos \left(\frac{2\pi z}{d} \right) \right) \tag{6.2}$$

where d is the spacing of the layers, ρ_0 is the average electron density and $\rho(z)$ is the electron density of the stack as a function of z. The lamellar order parameter (ψ) can be measured using X-ray, AFM, neutron and light scattering techniques (**Figure 6.7**). It is possible to theoretically predict the behaviour of the smectic and nematic order parameter described by equations (6.1) and (6.2) near to a critical point of a phase transition using a model due to Landau (**Section 6.2**). A further chiral order parameter is needed for a discussion of the phase behaviour of cholesterics and it will be introduced with an analysis of the elasticity of nematics that become twisted.

The *isotropic–nematic phase transition* for small rigid biological molecules in solution is now fairly well understood. Both analytic models and simulations predict the onset of a nematic phase for a particular molecular geometry and are in good agreement with experimental systems in which intermolecular potentials approximate to a hard sphere interaction. For short rigid rods in solution Lars Onsager analytically determined the phase diagram for nematic liquid crystals and found that it could be simply predicted as a function of the aspect ratio (L/D) of the molecules (length (L) and diameter (D)) and the volume fraction (ϕ) (**Figure 6.8**). Below a critical value of the product $\phi L/D$, the solution adopts an isotropic liquid structure,

$$\text{Isotropic ordering } \frac{\phi L}{D} < 3.34 \tag{6.3}$$

whereas above an upper bound the rods adopt a perfect nematic ordering,

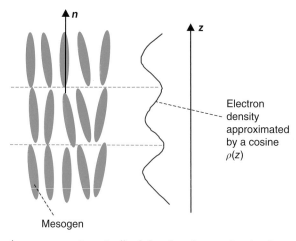

Mesogen

Figure 6.7 *The smectic order parameter is typically defined as the amplitude of a sinusoidal expansion (e.g. the amplitude of a cosine ρ(z), red) of the electron density of the lamellar stack along a direction perpendicular to the layers (z).*

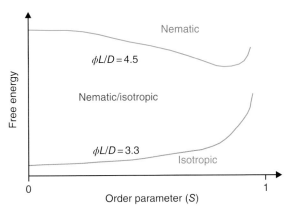

Figure 6.8 *The free-energy diagram of a solution of hard rods as a function of the nematic order parameter (S) shows the stable regions of nematic, nematic/isotropic, and isotropic phases. The critical parameter that determines the phase behaviour is the volume fraction (ϕ) multiplied by the aspect ratio (L/D).*

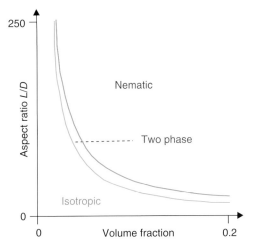

Figure 6.9 *Phase diagram for a solution of hard rods which shows the aspect ratio (L/D) as a function of the volume fraction. The phase behaviour is determined from the free energy shown in **Figure 6.8**.*

$$\text{Nematic ordering } \frac{\phi L}{D} > 4.49 \qquad (6.4)$$

where ϕ is the volume fraction, and L/D is the length/diameter of the rods (the aspect ratio). There is coexistence of the isotropic and nematic phases between the two critical values (**Figure 6.9**) described by equations (6.3) and (6.4). Nematic liquid-crystalline ordering from an isotropic fluid phase is a first-order phase transition.

Liquid-crystalline phases can also be adopted by long *semiflexible polymeric molecules* such as DNA, collagen and carrageenan. The behaviour is more complicated than that displayed with simple rod-like molecules, because the internal degrees of freedom of the polymer must be considered. The adoption of liquid-crystalline phases in polymers (e.g. DNA, collagen, etc.) is related to their persistence lengths (**Section 10.1**). The

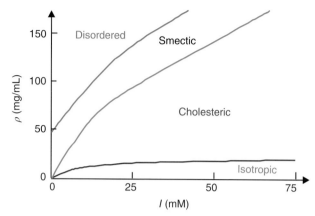

Figure 6.10 *Experimentally determined phase diagram for fD virus solutions (chiral semiflexible polymeric molecules) where the virus density (ρ) is shown as a function of the ionic strength (I). [Reproduced from S. Fraden. in 'Observation, Prediction and Simulation of Phase Transitions in Complex Fluids', ed M. Baus, L.F. Rull, J.P. Ryckaert, Kluwer Acdaemic Press, NATO ASI Series C, Vol. 460, 1995. With permission from Springer ScienceSpringer Science + Business Media.]*

isotropic/nematic phase diagram is qualitatively similar to that of small molecules (**Figure 6.9**) when semi-flexible chains are considered, with the phase boundaries scaled by the magnitude of the persistence length. As the orientational ordering grows in a solution of persistent chains, so does the mean size of the chains along the ordering axis. Thus, the internal conformation of the molecules is coupled to the phase behaviour, unlike the case of short rod-like molecules. **Figure 6.10** shows the phase diagram of a solution of fD viruses. These virus molecules are an ideal experimental system to examine semiflexible liquid crystals, since they form an optically observable microscopic smectic phase (**Figure 6.11**), and are perfectly monodisperse, since the protein sequence is genetically determined.

For an elastic Hookean spring in one dimension the energy (E) stored in the system is given by the familiar expression

$$E = \frac{1}{2}Kx^2 \tag{6.5}$$

where x is the extension, and K is the spring constant (**Figure 6.12**). The factor of a half is included so that upon differentiation the restoring force on the spring is given by Hookes law, $F = Kx$. In three dimensions (**Chapter 13**) the elasticity of an arbitrarily chosen material is much more complicated and a compliance tensor (a matrix of 81 numbers) is needed in place of the single constant, K. Fortunately, liquid-crystalline materials have an elasticity that is dependent on only the orientation of the director and the distortion of the director field. Thus, simple nematic liquid crystals in three dimensions only require three elastic constants, K_1 (splay), K_2 (twist), and K_3 (bend), for a complete description. The elastic energy (or free energy) of a nematic is constructed using symmetry relations. A fair amount of mathematical effort is required for the derivation, but the free energy per unit volume (F_v) of a nonchiral nematic liquid crystal is found to be

$$F_v = \frac{1}{2}K_1[\nabla . n]^2 + \frac{1}{2}K_2[n.(\nabla \times n)]^2 + \frac{1}{2}K_3[n \times (\nabla \times n)]^2 \tag{6.6}$$

where K_1, K_2, and K_3 are spring constants with units Joules m^{-3} and n is the unit vector that describes the direction of orientation of the rod-like molecules (the director). This nematic free energy (equation (6.6))

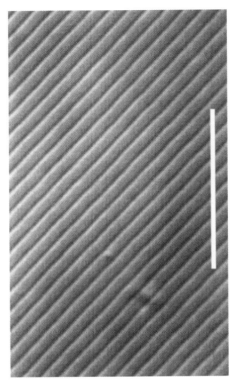

Figure 6.11 *Differential interference contrast image from smectic phases of fD viruses whose phase diagram is shown in* **Figure 6.10**. *The white scale bar is 10 μm and the smectic periodicity is 0.92 μm. [Reprinted with permission from Z. Dogic, S. Fraden, Physical Review Letters, 1997, 78, 12, 2417–2420. Copyright 1997 by the American Physical Society.]*

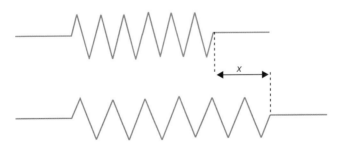

Figure 6.12 *Schematic diagram of the extension of a Hookean spring in one dimension. The spring is extended by a distance x and it stores an elastic energy,* $E = \frac{1}{2}kx^2$.

has the same form as that of a single elastic spring (equation (6.5)), except three elastic constants are now required to describe the material. A typical value of K_3 for a short molecular liquid crystal is 10^{-11} N m^{-2} and it is normally two to three times bigger than K_1 and K_2.

To physically understand the formula for the free energy (F_v) it is useful to simplify the vector calculus. The director is assumed to lie along the z direction so all of the derivatives of the z component of the director (n_z) are equal to zero,

$$\frac{\partial n_z}{\partial x} = \frac{\partial n_z}{\partial y} = \frac{\partial n_z}{\partial z} = 0 \tag{6.7}$$

The individual terms in equation (6.6) can then be calculated using equation (6.7),

$$[\nabla . n]^2 = \left[\left(\frac{\partial n_x}{\partial x}\right)_{y,z} + \left(\frac{\partial n_y}{\partial y}\right)_{x,z}\right]^2 \tag{6.8}$$

$$[n \times (\nabla \times n)]^2 = \left(\frac{\partial n_y}{\partial z}\right)^2_{x,y} + \left(\frac{\partial n_y}{\partial z}\right)^2_{x,y} \tag{6.9}$$

$$[n.(\nabla \times n)]^2 = \left[\left(\frac{\partial n_y}{\partial x}\right)_{y,z} - \left(\frac{\partial n_x}{\partial y}\right)_{x,z}\right]^2 \tag{6.10}$$

Figure 6.13 shows the director field when only the first terms on the right-hand side of equations (6.8), (6.9) and (6.10) are nonzero. This figure motivates a useful intuitive picture for the meaning of splay, twist and bend distortions.

The Frank's free energy (equation (6.6)) can be extended to describe a cholesteric liquid crystal. To model a chiral nematic (cholesteric) phase aligned perpendicular to the x-axis, the director needs to follow a helical path (**Figure 6.14a**) given by

$$n_x = 0 \tag{6.11}$$

$$n_y = -\sin\left(\frac{2\pi x}{P}\right) \tag{6.12}$$

$$n_z = \cos\left(\frac{2\pi x}{P}\right) \tag{6.13}$$

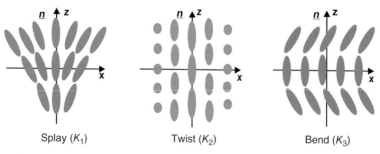

Splay (K_1)   Twist (K_2)   Bend (K_3)

Figure 6.13 *Visualisation of the splay, twist and bend constants for the director field of a nematic liquid crystal aligned along the z-axis.*

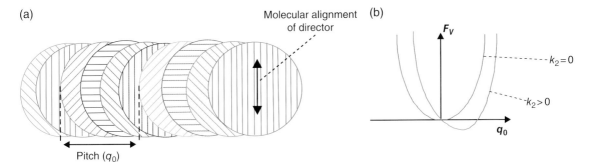

Figure 6.14   *(a) The helical pitch (q_0) of a cholesteric liquid crystal. It is the separation distance over which the directors are aligned. (b) The free energy density (F_v) for nonchiral ($k_2 = 0$) and chiral liquid crystals ($k_2 > 0$) as a function of the helical pitch (q_0).*

The cholesteric director thus twists around the *x*-axis with a characteristic length scale called the pitch (*P*). The div and curl of the director are required to construct the free energy in equation (6.6) and they are calculated as

$$\nabla . n = 0 \tag{6.14}$$

$$(\nabla \times n)_y = \frac{2\pi}{P} \sin \frac{2\pi x}{P} \tag{6.15}$$

$$(\nabla \times n)_z = -\frac{2\pi}{P} \cos \frac{2\pi x}{P} \tag{6.16}$$

Therefore, each of the three terms in equation (6.6) are

$$[\nabla . n]^2 = 0 \tag{6.17}$$

$$[n.(\nabla \times n)]^2 = \left(\frac{2\pi}{P}\right)^2 \tag{6.18}$$

$$[n \times (\nabla \times n)]^2 = 0 \tag{6.19}$$

And the total free energy only has a contribution from the twist term

$$F_v = \frac{1}{2} K_2 \left(\frac{2\pi}{P}\right)^2 = \frac{1}{2} K_2 a^2 \tag{6.20}$$

where $a = 2\pi/P$. Twist is the only distortion present in this simple example of a cholesteric liquid crystal. In a nematic liquid crystal (with no chirality) a helical distortion is unstable and a twisted nematic will relax the twist distortion to minimise its free energy, and from equation (6.20), $a = 0$, and the pitch becomes infinite. The free energy for a chiral nematic therefore needs a linear term to be added to equation (6.6) to stabilise its free energy,

$$K_4 [n.(\nabla \times n)] \tag{6.21}$$

where K_4 is a new elastic constant, which measures the degree of chirality. K_4 corresponds to an intrinsic chirality of the mesogens. Such chirality is common in biology (e.g. double-helical DNA, helices in proteins and carbohydrates, beta sheets in proteins, and lipids all have an intrinsic chirality) and in principle can be calculated from the molecular details of the mesogens.

Substitution of the helical director field equations (6.11–6.13) into the Frank's free energy with the additional chiral term (equation (6.21)) gives

$$P = \frac{2\pi K_2}{K_4} \tag{6.22}$$

$$a_0 = \frac{K_4}{K_2} \tag{6.23}$$

The pitch (P) is related to the ratio of the two elastic constants (a_0), the chiral term divided by the twist term. The free energy per unit volume for a cholesteric liquid crystal is given by

$$F_v = -\frac{K_4^2}{2K_2} \tag{6.24}$$

The free energy as a function of cholesteric pitch is shown in **Figure 6.14b**. The pitch relaxes to zero if there is no molecular chirality and the corresponding elastic constant (K_4) will be zero.

The Frank's free energy equation (6.6) can be used to calculate the free energy of a nematic liquid crystal around a defect. Let the director field parallel to the z-axis be

$$n_x = \cos\left[\theta(x,y)\right] \tag{6.25}$$

$$n_y = \sin\left[\theta(x,y)\right] \tag{6.26}$$

$$n_z = 0 \tag{6.27}$$

where $\theta(x,y)$ is the radial director of the mesogens. Assume that all the elastic constants are equal ($K_1 = K_2 = K_3$), then the free energy for an axial disclination can be calculated as

$$F_v = \frac{1}{2}K\frac{m^2}{\rho^2} \tag{6.28}$$

where m is the strength of the disclination ($m = \pm 1/2, \pm 1, \pm 3/2 \ldots$) and ρ is the radius of the core of isotropic material at the centre of the defect. The free energy diverges as the radius of the disclination reduces to zero ($\rho \to 0$) (**Figure 6.15**) and thus the nematic ordering becomes frustrated at the centre of a defect, i.e. real nematic defects have amorphous liquid-like ordering near to the centre of the defects.

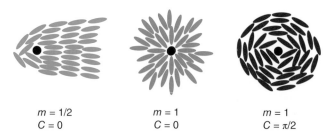

$$m = 1/2 \qquad\qquad m = 1 \qquad\qquad m = 1$$
$$C = 0 \qquad\qquad\quad C = 0 \qquad\qquad C = \pi/2$$

Figure 6.15   *Schematic diagram of some of the defect textures encountered with nematic liquid crystals. m is the defect strength and C is a constant in the relationship $\theta = m\phi + C$, where θ is the director angle and ϕ is the azimuthal angle.*

6.2 Liquid Nematic–Smectic Transitions

The *Onsager theory* allows the shape of the phase diagram for an isotropic–nematic transition to be calculated as a function of the aspect ratio of the molecules and the volume fraction of the solution (equations (6.3) and (6.4)). The theory allows both the form of the isotropic–nematic–smectic phase diagram to be motivated and provides quantitative predictions for the value of the order parameters near to the critical point of the phase transition.

The *Landau theory* for the isotropic–nematic phase transition develops on the ideas introduced by Onsager (**Figure 6.16**). The Landau theory assumes that the nematic order parameter (S) is small for the nematic phase in the vicinity of the isotropic/nematic transition and the difference between the free energy per unit volume of the isotropic and nematic phases ($G(S,T)$) can be expanded in powers of the nematic order parameter (S),

$$G(S,T) = G_{iso} + \frac{1}{2}A(T)S^2 + \frac{1}{3}BS^3 + \frac{1}{4}CS^4 + \tag{6.29}$$

where G_{iso} is the free-energy change for the isotropic material and A, B and C are the expansion coefficients. The theory only gives accurate information on the scaling of the phase behaviour near critical points. The term is S^1 in equation (6.29) has been neglected due to symmetry requirements. $A(T)$ is the most important parameter for the determination of the free-energy change during the phase transition and can be given a simple form (**Figure 6.17**) as a first approximation,

$$A(T) = A_0(T - T^*) \tag{6.30}$$

where T^* is the critical temperature for the transition. It is then possible to study the stability of the free energy using this functional form. The solutions of equation (6.29) are given by

$$S = 0 \, (\text{isotropic}) \tag{6.31}$$

And when the free-energy change is minimised as a function of the orientation,

$$\frac{\partial G}{\partial S} = A(T)S + BS^2 + CS^3 = 0 \tag{6.32}$$

there is the solution

$$S = \frac{-B \pm \sqrt{B^2 - 4AC}}{2C} \tag{6.33}$$

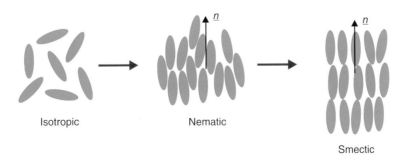

Figure 6.16 *Schematic diagram of the isotropic–nematic and nematic–smectic phase transitions in a molecular liquid crystal, e.g. the phase transition could be induced by a decrease in the temperature from left to right.*

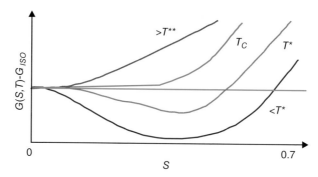

Figure 6.17 *Schematic diagram of the free-energy change (G(S,T)-G_iso) for a nematic phase compared with a liquid (isotropic) phase as a function of the order parameter (S) for the Landau model. The temperatures shown are T* for an isotropic phase, and T** for a nematic phase. T_c is the critical temperature for the nematic–isotropic phase transition.*

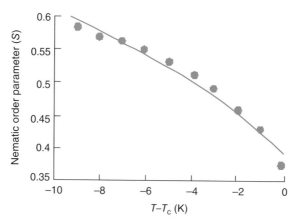

Figure 6.18 *Experimentally determined nematic order parameter near the isotropic-nematic phase transition as a function of the shifted temperature (T–T_c) compared with a fit of the Landau de Gennes theory. [Reprinted with permission from P.J. Collings, M.J. Hird, Introduction to Liquid Crystals, 1997. Copyright 1997, Taylor and Francis.]*

for the local maximum and minimum of G as a function of S. Such theories are in reasonable agreement with experiment (**Figure 6.18**).

For the *nematic–smectic transition* it is also possible to construct a similar theory for the free-energy change. The order parameter (ψ) is now the amplitude of the layered structure (equation (6.2), **Figure 6.7**). From the translational symmetry of the layered stack the Landau free-energy change can be constructed as

$$G(|\psi|,T) = G_{\text{nem}} + \frac{1}{2}\alpha(T)|\psi|^2 + \frac{1}{4}\beta|\psi|^4 + \frac{1}{6}\gamma|\psi|^6 \tag{6.34}$$

where G_{nem} is the free energy per unit volume of the nematic phase and α, β, and γ are characteristic constants. This free energy can be used to create a phase diagram for the nematic/smectic transition of a material and is in reasonable agreement with experiment.

6.3 Defects

Defects are an important facet of the structure of both solid and liquid-crystalline biological materials. Theory and experiment aim to explain a whole series of complex phenomena, e.g. how helices pack together in soft solids, the melting of mesophases and how chromosomes are constructed.

In solid materials which exhibit *lamellar* (a one-dimensional lattice) and *columnar* (a two-dimensional lattice) ordering, defect structures always occur. The *Landau–Peierls theorem* states that the geometry of one and two dimensional lattices is unable to constrain fluctuations in the positions of the molecules and they must display defect structures on large length scales, e.g. lamellar and columnar solids must be semicrystalline or liquid crystalline. Another important biological example of solid defect structures is the ordering of helical molecules on a hexagonal lattice that is commonly the case with solid biopolymers. Each helix has six identical neighbours and it is not possible for the helices to align with all of their neighbours simultaneously unless the helices have a perfect intrinsic six-fold symmetry, which is rarely the case. There must be frustration in these helical crystals and they must contain screw defects (**Figure 6.19**). This is an important concern in structure determination using X-ray fibre diffraction.

Generally, defects in solids consist of two main categories. *Point imperfections* are vacancy interstitials that involve an atom taken from a surface and inserted in an interior site not normally occupied. *Line imperfection* are defects localised along a continuous curve that passes through the ordered medium. The Volterra process is a geometrical method for the creation of dislocations in solids shown in the **Figure 6.20** and is therefore useful for the classification of dislocation structures.

There are two distinct categories of line defects; *dislocation line defects* involve translation of one part of a crystal with respect to another part and *disclination line defects* involve rotation of one point of the material relative to another part (**Figure 6.20**). The energy of disclinations is provided by the elastic energy associated with long-range distortions of the director field and can be calculated using generalised theories of elasticity, such as those discussed earlier (equation (6.6)). The strength of a disclination is determined by tracing a closed path that surrounds the disclination core, while the orientation of the director field is tracked (**Figure 6.21**). The

Figure 6.19 *Defect structure in the packing of helices on a hexagonal lattice. The arrows indicate the orientation of the interhelix potential in the horizontal cross section. The interhelix potential is frustrated by the hexagonal symmetry of the lattice.*

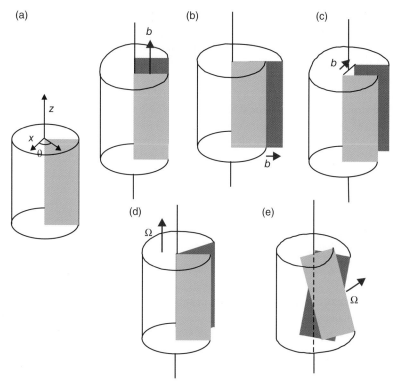

Figure 6.20 Dislocations *and* disclinations *in a cylindrical section of an elastic medium created using the volterra process. (a) Screw dislocations, (b) and (c) edge dislocations, (d) wedge disclinations and e) twist disclinations can occur.*

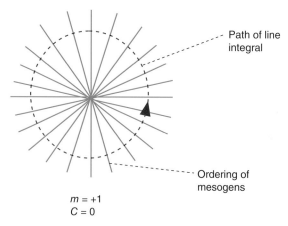

Figure 6.21 *To calculate the strength (m) of a defect, a line integral is used to add up the changes in direction. A hedgehog disclination is shown with m = +1 and C = 0.*

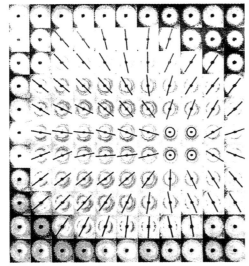

"Disclination" map of a single potato starch granule, based on the orientation of the (100) interhelix reflections. Each box samples a 5 μm by 5 μm region. The directors (lines) tend to point towards an eccentric region within the granule with no orientation, believed to be the hilum (circles). This map is physically analogous to a hedgehog disclination in liquid crystals, of strength $s = +1$.

Figure 6.22 *Single hedgehog defects naturally occur for the orientation of helical mesogens in carbohydrate granules. The disclination map was created using the X-ray scanning microfocus diffraction technique,* **Section 19.10**. *[Reproduced with permission from T.A. Waigh, K.L. Kato, A.M. Donald, M.J. Gidley, C.J. Clarke, C. Riekel, Starch, 2000, 52, 12, 450. © 2000 Wiley-VCH Verlag GmbH, Weinheim, Germany.]*

disclination strength (m) is defined as the normalised total angle (ϕ_{total}) or director reorientation in a complete circuit around the defect, i.e. divided by 2π,

$$m = \frac{1}{2\pi}\oint \frac{d\phi(r)}{d\theta} d\theta = \frac{\phi_{total}}{2\pi} \tag{6.35}$$

where $\phi(r)$ is the angle of the director at a position r, and θ is the angle that r makes with the positive horizontal axis. $\oint$ is the line integral around a complete circuit.

An illustration of a strength one hedgehog disclination found in a polymeric biological liquid crystal is shown in **Figure 6.22**. The diffraction patterns from small micrometre-sized elements in the biopolymer provide the direction of the helical mesogens and they are mapped across a single starch granule (~60 μm). This diffraction technique is a direct molecular probe of the structure of micrometre-sized disclinations in the material. Another method to characterise disclinations in a liquid-crystalline material is with polarised optical microscopy (**Figure 6.23**). The number of brushes that emanate from a defect (N) allows the strength of the defect to be calculated ($m = N/2$). The sign of the defect ($+/-$) can be determined from the direction of rotation of the brushes when one of the polarisers in the microscope is rotated.

From the generalised theory of elasticity the elastic energy per unit length (E) of a solid disclination located along the central line of a cylinder of radius R can be calculated as

$$E = 2\pi K m^2 \ln\left(\frac{R}{r_c}\right) + E_{core} \tag{6.36}$$

where K is the average elastic constant, m is the strength of the defect, r_c is the radius of the core and E_{core} is the core energy. The core of the line defect is on the order of the molecular size and it is assumed to contain an

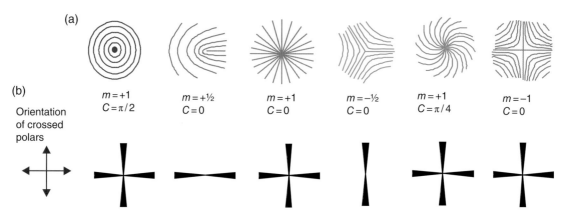

Figure 6.23 *(a) Examples of the orientation of the disclination fields of liquid crystals. (b) The defect textures that correspond to (a) when the liquid crystals are observed under a polarising microscope with crossed polarisers. m is the strength of the disclination and C is a constant in the relationship* $\theta = m\phi + C$, *where* θ *is the director angle and* ϕ *is the azimuthal angle.*

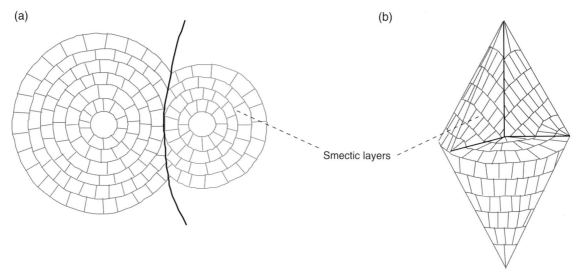

Figure 6.24 *Schematic diagram of defect textures observed with smectic liquid crystals, (a) Dupin cyclides and (b) focal conics. At small length scales the layered structures approximate to those for a standard smectic, i.e.* **Figure 6.7**.

isotropic material. The force (f_{12}) between a pair of straight parallel disclinations with strengths m_1 and m_2 separated by a distance r_{12} is

$$f_{12} = -2\pi K m_1 m_2 \frac{r_{12}}{(r_{12})^2} \tag{6.37}$$

When the strengths of two defects are equal, but of opposite sign, f_{12} is attractive and the defects can combine and annihilate.

For smectic materials step-like terraced defects are observed in optical and atomic force microscope images (Grandjean terraces). Optical polarising microscopy indicates there are also large-scale defect patterns specific to smectic liquid crystals, such as Dupin cyclides and focal conic domains (**Figures 6.24**).

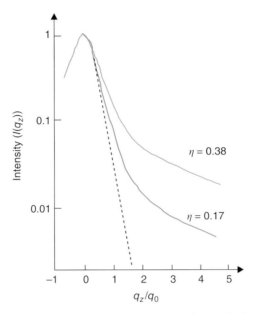

Figure 6.25   *The intensity from an X-ray diffraction experiment (with very high angular resolution) is shown as a function of the reduced momentum transfer perpendicular to a smectic stack of a liquid crystal. Bragg peaks are broadened into power-law cusps by the thermal fluctuations of smectic layers. The dashed line is a fit to $I \sim |q_z - q_0|^{-2+\eta}$ that allows the bending rigidity of the layers to be calculated.*

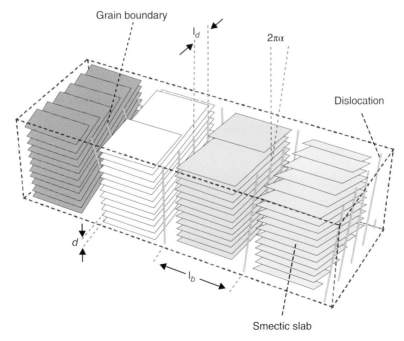

Figure 6.26   *Schematic diagram of a twisted grain-boundary phase in a smectic liquid-crystalline material. The chirality of the mesogens causes this unusual defect phase, which has been observed with DNA double helices. [Reproduced with permission from P.M. Chaikin and T.C. Lubensky, Principles of Condensed Matter Physics, 1995. Copyright Cambridge University Press, 1995.]*

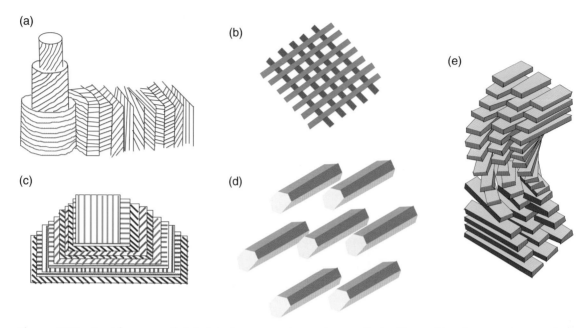

Figure 6.27 *A wide range of defect textures are observed naturally in fibrous biopolymers. These include (a) cylindrical helicoidal, e.g. bone, (b) orthogonal defects, e.g. basement lamellar in vertebrates, (c) twisted orthogonal defects, e.g. vertebrate cornea, (d) parallel defects, e.g. tendons, and (e) pseudo-orthogonal defects, e.g. endocuticle of beetles. [Reproduced with permission from A.C. Neville, Biology of Fibrous Composites, 1993. Copyright Cambridge University Press, 1993.]*

X-ray scattering is a standard technique for the characterisation of smectic liquid-crystalline materials (**Figure 6.25**). Due to the Landau–Peierls instability Bragg peaks are broadened into power-law cusps with X-ray scattering from smectics, which allow the bending rigidity of smectic stacks to be measured.

Twisted grain-boundary defects in smectic/cholesteric materials such as DNA (**Figure 6.26**) have experienced theoretical research, since there is an analogy to the phase behaviour of superconductors. Both processes can be described by the Landau equation (6.34) for the free energy of the nematic–smectic transition. Many other liquid-crystalline defect textures have been observed in naturally occurring soft solid biological materials (**Figure 6.27**). A sophisticated example of a biological defect structure is the blue phase in collagen observed in the skin of some fish (**Figure 6.28**), where the chirality of the collagen molecules are intimately connected with the defect texture observed.

6.4 More Exotic Possibilities for Liquid-Crystalline Phases

Biological polymeric liquid crystals are often induced by the increased persistence length produced by a *helix–coil transition* (e.g. DNA, and collagen) and conversely liquid-crystalline phases can induce increased persistence lengths in polymers due to steric constraints. Thus, helicity is seen to be intimately involved with the appearance of liquid-crystalline phases.

Starch, a storage polysaccharide, is a naturally occurring example of a *side-chain liquid-crystalline polymer* (**Figure 6.29**). In this material, additional order parameters are required to simultaneously describe the

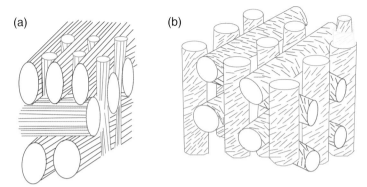

Figure 6.28 *(a) The solid cholesteric liquid-crystalline blue phase observed in the skin of fish is very similar in morphology to (b) the liquid cholesteric liquid-crystalline phase in synthetic small molecule chiral mesogens. [Reproduced with permission from M.M. Giraud, J. Castanet, F.J. Meunier, Tissue and Cell, 1978, 10, 671–686, with permission from Elsevier.]*

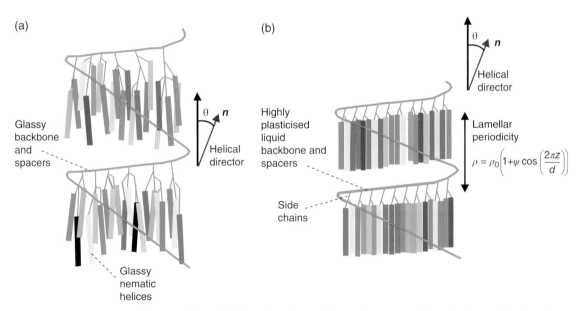

Figure 6.29 *Schematic diagram of a side-chain liquid-crystalline polymer (amylopectin in starch) that shows the process of self-assembly that occurs upon the addition of water. (a) The dry glassy nematic structure and (b) the hydrated side-chain liquid-crystalline smectic structure are shown. [Reprinted with permission from T.A. Waigh, I. Hopkinson, A.M. Donald, M.F. Butler, F. Heidelbach, C. Riekel, Macromolecules, 1997, 30, 3813–3820. Copyright © 1997, American Chemical Society.]*

nematic, cholesteric and smectic phases of both the backbone and the side chains. A wide range of distinct mesophases are therefore possible due to all the possible permutations of the order parameter values and the intricate interrelationships of the order parameters for the backbone and side chains directly relates to the macroscopically observed physical behaviour of these materials.

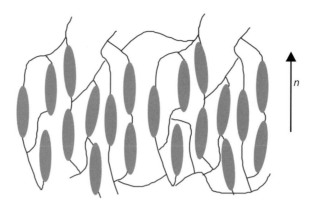

Figure 6.30   *Liquid-crystalline elastomers, where* n *is the director for the nematic ordering. The mesogens are crosslinked in a rubbery network to form a solid liquid-crystalline phase.*

It is possible for the orientational order of liquid crystals to persist into the solid state without the formation of a fully crystalline lattice, in so called *solid liquid crystals*. Solid nematic elastomer phases are formed by crosslinking mesogens with flexible (rubbery) polymeric chains. Liquid-crystalline elastomers models have been proposed for solid biopolymer elasticity (**Figure 6.30**), e.g. the soft anisotropic elasticity observed in collagen networks and spider silks.

Suggested Reading

If you can only read one book, then try:

Top recommended book: Collings, P.J. & Hird, M. (1997) *Introduction to Liquid Crystals*, Taylor and Francis. Provides a good introduction to the physics and physical chemistry of small molecule liquid crystals.

Allen, S.M. & Thomas, E.L. (1999) *The Structure of Materials*, Wiley. Introduction to materials science including defects.
Chaikin, P. & Lubensky, T. (1995) *Condensed Matter Physics*, Cambridge University Press. Mathematically advanced treatment of liquid crystals.
De Gennes, P.G. & Prost, J. (1995) *The Physics of Liquid Crystals*, 2nd edition, Oxford University Press. Tour of some key mathematical results in the physics of liquid crystals.
Kleman, M. & Lavrentovich, O.P. (2003) *Soft Matter Physics*, Springer. Similar in level to Chaikin and Lubensky's treatment.
Neville, A.C. (1993) *Fibrous Composites*, Cambridge University Press. Provides a large range of examples of biological liquid crystals.
Schey, H.M. (2005) *Div, Grad, Curl and All That*, W.W.Norton. A booster course in vector calculus.

Tutorial Questions 6

6.1 Hydrated potato starch consists of smectic ordering of the helical crystallites (**Figure 1.23** and **Figure 6.31**). If the starch sample is heated to create a chip there are three different order parameters that determine the manner in which

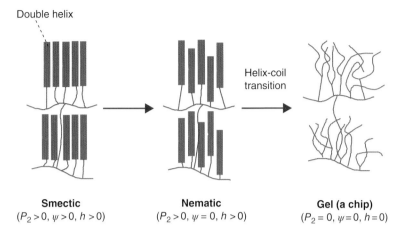

Figure 6.31 *A possible sequence of phase transitions that occurs during the break down in structure of a hydrated potato starch granule when heated. P_2 is the nematic order parameter, ψ is the smectic order parameter and h is the helicity order parameter. [Reproduced from T.A. Waigh, M.J. Gidley, B.U. Komanshek, A.M. Donald, Carbohydrate Research, 2000, 328, 165–176, with permission from Elsevier.]*

it cooks; the helicity (h), the nematic order parameter (P_2), and the smectic order parameter (ψ). Describe the sequence of phase transitions that are possible as the chip cooked (a possible scenario is shown in **Figure 6.31**). Define the physical processes that cause coupling between the three order parameters.

6.2 Calculate the nematic order parameter $P_2(\cos\theta)$ for a sample of mucin molecules in the trail of a slug if their orientational probability ($p(\theta)$) follows a top hat function given by

$$p(\theta) = \frac{2}{\pi} \qquad \frac{\pi}{4} < \theta < \frac{3\pi}{4}$$

$$p(\theta) = 0 \qquad \text{Otherwise}$$

In a particular trail of slime the mucin molecules have point defects of strength 1/2. Calculate the number of brushes (the black lines in the image) that would emanate from the defects under a polarising microscope.

6.3 A cylindrical virus is 200 nm in length and 10 nm in diameter. Calculate the critical volume fraction of virus particles to observe a nematic liquid-crystalline phase according to the Onsager theory.

6.4 The addition of flexible side chains to a flexible protein backbone induces the protein backbone to become more rigid and can result in the creation of a nematic phase. Suggest a reason for the induced rigidity.

7

Motility

Both living and inanimate microscopic objects are subject to *thermal fluctuations*, which causes them to jiggle about incessantly when viewed under an optical microscope. Many biological organisms modify these thermal fluctuations to facilitate the transport of the molecules inside their bodies and also to move themselves through their environment. Motility in biological systems is crucially important in a wide range of biological processes that include the transcription of DNA, the packaging of DNA in viruses, the propulsion of bacteria as they search for food and striated muscle as it is exercised (when a dumb-bell is lifted or the heart contracts).

Initially, an understanding of the undriven process of passive diffusion due to thermal energy will be developed, and this will then be extended to the analysis of motions produced by molecular motors (also see **Chapter 16** on molecular motors). Nanoparticles move as if caught in a randomly fluctuating hurricane and nanomotors are presented with a series of challenges to perform directed motion in this stochastic hurricane. Even more curious are the fluid mechanics effects due to the small length scales involved. Fluid dynamics occur at low Reynolds number, so inertial forces are negligible. Thus, the directed motion of biological molecules in the stochastic hurricane occurs in a watery environment that acts like treacle on the nanoscale.

Due to the importance of motility for the determination of biological processes, a series of noninvasive methods have been developed to measure molecular motility (**Chapter 19**), which include fluorescence correlation spectroscopy, pulsed femtosecond laser techniques, dynamic light scattering, neutron/X-ray inelastic scattering, video particle tracking, and nuclear magnetic resonance spectroscopy. Based on this wide range of dynamic techniques the field of biological motility has been provided with firm experimental foundations. The time scales that are now routinely experimentally probed range from femtoseconds (10^{-15} s) with biomolecular liquids, all the way up to the aging processes of biopolymer glasses, which are on the order of many years.

7.1 Diffusion

Diffusion is the process by which molecules jiggle around at small length scales due to thermal collisions with their neighbours, and equivalently, diffusion can be used to explain how macroscopic concentration gradients in materials evolve with time. Thus, a food dye injected into water eventually colours the whole vessel as

The Physics of Living Processes: A Mesoscopic Approach, First Edition. Thomas Andrew Waigh.
© 2014 John Wiley & Sons, Ltd. Published 2014 by John Wiley & Sons, Ltd.

the dye diffuses throughout the specimen; the jiggling motion at the nanometre scale produces a global redistribution of the dye molecules at the macroscale. To obtain a quantitative understanding of the process of diffusion, it will first be described in a statistical way at short length scales; the phenomenon of *Brownian motion*. At the macroscopic level an equivalent description is provided by *Fick's laws* for the concentration of a diffusing species.

As a first step it is useful to examine the statistical form of translational diffusion in one dimension, since it simplifies the analysis. A particle takes random steps to the left and to the right. In one dimension the displacement ($x_i(n)$) of a single diffusing particle as a function of the position of the previous random displacement ($x_i(n-1)$) after n steps is

$$x_i(n) = x_i(n-1) \pm \delta \tag{7.1}$$

where δ is the step size, which is assumed constant (this assumption can be relaxed using a Gaussian distribution of step sizes, but does not affect the final result), and n is the number of steps. The average of the displacements (x_i) is zero ($<x_i(n)> =0$), so the square of this quantity needs to be used to create a meaningful measure of the particle's motion,

$$x_i^2(n) = x_i^2(n-1) \pm 2\delta x_i(n-1) + \delta^2 \tag{7.2}$$

Next the mean square value of the displacement can be constructed and the second term on the right hand side of equation (7.2) averages to zero, since $< x_i(n-1) > = 0$. Therefore, the mean square displacement ($<x^2(n)>$) is given by

$$\left\langle x^2(n) \right\rangle = \frac{1}{N} \sum_{i=1}^{n} x_i^2(n) = \left\langle x^2(n-1) \right\rangle + \delta^2 \tag{7.3}$$

where N is the number of particles considered.

This expression for the mean square displacement is an iterative equation that relates the mean square displacement at step n ($<x_i^2(n)>$) to that of the previous step ($<x_i^2(n-1)>$). The application of equation (7.3) can be iterated all the way down to the first step of the motion ($n = 1$) and it is seen that the mean square displacement scales as the number of time steps (n), i.e.

$$\left\langle x^2(n) \right\rangle = n\delta^2 \tag{7.4}$$

where n is proportional to the time (t). This linear scaling of the mean square displacement with the time is a basic characteristic of diffusive motion. The number of time steps is related to the time (t) and the step size (τ). Thus, the number of steps is given by $n = t/\tau$ and this expression can be substituted in equation (7.4) to give

$$\left\langle x^2(t) \right\rangle = \left(\frac{\delta^2}{\tau} \right) t \tag{7.5}$$

The *diffusion coefficient* (D) is then defined to quantify the magnitude of the particle's mean square displacement fluctuations,

$$D = \frac{\delta^2}{2\tau} \tag{7.6}$$

Particles with large diffusion coefficients fluctuate a lot and *vice versa*. The factor of a ½ in equation (7.6) is used to tidy up Fick's equation in the corresponding macroscopic continuum description (equation (7.17)). Combination of equations (7.5) and (7.6) gives an expression that relates the diffusion coefficient to the mean square fluctuations of displacement in *one dimension*,

$$\left\langle x^2(t) \right\rangle = 2Dt \tag{7.7}$$

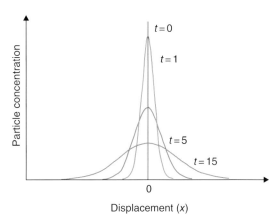

Figure 7.1   *The particle concentration as a function of displacement (x) for freely diffusing particles. The Gaussian probability distribution evolves with time (t). All of the particles start at x = 0, when t = 0.*

Diffusion in one dimension statistically corresponds to the probability distribution of the particle positions broadening with time (**Figure 7.1**). A well-localised point-like distribution of particles at the first time step ($t = 1$) evolves into a broad distribution ($t = 15$) as the diffusive motion takes place. Units of diffusion are $m^2 s^{-1}$, quite different from those of velocity, $m\ s^{-1}$. A velocity is not defined, because its value would depend on the time scale of observation, and is thus not a particularly meaningful quantity.

For a small molecule in water at room temperature a typical diffusion coefficient (D) is $10^{-5}\ cm^2\ s^{-1}$. The characteristic time for this molecule to diffuse the length of a bacterium ($10^{-4}\ cm$) is then $t \approx x^2/2D = 5 \times 10^{-4}\ s$ from equation (7.7).

For *two* or *three dimensions* the extension of the definition of the diffusion coefficient is fairly simple,

$$\langle r^2 \rangle = 2pDt \tag{7.8}$$

where r is the displacement in p dimensions, t is the time and D is the diffusion coefficient. An example of a two-dimensional random walk is shown in **Figure 7.2** for a polystyrene sphere moving in water and a more viscous glycerol solution. Clearly an increase in the viscosity of the solution decreases the amplitude of the fluctuations of the displacement of the polystyrene spheres. The decrease in the spheres' fluctuations is explained by the increased friction experienced by the particle in the more viscous fluid. The diffusion coefficient is related to the force of dissipation using the Einstein relationship (the *fluctuation–dissipation theory*),

$$D = \frac{kT}{f} \tag{7.9}$$

where kT is the thermal energy, D is the diffusion coefficient and f is the frictional coefficient. Something that dissipates a lot fluctuates very little and *vice versa*. The generalisation of this expression to viscoelastic materials is considered in **Chapter 15** and an electrical analogue appears in **Chapter 23**.

Newton's law of viscosity states that viscosity quantifies the effect of the shear on the velocity gradient in the fluid, i.e.

$$\frac{F}{A} = \eta \frac{v}{d} \tag{7.10}$$

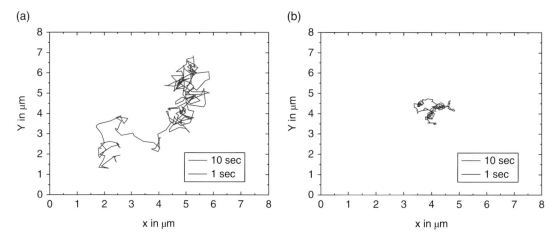

Figure 7.2 *Particle-tracking experiments on translational diffusion using video optical microscopy. Tracks of a 0.5-μm colloidal sphere in (a) water and (b) glycerol for two different time periods (1 s and 10 s) are shown. Glycerol is more viscous than water. [Reproduced with permission from A. Papagiannopoulos, PhD Thesis, University of Leeds, 2005.]*

Thus, viscosity (units Pa s) is the constant of proportionality between the shear stress (*F/A*, where *F* is the force and *A* is the area), and the velocity gradient (*v/d*, where *v* is the velocity and *d* is the plate separation, **Chapter 14**).

For a sphere in a fluid, if nonslip boundary conditions are assumed, the frictional coefficient (*f*) can be calculated from Navier–Stokes equations (**Chapter 14**) and is given by the Stoke's relationship,

$$f = 6\pi\eta a \tag{7.11}$$

where η is the viscosity of the fluid and *a* is the particle radius. Friction coefficients are known (or can be numerically calculated) for a wide range of rigid microscopic objects in solution. The *Stokes–Einstein* equation for a sphere combines equations (7.9) and (7.11) to give

$$D = \frac{kT}{6\pi\eta a} \tag{7.12}$$

Thus, measurement of the fluctuations in a particle's position as a function of time (or equivalently the diffusion coefficient) allows the size of the particle to be calculated.

It is important to realise that there is a difference between *mutual diffusion* and *self-diffusion*. With mutual diffusion the fluctuating rearrangement of particles with respect to their neighbours is considered, whereas with self-diffusion it is the rearrangement of individual particles relative to the laboratory that is important. Experimental techniques are often sensitive to one or other of the two types of diffusion. The previous discussion was centred on *translational self-diffusion*, e.g. measured using video particle tracking experiments. Photon correlation spectroscopy experiments often measure translational mutual diffusion.

Particles in solution experience fluctuations in their *rotational motion* in much the same way as with translational motion. The particles are constantly being buffeted by the surrounding solvent molecules, which impart angular momentum to them. A similar statistical analysis is possible for their angular motion as that for translational motion, equations (7.1)–(7.9). The mean square angular rotation ($<\theta^2>$) for small angular rotations (a small angle approximation is used in the derivation) is found to be related to the time (t) through the *rotational diffusion coefficient* (D_θ),

$$\langle \theta^2 \rangle = nD_\theta t \tag{7.13}$$

The fluctuation dissipation theory can again be used and in this case it relates the rotational diffusion coefficient to the thermal energy (kT) and the frictional coefficient (f_θ) for rotational motion,

$$D_\theta = \frac{kT}{f_\theta} \tag{7.14}$$

For a sphere the frictional coefficient for rotational motion is given by

$$f_\theta = 8\pi\eta a^3 \tag{7.15}$$

where a is the radius and η is the solvent viscosity. The two equations (7.14) and (7.15) can be combined to provide an expression for the rotational diffusion coefficient of a sphere,

$$D_\theta = \frac{kT}{8\pi\eta a^3} \tag{7.16}$$

There is a strong dependence of the rotational frictional coefficient on the particle radius (a), and thus large particles rotate very slowly.

There is a macroscopic description of *translational diffusion* that uses Fick's laws and is equivalent to the microscopic approach for translation on small length scales. *Fick's first equation* relates the flux of particles (J_x) that diffuse to the gradient of the particle concentration ($\partial c/\partial x$),

$$J_x = -D\frac{\partial c}{\partial x} \tag{7.17}$$

where the particle concentration (c) is in moles per cm^3, and the flux is in particles/cm^2 s. In words, equation (7.17) states that the net diffusive flux (at both position (x) and time (t)) is proportional to the slope of the concentration function (at both x and t). The constant of proportionality is the negative of the diffusion coefficient ($-D$). Diffusion occurs down concentration gradients.

Fick's second equation is

$$\frac{\partial c}{\partial t} = D\frac{\partial^2 c}{\partial x^2} \tag{7.18}$$

Again in words, the time rate of change in concentration (at x and t) is proportional to the curvature of the concentration function (at x and t), and the constant of proportionality is again the diffusion coefficient (D). A nonuniform distribution of particles redistributes itself in time according to Fick's two laws. This is pictured in **Figure 7.3**, where a sharp concentration gradient in dye molecules is reduced over time by inter-diffusion of the dye and solvent molecules.

In three dimensions Fick's two laws, equations (7.17) and (7.18), can be written in vector notation as

$$\underline{J} = -D\underline{\nabla}c \tag{7.19}$$

$$\frac{\partial c}{\partial t} = D\nabla^2 c \tag{7.20}$$

For a quick solution to these partial differential equations in a particular geometry, the most efficient strategy is to look them up in a specialist applied mathematics text book. Solution of such diffusion

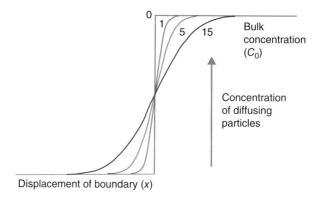

Figure 7.3   *The development of the concentration gradient of a step concentration profile (height c_0) as a function of distance (x) at the times 0, 1, 5, and 15 seconds, due to diffusion.*

problems often requires sophisticated mathematical methods, i.e. Green's functions. However, some solutions to diffusion in a range of specific geometries will be briefly quoted to give a flavour of the basic principles involved.

a. Diffusion from a *point source*. Consider the injection of some fluorescent dye from a micropipette. The diffusion of the dye is found to be well described by the equation

$$c(r,t) = \frac{N}{(4\pi Dt)^{3/2}} \exp\left(\frac{-r^2}{4Dt}\right) \qquad (7.21)$$

where $c(r,t)$ is the concentration of the dye molecules as a function of time (t), r is the distance from the point of injection, and N is the total number of dye molecules. It agrees with the expected result for the microscopic mean square displacement (equation (7.8)), i.e.

$$\langle r^2 \rangle = \frac{\int\limits_{-\infty}^{\infty} c(r,t)r^2 dx}{N} = 2Dt \qquad (7.22)$$

The flux of dye molecules can then be calculated from Fick's first law, equation (7.19). Experimentally the dye molecules under the microscope first appear as a bright spot upon injection that spreads rapidly outwards, and then fades away as the concentration becomes homogenised at a low value (**Figure 7.1**).

b. Diffusion to a *spherical absorber*. It is assumed that every particle that reaches the surface of a sphere is gobbled up (**Figure 7.4a**). These boundary conditions are slightly artificial (perhaps a good model for a stationary feeding bacterium), but mathematically the concentration at the surface of the sphere ($r = a$) is assumed to be zero and at a long distance from the sphere ($r = \infty$) it is c_0. The solution for the concentration of diffusing particles is found to be

$$c(r) = c_0\left(1 - \frac{a}{r}\right) \qquad (7.23)$$

where a is the radius of the sphere. The flux of diffusive particles can be then calculated from Fick's first law (equation (7.19)),

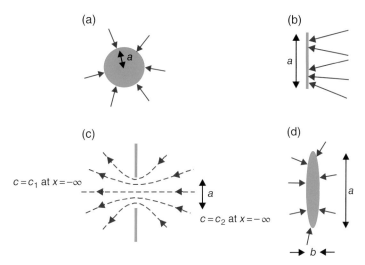

Figure 7.4 *Morphologies of different absorbers for diffusing particles. (a) A sphere, (b) a disk, (c) a circular aperture and (d) an ellipsoid. Red arrows indicate the motion of the diffusing species that is absorbed.*

$$J_r = -D\frac{\partial c}{\partial r} = -Dc_0\frac{a}{r^2} \tag{7.24}$$

The particles are absorbed by the sphere at a rate (I) equal to the area ($4\pi a^2$) times the inward flux ($-J_r(a)$) given by equation (7.24),

$$I = 4\pi Dac_0 \tag{7.25}$$

The adsorption rate (I) is the *diffusion current of particles* per second, and c_0 is the particle concentration per cm^3. Similar results with the current proportional to the size of the particles ($I \sim a$) are found for a wide range of different absorption geometries and thus the rate of capture is relatively independent of the geometry. This implies a wide range of efficient mechanisms are possible in Nature for the absorption of biomolecules and this is indeed observed with the feeding strategies of micro-organisms. Such considerations with the diffusion equation are also important for a range of reaction diffusion problems (e.g. the Turing model for morphogenesis in **Section 22.9**) and three more results will be quoted for completeness.

c. For diffusion to a *disk-like absorber* the adsorption rate (**Figure 7.4b**) is

$$I = 4Dac_0 \tag{7.26}$$

where a is the diameter of the aperture.

d. For diffusion through a *circular aperture* (**Figure 7.4c**) from a particle concentration of c_1 on one side of the aperture to c_2 on the other side of the aperture, the current is

$$I_{2,1} = 2Da(c_2 - c_1) \tag{7.27}$$

The currents are not proportional to the area of the disk, but instead depend on its radius (s).

e. For diffusion to an *ellipsoidal adsorber* (**Figure 7.4d**) the concentration at the surface of the ellipsoid is zero and the concentration at a large distance of separation ($r = \infty$) is c_0. The length of the major axis of the ellipsoid is a and its minor axis is b. If $a^2 >> b^2$ the diffusive current is

$$I = \frac{4\pi Dac_0}{\ln(2a/b)} \tag{7.28}$$

Again the current is roughly proportional to the length (a).

Often, diffusion is not sufficiently fast to transport cargoes inside large cells. Motor proteins are required in this case to reduce the transit times. Even just a partial breakdown of motor protein transport can have serious consequences for the organisms involved, e.g. motor neuron disease occurs when the dyneins malfunction that are required to move neurotransmitters along nerve cells (specifically dynactin, a protein that improves dynein's processivity, is implicated in some human motor neuron diseases).

7.2 Low Reynolds Number Dynamics

The Reynolds number turns up in a number of different guises in fluid mechanics, because a wide range of phenomena depend on the relative importance of frictional effects to inertial effects. Reynolds originally introduced his dimensionless ratio to describe the onset of turbulent flow of a fluid in a pipe. *Laminar flow* occurs at high viscosities (low Reynolds numbers). *Turbulent flow* occurs at low viscosities (high Reynolds numbers). When a fluid is stirred at low Reynolds number it produces the least possible disturbance (laminar flow) and the flow stops immediately after the external force stops. When the Reynolds number is large, inertial effects can dominate, e.g. coffee will continue to swirl after stirring is stopped and the flow is turbulent.

In molecular biophysics diffusion predominantly occurs under *low Reynolds number conditions*. This provides some counterintuitive results, since viscous effects dominate the motion and particle inertia is negligible, but the good news is that low Reynolds number conditions greatly simplify the mathematics required to understand the motion of biological molecules at small length scales.

The *Reynolds number* (Re, a dimensionless ratio) of a particle that moves at a velocity (v) in a fluid is defined by

$$\text{Re} = \frac{vL\rho}{\eta} \tag{7.29}$$

where L is the size of a particle, ρ is the specific gravity (density) of the fluid and η is the viscosity. The utility of the Reynolds number is found through an analysis of Navier–Stokes equations; the equations that predict the general motion of fluids, **Chapter 14**. Navier–Stokes equation for an incompressible fluid is

$$-\nabla p + \eta \nabla^2 v = \rho \frac{\partial v}{\partial t} + \rho(v.\nabla)v \tag{7.30}$$

where p is pressure, v is velocity, η is viscosity, and ρ is density. If Re $<<1$, the time dependence can be neglected in equation (7.30) and the inertial terms on the right can be neglected. The pattern of motion is the same, whether slow or fast, whether forward or backward in time. Practically it is found that when Re < 1 inertial forces (mdv/dt, where m is the mass) can be neglected at reasonably long time scales (>0.001 s). Furthermore, there is no turbulent flow in the system at low Reynolds number. For a salmon that travels at a velocity (v) of 10^2 cm s^{-1}, with a length (L) of 10 cm, specific density (ρ) of 1 g cm^{-3} and water viscosity (η) of 10^{-2} g cm^{-1} s^{-1}

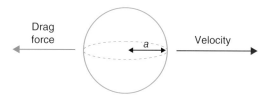

Figure 7.5 *A bacterium, which approximates to a spherical colloid (radius a), experiences a drag force (6πηav) from the viscosity of the surrounding water that rapidly decelerates its motion. The gliding distance when no additional motile force is applied can be as small as 0.04 Å.*

the Reynolds number is 10^5 (large Reynolds number dynamics). However, for a bacterium travelling at $v \approx 10^{-3}$ cm s^{-1}, $L = 10^{-4}$ cm, $\rho \approx 1$ g cm^{-3}, $\eta \approx 10^{-2}$ g cm^{-1} s^{-1} the Reynolds number is very small $R \approx 10^{-5}$ (small Reynolds number dynamics). Due to the relative importance of the inertial terms the fish and the bacterium have different strategies for swimming. The salmon propels itself by the acceleration of the water that surrounds it. A bacterium uses viscous shear to propel itself.

A useful example that emphasises the counterintuitive behaviour in low Reynolds number motility is to calculate the length a bacterium can coast before it comes to a stop (**Figure 7.5**). The mathematical analysis is very simple. Without any external forces in a purely viscous material, Newton's second law relates the acceleration (dv/dt) to the frictional force created by the relative motion of the surrounding water, equation (7.11). The bacterium is assumed spherical and therefore

$$-m\frac{dv}{dt} = 6\pi\eta av \tag{7.31}$$

where m is the mass of the particle, v is the velocity, η is the viscosity of water and a is the particle radius. There is a velocity on both sides of this equation and it can be integrated by separation of variables,

$$\frac{dv}{v} = -\frac{6\pi\eta a}{m}dt \tag{7.32}$$

Solution of this equation indicates that the velocity relaxes to zero with a characteristic time constant (τ),

$$v(t) = v(0)e^{-t/\tau} \tag{7.33}$$

$$\tau = \frac{2a^2\rho}{9\eta} \tag{7.34}$$

where ρ is the density of the bacterium. This result for the velocity can be integrated once more to provide the distance coasted (d) by the bacterium before it comes to a halt,

$$d = \int_0^\infty v(t)dt = v(0)\tau \tag{7.35}$$

Through substitution of typical values for v, a, ρ and η, the coasting distance of a bacterium is found to be very small, it is 0.04 Å, smaller than the size of a single covalent bond. The stopping distance scales as the square of the particle radius, and thus small particles coast much smaller distances than large particles.

A range of microfluidic experiments depend on the low Reynolds number approximation to interpret the data, e.g. microrheology apparatus and optical/magnetic tweezer force measurements at low frequencies

(**Chapter 19**). The *Langevin equation* is a useful method for understanding the displacement spectrum of thermally driven motion of a particle. Practically, the Langevin equation is encountered in situations such as the fluctuations in the position of an AFM tip or the high-frequency fluctuations of the bead position trapped using optical tweezers. The equation for the displacement fluctuations of a particle's motion can be written using Newton's second law,

$$m\frac{d^2x(t)}{dt^2} + \gamma\frac{dx(t)}{dt} + \kappa x(t) = F(t) \tag{7.36}$$

where $x(t)$ is the particle displacement as a function of time, md^2x/dt^2 is the inertial force that acts on the particle, $\gamma dx/dt$ is the drag force, κx is the elastic force and $F(t)$ is the random force that causes the motion of the particle, e.g. driven by thermal energy. It is the introduction of the fluctuating random force that complicates the analysis and gives rise to a separate designation for equation (7.36) as the *Langevin equation*.

The *autocorrelation function* of the displacement $(R_x(\tau))$ of the particle displacement is practically very useful in a range of spectral applications. The autocorrelation function is defined as

$$R_x(\tau) = \langle x(t)x(t-\tau) \rangle = \lim_{T \to \infty} \left\{ \frac{1}{T} \int_{-T/2}^{T/2} x(t)x(t-\tau)dt \right\} \tag{7.37}$$

This autocorrelation function satisfies the equation of motion (7.36) with the simplification that the right-hand side is zero, since the displacement (x) and the force (F) are uncorrelated by definition. Substitution of R_x is equation (7.32) gives

$$m\frac{d^2R_x(\tau)}{d\tau^2} + \gamma\frac{dR_x(\tau)}{d\tau} + \kappa R_x(\tau) = 0 \tag{7.38}$$

Such a second-order ordinary differential equation can be solved in the standard manner and the solution compared with experiments.

7.3   Motility of Cells and Micro-Organisms

The absence of inertial forces could at first sight appear to present an insurmountable barrier for a biological organism that needs to propel itself at the micrometre length scale. Reciprocal motion (e.g. waggling a paddle to and fro) does not lead to motility at the micrometre scale. It is similar to the case of a human who attempts to do the breast stroke in a swimming pool filled with tar; they move nowhere. Consideration of such behaviour led Ed Purcell to postulate his Scallop theorem (**Figure 7.6a**). At low Reynolds number a scallop with a single hinge can move nowhere. In contrast, the inclusion of an additional hinge on an organism allows the time symmetry of the motion to be broken and the organism can swim using the gait shown in **Figure 7.6b**.

Evolution has overcome the problem of low Reynolds number dynamics in a variety of ways. Propulsion mechanisms most commonly depend on the anisotropy of the drag force on a slender filament. The drag force is two times larger perpendicular to the filament than parallel to it. If a slender filament is moved in a non-reciprocal manner (the forward and backward strokes are different) often the organism is provided with a mechanism for propulsion. Flagellae (from the Latin for whip, where motor proteins at their base cause them to rotate) are used for bacterial propulsion, but another standard mechanism is to use cilia (from the Latin for eye lashes, where motor proteins are distributed along the cilium's length and cause active shape modifications). Cilia are used for micro-organism locomotion (e.g. with protists such as paramecium), but also in

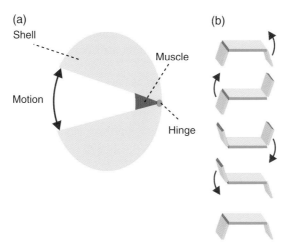

Figure 7.6   *(a) A scallop with a single hinge is unable to move forward at low Reynold's number due to the time reversal symmetry of the Navier–Stokes equation. (b) Purcell's hypothetical two hinge organism is able to propel itself forward using a series of conformational changes.*

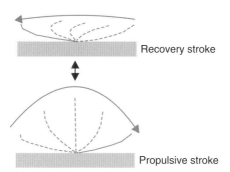

Figure 7.7   *Cilia (green) can cause propulsion by an alternation between propulsive and recovery strokes, e.g. in the lungs of humans or for the motility of paramecium. The motion is not reciprocal due to the different paths of the two strokes.*

the lungs and reproductive tracts of humans (**Figure 7.7**). A rod dragged along its axis at velocity $\underline{v}$ feels a resisting force proportional to $-\underline{v}$ (also directed along its axis). Similarly, a rod dragged perpendicular to its axis feels a resisting force also proportional to $-\underline{v}$ (directed perpendicular to its axis). However, the viscous friction coefficient for motion parallel to the axis is smaller than for perpendicular motion. The motion of the fluid created by a power stroke of a cilium is only partly undone by the backflow created by the recovery stroke and the organism can move. Other examples of motor protein motility are provided in **Chapter 16**.

A flagellated bacterium swims in a manner characteristic of the size and shape of the cell and the number and distribution of the flagellae, e.g. an E.coli of 1 µm in diameter and 2 µm in length has six flagellar filaments for propulsion. The flagellae are driven by a particularly elegant device for propulsion; the rotatory motor (**Chapter 16**). The motion of the bacteria is determined by the simultaneous action of the six flagellar filaments. When the flagellae turn counterclockwise they form a synchronous bundle that pushes the body steadily forward; the cell 'runs'. When the filaments turn independently the cell moves erratically and the cell 'tumbles'. The cells switch back and forth between the run and tumble modes of transport at random. The distribution of run (or tumble) intervals is exponential and the length of a given interval does not depend on the length of the intervals that precede it. Using this mechanism the bacteria can search for nutrients in its aqueous environment (**Figure 7.8**). This active motion has completely different statistics from the Brownian process depicted in **Figure 7.2**; its statistics are Poisson (**Chapter 3**).

Bacterial locomotion is thus achieved by the action of flagellar filaments. The thrust is produced from the component of viscous shear (F_v) of the helical filament on the surrounding water in the direction of motion (**Figure 7.9**).

The probability distribution (P) for a Poisson statistical process is given by equation (3.10). Therefore, the probability that a particle changes direction once in a time period between t and $t + dt$ is

$$P(t, \lambda) = \lambda e^{-\lambda t} dt \tag{7.39}$$

The probability that there is a change in direction per unit time is λ. The expectation time for the particle to change direction is

$$\langle t \rangle = \frac{1}{\lambda} \tag{7.40}$$

0.005 cm

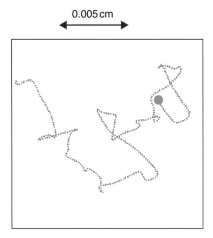

Figure 7.8   *Schematic diagram of the results of a particle tracking experiment with a bacterial cell in water under an optical microscope. The trajectory describes a motility process with Poisson statistics, e.g. equation (7.39). [Reprinted by permission from Macmillan Publishers Ltd: H.C. Berg, D.A. Brown, Nature, 1972, 239, 500–504. Copyright 1972.]*

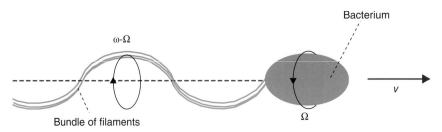

Figure 7.9   *Schematic diagram of the process of bacterial locomotion. The bacterium travels at velocity (v) and the forces that propel the bundle of filaments result from the component of the viscous shear.*

as expected for a Poisson process. The mean squared time interval is

$$\langle t^2 \rangle = \frac{2}{\lambda^2} = 2\langle t \rangle^2 \qquad (7.41)$$

And the standard deviation of the time interval is equal to the mean,

$$\left(\langle t^2 \rangle - \langle t \rangle^2 \right)^{1/2} = \langle t \rangle \qquad (7.42)$$

It is found for the Poisson distribution that describes bacterial motion that the apparent diffusion coefficient (D) is given by

$$D = \frac{v^2 \tau}{3(1-\alpha)} \qquad (7.43)$$

where α is the mean value of the cosine between successive runs, v is the velocity at which the bacterium propels itself and τ is the mean duration of the straight runs. If the mean angle between successive runs is zero ($\alpha = 0$) the apparent diffusion coefficient is $D = v^2\tau/3$, which is identical to the result for unbiased translational diffusion.

Figure 7.10 shows the range of velocities at which cells can propel themselves at. These motile processes fulfil a diverse range of functions such as to search for food or to explore a new ecological niche, but also many other

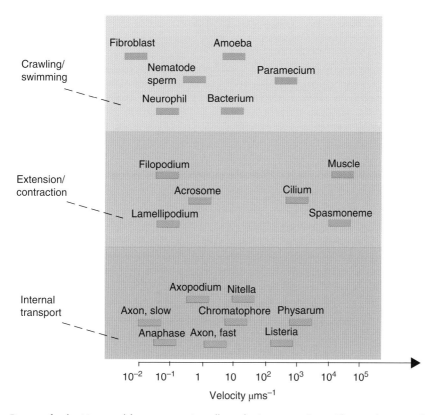

Figure 7.10 *Range of velocities used for transport in cells and micro-organisms. The mechanism of motility can be broadly classified as* crawling/swimming, extension/contraction *and* internal transport. *[Copyright 2000 from Cell Movements: From Molecules to Motility by Bray. Reproduced by permission of Garland Science/Taylor & Francis LLC.]*

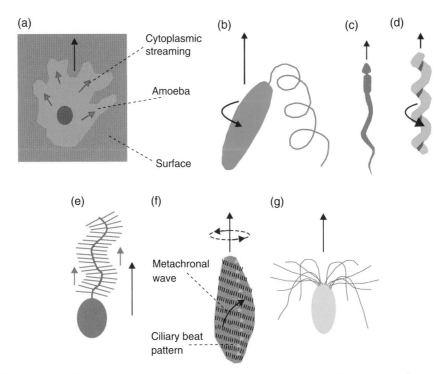

Figure 7.11 *Range of different strategies used by micro-organisms for motility. (a)* Amoeba *use cytoplasmic streaming to crawl across a surface, (b) euglena have a single flagellum and the cell body acts as a propeller, (c)* sperm *cells use flagellae to swim, (d)* spirochetes *swim using a cork screw motion, (e) chrysophytes (golden algae) have a hairy flagellum for propulsion, (f) paramecium use the coordinated beating of cilia in metachronal wave to propel themselves, and (g) chlamydomonas (green algae) swim with multiple flagellae.*

more sophisticated tasks require motility such as when nerve cells are plugged into muscle tissue during morphogenesis or sound waves in the cochlea of the mammalian ear are detected by active cilial oscillations.

Cells employ a range of unusual strategies and gaits to propel themselves around (**Figure 7.11**). Ameoboid-type motion is observed with many mammalian cells at surfaces (an active gel on the cell periphery causes protrusions of the membrane that results in global motion), cilia rotate around in helical waves (metrachronal waves) along some micro-organisms and similarly the internal area of many mammalian organs is covered with cilia to clear airways, move embryos and pump fluids. Some bacteria are helical in shape and follow a corkscrew motion. Algae can have a comb-like appendage that they hold in front of themselves and are unusual in that they can pull themselves through the fluid (E.coli in contrast are pushers). Crucial motile processes occur during cell division as the nucleus divides and the mitotic spindle contracts (**Chapter 16**). Other examples of cellular motility include the gas-vesicle-mediated motion of bacteria in the sea. Gas vesicles allow them to modify their buoyancy and thus to move vertically. Pili allow some bacteria to crawl across surfaces in an analogous manner to tanks on tank treads.

7.4   First-Passage Problem

The first-passage problem is a basic question in the statistical physics of biological processes. It asks, what is the time for a particle to travel a certain distance for the first time? Events that exceed the threshold distance and then return to it are not included in the calculation. It is particularly important for transport processes involved in chemical reactions, since the reaction rate is determined by the first-passage time, but it also provides a useful statistical measure to quantify motility processes in addition to the mean square displacement, and the kurtosis of the displacement (**Section 7.6**). The diffusion limited rate of a first passage process is thus the reciprocal of the first-passage time. For example this could be the time for a particle released at the origin ($x = 0$) to be absorbed at a boundary at $x = x_0$ (**Figure 7.12**). The mean time to capture (t_0) is given by

$$t_0 = \frac{1}{j(x_0)} = \frac{1}{j(x)} \tag{7.44}$$

where $j(x)$ is the concentration flux, as described in **Section 7.1**.

The solution to the reflecting wall and absorbing wall problem is plotted in **Figure 7.13**. In the absence of an external force the mean first-passage time (t) is approximately just the time to diffuse in one dimension (from equation (7.7)),

$$t \approx \frac{x_0^2}{2D} \tag{7.45}$$

More accurate expressions include the corrections for particles that arrive for the 2nd, 3rd, 4th,... times (see Sidney Redner's book on first-passage processes).

As an example, consider a 500-nm diameter protein that diffuses near a fibrin fibre. The first-passage time to diffuse the distance between adjacent fibrin monomers (45 nm) is found to be 2.33×10^{-3} s, if an estimated

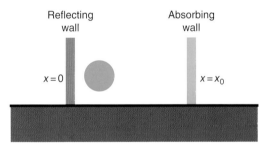

Figure 7.12 *Schematic diagram of the geometry used to calculate the first-passage time for a freely diffusing particle which travels from a reflecting wall at* x = 0 *to an absorbing wall at* x = x₀.

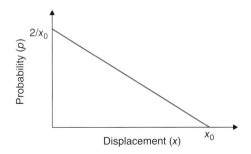

Figure 7.13 *Solution of the first-passage problem of a freely diffusing particle in one dimension that is placed between a reflecting wall and an absorbing wall. The probability (p) is shown as a function of displacement (x) for the situation illustrated in* **Figure 7.12***.*

diffusion coefficient from the Stokes–Einstein equation is used (equation (7.11), $D \approx 4.35 \times 10^{-13} \, \mathrm{m^2 \, s^{-1}}$). A similar calculation is useful to understand the mechanism of motion in the lac repressor along a DNA chain, i.e. the time to explore the one dimensional length of a DNA chain by an enzyme (**Chapter 18**).

A more sophisticated problem is how long a molecule that is initially at the origin ($x = 0$) takes to diffuse over an energy barrier placed a certain distance away (x_0). For a constant force (F) the potential (U) as a function of distance (x) is given by

$$U(x) = -Fx \tag{7.46}$$

The first-passage time (t) for the motion is found to be

$$t = 2\left(\frac{x_0^2}{2D}\right)\left(\frac{kT}{Fx_0}\right)^2\left\{e^{-Fx_0/kT} - 1 + \frac{Fx_0}{kT}\right\} \tag{7.47}$$

Detailed examination of this expression indicates that when the diffusion is steeply downhill, the force is large and positive, and the first-passage time approaches the distance divided by the average velocity (x_0/v). When the diffusion is uphill the first-passage time increases approximately exponentially as the opposing force is increased.

For the diffusion of a particle in a parabolic potential well (**Figure 7.14**), the first-passage time (t_k, the Kramer's time) now becomes

$$t_k = b\sqrt{\frac{\pi}{4}}\sqrt{\frac{kT}{U_0}}e^{U_0/kT} \tag{7.48}$$

where $b = \gamma/\kappa$ is the drag coefficient divided by the spring constant (**Figure 7.15**) and U_0 is the height of the parabolic potential. The shape of the potential in which a particle diffuses thus sensitively affects the functional form of the first-passage time.

An important example of the first-passage problem considers the rate at which a protein changes its conformation. A first possible solution is provided by the *Arrhenius equation* (**Figure 7.16**), which describes the

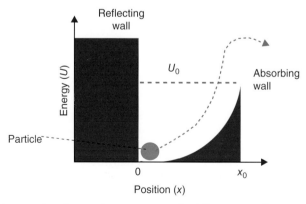

Figure 7.14   *Schematic diagram that depicts the energy (U) of a diffusing particle in one dimension as a function of position (x) that moves out of a parabolic potential well of height U_0, placed next to a reflecting wall.*

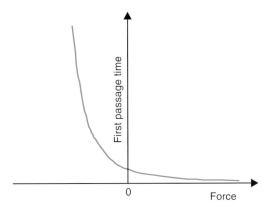

Figure 7.15 *First-passage time for the diffusion of a particle in one dimension as a function of the applied force, if the particle is attached to an elastic element* $U(x) = (1/2)\kappa x^2$ *(a harmonic potential).*

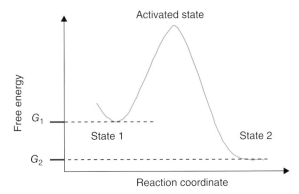

Figure 7.16 *The free energy as a function of reaction coordinate for the conformational change of a molecule from a state 1 (free energy G_1) to a state 2 (free energy G_2).*

transition of a protein between two states of free energy. The probability that the protein is in an activated state is given by a Boltzmann distribution. The rate constant (k_1) is therefore

$$k_1 = Ae^{-\Delta G_{al}/kT} \tag{7.49}$$

where $\Delta G_{al} = G_a - G_l$ is the difference in energy between the two transition states (a and l). The Arrhenius equation gives no information on the constant prefactor (A) and additional assumptions are required to calculate this coefficient. In the *Eyring rate theory* the reaction constant (A) corresponds to the breakage of a single quantum mechanical vibrational bond and is typically of the order of $kT/h \approx 6 \times 10^{12}\,\mathrm{s}^{-1}$. This approximation applies equally well to the breakage of covalent bonds, but it is not useful for global conformational changes of protein chains.

For global protein conformational changes the *Kramers rate theory* is a more realistic calculation of the prefactor A in equation (7.45). This includes the concept that diffusive fluctuations that determine the reaction rate,

$$k_1 = \frac{\varepsilon}{\pi\tau}\sqrt{\frac{\Delta G_{al}}{kT}}e^{-\Delta G_{al}/kT} \tag{7.50}$$

$$\Delta G_{al} = G_a - G_l \tag{7.51}$$

Proteins diffuse into the transition state with a rate equal to the reciprocal of the diffusion time. The efficiency factor (ε) for the transition rate is equal to the probability that the conformational transition is made when in the transition state. τ is the time over which the protein's shape becomes uncorrelated. Kramer's theory indicates that the frequency factor (τ^{-1}) is approximately equal to the inverse of the relaxation time ($\tau^{-1} = \kappa/\gamma$, where κ is the elastic constant and γ is the dissipative constant for the protein).

7.5   Rate Theories of Chemical Reactions

The rates of many biochemical processes are determined by the combined diffusion of the reactants (**Chapter 20**). It is assumed that the only interaction between biomolecules A and B in a mixture is during a collision (**Figure 7.17**). The flux of matter due to molecule A, if B is at rest, is given by J_A. Fick's first equation (7.14) can be written for the flux A,

$$J_A = -D_A \frac{\partial c_A}{\partial r} \tag{7.52}$$

where D_A is the diffusion coefficient of species A.

There is an excluded region around the two particles A and B equal to the sum of their two radii ($r_0 = r_A + r_B$). Therefore, for separation distances less than the radius ($r < r_0$) the concentration of A is equal to zero ($c_A = 0$). For large separation distances ($r \to \infty$) the concentration of A approaches the bulk concentration ($c_A \to c_A^0$).

The total current of A that flows towards B is the flux multiplied by the area of the surface around the particle that is considered. Let dq_A/dt be the current of particles of type A over the surface area of a sphere of radius (r),

$$\frac{dq_A}{dt} = -4\pi r^2 J_A = 4\pi r^2 D_A \frac{\partial c_A}{\partial r} \tag{7.53}$$

where equation (7.52) has been used for the flux of A particles. This expression can be simplified further by the addition of all the contributions over space to the current,

$$\frac{dq_A}{dt} \int_{r_0}^{\infty} \frac{dr}{r^2} = 4\pi D_A \int_0^{c_A^0} dc_A \tag{7.54}$$

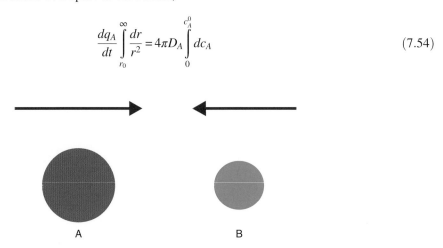

Figure 7.17   *Schematic diagram of two biochemical species (A and B) that experience a collisional diffusive reaction.*

where r_0 is the excluded region between the particles. Integration of this variables-separable equation gives

$$\frac{dq_A}{dt} = 4\pi D_A r_0 c_A^0 \tag{7.55}$$

To calculate the total amount of complex formed per second an additional term for B is required for the total current (dq_A/dt),

$$\frac{dq_{AB}}{dt} = 4\pi r_0 (D_A + D_B) c_A^0 c_B^0 \tag{7.56}$$

The rate constant (k in $\mathrm{mol^{-1}s^{-1}}$) for the reaction rate of molecules A and B that is purely due to diffusion is therefore

$$k = 4\pi f (D_A + D_B) 10^3 N r_0 \tag{7.57}$$

where N is Avogadro's number and f is a factor that measures the contribution of non-collisional interactions ($f \approx 1$). This expression will be used in **Chapter 18** to understand the interaction of the lac repressor with DNA.

7.6   Subdiffusion

Inside cells, membrane pores, colloidal suspensions and generally in congested soft-matter environments, thermally driven motion often does not follow the Einstein prediction for the mean square displacement on the lag time, i.e. $<r^2>$ does not scale linearly as t^1, equation (7.8). Instead a subdiffusive power law is observed at short times, $<r^2> \sim t^\alpha$ where $\alpha < 1$ (**Figure 7.18**). The exact physical origin of the subdiffusive motions is still being investigated. It is observed in a wide range of systems and it may have a number of physical origins. Inside cells it is likely due to a combination of diffusion in a congested environment

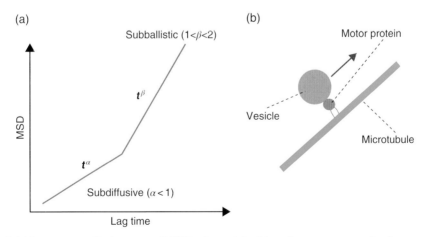

Figure 7.18   *(a) Mean square displacement (MSD) of a vesicle driven by a motor protein along a microtubule as a function of lag time. The motion is subdiffusive at short times ($\alpha < 1$) and subballistic at long times ($1 < \beta < 2$). (b) Schematic diagram of a motor protein (dynein or kinesin) that moves a vesicle along a microtubule. [Reproduced from A. Harrison, et al, Physical Biology, 2013, 10, 36002. © IOP Publishing. Reproduced with permission. All rights reserved.]*

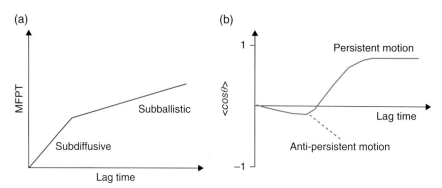

Figure 7.19   *(a) Mean first-passage time and (b) mean cosine angle between consecutive displacements, as a function of lag time for a vesicle transported by motor proteins along a microtubule. The corresponding mean square displacement is shown in **Figure 7.18a** and a schematic of the vesicle is shown in **Figure 7.18b**.* [Reproduced from A. Harrison, et al, Physical Biology, 2013, 10, 36002. © IOP Publishing. Reproduced with permission. All rights reserved.]

(e.g. caging of particles by surrounding molecules), and the internal fluctuating displacements of soft molecules and aggregates, e.g. the internal conformational fluctuations of polymers and membranes. Specifically, internal motions of flexible polymer chains are subdiffusive (**Chapter 10**) which gives a molecular motivation for some of the subdiffusive motion phenomena observed, e.g. with semiflexible cytoskeletal polymers such as actin and microtubules $<r^2> \sim t^{3/4}$. Subdiffusive motion is also observed for probes embedded in viscoelastic materials in particle-tracking experiments (**Section 19.16**). An alternative model for subdiffusive motion is the continuous time random walk in which the wait time between each step of the motion is chosen from a fat-tailed distribution, e.g. equation (3.13). Such fat-tailed distributions can be created by cage hopping of particles in congested solutions.

Confusingly, when intracellular cargoes are driven by motor proteins, the activity of the motor is often superposed on the underlying thermal subdiffusive motion. A schematic diagram of the mean square displacement as a function of lag time for the driven motion of a cargo (say a vesicle propelled by kinesin or dynein along microtubules) is shown in **Figure 7.18**. At short times the motion is subdiffusive, and it switches to directed (ballistic) motion at long times.

There are a range of alternative statistical measures to analyse data in addition to the mean square displacement that provide complementary information to dissect the exact origin of subdiffusive motion. The mean first-passage time (MFPT) is a useful unbiased measure of velocity (**Figure 7.19a**). Furthermore, the average cosine angle between successive displacement ($<\cos\theta>$) is also useful to quantify caging and directed motion, since it includes directional information, unlike the MSD or MFPT. In **Figure 7.19b** $<\cos\theta>$ is shown for the same system as **Figure 7.18**. At short times, the motion is diffusive ($\langle\cos\theta\rangle = 0$), then it becomes antipersistent ($\langle\cos\theta\rangle < 0$) due to motion of the vesicle on the motor protein tether, and finally it becomes persistent ($\langle\cos\theta\rangle > 0$) due to continued motion in a single direction. The kurtosis of displacement fluctuations (equation (3.5)) allows the non-Gaussian nature of displacement fluctuations to be probed, which is particularly useful for the analysis of constrained diffusive motion or glassy behaviour. Particle tracking can be performed with single molecules or single organisms, due to the availability of high-resolution microscopes and fast sensitive cameras, which allows accurate statistical theories to be tested for many types of biomolecular and micro-organism motility through the calculation of MSDs, MFPTs, angular correlations and kurtosises.

Suggested Reading

If you can only read one book, then try:

Top recommended book: Howard, J. (2001) *Mechanics of Motor Proteins and the Cytoskeleton,* Sinauer. Very clear modern account of motility.

Ben Avrahan, D. & Havlin, S. (2000) *Diffusion and Reaction in Fractals and Disorder Systems*, Cambridge University Press. More mathematical models of diffusion.

Berg, H. (1993) *Random Walks in Biology*, Princeton University Press. Classic text on biomolecular motion from an expert in the field. Much of the current chapter draws heavily on this clear exposition, which is readily accessible.

Bray, D. (2000) *Cellular Motility: From Molecules to Motility*, Garland Science. Classic text on cellular motility, focusing on zoological and molecular aspects.

Dusenbery, D.B. (2009) *Swimming at Microscale*, Harvard. A popular account of cellular motility.

Hughes, B.D. (1996) *Random Walks and Random Environments*, Oxford University Press. Extremely broad coverage of models for Brownian motion.

Klafter, J. & Sokolov, I.M. (2011) *First Steps in Random Walks*, Oxford University Press. Introduction to modern mathematical techniques for random walks, e.g. continuous time random walks.

Mazo, R.M. (2002) *Brownian Motion: Fluctuations, Dynamics and Applications*, Oxford University Press. Another useful introduction to Brownian motion.

Nelson, P. (2007) *Biological Physics*, W.H.Freeman. Interesting introduction to statistical mechanics tools in motility.

Powers, T. & Lauga, E. (2009) The Hydrodynamics of Swimming Microorganisms, *Reports on Progress in Physics*, 72, 1. Excellent comprehensive discussion of the fluid mechanics of micro-organism motility.

Redner, S. (2001) *A Guide to First Passage Processes*, Cambridge University Press. Section 7.4 only considers approximate expressions for the first passage probability. More accurate results can be found in this text.

Vogel, S. (1994) *Life in Moving Fluids: The Physical Biology of Flow*, 2nd edition, Princeton. Classic popular account of biological hydrodynamics.

Tutorial Questions 7

7.1 A mosquito of length 10^{-3} m flies at a speed of 10^{-1} m s^{-1}. Calculate its Reynolds number, given that the density and dynamical viscosity of air are 1.3 kg m^{-3} and 1.8×10^{-5} N s m^{-2}, respectively.

7.2 The flow of sodium ions in a cell is assumed to satisfy Fick's law and the diffusion equation (the diffusion coefficient is 1.35×10^{-9} m^2 s^{-1}). The flux of sodium is used for signalling inside an organism. Calculate how long it would take for the sodium to diffuse the length of a neuron (2.7 mm). Determine whether this is a practical mechanism of signalling.

7.3 Consider the rotational diffusion of a spherical virus. Write down an equation that relates the mean square fluctuation ($<\theta^2>$) of the virus's rotational angle to the rotational diffusion coefficient. Estimate the time for the virus to fluctuate by 90° if the thermal energy (kT) is 4.1×10^{-21} J, the viscosity is 0.001 Pa s and the virus can be approximated by a sphere of radius 2 μm. Calculate the time for a point on the circumference of the virus to rotate by diffusion through a distance $2\pi a$ and compare it with the time to translate by $2\pi a$.

7.4 A Poisson distribution can be used to describe the motion of a bacterium. The apparent diffusion coefficient of the bacterium is 4×10^{-6} cm^2 s^{-1}. Calculate the average value of the cosine of the angle between consecutive runs if the cell swims at a constant speed of 1×10^{-3} cm s^{-1} and the mean duration of the straight runs is one second.

8

Aggregating Self-Assembly

Biological complexes are often extremely complicated, and it was an important advance when many were found to *self-assemble* on a molecular level from their raw ingredients. The molecules can arrange themselves spontaneously into aggregates without any outside assistance. If the components, the solvent, the pH and the temperature are correctly chosen, the system will minimise its free energy in equilibrium and organise itself in the correct manner. Such strategies for self-assembly have been created countless times during biological evolution and appear intimately connected with life itself.

There are a diverse range of examples of self-assembling biological systems. In the construction of tobacco mosaic virus, an RNA chain attaches itself onto pie-shaped coat proteins, which produces a rod-like helical virus that is pathogenic to tobacco plants. Similarly, many globular enzymes can self-assemble from their primary structure into fully functioning chemical factories. This is the extremely complicated Levinthal problem referred to in the discussion of the globule–coil transition in **Section 5.3**. Actin, tubulin and flagellin can self-assemble to provide a force for cellular locomotion. The gelation of self-assembled haemoglobin fibres in the interior of red blood cells can disrupts their function and gives them a characteristic sickle shape (**Figure 8.1**); a first example of a self-assembling disease, sickle cell anaemia. A further medical condition that is currently the subject of intense research is amyloidosis in prion diseases. Self-assembled beta-sheet amyloid plaques are implicated in a large range of diseases including Alzheimer's, mad cow disease (Bovine Spongiform Encephalopathy), and Parkinson's disease (**Figure 8.2**). One of the most compelling demonstrations of self-assembly is that of cardiac thin filaments. Not only do the actin filaments self-assemble to form fibres from the soup of molecules found in a heart cell, all of the associated regulatory proteins (troponin and tropomyosin) lock themselves perfectly into place with the correct periodicity. This mechanism has also been demonstrated in live cardiac muscle and provides a fascinating example of heart repair at the molecular level. Historically, lots of work on self-assembly has been performed on soap that is economically very important to humans, since it removes dirt and kills bacteria. Dirt is removed by self-assembled soap micelles, which help solubilise the dirt particles. Soap also has a depletion interaction that enables surfactants to prise adsorbed molecules off dirty surfaces, e.g. scrambled eggs can be removed from cooking pans.

Cell membranes are found to self-assemble from their raw components, driven by finely balanced thermodynamic processes. Bilayers are readily created *in vitro* from purified lipid molecules and the bilayers

The Physics of Living Processes: A Mesoscopic Approach, First Edition. Thomas Andrew Waigh.
© 2014 John Wiley & Sons, Ltd. Published 2014 by John Wiley & Sons, Ltd.

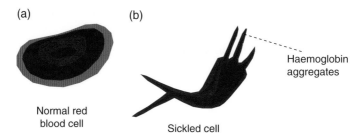

Figure 8.1   *Schematic diagram of (a) a normal human red blood cell and (b) a red blood cell from someone with sickle cell anaemia. Sickle-shaped cells are formed when fibrous aggregates of misfolded haemoglobin molecules self-assemble within cells and produce elongated structures on the membrane surfaces. The flow of sickled cells is impaired in the circulation system and this reduces the transport of oxygen in the body (an anaemia).*

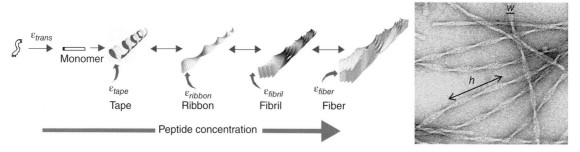

Figure 8.2   *Giant self-assembled beta sheet aggregates are thought to be responsible for a range of prion diseases. A hierarchy of structures are found with model peptide systems as a function of peptide concentration. The diameters of the resultant fibres are controlled by the chirality of the peptide monomers. h is the pitch of the self-assembled fibre and w is the width of a fibre. [Ref. A. Aggeli, I.A. Nyrkova, M. Bell, R. Harding, L. Carrick, T.C.B. McLeish, A.N. Semenov, N. Boden, PNAS, 2001, 98, 21, 11857–11862.]*

can spontaneously arrange themselves into vesicles. *In vivo* cell membranes often follow more complicated schemes of construction (their structures include intramembrane proteins and scaffolding, **Figure 8.3**), however, the underlying scheme of self-assembly is still thought to hold. Carbohydrates can also experience a process of self-assembly; the double helices of starch in plant storage organs are expelled into smectic-layered structures when the carbohydrate is hydrated. An important consideration for the next time the reader cooks a chip or makes custard.

A distinction is made between examples of *aggregating self-assembly* (e.g. micellisation of lipids) and *nonaggregating self-assembly* (e.g. folding of globular proteins). Aggregating self-assembly has some conceptually sophisticated universal thermodynamic features (e.g. a critical micelle concentration) that will be considered in detail in the present chapter. Nonaggregating self-assembly usually describes the behaviour of a system that moves between some hidden free energy minima, e.g. the subtle molecular origami involved in the folding of globular proteins. Examples of such phase transitions were covered in **Chapter 5**.

Other more general examples of self-assembly exist in soft condensed-matter physics such as the morphologies produced in the phase separation of liquids, liquid crystals, polymers and block copolymers. All of these have analogues in molecular biophysics. *Self-organisation* is another closely related field of pattern formation in molecular biology and typically is used to describe the results of nonequilibrium thermo-dynamic processes, e.g. morphogenesis during cell division (how the leopard got its spots, **Section 22.9**).

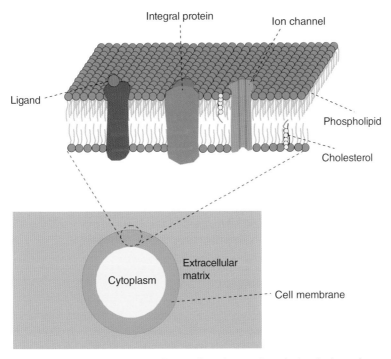

Figure 8.3 *A schematic diagram of a cell membrane that shows phospholipids, ligands, proteins, cholesterol, and ion channels. The cell membrane separates the cytoplasm from the extracellular matrix. [Copyright 2008 from Molecular Biology of the Cell, Fifth Edition by Alberts et al. Reproduced by permission of Garland Science/Taylor & Francis LLC.]*

Nonequilibrium self-assembly is also observed with motor proteins and is described in **Chapter 16**, where ATP and GTP are used to drive the process of self-assembly.

To use thermodynamics with processes of aggregating self-assembly the change in free energy (G) of a system due to the exchange of one of its components needs to be considered. The partial molar Gibbs free energy of a biomolecular system with a number of components is given by the symbol μ (the chemical potential). The chemical potential (μ_i) with respect to one of the species (**Sections 3.4** and **20.3**) is defined as

$$\mu_i = \left(\frac{\partial G}{\partial n_i}\right)_{T,P,n_{i \neq j}} \tag{8.1}$$

where n_i is the number of species of type i in the system. The subscripts T, P and $n_{i \neq j}$ imply the temperature, pressure and number of other species are held constant.

The total Gibbs free energy (G) of a biomolecular system is the sum of the partial free energies of each of its components,

$$G = \sum_{i=1}^{N} n_i \mu_i \tag{8.2}$$

where μ_i are the potentials that drive the chemical reactions, n_i are the number of molecules of type i involved in the reaction and N is the total number of molecular species.

Generally, the processes of aggregating self-assembly in molecular biophysics have a number of common themes; there exists a critical micellar concentration, which is a value of the concentration of subunits above which self-assembly occurs (the free monomer concentration is pinned at a single value above this concentration), the entropy change is positive on assembly (globally the entropy is still maximised due to the increased randomisation of associated solvent molecules), hydrogen bonding and hydrophobicity are often an important driving factor, and the surface free energy is minimised as the self-assembly proceeds.

The general features of self-assembly also depend on the dimensionality of the system. Self-assembly in one dimension produces highly polydisperse polymeric aggregates. In two dimensions the equilibrium aggregate is a single raft and in three dimensions the aggregate is a single micelle or crystal. Self-assembly is driven by the minimisation of the surface free energy. In one dimension the reduction in free energy is independent of polymer length and thus polydisperse aggregates are formed in the self-assembly of single-stranded fibrous proteins and linear surfactant aggregates (e.g. worm-like micelles). Fusion of two surface rafts in two dimensions reduces the surface area and drives a coarsening of the raft morphology. Similarly in three dimensions, the process of Ostwald ripening causes a gradual increase in aggregate size as small micelles are subsumed by their larger neighbours as they minimise their surface free energy (**Chapter 9**).

8.1 Surface-Active Molecules (Surfactants)

The essential framework of biological membranes in cells is provided by lipids that spontaneously aggregate to form bilayer vesicles (**Figure 8.3**). The bilayer encapsulates an internal cavity in which the environment for a living cell is maintained; its osmotic pressure, salt concentrations and pH. Amphiphilic molecules such as surfactants, lipids, copolymers and proteins can spontaneously associate into a wide variety of structures in aqueous solutions. With naturally occurring lipids, the critical micelle concentrations (CMCs) occur at extremely low concentrations, which allow stable bilayers to be formed from globally very low concentrations of subunits. Critical micelle concentrations are typically in the range 10^{-2}–10^{-5} M and 10^{-2}–10^{-9} M for single- and double-chained phospholipids, respectively.

A graphical illustration of surfactant self-assembly is shown in **Figure 8.4**. Surfactants are partitioned between micelles and solutions above the CMC (**Figure 8.5**). The unusual phenomenon above the CMC is that the monomers are pinned at a fixed concentration due to the thermodynamics of the assembly process and this process will be motivated theoretically in the following section.

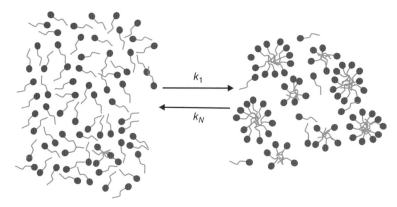

Figure 8.4 *Schematic diagram of the self-assembly of surfactants into micellar structures. Monomers on the left assemble into multimer aggregates (on the right) above the critical micelle concentration. The process is one of dynamic equilibrium.* k_1 *is the rate constant for association and* k_N *is the rate constant for dissociation.*

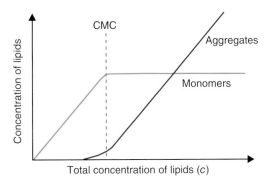

Figure 8.5 *Monomers and aggregate concentrations as a function of the total concentration of lipids. Above the CMC the concentration of lipid monomers is held fixed by the thermodynamics, whereas the concentration of lipids in aggregates increases exponentially.*

Surfactants are in dynamic equilibrium with their aggregates during micellar assembly. There is a constant interchange between lipids in the micelles and those free in solution (**Figure 8.4**). The morphology of the aggregates is determined by the geometry of the molecules (principally the head group area and length of the tails) and the hydrophobicity/hydrophilicity of both the head and tail groups.

There is a strong similarity between **Figure 8.5** that shows the CMC for amphiphiles and that for protein assembly. The similarity will be covered in a following section (**Figure 8.13**), and it indicates the universal thermodynamic processes in action. In equilibrium during surfactant self-assembly, thermodynamics requires that the chemical potential of all identical molecules in different aggregates are equal. The chemical potential for monomers, dimers, and trimers can thus be equated,

$$\mu = \mu_1 + kT \ln c_1 = \mu_2 + kT \ln \frac{c_2}{2} = ... \tag{8.3}$$

and more generally for an *N*-mer

$$\mu = \mu_N + \frac{kT}{N} \ln \left(\frac{c_N}{N}\right) = \text{const} \tag{8.4}$$

where N is the aggregation number, kT is the thermal energy, c_N is the concentration of species N and μ_N is the standard part of the chemical potential. The first term on the right-hand side of equation (8.4) is the internal energy per monomer, whereas the second term is the entropic contribution per monomer. The rate of association is $k_1 c_1{}^N$ and the rate of dissociation is $k_N (c_N/N)$, where k_N is the reaction rate for the *N*th-order association process and K is the dissociation constant for the equilibrium process. The equilibrium dissociation constant is given by

$$K = \frac{k_1}{k_N} = e^{-N\left(\mu_N^0 - \mu_1^0\right)/kT} \tag{8.5}$$

Solute molecules self-assemble in solution to form clusters of aggregation number (*N*) per cluster. The reaction between monomers (*A*) and aggregates (*B*) in solution can be expressed as

$$A + A + A + ... = B \tag{8.6}$$

where c_A and c_B are defined as the concentrations of *A* and *B* in mole fraction units, respectively. A general relationship can be obtained between K, N, c and c_A. c is the total concentration of solute molecules. For large values of the dissociation constant ($K \gg 1$) and large micellar aggregates ($N \gg 1$) it will be shown that the concentration of monomers c_A can never exceed $(NK)^{-1/N}$. By definition the equilibrium dissociation constant is

$$K = \frac{c_B}{c_A^N} \tag{8.7}$$

The total concentration of species equals the sum of the concentration of the components,

$$c = c_A + N c_B \tag{8.8}$$

Substitution of equation (8.7) gives

$$K = \frac{(c - c_A)}{N c_A^N} = \text{const} \tag{8.9}$$

Rearrangement gives an equation for c_A,

$$c_A = \left[\frac{c - c_A}{NK} \right]^{1/N} \tag{8.10}$$

The maximum possible value of $c - c_A$ is 1, since the calculation is in fractional molar units, i.e. when $c_A = 0$, $c = 1$. Therefore, c_A cannot exceed $(NK)^{-1/N}$. This is shown graphically in **Figure 8.5**. When a large value is taken for the dissociation constant ($K = 10^{80}$) and a reasonable value for the aggregation number ($N = 20$), the critical concentration (c_A) is 0.86×10^{-4}. Substitution in equation (8.10) gives

$$c_A = 10^{-4} \left[\frac{(c - c_A)}{20} \right]^{1/20} \tag{8.11}$$

Detailed analysis of this equation shows for $c \ll 10^{-4}$, $c_A \approx c$, whereas for $c \approx 10^{-4}$, $c_A \approx N c_B$, and there is an equal partition between micelles and unimers. Thus, $(NK)^{-1/N}$ is the *critical micelle concentration* for self-assembly.

The process of *one-dimensional aggregation* can be considered in detail, e.g. creation of worm-like micelles or peptide fibres. $\alpha k T$ is the monomer–monomer 'bond' energy of the linear aggregate relative to isolated monomers in solution. The total interaction free energy ($N \mu_N$) of an aggregate of N monomers is therefore (terminal monomers are unbonded),

$$N \mu_N = -(N - 1) \alpha k T \tag{8.12}$$

This can be rearranged as

$$\mu_N = -\left(1 - \frac{1}{N} \right) \alpha k T \tag{8.13}$$

and this can be written in the equivalent form,

$$\mu_N = \mu_\infty + \frac{\alpha k T}{N} \tag{8.14}$$

Thus, μ_N decreases asymptotically towards μ_∞, the bulk free energy of an extremely large aggregate ($N \to \infty$).

In *two-dimensional aggregation* the number N of molecules in a disc is proportional to its area, πR^2. The number of unbonded molecules in the rim is proportional to the circumference $2 \pi R$ and hence $N^{1/2}$. This implies that the free energy of an aggregate is therefore

$$N \mu_N = -\left(N - N^{1/2} \right) \alpha k T \tag{8.15}$$

and rearrangement of this equation gives the free energy per molecule in an aggregate as

$$\mu_N = \mu_\infty + \frac{\alpha kT}{N^{1/2}} \qquad (8.16)$$

where μ_∞ $(=-\alpha kT)$ is again the free energy per particle of an infinitely large aggregate.

For spherical, *three-dimensional aggregates*, N is proportional to the volume $((4/3)\pi R^3)$ and the number of unbonded molecules is proportional to the area $(4\pi R^2)$, and hence $N^{2/3}$. Therefore, the total free energy of an aggregate is

$$N\mu_N = -\left(N - N^{2/3}\right)\alpha kT \qquad (8.17)$$

and therefore the free energy per particle is

$$\mu_N = \mu_\infty + \frac{\alpha kT}{N^{1/3}} \qquad (8.18)$$

where μ_∞ $(= -\alpha kT)$ is the free energy per particle of an infinitely large aggregate.

It can be shown that for a spherical micelle the proportionality constant (α) is related to the surface tension (γ) and the size of the aggregate (R),

$$\alpha = \frac{4\pi R^2 \gamma}{kT} \qquad (8.19)$$

In general, the critical micelle concentration (CMC) for the aggregation of surfactants is given by the exponential of the difference in chemical potentials for the monomer and an aggregate,

$$\text{CMC} \approx e^{-(\mu_1 - \mu_N)/kT} \qquad (8.20)$$

Therefore, for three-dimensional aggregates the CMC is

$$\text{CMC} \approx e^{-\alpha/N^{1/3}} \qquad (8.21)$$

where α is given by equation (8.19).

An experimental plot of the osmotic pressure or the surface tension of surfactant solutions as a function of monomer concentration shows a clear discontinuity at the CMC concentration. A huge range of evidence for the micellar morphologies created above the CMC has been provided by scattering and microscopy techniques.

8.2 Viruses

The self-assembled geometrical structure of hepatitis B is shown in **Figure 8.6** as determined by X-ray crystallography measurements. This virus is pathogenic to humans, and the self-assembly (reproduction) of such parasites is important for medical science. The general process of self-assembly in viruses is thought to be explained by an interplay between short-range hydrophobic and long-range electrostatic forces. However, the details of the mechanism of self-assembly can be very complicated and are often specific to the particular variety of virus that is considered.

Tobacco mosaic virus (TMV, **Figure 8.7**) is one of the simplest helical viruses known and consists of a single strand of RNA surrounded by ~2000 identical pie-shaped protein subunits. TMV has been a favourite

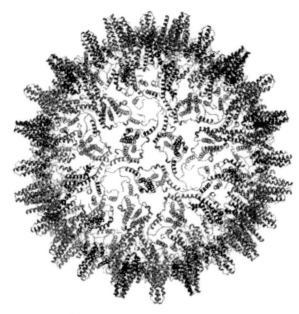

Figure 8.6 *Graphical representation of hepatitis B viruses based on X-ray crystallography data. The virus has a tessellated capsid structure.*

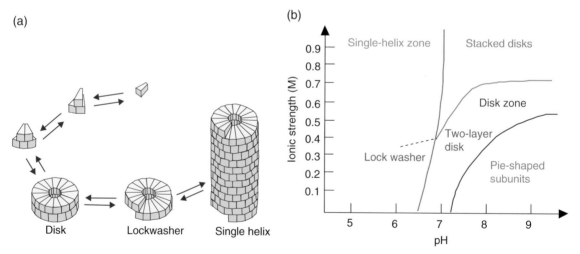

Figure 8.7 *(a) Tobacco mosaic virus self-assembles from 'lock washer' units that become attached to a central RNA core (not shown). (b) The phase diagram of TMV as a function of ionic strength and pH. The ionic strength and pH can switch the process of self-assembly on and off. The self-assembly of a complete TMV virus is favoured at low pH values.*

topic of research for physicists, since it is not pathogenic to man and presents an ideal monodisperse rod-like system with which to study liquid-crystalline phases. Assembly of TMV *in vitro* can occur with and without a chain of RNA. *Without RNA*, the protein monomers of TMV first form double disks that contain 17 monomer units. The disks contain holes at their centres. If the pH is changed appropriately, which modulates the electrostatic interactions of the disks, they slip join together and aggregate. The protein disk-like subaggregate units have a 'lock washer' morphology and slowly stack upon each other to form rods with a high polydispersity in their length. *With RNA* the nucleic acid directs aggregation of the disks and a monodisperse virus is formed; the RNA dictates the length of the helical virus.

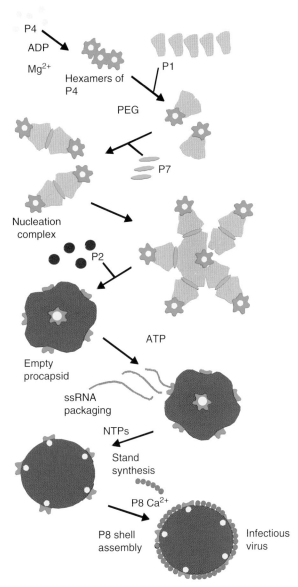

Figure 8.8 *Schematic diagram of the sophisticated process of self-assembly found in the φ6 NC double-stranded bacterial virus. Self-assembly occurs, but there are a number of biochemical steps that are needed to help the construction process. [Reproduced from D.H. Bamford, Philos. Trans. R. Soc. Lond. A, 2003, 361, 1187–1203, by permission of The Royal Society.]*

Many other viruses consist of a nucleic acid core surrounded by a symmetrical shell assembled from identical protein molecules (icosahedra viruses, **Figure 8.6**). There are geometrical selection rules for the symmetry of the arrangement of the coat proteins. The process of self-assembly in these cases is often much more complicated than with TMV. The self-assembly is sometimes directed by chaperone proteins that guide the process. The self-assembling process involved in the life cycle of $\phi6$ NC double-stranded RNA bacterial virus is shown in **Figure 8.8**.

The T4 bacteriophage is one of several viruses whose self-assembly is directed by DNA. They can infect E.coli bacteria (**Figure 8.9**) and the process of self-assembly is again more sophisticated than with TMV. Separate sections of this virus have been observed to self-assemble from their constituent components. The tail tube forms spontaneously from the core proteins and purified base plates (**Figure 8.10**). Using only

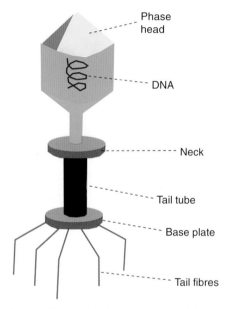

Figure 8.9 *Schematic diagram of the self-assembled structure of T4 bacterophage. The tail tube is found to spontaneously self-assemble* in vitro *(**Figure 8.10**).*

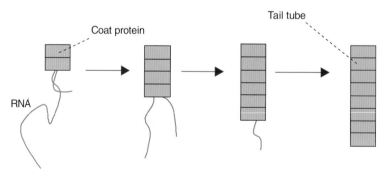

Figure 8.10 *The self-assembly of the coat proteins of T-4 bacteriophage (**Figure 8.9**) on a single RNA chain template.*

purified base plates and core protein monomers, the tail tube can self-assemble in vitro to the length of ~100 nm which is observed in the intact pathogenic virus.

8.3 Self-Assembly of Proteins

Two important examples of aggregating protein self-assembly in medical conditions are the aggregation of proteins in amyloid diseases (Alzheimer's, BSE, etc., **Figure 8.2**) and the aggregation of haemoglobin molecules in sickle cell anaemia (**Figure 8.1**). There are a series of unusual features found *in vitro* assembly of filamentous proteins that are indicative of a process of self-assembly, as opposed to a conventional chemical reaction. The temperature dependence of the process of polymerisation is very different from that of standard inorganic reactions. At low temperatures there is no polymerisation and at high temperatures polymerisation occurs; normally with synthetic polymerisation reactions the reverse is true. A rise in pressure causes depolymerisation; behaviour opposite to that normally found with covalent bonds, e.g. polymerisation of polyethylene. Polymerisation of self-assembling proteins only occurs above a critical initial monomer concentration, the CMC of the self-assembly process, as in the case of surfactant self-assembly. The kinetics of protein polymerisation are characterised by a long lag period followed by rapid formation of polymers. This is of particular concern in amyloid diseases, since the resultant self-assembled amyloid aggregates are implicated in these fatal conditions and nucleation during the long lag time may go unnoticed.

The change in Gibbs free energy (ΔG) during polymerisation is given by the standard thermodynamic equation (equation (3.18)),

$$\Delta G = \Delta U + P \Delta V - T \Delta S \qquad (8.22)$$

where ΔU is the change in free energy, P is the pressure, ΔV is the change in volume, T is the temperature and ΔS is the change in entropy. The polymerisation reaction is found to proceed more favourably at high temperatures. From equation (8.22) this signifies that the entropy change occurring upon polymerisation is positive and is provided by the increased entropy of the associated solvent molecules (the entropy of the protein aggregates actually decreases as they self-assemble into ordered structures).

The mechanism of nonaggregating self-assembly that forms the internal structure of globular proteins has been covered in **Chapter 5**. It is intimately associated with the globule–coil, helix–coil and beta sheet–coil phase transitions. With natural proteins there is the additional complication of frustration during folding that exists due to the large number of closely spaced local minima explored by the protein on its free energy landscape (Levinthal's paradox). This frustration can lead to misfolded proteins if they are improperly chaperoned to their final active states or unfolded by chemical/physical denaturants. The misfolding of globular proteins is thought to be the nucleation step in a number of amyloid diseases and constitutes a rate-limiting step in the development of such conditions.

8.4 Polymerisation of Cytoskeletal Filaments (Motility)

The polymerisation of cytoskeletal polymers is another important example of self-assembly, since it can lead to cellular motility. Actin and microtubule polymerisation allow forces to be applied to the cell's environment and have thus been intensively researched. Single-stranded polymerisation is found to be unlikely (**Figure 8.11**) and actin circumvents this problem by using the interaction of two (double-helical) strands. Similar multistranded schemes hold for a range of other helical cytoskeletal filaments, e.g. with tubulins the multiple strands are arranged in a tube.

Figure 8.11 *Self-assembly of filamentous proteins. The equilibrium constant is the same for each unit added to the polymer for a single filament that consists of symmetric subunits.*

The addition of a monomer unit to an actin fibre is a process of self-assembly and equilibrium is reached through a balance between the rate of addition and dissociation of the subunits (k_{on} and k_{off}),

$$A_n + A_1 \underset{k_{off}}{\overset{k_{on}}{\rightleftharpoons}} A_{n+1} \tag{8.23}$$

There is a dissociation constant (K) attributed to this process of monomer addition on an n-mer aggregate,

$$K = \frac{c_n c_1}{c_{n+1}} = \frac{k_{off}}{k_{on}}, \quad n \geq 1 \tag{8.24}$$

where c_1, c_n, and c_{n+1} are the molar concentrations of monomer, n-mer and $(n + 1)$-mer aggregates, respectively.

It is assumed that all the individual subunit addition reactions that determine the length of an n-mer have the same dissociation equilibrium constant (K), dissociation rate constant (k_{off}), and second order association constant (k_{on}). However, the dissociation constants (K) are not held fixed and change with the length of the polymer. The dissociation equilibrium constant is given by the standard thermodynamic relationship (equation 20.18),

$$K = e^{\Delta G_0 / kT} \tag{8.25}$$

The dissociation equilibrium constant (K) is thus associated with the standard free energy change (ΔG_0) of the reaction. ΔG_0 is the sum of the internal energy (negative) associated with formation of a monomer–protein bond and an entropy change (positive) associated with the loss of translational and rotational entropy as the subunits go from the standard 1 M concentration in solution to a bound state in a polymeric aggregate (equation (8.22), at constant volume and pressure $\Delta V = 0$).

As argued qualitatively in **Section 8.1**, it is found that *single-stranded filaments are polydisperse and short.* This result can be deduced more formally by the calculation of how the total concentration of subunits changes with the average length of the polymers. The dissociation constant for the reaction is given by equation (8.24). The concentration of different lengths of polymers (c_n) are assumed to be exponentially distributed[*],

$$c_n = K e^{-\frac{n}{n_0}} \tag{8.26}$$

where K is the equilibrium constant defined by equation (8.25), n is the length of the polymer, and n_0 is defined as

$$n_0 = -\frac{1}{\ln a_1} \tag{8.27}$$

where a_1 is given by

$$a_1 = \frac{c_1}{K} \tag{8.28}$$

---

[*] The exponential distribution is commonly assumed for all self-assembled polymers and was found to be a reasonably good approximation in an experimental study by our group on the self-assembly of cardiac thin filaments [M. Tassieri *et al,* Biophysical Journal, 2008, 94, 2170–2178]. However at small polymer lengths a peak is found in the distribution, which is attributed to additional entropic effects during the assembly of short rods.

The exponential distribution can be shown to be a solution of the problem by substitution in equation (8.24), but to calculate the average length of a filament (n_{av}) requires more effort. It is

$$n_{av} \approx \sqrt{\frac{c_t}{K}} \qquad (8.29)$$

where c_t is the total concentration of monomers and K is the equilibrium constant. The contribution from the monomer lengths is neglected, and the definition of an independent probabilistic average (equation (3.1)) gives

$$n_{av} = \sum_2^\infty n p_n = \sum_2^\infty n \frac{a_n}{\sum_2^\infty a_n} = 1 + \frac{1}{1 - a_1} \qquad (8.30)$$

where a_n are the statistical weights, $a_n = c_n/K$, and from equation (8.26) these are given by

$$a_n = a_1^n = e^{-n/n_0} \qquad (8.31)$$

The total number of subunits (a_t) in the solution is the sum of the numbers in each variety of aggregate,

$$a_t = \sum_1^\infty n a_n = \frac{a_1}{(1 - a_1)^2} \qquad (8.32)$$

$$a_1 = 1 + \frac{1}{2a_t} - \sqrt{\frac{1}{a_t} + \frac{1}{4a_t^2}} \qquad (8.33)$$

when $a_t \gg 1$ the expression for a_1 simplifies to $a_1 \approx 1 - a_t^{-1/2}$ from equation (8.27) and using the Taylor expansion $ln(1 + x) = x - x^2/2 + x^3/3$.... gives

$$n_0 \approx \sqrt{a_t} \qquad (8.34)$$

and then from equation (8.30) the average number of monomers in an aggregate is

$$n_{av} \approx 1 + \sqrt{a_t} \qquad (8.35)$$

This equation gives equation (8.29) when $\sqrt{a_t} \gg 1$. The equation predicts that the single stranded filaments are relatively short, as expected. From a similar type of analysis it can be shown that *multistranded filaments tend to be very long*. The problem is slightly more complicated, because the geometry of the fibre imposes three separate dissociation constants K, K_1 and K_2 on the process of assembly (**Figure 8.12**). The average length of multistranded filaments is found to be given by

$$n_{av} \approx \sqrt{\frac{K_1}{K_2}} \sqrt{\frac{c_t}{K}} \qquad (8.36)$$

where c_t is the total monomer concentration.

Filaments of actin and microtubules can polymerise and are sufficiently long for their roles in motility, because they are multistranded. Calculations show that the lengths of multistranded polymers are again distributed exponentially, but the average length is much greater than predicted by equation (8.35), in agreement with equation (8.36).

The rate of elongation (dn/dt) for filamentous self-assembly is given by the Oosawa equation,

$$\frac{dn}{dt} = k_{on} c_1 - k_{off} \qquad (8.37)$$

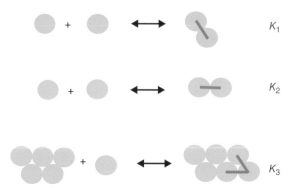

Figure 8.12 *Three association constants are required to describe the self-assembly of a two-stranded filament formed from symmetric subunits (K$_1$, K$_2$ and K$_3$).*

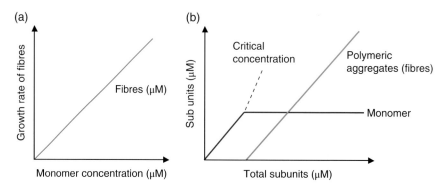

Figure 8.13 *The generic self-assembly of filamentous protein aggregates. (a) The growth rate of aggregates as a function of the monomer concentration and (b) the concentration of subunits of monomers that are free in solution and those in fibrous aggregates. Note the similarity to **Figure 8.5** for surfactant assembly.*

The graphical solution is shown in **Figure 8.13b** and the equation will be examined in more detail in **Section 16.1**. The behaviour of the CMC for one-dimensional self-assembly of the cytoskeletal fibres is shown in **Figure 8.13**. There is a close resemblance to the equivalent diagram (**Figure 18.5**) for surfactant self-assembly which emphasises the common features involved.

Extensive theoretical development of the assembly of multistranded fibres show that multistranded filaments grow and shrink at both their ends, and there again exists a critical concentration for self-assembly (**Figure 8.14**). Multistranded filaments are stable to fibre breakage, whereas single-stranded filaments continually break and reform. Other biological examples that follow the same trends as with actin (**Figure 8.13**) are sickle cell haemoglobin aggregates (double-helical fibres) and amyloid aggregates (twisted chiral multitape fibres).

The critical concentration (c_c) for self-assembly is given by the minimum in the rate of addition ($dn/dt = 0$) and substitution in the Oosawa equation (8.37) gives

$$c_c = \frac{k_{off}}{k_{on}} = K \tag{8.38}$$

This equation provides a method to determine the dissociation constant experimentally from a plot of the monomer concentration against the total concentration of subunits.

(a)

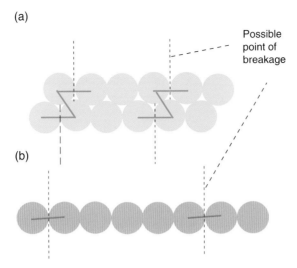

Possible
point of
breakage

(b)

Figure 8.14 *Multistanded polymeric filaments (a) are stable against thermal breakage, whereas single-stranded filaments (b) continually break and reform along their length.*

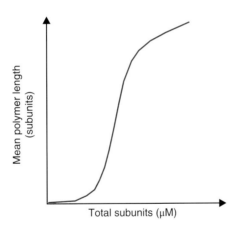

Figure 8.15 *The mean length of a self-assembled two-stranded filamentous aggregate chains is an S-shaped function of the total number of subunits.*

When the concentration of nuclei are very small for the self-assembly of multistranded fibres, the ends of the filaments are blunt due to the extra stability of a snug fit and the mean lengths of the filaments increase very steeply above the critical concentration for self-assembly. Thus, slight changes in free monomer concentration give a large change in polymer length for multistranded filaments (**Figure 8.15**).

Many fibrous biopolymers are self-assembled through the intermediate step of protofilaments, in contrast to the direct mechanism of addition of globular proteins monomers seen with actin and haemoglobin. Examples of protofilament assembly are shown in **Figure 8.16**. Such mechanisms are found to be important in a range of proteins that include the collagens, and the intermediate filaments such as the keratins and desmins. Protofilament assembly provides another mechanism for these systems to circumvent the problems of

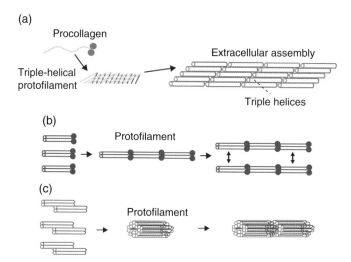

Figure 8.16 *(a) Collagen, (b) lamin and (c) vimentin fibres form from extended protofilament units, in contrast to actin and tubulin fibres that self-assemble directly from small spherical subunits. [Reproduced from D.A.D. Parry, S.V. Strelkov, P. Burkhard, U. Aebi, H. Herrmann, Experimental Cell Research, 2007, 313, 2204–221. With permission from Elsevier.]*

one-dimensional self-assembly, such as single-filament polydispersity and small size, which enables the construction of giant fibrous networks.

Suggested Reading

If you can only read one book, then try:

Dill, K. & Bromberg, S. (2010) *Molecular Driving Forces: Statistical Thermodynamics in Biology, Chemistry, Physics and Nanoscience*, Garland Science. Good clear introduction to statistical mechanics and thermodynamics of self-assembly.
Howard, J. (2001) *Mechanics of Motor Proteins and the Cytoskeleton*, Sinauer. The self-assembly of motor proteins (actin and microtubules) is well covered in this text.
Israelivichi, J.N. (2011) *Intermolecular and Surface Forces*, 3rd edition, Academic Press. A classic text with a useful account of the self-assembly of lipid molecules.

Tutorial Questions 8

8.1 Calculate the critical micelle concentration for spherical lipid micelles if the number of lipid molecules in a micelle is 1000, the surface tension is $20\,mJ\,m^{-2}$ and the micellar radius is 2 nm.
8.2 Calculate the critical subunit concentration for self-assembly (c_c) of a linear fibrous aggregate if the free-energy change on the addition of a subunit (ΔG) is $-4.6\,kT$. Also, calculate the average filament length if the monomer size is 5 nm and the total monomer concentration is 1 M.
8.3 Explain why the process of self-assembly depends on the dimensionality of the system considered.

9

Surface Phenomena

The physical phenomena associated with surfaces are a large emergent area of interest in molecular biophysics. A wide range of biophysical problems are currently being examined in relation to surfaces, many of which could have important industrial applications. Relatively small numbers of molecules are typically needed to cover the surfaces of soft-matter systems (such modifications are economical), but they have a huge impact on their behaviour, since they modulate the forces experienced. Examples of surface biological phenomena include the adhesive hairs on insect legs that allow them to hang upside down (**Figure 9.1**), the self-cleaning ability of lotus leaves (hydrophobic surface coatings, **Figure 9.2**), how slugs travels over razor blades (they slide on trails of liquid crystalline polyelectrolytes), how the skin of sharks suppresses short-wavelength turbulence for efficient locomotion (riblet patterned surfaces, **Figure 9.3**), how molluscs attach themselves to rocks (protein cements that act in a highly hydrated environment) and how dental fillings are glued to hydroxyapatite (the physics of composite materials, their adhesion and fracture).

9.1 Surface Tension

The work done (w) during the creation of a new surface is proportional to the number of molecules that are transported to the surface and thus to the area of the new surface,

$$w = \gamma \Delta A \tag{9.1}$$

where γ is a constant of proportionality (the surface tension) and ΔA is the change in surface area. γ has two equivalent definitions as the free energy per unit area (units of [energy]/[length]^2) or as a surface tension (units of [force]/[length]).

The Physics of Living Processes: A Mesoscopic Approach, First Edition. Thomas Andrew Waigh.
© 2014 John Wiley & Sons, Ltd. Published 2014 by John Wiley & Sons, Ltd.

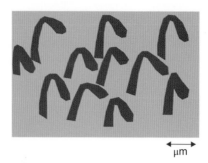

Figure 9.1 *Schematic diagram of the hairy appendages found on the surface of the legs of a range of insects. The hairs encourage adhesion due to van der Waals attractive forces.*

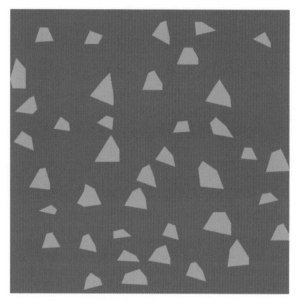

Figure 9.2 *Schematic diagram of the textured hydrophobic surfaces that are found on a range of plants and animals. The wax crystalloids on a plant leaf are shown (µm sized). They help to keep the surfaces clean.*

Figure 9.3 *Riblets patterns (µm grooves) along the shark's skin are thought to decrease the amount of frictional dissipation due to small wavelength turbulence. The arrows indicate the direction the riblets follow on the skin of the shark.*

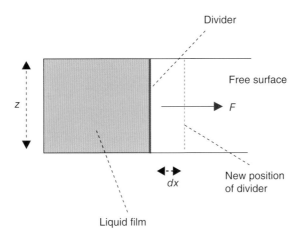

Figure 9.4   *The increase in surface area and consequently the surface free energy, is directly related to the amount of force (*F*) placed on the divider (and thus the amount of work it does on the liquid surface as it moves through a distance* dx*). z is the length of the divider.*

To understand the concept of surface free energy in more detail, consider a wire loop that encloses a liquid film (**Figure 9.4**). The force operates along the entire surface of the film and the resultant force on a slide wire varies linearly with its length. For static equilibrium when a force (F) is applied to a slide wire Newton's laws require that the forces must balance,

$$\gamma = \frac{F}{2x} \tag{9.2}$$

The factor of 2 is included because the force acts on the two sides of the film. The change in work (dw) associated with the expansion of the interfacial energy due to a movement of the slide wire by a small distance (dx) is given by the force multiplied by the distance moved and is therefore

$$dw = Fdx = \gamma 2xdx = \gamma dA \tag{9.3}$$

where dA is a small change in area. This work is equal to the change in Gibbs free energy (dG), at constant temperature and pressure. The change in Gibbs free energy is thus given by

$$dG = \gamma dA \tag{9.4}$$

And hence at constant temperature and pressure an expression for the surface free energy is

$$\gamma = \left(\frac{\partial G}{\partial A}\right)_{T,P} \tag{9.5}$$

The surface tension is therefore the *increase in Gibbs free energy per unit increment in area* and is equivalent to its definition as the *force per unit length*.

9.2   Adhesion

Adhesion relates to the phenomena involved in the cohesion of biological materials. It is useful to estimate the *interfacial fracture energy* (**Figure 9.5**), to provide a quantitative analysis of adhesion phenomena. Consider an interface between surfaces A and B which have an interfacial tension, γ_{AB}. If these two surfaces (surface

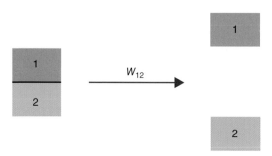

Figure 9.5 *The interfacial fracture energy during the separation of two phases (1 and 2) defines the total work of adhesion between the two phases.*

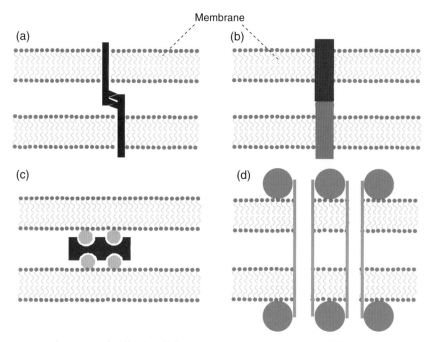

Figure 9.6 *A range of organised adhesive links can occur between two cellular membranes. Nonjunctional adhesions are possible such as (a) homophilic, (b) heterophilic and (c) extracellular protein linked to four ligands. Junctional adhesions are also important, such as (d) gap junctions that facilitate ion conductivity between cells, e.g. cardiac cells. [Reproduced with permission from G. Forgacs, S.A. Newman, 'Biological Physics of the Developing Embryo', 2005. Copyright Cambridge University Press, 2005.]*

tensions γ_A and γ_B in a vacuum) are separated in a thermodynamically reversible manner the total work of adhesion (W_{AB}) is the difference in surface tension before and after they are separated,

$$W_{AB} = \gamma_A + \gamma_B - \gamma_{AB} \tag{9.6}$$

However, experimentally it is found that fracture energies can be orders of magnitude greater than the value predicted by equation (9.6). The reason for this shortfall is that materials possess the ability to dissipate energy by a number of mechanisms not accounted for in the equation, e.g. plastic deformation at the surface. Indeed, this is the function of the soft filler in many biological composite structures; to maximise the dissipation of energy during impacts through plastic flow (**Section 13.2**).

In molecular biophysics adhesion is often carefully controlled by specialised structures at the nanoscale. For example, organised adhesive links form as cells mature, which guide their arrangement into tissues and drive the process of morphogenesis. The adhesive links can act as both junctions and switch yards for cytoskeletal components. To bind membranes, there are homophilic, heterophilic, and extracellular proteins that are used to create linkages (**Figure 9.6**). The proteins used in cell adhesion include the cadherins, integrins and selectins, and their binding energies are typically in the range 5–30 kT. Experiments to study such interactions have been made using AFM, optical tweezers and surface-force apparatus. Many experiments also consider the adhesion of pure bilayers by the measurement of the contact angle when two bilayers are pressed together. Micromanipulation techniques have proven particularly successful in this area. **Figure 9.7** illustrates an experiment based on the compression of two spherical bilayers. The contact angle (ϕ) can be related to the force of adhesion between the cells.

9.3   Wetting

Consider a liquid droplet that is sat (sessile) on a solid surface. Young's equation for the contact angle (θ) on the solid surface (**Figure 9.8**) is given by

$$\gamma_{sg} = \gamma_{sl} + \gamma_{lg} \cos\theta \qquad (9.7)$$

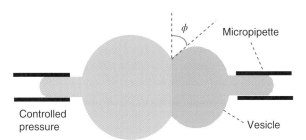

Figure 9.7   *Adhesive forces between two vesicles can be measured using micropipettes. The contact angle (ϕ) is related to the pressure exerted on the vesicles.*

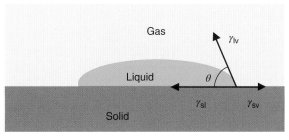

Figure 9.8   *A liquid drop sat on a solid surface in a gas. In equilibrium the components of the surface tension parallel to the surface of the solid balance to give Young's equation (9.7).*

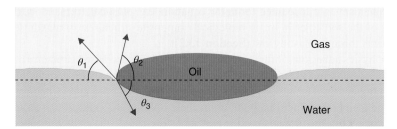

Figure 9.9   *An oil (liquid) drop placed on a liquid/gas interface requires three contact angles (θ_1, θ_2 and θ_3) to describe the position of the triple line of contact. The angles are described by a modified Young's equation (9.8).*

where γ_{sg}, γ_{sl}, and γ_{lg} are the surface tensions of the solid/gas, solid/liquid and liquid/gas interfaces, respectively. Young's equation is deduced from the force balance at the three phase contact line (Newton's law), and the components of the forces are taken in the direction parallel to the surface.

Similarly, it is possible to perform a force balance to deduce the structure of liquid/liquid/gas interfaces (**Figure 9.9**), e.g. oil on water exposed to the air. However, all of the components are fluids and can be deformed in this case. Three angles (θ_1, θ_2, θ_3) must be introduced to calculate the static conformation,

$$\gamma_{wa} \cos\theta_1 = \gamma_{oa} \cos\theta_2 + \gamma_{wo} \cos\theta_3 \tag{9.8}$$

where the subscripts w, a and o refer to interfaces of water, air, and oil, respectively.

The Young's equations allow the subject of *wetting* to be approached. At equilibrium in a phase-separated system of a surface immersed in a solution of molecules (say oil and water), the surface is expected to be coated by a macroscopic thick layer of the phase of lower surface energy that is called a *wetting layer*, e.g. a film of oil can form a film on the surface of crockery. The layer is macroscopically thick and the interface between the surface layer and the bulk is identical to that between the two bulk coexisting phases. It is therefore possible to write down an inequality for the surface tension of the two phases and the interfacial tension between the phases. This inequality can be used to understand how molecules spread on a surface in terms of a phase transition (a *wetting transition*). The wetting transition has a range of practical applications that include an understanding of biofouling on the membranes used in kidney dialysis machines and how agrochemicals adhere to waxy plant leaves.

To have *perfect wetting* the Young equation (9.7) must not have a solution that corresponds to a finite contact angle (**Figure 9.8**). Without a finite contact angle, a liquid will completely coat a solid surface with a macroscopically thick layer. When a liquid (l) only *partially wets* a surface (s) a finite contact angle occurs and the surfaces are partially coated with droplets. The surface energies of the two phases can be written in the form of equation (9.7) and an inequality must hold for stable equilibrium of the forces at a point of three phase coexistence,

$$\gamma_{sg} > \gamma_{lg} + \gamma_{sl} \tag{9.9}$$

Young's equation (9.7) can be rearranged to form the wetting coefficient (k),

$$k = \frac{\gamma_{sg} - \gamma_{sl}}{\gamma_{lg}} = \cos\theta \tag{9.10}$$

When the wetting coefficient equals one ($k = 1$), the contact angle of the drop is zero and the solid is *completely wetted* as in **Figure 9.10**. In the intermediate regime the wetting coefficient is between zero and one ($0 < k < 1$)

Figure 9.10   *Complete wetting of the liquid (β) on the solid surface (γ) is observed with a uniform macroscopically thick film of the liquid β.*

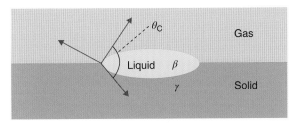

Figure 9.11   *Partial wetting of a liquid (β) on the solid surface (γ) causes the solid surface to be inhomogeneously wet.*

and the contact angle (θ) lies between 0 and $\pi/2$. The solid is *partially wetted* by the liquid (**Figure 9.11**). Again when the coefficient is minus one ($k = -1$) the contact angle is π and the solid is completely wetted. In the range of wetting coefficients between minus one and zero ($-1 < k < 0$) the solid is unwetted by the liquid (experimentally the liquid would form a tight ball with a high contact angle and roll off the surface under gravity).

Once the stability of equilibrium states has been examined, it is possible to understand the dynamic development of surface morphologies. From equation (9.9) no equilibrium position of the contact line exists when the spreading coefficient (s) is greater than zero,

$$s = \gamma_{sg} - \gamma_{sl} - \gamma_{lg} > 0 \tag{9.11}$$

Thus, the angle of contact will change when the spreading coefficient is greater than zero, a measure of the driving force for this nonequilibrium phenomena. The *spreading coefficient* (s) is related to the wetting coefficient (k) and is defined as

$$k = 1 + \frac{s}{\gamma_{lg}} = \cos\theta \tag{9.12}$$

If the spreading coefficient is greater than zero ($s > 0$) then the wetting coefficient is greater than one ($k > 1$). The larger the positive value of the spreading coefficient (s) the better the spreading over the solid surface. The transformation from a partially to a totally wet condition has the form of a phase transition (**Section 5.1**). This is normally a first-order process, but can become continuous with the correct choice of interaction potential and state variable (temperature, pressure, etc.).

Consider the wetting of a protein on an aqueous surface. Initially the spreading coefficient is found to be positive ($s_{\text{inital}} = 13 \text{ mJ/m}^2$) and at equilibrium s is negative ($s_{\text{final}} = -2 \text{ mJ/m}^2$). This protein is therefore strongly surface active and it leads to an important reduction of the surface tension. Trace impurities on a surface, reduce its free energy and can cause a dramatic change in the spreading behaviour, e.g. surface active proteins will often spread on a clean water surface, but not on a contaminated one. This process of wetting

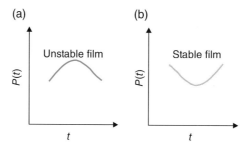

Figure 9.12   *The stability of surface-absorbed films depends on the form of the surface potential (P(t)) on the thickness of the film (t). (a) A positive curvature of P(t) leads to an unstable film, whereas (b) has a negative curvature that provides film stability.*

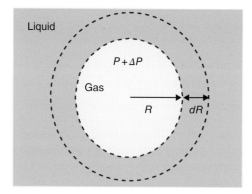

Figure 9.13   *The expansion of a spherical gas bubble in a liquid is controlled by its surface tension. A change in pressure (ΔP) causes the bubble to increase its radius (R) by an amount dR.*

can also be considered in terms of the form of the energy of the protein/surface interaction (P, **Figure 9.12**) on the thickness of the film (t). A positive curvature for the functional form of the energy of the film's thickness ($P(t)$) give an unstable film, whereas a negative curvature gives a stable film.

9.4   Capillarity

The *Young–Laplace* equation can be used to relate the pressure difference across a surface to its curvature. For an isolated particle the surface tension is balanced by the stresses within the particle.

Consider a *spherical gas bubble* suspended in a liquid as shown in **Figure 9.13**. The bubble is expanded infinitesimally by a radial distance dR and the change in surface area (dA) is given by

$$dA = 4\pi \left\{ (R + dR)^2 - R^2 \right\} \approx 8\pi R dR \tag{9.13}$$

The corresponding change in volume (dV) is

$$dV = \frac{4\pi}{3} \left\{ (R + dR)^3 - R^3 \right\} \approx 4\pi R^2 dR \tag{9.14}$$

The free-energy change by the expansion of the bubble is $dw = \Delta p dV$, simply the force multiplied by the distance moved, where Δp is the change in pressure. The corresponding free energy due to the surface tension from equation (9.4) is

$$dG = \gamma 8\pi R dR \qquad (9.15)$$

where the geometric relationship for the change in surface area has been substituted. The work done by the increase of the surface area is balanced by the pressure volume work that gives

$$\gamma 8\pi R dR = \Delta p 4\pi R^2 dR \qquad (9.16)$$

Therefore, the pressure difference across the bubble is given by

$$\Delta p = \gamma \frac{2}{R} \qquad (9.17)$$

For *cylindrical bubbles* the physical behaviour during expansion is very similar. The bubble has the morphology of a cylinder of length L (**Figure 9.14**). The area change (dA) for a small variation in the radius is given by

$$dA = 2\pi L (R + dr - R) = 2\pi L dR \qquad (9.18)$$

And the volume change (dV) is

$$dV = \pi L \left\{ (R + dR)^2 - R^2 \right\} \approx 2\pi L R dR \qquad (9.19)$$

A balance between the surface and pressure volume work now gives the pressure drop to be

$$\Delta p = \frac{\gamma}{R} \qquad (9.20)$$

The pressure drop for the cylindrical bubble (equation (9.20)) is a factor of two smaller than the spherical bubble (equation (9.17)), since there is half the curvature.

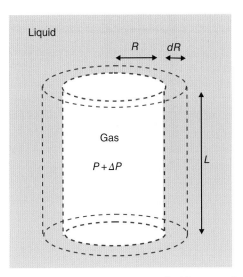

Figure 9.14 *The expansion of a cylindrical gas bubble is controlled by its surface tension. A change in pressure (ΔP) causes the radius (R) of the bubble to increase by an amount* dR.

For interfaces of *arbitrary shape* a general expression is found that relates the pressure drop across the surface to its curvature,

$$\Delta p = \gamma \left(\frac{1}{R_1} + \frac{1}{R_2} \right) = \gamma 2H \tag{9.21}$$

where R_1 and R_2 are the radii of curvature and H is the mean curvature given by

$$H = \frac{1}{2} \left(\frac{1}{R_1} + \frac{1}{R_2} \right) \tag{9.22}$$

Equations (9.20) and (9.17) for spherical and cylindrical bubbles are both then special cases of equation (9.21).

For simple liquids such as cyclohexane, the macroscopic approach to surface tension has been shown to be accurate down to length scales of seven times the diameter of the molecules using a surface-force apparatus. This is exceedingly good agreement for such a simple continuum theory. This lends confidence for the utility of equations such as (9.21) to gauge the strength of capillary forces on the nanoscale.

Surface tension can govern the rise of a liquid in a *capillary tube*. The capillary forces that drive this motion are important in a number of biological processes such as the rise of sap in plants, mucin in the trachea of the lungs and urine in the kidneys. Although it is a classic example, it needs to be stressed that the main driving force for the motion of sap in plants is a reduction in pressure due to the evaporation through the leaves. Capillary forces alone are unable to drive a fluid up the hundred metres of trunk in a giant redwood.

For the rise of a fluid in a capillary a simple relation exists between the surface tension (γ), the height of capillary rise (h), the capillary radius (r) and the contact angle (θ) on the surface of a capillary. This expression can be used to measure the surface tension of simple liquids (**Figure 9.15**). To perform these experiments all

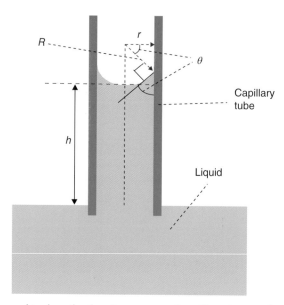

Figure 9.15   *The geometry used to describe the phenomena of capillary rise in cylindrical tubes. h is the height of capillary rise, R is the radius of curvature of the liquid meniscus, r is the radius of the tube and θ is the contact angle of the liquid on the wall of the capillary tube.*

that is required is a capillary tube with a well-defined clean surface. From the capillary geometry the radius of curvature of the meniscus is related to the contact angle and the radius of the capillary tube,

$$R\cos\theta = r \tag{9.23}$$

There is a Laplace pressure across the interface of the cylindrical tube due to the curvature of the liquid's surface and it is given by equation (9.17),

$$\Delta p = \frac{2\gamma}{R} = \frac{2\gamma\cos\theta}{r} \tag{9.24}$$

In equilibrium, the Laplace pressure is balanced by the hydrostatic pressure due to the height of the fluid (h) above the level of the reservoir. This hydrostatic pressure (Δp) is linearly related to the height of the fluid,

$$\Delta p = \rho g h \tag{9.25}$$

where ρ is the density of the fluid and g is the acceleration due to gravity. Combination of equations (9.24) and (9.25) provides an expression for the surface tension,

$$\gamma = \frac{\rho g h r}{2\cos\theta} \tag{9.26}$$

This equation is particularly simple to apply when the fluid wets the surface of the capillary, since $\cos\theta = 1$.

Capillary condensation is known to be important for beetles (beetles exude a fluid on their legs) and tree frogs, since it helps them enhance their adhesion on to surfaces. In contrast, for Geckos and flies, van der Waals interactions are thought to be the predominant force of adhesion.

9.5 Experimental Techniques

The *Wilhemy plate method* is a simple robust technique for the measurement of the surface tension of liquid films (**Figure 9.16**). The apparatus uses the balance of torques on a pivoted beam to measure the force due to the tension of a liquid surface. The apparatus is balanced before it is brought into contact with the liquid, and

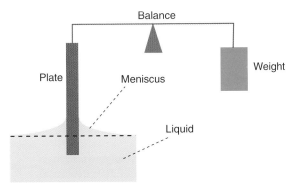

Figure 9.16 *Schematic diagram of a Wilhemy plate used to measure the surface tension of a liquid. The weight required to balance the surface tension of the entrained meniscus is measured.*

the increase in weight (w) is a result of the surface tension of the entrained meniscus. The weight is therefore given by

$$w = \gamma x \cos \theta \tag{9.27}$$

where x is the width of the plate, θ is the contact angle and γ is the surface free energy. **Table 9.1** lists a range of methods that can be used for the measurement of surface tensions.

9.6 Friction

The coefficient of friction (μ) between two solids is defined as the ratio of the frictional force (F) to the normal force to the surface (W, $\mu = F/W$). *Amonton's law* states that the coefficient of friction is independent of the apparent area of contact (a surprising counterintuitive result). Amonton's law is an empirical result that has been shown to hold for a wide variety of materials. A corollary is that the coefficient of friction is independent of load. Furthermore if two objects of equal weight ($W_1 = W_2$) are made of the same material, then their frictional forces are equal ($F_1 = F_2$) (**Figure 9.17**). To illustrate the bizarre nature of this law consider a rectangular block of wood initially on an end with a small surface area (**Figure 9.18**). Toppling the block on its side results in a large increase in the area of contact. However, this *does not change* the horizontal frictional force that opposes motion, since the total normal reaction force to the block's weight is unchanged.

A qualitative microscopic explanation can be made for Amonton's law. As the two surfaces are brought together the pressure is initially extremely large at the first few points of contact that develop and deformation immediately occurs to allow more and more contacts to be formed (**Figure 9.19**). This plastic flow continues until the total area of contact is such that the local pressure has fallen to a characteristic yield pressure (P_m) of the softer material. The actual contact area (A) is determined by the characteristic yield pressure of the material,

$$A = \frac{W}{P_m} \tag{9.28}$$

Table 9.1 *Methods for the measurement of the surface tension of liquids.*

Method	Principle
Capillary height	Capillary rise
Wilhemy plate	Capillary force on a plate
Drop profile	Digital analysis of drop geometry
DuNouy ring	Capillary forces on a wire ring
Spinning-drop tensiometer	Digital analysis of drop geometry

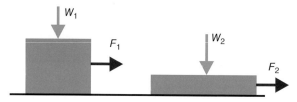

Figure 9.17 *The frictional forces on two objects made from the same material, placed on the same surface, are equal if the objects have the same weights and obey Amonton's law, i.e. if $W_1 = W_2$ then $F_1 = F_2$.*

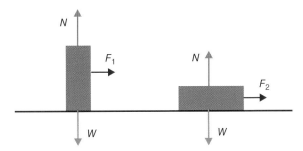

Figure 9.18 *Topple a solid block on its side and Amonton's law gives the same frictional force that opposes the motion, F$_1$ = F$_2$, since the normal force of reaction (N) is unchanged (F$_1$ = F$_2$ = μN = μW, where W is the weight).*

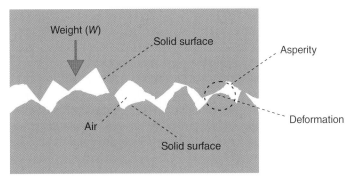

Figure 9.19 *Schematic diagram of the microscopic contacts (asperities) between two solid surfaces that give rise to the frictional force that opposes lateral motion.*

where W is the weight of the upper solid object. By definition the force to shear the junctions at the point of contact (F) is

$$F = As_m \tag{9.29}$$

where s_m is the shear strength per unit area. As an approximation the contact area (A) can be eliminated to give the result

$$F = W \frac{s_m}{P_m} \tag{9.30}$$

or alternatively

$$\mu = \frac{s_m}{P_m} = const \tag{9.31}$$

which is Amonton's law and is a useful starting point for the study of frictional forces. However, many non-Amonton's law materials have evolved in Nature to provide carefully designed frictional properties that do not follow equation (9.31).

Dynamic frictional effects are also an important, but extremely complex area. From detailed studies of lubrication in automotive applications (a billion-pound industry that has invested considerably into lubricant research) the typical forms of the frictional force as a function of shear rate are well known (**Figure 9.20**).

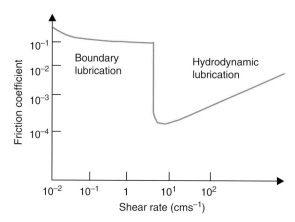

Figure 9.20   *Schematic diagram of a Stribeck curve that illustrates the regions of hydrodynamic and boundary lubrication for liquid films held between two solid surfaces. The frictional coefficient is shown as a function of the relative shear rate of the two identical solid surfaces.*

Such Stribeck curves have two important regimes: *hydrodynamic lubrication* at high shear rates in which the details of the lubricant are not critically important and *boundary lubrication* at low shear rates that is very sensitive to the molecular properties of the lubricant. Engine oils are specifically designed to modify the region of low shear rate lubrication and it is believed that proteoglycans and glycoproteins have evolved naturally to reduce frictional effects in this regime for a wide variety of biological processes, e.g. ocular mucins in the eye to reduce wear and damage.

   An important example of lubrication is the ultralow friction of cartilage on cartilage in vertebrate joints. Cartilage has an extremely low friction coefficient, lower than teflon on teflon and there are large deviations from Amonton's law for the interactions of cartilage surfaces in articulated joints (**Section 17.1**).

   Anisotropic frictional behaviour has evolved in the textured skin of snakes. This provides low frictional resistance to forward motion, but allows the skin of the snake to grip the surface upon muscular retraction to produce the explosive force needed to strike the snake's prey. The frictional losses of fibres (actin and myosin) in striated muscle, when they slide past one another, are another important area (**Section 16.2**). Non-Amonton's law friction occurs again in this system and frictional losses are reduced by a combination of the electrostatic stabilisation of the fibrillar array and the hydrodynamic lubrication of fluids inside the striated muscle.

9.7   Adsorption Kinetics

The processes that determine the rate and extent of adsorption of molecules onto solid and liquid surfaces are of wide interest, e.g. hCG hormones that bind to antigens in pregnancy kits or pesticides that adsorb to the leaves of crops. Sometimes the processes depend sensitively on the internal degree of freedom of the molecules (e.g. specific conformations of flexible polymers), but often molecular adsorption can be understood by reference to a general framework developed for rigid spherical particles. The Langmuir adsorption rate assumes no inter-action between molecules adsorbed to a surface (**Section 3.5**) and that they have no internal degrees of freedom (they act as hard spheres). A general equation for the rate of adsorption at a surface is

$$\frac{dM}{dt} = k_a c_b \phi(t, M) - k_d(t, M) M \qquad (9.32)$$

where M is the adsorbed mass, c_b is the bulk molecular concentration, $\phi(t,M)$ is the uncovered surface fraction, k_a is the chemical adsorption rate constant and k_d is the desorption rate constant. For Langmuir adsorption this equation simplifies to a simple linear dependence on the bulk concentration

$$\frac{dM}{dt} = k_a c_b \phi(t) \tag{9.33}$$

where $\phi(t) = 1 - \theta(t)$ and $\theta(t)$ is the fraction of surface covered by molecules. However, due to the excluded volume, jamming can occur on the surface at high volume fractions (think of trying to park a car on a road where there are already many other cars randomly parked). A better model for higher surface concentrations is given by the *random sequential adsorption* model (which includes jamming phenomena), and the uncovered surface fraction in equation (9.33) is replaced by

$$\phi = 1 - 4\theta + \frac{6\sqrt{3}}{\pi}\theta^2 + b\theta^3 \tag{9.34}$$

where b is a constant. At high surface volume fractions, jamming can also affect the kinetic behaviour, as well as the numbers of molecules adsorbed, and the rate follows a power law decay with time. Simulations and scaling theory give the jammed volume adsorption kinetics as

$$M_\infty - M \sim t^{-1/d_f} \tag{9.35}$$

where M_∞ is the adsorbed mass at long times and d_f is the number of degrees of freedom, e.g. it is 2 for two dimensional surfaces.

   Figure 9.21 shows the dynamics of adsorption of the globular protein BSA on to a silicon surface. BSA is used as a blocker to avoid false positives in antigen-based pregnancy kits. For low bulk BSA concentrations the dynamics follow the expectation for Langmuir kinetics in dual polarisation interferometry experiments, but at higher BSA concentrations a characteristic power law for jamming is observed (equation (9.35) with $d_f = 2$).

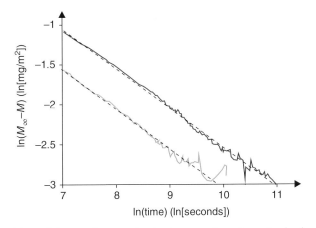

Figure 9.21 *Natural logarithm of the maximum adsorbed mass minus the adsorbed mass ($M_\infty - M$) of globular proteins attached to a solid silicon surface as a function of time. The adsorption of the globular proteins BSA (orange) and transferrin (blue) were measured using dual polarisation interferometry. The gradient of both curves is –0.5 in good agreement with the predictions of the kinetics of a jamming transition, equation (9.35). [Reproduced from B.J. Cowsill, P.D. Coffey, M. Yaseen, T.A. Waigh, N.J. Freeman, J.R. Lu, Soft Matter, 2011, 7, 7223–7230 by permission of The Royal Society of Chemistry.]*

9.8 Other Physical Surface Phenomena

When two surfaces approach one another surface energy effects can cause a liquid to condense on a surface prior to saturation in the bulk phase. This effect can cause a large change in the adhesive energies and is termed *capillary condensation*.

Ostwald ripening is the process by which surface free energies govern the growth of colloidal particles. Small crystals are subsumed by larger ones during their growth in order to minimise the surface free energy. Such behaviour is important with the production of ice cream (large crystals are considered unpalatable), and in the creation of large defect-free protein crystals for crystallographic structure determination (**Section 5.4**).

During the *nucleation* of a new phase of material the surface free energies oppose the nucleation of a new droplet or microcrystal. Thus, the surface forces are the critical factor that determines the dynamics of phase separation (**Section 5.4**).

Gradients in surface tension can drive diffusion in mixed liquids and this process is called the *Marangoni effect*. The effect can be easily observed in the production of tear drops of brandy or wine, if a thin meniscus of the alcohol/water mixtures are deposited on the side of a glass and the alcohol begins to evaporate.

There are a range of surface effects that relate to *liquid crystallinity*. Phase transitions can be induced by the interaction of the liquid-crystalline order parameters with a surface, e.g. smectic terraces often form at solid/nematic interfaces whereas the bulk phase remains in the nematic state.

The dynamics of wetting are affected by the motion of the triple line (i.e. the line of contact between the liquid, solid and gas phases). Many soft matter processes are heavily influenced by the dynamics of this contact line e.g. skidding of an elastic material on a thin aqueous film.

Suggested Reading

If you can only read one book, then try:

Top recommended book: Israelivichi, J.N. (2011) *Intermolecular and Surface Forces*, 3rd edition, Academic Press. A classic readable text on surface forces.

Adamson, A.W. & Gast A.P. (1997) *Physical Chemistry of Surfaces*, Wiley. Extensive coverage of the physical processes at surfaces.
Berg, J.C. (2010) *An Introduction to Interfaces and Colloids: The Bridge to Nanoscience*, World Scientific. Simple straightforward modern account of soft-matter surface physics.
De Gennes, P.G., Quere, D. & Brochard-Wyart, F. (2004) *Capillarity and Wetting Phenomena*, Springer. Clear insightful theoretical account of physical phenomena at surfaces.
Krapivsky, P.L., Redner, S. & Ben-Naim, E.A (2010) *A Kinetic View of Statistical Physics*, Cambridge University Press. Describes modern results on the effect of jamming on surface adsorption.
Persson, B.N.J. (1998) *Sliding Friction: Physical Principles and Applications*, Springer. Considers the fundamental origins of frictional forces.
Scherge, M. & Gorb, S. (2001) *Biological Micro and Nano Tribology*, 2nd edition, Springer. Fascinating discussion of the biological implications of surface forces.

Tutorial Questions 9

9.1 A water droplet sits on a lotus leaf, which is hydrophobic. Calculate the equilibrium contact angle of the drop if the surface free energy of the leaf/air, leaf/water and air/water interfaces are $10 \, \text{mJ m}^{-2}$, $73.2 \, \text{mJ m}^{-2}$ and $72 \, \text{mJ m}^{-2}$, respectively. Also calculate the wetting coefficient of the system. Explain whether the surface is unwetted, partially wetted or completely wetted by the droplet.

10

Biomacromolecules

Macromolecules (polymers) are long-chain molecules built of repeated subunits (monomers). Proteins, nucleic acids and carbohydrates are polymeric, and polymers are therefore thoroughly discussed in the current chapter. Synthetic polymers have been extensively investigated for their diverse range of industrial applications over many years, e.g. polyethylene, polyacrylamide, polystyrene, etc. This provides the field of biomacromolecules with a rich variety of quantitative models that can be readily transported from their original synthetic origins into the analysis of biological problems.

10.1   Flexibility of Macromolecules

The *persistence length* of a polymer is a standard method to measure its flexibility. There are three standard classifications of single polymer chains with respect to their *persistence length*; *flexible*, *semiflexible* and *rod-like*. Each of these different classes of polymers requires a separate theoretical model to develop an understanding of their structure and dynamics in solution. *Rigid* polymeric rods form liquid-crystalline phases and have no internal dynamic modes. At the opposite extreme the structures of *flexible* polymers are adequately described by blob models (blobs of monomers connected end-on-end in a chain) and the chain conformations are dominated by thermal fluctuations inside the blobs at small length scales and the solvent quality (chain–solvent interaction) at large length scales. The internal dynamic modes of flexible polymers in solution are well described in terms of the Zimm model, which describes the coupled hydrodynamic interactions of the monomers. The intermediate *semiflexible* class of polymers has been subject to a series of recent developments. The role of both transverse and longitudinal fluctuations of the filaments is highlighted in the theoretical models, which determine the filaments' structures and hydrodynamic modes. It is possible for a long polymeric chain to exhibit dynamics from all three regimes when it is considered on a series of different length scales, e.g. *rod-like* (<1–10 Å), *semiflexible* (~1–100 nm) and *flexible* (>100 nm).

The *persistence length* of a macromolecule can be measured using a host of techniques such as dynamic light scattering, electron microscopy, optical microscopy (optical tweezers), small-angle X-ray scattering,

The Physics of Living Processes: A Mesoscopic Approach, First Edition. Thomas Andrew Waigh.
© 2014 John Wiley & Sons, Ltd. Published 2014 by John Wiley & Sons, Ltd.

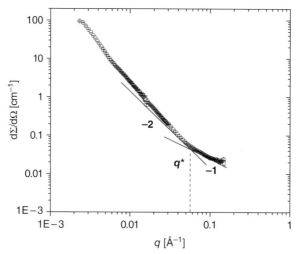

Figure 10.1   *Small-angle neutron scattering experiment to study the elasticity of titin, a giant semi-flexible protein in striated muscle. The scattered intensity is shown as a function of the momentum transfer (q). q* corresponds to the persistence length (l$_p$) of the chains, i.e. q* = 2π/l$_p$. [Reproduced with permission from E. Di Cola, T.A. Waigh, J. Trinick, L. Tskhovrebova, A. Houmeida, W. Pyckhout-Hintzen, C. Dewhurst, Biophysical Journal, 2005, 88, 4095–4106, with permission from Elsevier.]*

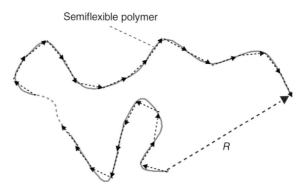

Figure 10.2   *Schematic diagram of the conformation of a semiflexible polymeric chain. R is the end-to-end distance (red). The black dashed arrows indicate the direction cosines tangentially aligned along the chain.*

static light scattering, small-angle neutron scattering, atomic force microscopy and fluorescence measurements. **Figure 10.1** shows the effect of the persistence length of giant protein molecules (titin) in small-angle neutron scattering experiments (q is the momentum transfer, equation (19.51)). The crossover from q^{-1} to q^{-2} scaling dependence of the scattered intensity corresponds to the change in chain conformation from a rigid rod on small lengths scales (<10 nm) to flexible conformations on larger length scales (>10 nm).

The geometrical construction of the persistence length of a macromolecule is shown in **Figure 10.2**. The persistence length (l_p) is the decay length of the cosines between the tangent vectors along the monomers of the chain. It is found that the correlation decays exponentially with contour length (s) along the chain,

$$\langle \underline{t}(0).\underline{t}(s)\rangle \sim e^{-s/l_p} \tag{10.1}$$

For a rigid rod the persistence length is infinite ($l_p \to \infty$), a flexible chain has a small value of persistence length ($l_p \to 0$), and with a real semiflexible chain, the persistence length takes intermediate values between the two extremes. The global conformations of chains can be classified by comparison of the persistence length (l_p) with the contour length (L, equal to the number of monomers multiplied by the size of each monomer). Thus, chain conformations are defined as

$l_p << L$	*Flexible chain*
$l_p \sim L$	*Semiflexible chain*
$l_p >> L$	*Rigid chain*

It is useful to connect the persistence length with the global conformation of a full polymeric chain. The average end-to-end distance ($<R>$) of a semiflexible chain is the sum of the cosine projections of the tangent vector along the chain,

$$\langle R \rangle = a \sum_{k=0}^{N} x^k = \frac{a(1-x^N)}{1-x} \tag{10.2}$$

where a is the monomer length, x is $\cos\theta$ and the N segments make an average angle θ with their adjacent segments. The persistence length (l_p) can be defined as the limiting value of $<R>$ as the chain becomes infinitely long ($N \to \infty$),

$$l_p = \frac{a}{1-x} \tag{10.3}$$

Since θ is small for semiflexible rods, the small-angle expansion can be used for the cosine. If only the first two terms are kept

$$\cos\theta \approx 1 - \frac{\theta^2}{2} \tag{10.4}$$

Substitution of equation (10.4) into equation (10.3) gives

$$l_p = \frac{2a}{\theta^2} \tag{10.5}$$

x^N can be written in the form of an exponential using the mathematical identity

$$x^N = \left(1 - \frac{\theta^2}{2}\right)^N \approx e^{-N\theta^2/2} \tag{10.6}$$

The contour length is equal to the number of segments (N) multiplied by the segment length (a),

$$L = Na \tag{10.7}$$

Equations (10.5) and (10.7) allow equation (10.6) to be re-expressed as

$$x^N = e^{-L/l_p} \tag{10.8}$$

From equations (10.2) and (10.3) it follows that

$$\langle R \rangle = l_p\left(1 - e^{-L/l_p}\right) \tag{10.9}$$

As the length of the chain becomes very large ($L \rightarrow \infty$) the persistence length becomes equal to the end-to-end distance ($<R> \rightarrow l_\mathrm{p}$). Furthermore if the persistence length is very large ($l_\mathrm{p} >> L$) the end-to-end distance becomes equal to the contour length ($<R> \approx L$ for a rigid rod) as expected.

The mean square end-to-end distance is another useful quantity for sizing chains, since the average chain size is zero ($<R> = 0$) for completely flexible chains. The mean square end-to-end distance is the expected value of R^2 ($<R^2>$) and can be calculated by integration of the infinitesimal sections,

$$d\langle R^2 \rangle = d\langle R.R \rangle = 2\langle R.dR \rangle = 2\langle R \rangle ds \tag{10.10}$$

where ds is a small increment along the chain's contour length. Therefore, integration of this expression combined with equation (10.9) gives

$$\langle R^2 \rangle = \int_0^L d\langle R^2 \rangle = 2l_\mathrm{p} \int_0^L \left[1 - \exp\left(-L/l_\mathrm{p}\right)\right] ds \tag{10.11}$$

The integral can be calculated analytically to give the expression

$$\langle R^2 \rangle = 2l_\mathrm{p} \left(L - l_\mathrm{p}\left(1 - e^{-L/l_\mathrm{p}}\right)\right) \tag{10.12}$$

In the limit of very long chains ($L \rightarrow \infty$) the radius of gyration of a semiflexible chain is given by

$$\langle R^2 \rangle = 2Ll_\mathrm{p} \tag{10.13}$$

where $2l_\mathrm{p}$ is the length of a rigid monomer in an equivalent flexible chain (the segment length in a freely jointed chain model). Equation (10.13) is a very useful method to estimate the size of biopolymers in solution. For example, with a B-type variety of DNA chain the persistence length (l_p) is 45 nm, the contour length is 80 nm, and the root mean square end-to-end distance is then calculated to be 60 nm. The chain is fairly compact and the calculation highlights the significant impact of its flexibility on the overall conformation of DNA chains.

The *Kratky–Porod model* allows the connection to be made between the persistence length of a polymeric material and its bending rigidity (an intrinsic property of the material). The infinitesimal change in free energy (dG) of a semiflexible chain for small values of the curvature (ds) is

$$\frac{dG}{ds} = \frac{dG\,d\theta}{d\theta\,ds} + \frac{1}{2}\left(\frac{d^2G}{d\theta^2}\right)\left(\frac{d\theta}{ds}\right)^2 \tag{10.14}$$

In the absence of a permanent bending moment for the chain

$$\frac{dG}{d\theta} = 0 \tag{10.15}$$

And therefore

$$\frac{dG}{ds} = \frac{\kappa}{2}\left(\frac{d\theta}{ds}\right)^2 \tag{10.16}$$

where the bending rigidity (κ) is defined as

$$\kappa = \frac{d^2G}{d\theta^2} \tag{10.17}$$

The total energy to bend a finite chain (ΔG) is therefore the integral of equation (10.16) over the length of the chain (L),

$$\Delta G = \frac{\kappa}{2}\int_0^L \left(\frac{d\theta}{ds}\right)^2 ds \qquad (10.18)$$

For small displacements of the chain the angle of deviation is proportional to the size of a step along the contour length ($\theta = ks$ when k is a constant),

$$\Delta G = \kappa k^2 \int_0^L \frac{ds}{2} \qquad (10.19)$$

where k is defined as

$$k = \frac{d\theta}{ds} \qquad (10.20)$$

If θ_L is the total angle between the ends of the chains, it can be found by integration of equation (10.20),

$$\theta_L = \int_0^L k\,ds = kL \qquad (10.21)$$

where L is the total contour length. Therefore, if the probability of bending the chain is characterised by the energy ΔG, combination of equation (10.19) and equation (10.21) gives

$$\Delta G = \frac{\kappa \theta_L^2}{2L} \qquad (10.22)$$

The bending coefficient (κ) is therefore equal to twice the energy required to bend a unit length of polymer chain through one radian. The mean square value of the bending angle of the chain can be calculated with a Boltzmann distribution (equation (3.22)) for the chain-bending energies, divided by the necessary normalisation factor for the probability distribution,

$$\langle \theta_L^2 \rangle = \frac{\int_0^\pi e^{-\Delta G/kT}\theta_L^2 d\theta_L}{\int_0^\pi e^{-\Delta G/kT} d\theta_L} = \frac{LkT}{\kappa} \qquad (10.23)$$

There are two transverse modes for bending the semiflexible chain (y and z on **Figure 10.3**), so equation (10.23) must be multiplied by a factor of 2,

$$\langle \theta_L^2 \rangle = \frac{2LkT}{\kappa} \qquad (10.24)$$

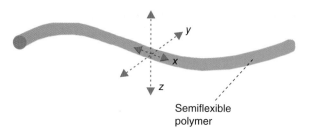

Semiflexible
polymer

Figure 10.3 *Two modes of transverse fluctuations (z and y) are possible with semiflexible polymers. Longitudinal fluctuations are supressed by the compressibility of the chain.*

It is then possible using $\langle h \rangle = \int_0^L \langle \cos\theta_L \rangle ds$ and by comparison with equation (10.9), to relate the persistence length to the bending rigidity,

$$L_P = \frac{\kappa}{kT} \tag{10.25}$$

This is a successful theory for the conformation of semiflexible chains and only requires a single elasticity constant, the bending coefficient (κ), to describe many of the experimentally observed phenomena.

It is revealing to connect the statistics of the conformations of random polymeric chains (monomer displacements as a function of *distance*) to the calculations of Brownian motion considered in **Chapter 5** (particle displacements as a function of *time*). The analysis provides a *Gaussian probability distribution* model for polymer conformations. Such random-walk models for macromolecules view them as rigid segments connected by freely jointed hinges. Thus, DNA can be modelled as a collection of rigid elements of length two times the persistence length. In random-walk models of polymers every macromolecular configuration is equally probable. The expected total end-to-end length ($<R>$) of the macromolecule is related to the sum of the individual monomeric displacements (x_i),

$$\langle R \rangle = \left\langle \sum_{i=1}^{N} x_i \right\rangle \tag{10.26}$$

where N is the number of monomers in a chain. However, this is not a useful measure (in common with the analysis of Brownian motion), since the molecule steps forward and backward with equal probability and the ensemble average is zero ($<R> = 0$). The mean square end-to-end distance is a better measure of the chain's size,

$$\langle R^2 \rangle = \left\langle \sum_{i=1}^{N} \sum_{j=1}^{N} x_i x_j \right\rangle \tag{10.27}$$

The sums can be separated to give

$$\langle R^2 \rangle = \sum_{i=1}^{N} \langle x_i^2 \rangle + \sum_{i \neq j=1}^{N} \langle x_i x_j \rangle \tag{10.28}$$

Adjacent steps are not correlated, so $<x_i x_j> = 0$ when $i \neq j$. Therefore

$$\langle R^2 \rangle = Na^2 \tag{10.29}$$

where a is the size of a step. Thus, the mean size of a random walk macromolecule ($<R^2>^{1/2}$) scales as the square root of the number of segments, $N^{1/2}$ (**Section 10.2**). This describes the conformational statistics of a phantom chain, i.e. a chain with no excluded volume effects, where the chain segments can pass through one another. A simple estimate of the size of a polymer in solution is therefore given by the root mean square end-to-end distance, from equation (10.13) and (10.29),

$$\sqrt{\langle R^2 \rangle} = \sqrt{2Ll_{\mathrm{p}}} = aN^{1/2} \tag{10.30}$$

since $N = L/2l_{\mathrm{p}}$ and $a = 2l_{\mathrm{p}}$ (the Kuhn segment length). The *radius of gyration* is perhaps a more precise measure of polymer size than the end-to-end distance, since it agrees better with experiments such as light/X-ray/neutron scattering. It is defined as the average square distance from the centre of mass of the chain (all the monomers have equal masses, so mass does not appear explicitly),

$$\langle R_G^2 \rangle = \frac{1}{N}\sum_{i=1}^{N} \left\langle \left(R_i - R_{\mathrm{CM}}\right)^2 \right\rangle \tag{10.31}$$

where N is the number of monomers in a chain and R_{CM} is the position of the centre of mass of the chain. It can be shown that

$$\sqrt{\langle R_G^2 \rangle} = \sqrt{\frac{Ll_{\mathrm{p}}}{3}} \tag{10.32}$$

which is similar to the expression of the root mean square end-to-end distance, equation (10.30), with a slight change in the prefactor.

For semiflexible polymers there are other contributions to the free energy, in addition to the bending energy. Hooke's law allows the extensional force (F) to be modelled (**Figure 10.4**),

$$F = -k\Delta L \tag{10.33}$$

where k is the elastic constant and ΔL is the extension. A slightly more general formalism for Hooke's law (**Chapter 13**) is to use the Young's modulus, which is independent of the specimen size,

$$\frac{F}{A} = E\frac{\Delta L}{L} \tag{10.34}$$

where ε is the strain ($\varepsilon = \dfrac{\Delta L}{L}$), ΔL is the extension, L is the length, and A is the cross-sectional area. The strain energy (W_{strain}) is calculated from the force by integration of equation (10.33) to give

$$W_{\mathrm{strain}} = \frac{1}{2}k(\Delta L)^2 \tag{10.35}$$

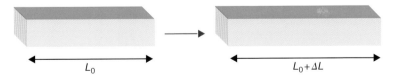

Figure 10.4   *Energy can be stored in an elastic material that is extended by a distance* ΔL *from an undeformed length* L_0.

More generally the calculation can be performed for a continuous fibrous material in which the degree of strain varies along the chain,

$$W_{strain} = \frac{EA}{2} \int_0^L \left(\frac{\Delta L}{L}\right)^2 dx \qquad (10.36)$$

where the Young's modulus (E) has been introduced into the expression. The energy to deform an elastic material is seen to be a quadratic function of the strain and a slightly more general equation uses the chain position ($u(x)$) to define the continuous extension along the chain,

$$W_{strain} = \frac{EA}{2} \int_0^L \left(\frac{du(x)}{dx}\right)^2 dx \qquad (10.37)$$

Detailed consideration of the three different modes of deformation of an elastic beam provides a more accurate model of a semiflexible polymer that includes *stretching*, *bending*, and *twisting* behaviour (**Figure 10.5**). All of these mechanisms can store energy. The Kratky–Porod model considered previously for semiflexible polymers only considered the bending energy (equation (10.25)). *Stretching* involves extension or compression (equation (10.37)), *bending* involves deflection with curvature, and *twisting* involves changes in torsion. For semiflexible biological polymers bending is often the dominant contribution, but stretching and, to a lesser extent, twisting, can play a role. The contribution of entropic elasticity to stretching will be considered in more detail in **Section 10.2**, since it provides a significant contribution to the elasticity of flexible chains.

Consider the *bending energy* of a semiflexible beam in more detail. A beam can be thought of as a collection of segments each of which is bent locally into an arc of a circle. The energy of extension ($W(\varepsilon)$) can be approximated as (Young's modulus E)

$$W(\varepsilon) = \frac{1}{2} E \varepsilon^2 = \frac{1}{2} E \left(\frac{\Delta L}{L_0}\right)^2 \qquad (10.38)$$

As before this is a quadratic function of the strain. A more precise statement for the total bending energy stored in a beam is

$$E_{bend} = L_0 \int_{\partial\Omega} \frac{E}{2R^2} z^2 dA \qquad (10.39)$$

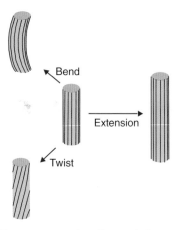

Figure 10.5   *Elastic fibres can store bending, twisting and extensional energy.*

where $\partial \Omega$ indicates that the integral should be made over the full cross-sectional area (A). This can be simplified as

$$E_{bend} = \frac{EIL}{2R^2} \tag{10.40}$$

where $I = \int_{\partial \Omega} z^2 dA$ is the momentum of inertia of the chain. For a chain bent into a loop, $L = 2\pi R$, and the total bending energy is therefore

$$E_{loop} = \frac{\pi EI}{R} \tag{10.41}$$

Thermal fluctuations tend to randomise the orientation of biological polymers. The thermal energy can be equated with the bending energy (equation (10.40)) to give

$$kT \approx \frac{EIL}{2R^2} \tag{10.42}$$

The persistence length is the length over which a polymer is roughly rigid, so $L = R \sim l_p$, and therefore

$$l_p \approx \frac{EI}{2kT} \tag{10.43}$$

This is again equation (10.25), where $\kappa = EI/2$ and this expression can be shown to be exact in a more rigorous calculation. More modern theories for semiflexible polymers often include chirality (spontaneous torsional twisting of the chains) and even twist–bend coupling, but they continue to be based on the Kratky–Porod model.

   The cytoskeleton in a cell can be viewed as a collection of elastic beams. Actin filaments can be tied in a knot by optical traps and thus torsional energy can be stored in such chains. In addition to their role in the cytoskeleton, actin filaments are important for the rigidity for microvilli and hair cells (among many other examples). An important aspect of the mechanical properties of actin in such roles is its ability to buckle. A beam subject to a large enough force will buckle, with a sudden, often catastrophic, increase in its curvature (**Figure 10.6**). Consider a long elastic beam with a force (F) applied to both the ends. The energy of the beam (W_{tot}) is the sum of the bending energy and the work done by the applied force,

$$W_{tot} = \frac{l_p kT}{2} \frac{L}{R^2} - F(L-x) \tag{10.44}$$

where $x = 2R\sin\frac{\theta}{2}$, the distance between the two ends of the beam and L is the contour length of the chain. A critical force is found from this expression when the beam transfers from the unbuckled ($\theta = 0$) to

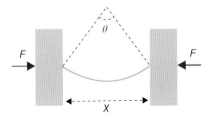

Figure 10.6   *The buckling transition of a semiflexible fibre compressed between two supports which are separated by a distance* x. *The critical force* (F = F_{crit}) *depends on the persistence length of the chain, equation (10.45).*

the buckled state ($\theta \neq 0$), i.e. it experiences a buckling transition. The critical force for buckling (F_{crit}) is found to be

$$F_{crit} = \frac{12kTl_p}{L^2} \tag{10.45}$$

10.2   Good/Bad Solvents and the Size of Flexible Polymers

Excluded volume is an important parameter for the determination of the configuration of flexible polymer chains. Steric interactions cause an expansion of a chain in solution when compared with a phantom model (a random walk of monomers with no excluded volume, equation (10.29)). The quality of a solvent for a polymer also affects its degree of expansion. There are typically three regimes of solvent quality that are defined; *good* (the conformation of a polymer chain expands as it tries to increase the number of contacts with the solvent), *bad* (the chain forms a compact globular conformation as it decreases the number of contacts with the solvent) and *theta* (the excluded volume interaction is exactly balanced by the attractive intermonomer potential) conditions. Variation of the solvent quality from good to bad can induce phase transitions as described in **Chapter 5**, such as the globule–coil transition and liquid-liquid phase separation. In most cases, chirality has a secondary impact on the size of flexible polymer chains and is neglected in what follows.

The root mean square end-to-end distance ($<R^2>^{1/2} = R_m$) for flexible polymer chains is defined in the same manner as for semiflexible chains and is a useful measure of the degree of expansion of the chains. With ideal flexible chains, where excluded volume interactions are neglected, accurate models for chain statistics have been known to synthetic polymer physicists for over fifty years. Flexible chains have random walk statistics in three dimensions, that lead to a characteristic scaling exponent (½) on the number of monomers (the same as that for diffusion in **Section 7.1**, equation (10.29)),

$$R_m \sim aN^{1/2} \quad \text{flexible phantom chain} \tag{10.46}$$

where a is the monomer length and N is the number of monomers in a chain. For a *rigid rod* the interpretation of the root mean square end-to-end distance (R_m) is also straightforward, its length is simply the number (N) of monomers multiplied by their size (a),

$$R_m \sim aN \quad \text{rigid rod} \tag{10.47}$$

For *self-avoiding walks* (with excluded volume and good solvent statistics) the calculation for the root mean square end-to-end distance is more complicated. The problem requires the renormalisation group technique, a sophisticated mathematical method from the theory of phase transitions. A similar scaling result to that of phantom and rigid rod chains is obtained for the root mean square end-to-end distance of the chains in a good solvent,

$$R_m \sim aN^{3/5} \quad \text{flexible good solvent} \tag{10.48}$$

At the *theta point* the attractive interchain forces due to the solvent and the repulsive forces due to the excluded volume of the chains balance. Phantom Gaussian statistics occur (as in equation (10.46)), but this time in a real experimentally realisable system,

$$R_m \sim aN^{1/2} \quad \text{flexible theta solvent} \tag{10.49}$$

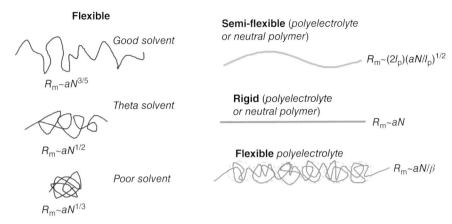

Figure 10.7 *Different scaling models for polymer-chain conformations as a function of solvent quality, charge and backbone rigidity. The root mean square end-to-end distance (R_m) is shown for each class of polymeric chain. β is a solvent quality-dependent parameter, l_p is the persistence length, a is the size of a monomer and N is the number of monomers in a chain.*

For a globule (a chain in a bad solvent) the chain forms a compact hard sphere whose radius is equivalent to a sphere constructed from N smaller spheres of volume $\frac{4}{3}\pi a^3$ and therefore

$$R_m \sim aN^{1/3} \quad \text{flexible bad solvent} \tag{10.50}$$

A range of possible root mean square end-to-end distance scalings with polymeric chains are summarised in **Figure 10.7**. Thus, all the scaling models for the size of a flexible chain correspond to the same general equation

$$R_m \sim aN^{\nu} \tag{10.51}$$

where ν is an exponent that depends on the type of the chain and the quality of the solvent for the chain's backbone chemistry.

A better understanding of the factors that affect the chain conformation as a function of solvent quality can be developed using the Flory calculation for the end-to-end distance. This is not a completely accurate calculation, indeed its success is in part due to the cancelation of two opposing misapproximated terms, but it is a useful starting point to understand the interplay between entropic forces and the monomer–monomer interaction that determines the conformation of a flexible chain. In this model the swelling of a polymer chain is due to a balance between the repulsion of the segments inside the coil (binary collisions) and the elastic forces that arise from the monomer entropy. The internal energy (U) due to the monomer–monomer collisions is given by

$$U(\alpha) \sim \frac{kTBN^{1/2}}{a^3\alpha^3} \tag{10.52}$$

where α is the expansion coefficient ($\alpha = R/R_0$, radius/initial radius), N is the number of monomers, a is the monomer length, and B is the second virial coefficient. The entropy of a chain is also related to the expansion coefficient and is given by

$$S(\alpha) = S_1 - \frac{3}{2}k\alpha^2 \tag{10.53}$$

where k is Boltzmann's constant and S_1 is a constant. The total free energy (F) of the chain can then be calculated as a function of the expansion coefficient by addition of the two terms given by equations (10.52) and (10.53),

$$F(\alpha) = U(\alpha) - TS(\alpha) = \text{const} + kT\frac{BN^{1/2}}{a^3\alpha^3} + \frac{3}{2}kT\alpha^2 \tag{10.54}$$

This can be solved graphically as shown in **Figure 10.8**. The minimum on the figure corresponds to the equilibrium conformation of the polymer chain. The renormalisation group technique is required for exact quantitative results on the size of a polymer, since the excluded-volume effect has not been accurately incorporated into the Flory model. However, the Flory calculation does qualitatively describe the experimentally observed behaviour for the size of flexible polymer chains.

Blob models are very useful tools for the determination of the physical properties of flexible polymers. Also, they are a useful vehicle for learning scaling ideas, which usually present the first best guess for a biophysical model, before a complete quantitative solution can be developed. A blob is a small section of a flexible polymeric chain that has some well-defined statistical properties (**Figure 10.9**). There are three varieties of blob that are typically used in polymer physics; *thermal blobs, electrostatic blobs* and *tension blobs*.

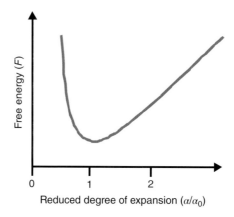

Figure 10.8 *Free energy of a single Flory chain as a function of the degree of expansion (α). α_0 is the equilibrium degree of chain expansion. The single minimum corresponds to the equilibrium conformation of the chain.*

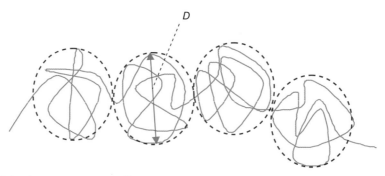

Figure 10.9 *A blob (diameter D) is a small section of a flexible polymer chain that has a free energy equal to kT.*

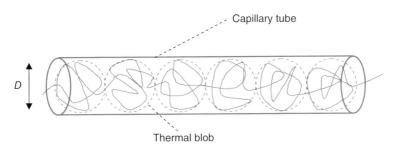

Figure 10.10   *A flexible polymer chain inserted into a capillary, e.g. a titin molecule between actin cages in striated muscle, tutorial question (10.3) or a flexible chain in a reptation tube. The blob size (D) is equal to the size of the capillary.*

A *thermal blob* (the simplest that will be encountered) is defined as a region of polymer chain in which the chain conformation is unperturbed from its thermal statistics. For example, consider a polymer chain in a capillary of diameter D (**Figure 10.10**). The chain is self-similar (fractal like) on a range of length scales, so the size of a small chain segment (R_s) scales as that of the whole chain

$$R_s \sim ag^\nu \tag{10.55}$$

where g is the number of monomers in a small segment and ν is an exponent, which depends on the quality of the solvent (compare with equation (10.51)). When a polymer chain is confined to a pore, the size of the chain segment (R_s) must be equal to the diameter of the pore (D) that contains it and from equation (10.55)

$$ag^\nu \sim D \tag{10.56}$$

Rearrangement of this equation gives the number of monomers in a thermal blob (g),

$$g \sim \left(\frac{D}{a}\right)^{1/\nu} \tag{10.57}$$

Thus, with very little mathematical effort the size of the chain in the capillary (R) can be calculated. The chain is decomposed into a string of equally sized blobs and is equal to the number of blobs multiplied by the size of each blob, i.e. $R = (N/g)D$.

The idea of thermal blobs can be extended into *semidilute solutions* (**Figure 10.11**) and calculations can thus be performed when flexible polymeric chains overlap. A semidilute solution of flexible polymer chains can be pictured as a close packed array of blobs. The number of contacts that a blob has with sections of another chain is approximately one and the free energy of the solution has by definition kT per blob, i.e. blobs are defined as chain regions with kT of free energy. Through a dimensional scaling argument it is possible to show that the semidilute correlation length (ξ), the mesh size of the polymer solution (the average size of the holes in a polymeric fish net), scales with concentration (c) as

$$\xi = ac^{-\gamma} \tag{10.58}$$

The exponent (γ) again depends on the quality of the solvent for the polymer chains. The exponent is 1/2 for the extreme limit of completely extended chains (found with polyelectrolytes and liquid crystalline polymers) and ¾ for a flexible chain in a good solvent. The predictions for the correlation length are in good agreement with scattering and atomic force microscopy experiments for a wide range of polymeric systems.

The concept of a *charged blob* is very useful for the calculation of chain statistics with weakly charged flexible polymeric chains (**Figure 10.12**). For the charged blob, by definition, the electrostatic repulsion of

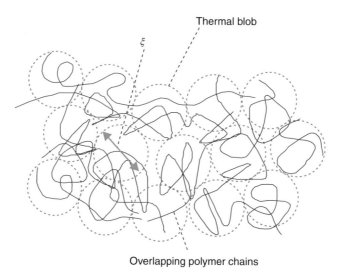

Figure 10.11 *Schematic diagram of a close-packed array of polymeric blobs in a semi-dilute solution of flexible polymers with an average correlation length, ξ.*

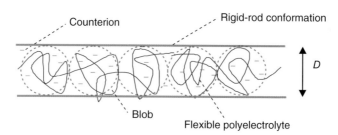

Figure 10.12 *Schematic diagram of a scaling model for a* flexible linear polyelectrolyte *that shows a rigid bayonet conformation (at large length scales) whose subsections are arranged in a series of electrostatic blobs (size D). At small length scales (inside a blob) the chain conformations are random walks.*

two neighbouring blobs in the chain is equal to the thermal energy (kT). Consideration of the Coulombic electrostatic energy of repulsion between charged monomers along a chain, gives the dominant term to be $g^2e^2/\varepsilon D$, where g is the number of monomers in a blob, e is the charge on each monomer, ε is the solvent dielectric and D is the blob size. A balance of the electrostatic energy to the thermal energy (kT), for a blob, gives

$$\frac{g^2e^2}{\varepsilon D} \sim kT \tag{10.59}$$

The size of an electrostatic blob is therefore

$$D \sim \frac{g^2e^2}{\varepsilon kT} \tag{10.60}$$

The conformation of charged polymers is a particularly hard problem due to the long-range nature of the electrostatic interactions and the large number of competing effects. Equation (10.60) is therefore just a first approximation. The conformation of a flexible polyelectrolyte chain can be approximated by a rod of electro-static blobs arranged end-to-end in a straight line (**Figure 10.7**). The topic is considered in more detail in **Chapter 11**.

Tension blobs are defined for a chain with an applied force (consider a single molecule that acts as an elastic spring, such as titin in striated muscle), which reduces the size of the lateral entropic fluctuations of the flexible chain (**Figure 10.13**). Such tension blobs (kT of stretching energy is stored per blob) are a useful tool for the description of deformed polymeric systems.

10.3   Elasticity

An analysis of the *elasticity of rubbery networks* is useful to understand both the static and dynamic properties of many types of naturally occuring biopolymers (**Figure 10.14**). Flexible chains are connected together at a series of junction sites to form a solid network. The elasticity of such networks is important in the function of a range of biological materials, e.g. resilin in the hinges of dragon flies wings and abductin in the hinges of clams. The behaviour of crosslinked flexible polymer chains is relatively simple and will be covered here, whereas semiflexible polymer chains exhibit more subtle phenomena related to the anisotropy of the networks (**Section 15.3**), and both entropic and bending mechanisms contribute to the elasticity.

To calculate the elasticity of a network of completely flexible chains an approach due to Flory will again be followed. An expression for the change in entropy of the polymer network on stretching is required. For a

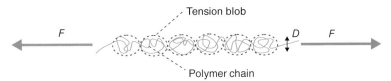

Figure 10.13   *Tension blobs (size D) for a stretched flexible polymer are determined by the size of the external force (F) and the quality of the solvent (the polymer/solvent interactions).*

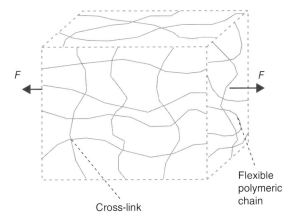

Figure 10.14   *Schematic diagram of a rubber network under a tension force (F). The extending force is resisted by the entropy of the chains of the flexible polymeric network.*

single polymer chain the chain entropy has been introduced previously (equation (10.53)) and the chain expansion factor is $\alpha = R/R_0$, the ratio of the root mean square (RMS) end-to-end distance to the initial RMS end-to-end distance. The entropy of a flexible Gaussian chain (unperturbed radius $R_0 = N^{1/2}a$) is therefore

$$S(R) = -k\left(\frac{3R^2}{2Na^2}\right) + \text{const} \tag{10.61}$$

where a is the monomer length, N is the number of monomers in a chain and k is Boltzmann's constant.

The change in entropy ($\Delta S(R)$) upon extension of a chain in three dimensions is the difference of the two entropies before and after stretching

$$\Delta S(R) = S(R) - S(R_0) \tag{10.62}$$

where R is the new chain radius and R_0 is the original chain radius. In three dimensions with Cartesian coordinates this entropy difference can be calculated as

$$\Delta S(R) = -\frac{3k}{2Na^2}\left[\left(R_x^2 - R_{0x}^2\right) + \left(R_y^2 - R_{0y}^2\right) + \left(R_z^2 - R_{0z}^2\right)\right] \tag{10.63}$$

It is useful to define the draw ratios $\lambda_x = R_x/R_{0x}$, etc., and this gives another equation equivalent to equation (10.63),

$$\Delta S(R) = -\frac{3k}{2Na^2}\left[\left(\lambda_x^2 - 1\right)R_{0x}^2 + \left(\lambda_y^2 - 1\right)R_{0y}^2 + \left(\lambda_z^2 - 1\right)R_{0z}^2\right] \tag{10.64}$$

The elasticity of a flexible polymer network is mainly entropic in Nature and the agreement of the prediction of the entropic model with experiment is very good with simple synthetic polymers, e.g. the contraction of a synthetic rubber band when heated. Summation of the changes in entropy over all the subchains gives the total entropy for stretching an entire polymeric network,

$$\Delta S = -\frac{3kvV}{2Na^2}\left[\left(\lambda_x^2 - 1\right)\left\langle R_{0x}^2\right\rangle + \left(\lambda_y^2 - 1\right)\left\langle R_{0y}^2\right\rangle + \left(\lambda_z^2 - 1\right)\left\langle R_{0z}^2\right\rangle\right] \tag{10.65}$$

where V is the volume of the sample and v is the concentration of subchains per unit volume. For a rubber network of ideal Gaussian chains the root-mean-square end-to-end distance ($<R_0^2>^{1/2}$) is proportional to the square of the number of monomers (equation (10.46)) and therefore

$$\left\langle R_0^2\right\rangle = \left\langle R_{0x}^2\right\rangle + \left\langle R_{0y}^2\right\rangle + \left\langle R_{0z}^2\right\rangle = Na^2 \tag{10.66}$$

where a is the size of a monomer in the chain. For isotropic Gaussian statistics of the chains, the size of the unperturbed chains in each of the draw directions are equal and therefore

$$\left\langle R_{0x}^2\right\rangle = \left\langle R_{0y}^2\right\rangle = \left\langle R_{0z}^2\right\rangle = \frac{Na^2}{3} \tag{10.67}$$

The chain entropy from equation (10.64) is thus

$$\Delta S = -\frac{kvV\left(\lambda_x^2 + \lambda_y^2 + \lambda_z^2 - 3\right)}{2} \tag{10.68}$$

The condition of uniaxial deformation along the x-axis for incompressible materials (volume conservation) introduces an additional condition on the draw ratios,

$$\lambda_y = \lambda_z = \lambda_x^{-1/2} = \lambda \tag{10.69}$$

Therefore, the change of entropy can be simplified

$$\Delta S = -kvV\frac{(\lambda^2 + 2)}{2(\lambda - 3)} \tag{10.70}$$

The free energy of the network (f) is proportional to the rate of change of entropy with extension ($\partial S/\partial \lambda$, **Chapter 3**) and therefore

$$f = -T\frac{\Delta S}{\Delta L_x} = -\frac{T}{L_{0x}}\frac{\partial S}{\partial \lambda} \tag{10.71}$$

where ΔL_x is the change in specimen length and L_{0x} is the original specimen length. The stress (σ) experienced by the elastic network is the force per unit area,

$$\sigma = \frac{f}{L_{0x}L_{0y}} = -\frac{T\partial S/\partial \lambda}{L_{0x}L_{0y}L_{0z}} = -\frac{T\partial S/\partial \lambda}{V} \tag{10.72}$$

where L_{0y} and L_{0z} are the original specimen width and height, respectively. A final expression for the stress in terms of the draw ratio is therefore

$$\sigma = kTv\left(\lambda - \frac{1}{\lambda^2}\right) \tag{10.73}$$

There is no dependence on the number of monomers (N) or on the size of the monomers (a) in this expression for the stress. The only parameters that are important for the determination of the response of the network to a stress is the extension ratio (λ) and the density of crosslinks (v). For small amounts of extension the draw ratio is approximately one ($\lambda \approx 1$) and therefore

$$\lambda - \frac{1}{\lambda^2} \approx 3(\lambda - 1) \tag{10.74}$$

By definition, the Young's modulus (E) is given by the ratio of the stress (σ) to the strain ($\Delta l/l$) (**Section 13.1**) and therefore

$$\sigma = E\frac{\Delta l}{l} \tag{10.75}$$

The draw ratio is simply related to the strain.

$$\lambda - 1 = \frac{\Delta l}{l} \tag{10.76}$$

And thus it is found that the Young's modulus is directly proportional to the number of crosslinks for a rubbery network,

$$E = 3kTv \tag{10.77}$$

where v is the density of cross links, and kT is the thermal energy. This expression for the modulus is compact, simple and very useful for the estimation of the elasticity of polymer networks. It can also be used to calculate the plateau in the shear modulus of viscoelastic networks of entangled polymer chains in which the crosslinks are topological in Nature (no chemical bonds contribute directly to v in this case). A complication for a detailed description of most biological systems is that polymeric chains can include rigid-rod or semiflexible sections (they are nematic elastomers) and the materials therefore act as biocomposites of their molecular components with unique strain-hardening mechanisms (**Chapter 15**), i.e. the approximation that the elasticity is completely entropic in origin is not a good one, due to significant alignment effects, as well as bending and twisting of the component chains.

10.4 Damped Motion of Soft Molecules

A first step in understanding the dynamics of polymers is through an inspection of the decay of forced vibrations of a rigid rod. This could be a fibre rigidly attached at one end (**Figure 10.15**), e.g. a hair in the cochlea of the ear (**Section 25.4**). The equation of motion for a damped spring is given by

$$m\frac{d^2x}{dt^2} + \gamma\frac{dx}{dt} + \kappa x = F \tag{10.78}$$

where x is the displacement of the end of the rod, γ is the viscous coefficient, κ is the elastic modulus and F is the external force. Through analysis of equation (10.78) it is found there are two distinctly different forms of motion for the fibre. When the damping is small

$$\gamma^2 < 4m\kappa \tag{10.79}$$

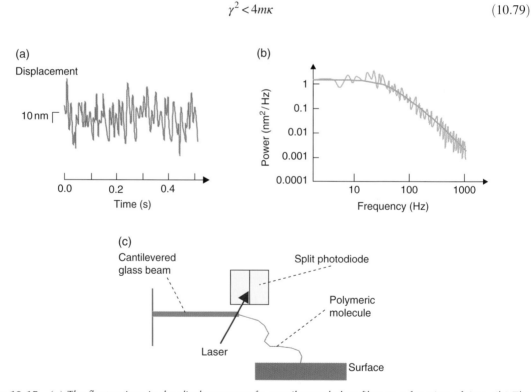

Figure 10.15 *(a) The fluctuations in the displacement of a cantilevered glass fibre as a function of time. (b) The resultant power spectrum as a function of frequency. (c) The experimental arrangement for the measurement of the displacement fluctuations of the glass fibre using a laser and a split photodiode. [Ref. E. Meyhofer, J. Howard, PNAS, 1995, 92, 574–578.]*

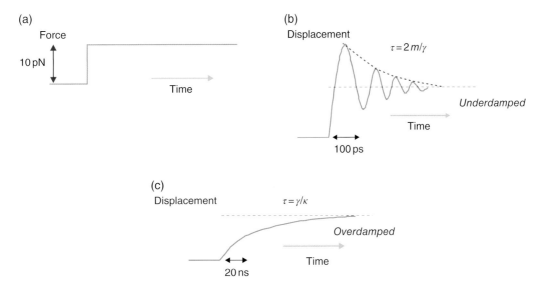

Figure 10.16 *(a) A mass (m) is attached to an elastic spring (elastic constant κ) embedded in a liquid with viscous dissipation (γ) is subjected to a force of 10 pN at time t = 0. The resultant motion of the mass can be (b) underdamped or (c) overdamped; the displacement is shown as a function of time. The time constant for the decay of the underdamped motion is $2\,m/\gamma$ and for the overdamped motion it is γ/κ.*

the motion is oscillatory and *underdamped* (**Figure 10.16b**). However, for large damping,

$$\gamma^2 > 4m\kappa \tag{10.80}$$

the motion is *overdamped* (**Figure 10.16c**). There are two time constants associated with the damped oscillatory motion. A fast time constant occurs for the mass to accelerate to the velocity of F/γ and a slow time constant is introduced for the relaxation time of the viscous forces.

   Global motions of small comparatively soft objects, such as proteins, polysaccharides and nucleic acids, in aqueous solutions are overdamped. This can be shown using a crude mechanical model of a soft biomolecule, e.g. a globular protein. The protein is pictured as a homogeneous isotropic cube of material with a given side length (L), density (ρ), Young's modulus (E) and viscosity (η). The mass (m) of the cube is

$$m = \rho V = \rho L^3 \tag{10.81}$$

The stiffness (κ) of the cube is related in the standard way to the length (L) and Young's modulus (E)

$$\kappa = EL \tag{10.82}$$

The drag coefficient on the cube (γ) is given by Stokes law (compare with equation (7.11)) as

$$\gamma = 3\pi\eta L \tag{10.83}$$

From equation (10.78), it is known that the motion of the cube is overdamped if the ratio $4m\kappa/\gamma^2$ is less than 1, and this can be related to the material properties of the protein,

$$\frac{4m\kappa}{\gamma^2} = \frac{4\rho L^3 EL}{(3\pi\eta L)^2} = \left(\frac{2}{3\pi}\right)^2 \frac{\rho E}{\eta^2} L^2 \tag{10.84}$$

How the characteristic ratio scales with the dimension L is particularly important. The smaller the physical dimension of the protein the higher the tendency for its motion to be overdamped.

The motions of protein domains of diameter less than the characteristic length (L_c) can be analysed. Equation (10.84) allows the characteristic length to be calculated, since $\eta \sim 1\,\text{mPa s}$, $\rho \sim 10^3\,\text{kg/m}^3$, $E \sim 1\,\text{GPa}$,

$$L_c \approx \frac{3\pi}{2}\left(\frac{\eta^2}{\rho E}\right)^{1/2} \approx 5\,\text{nm} \tag{10.85}$$

Small motions of globular proteins such as lysozyme are thus overdamped. The internal friction of the proteins will accentuate the degree of damping over and above the previous analysis. Elongated proteins are more highly damped than globular proteins, because as the aspect ratio increases the damping increases (increased L) and the stiffness decreases (E). Thus, the motion of the cytoskeleton is overdamped. Similarly, the motions of most extended biological polymers are overdamped.

A useful calculation is to solve a model for an oscillatory damped fibre by calculation of its power spectrum. Consider the motion of a fibre that undergoes thermal motion, e.g. the glass fibre is attached to a DNA chain (**Figure 10.15**). The autocorrelation function of the displacement of the fibre end ($R(\tau)$) is introduced to help solve the equation of motion and it is the time-dependent quantity that is often measured experimentally (equation (7.37)). The autocorrelation function is defined as

$$R(\tau) = \langle x(t)x(t-\tau)\rangle = \lim_{T \to \infty}\left\{\frac{1}{T}\int_{-T/2}^{T/2} x(t)x(t-\tau)dt\right\} \tag{10.86}$$

where T is the time over which the function is sampled and $x(t)$ is a spatial coordinate as a function of time,, e.g. the position of the fibre end. The autocorrelation function of the Langevin equation (10.78) can be calculated and it is found that the fibre fluctuates around its average position with a characteristic spectra

$$R(\tau) = \frac{kT}{\kappa}e^{-|\tau|/\tau_0} \tag{10.87}$$

$$\tau_0 = \frac{\gamma}{\kappa} \tag{10.88}$$

Typically, a power spectrum is experimentally used to analyse fluctuations in the displacement of a probe and is defined as the Fourier transform of the autocorrelation function ($G(f) = \Im(R(\tau))$). The power spectrum of the position of the glass fibre is found to be

$$G(f) = \frac{4kT\gamma}{\kappa^2}\frac{1}{1+(2\pi f\tau_0)^2} \tag{10.89}$$

This theoretical power spectrum can then be compared with that calculated using a numerical Fourier transform of the raw experimental data. It is found that the mean square displacement of the fibre end is $\langle x^2\rangle = kT/\kappa$ and the correlation time is $\tau = \gamma/\kappa$ (this is again related to the cut-off frequency of the power spectrum, $f_0 = (2\pi\tau)^{-1}$,

Section 7.3). The elastic (κ) and dissipative (γ) constants for a biological molecule can thus be directly calculated from the experimental data.

10.5 Dynamics of Polymer Chains

Consider an approximation to a polymer chain that consists of a spherical test particle of radius (a) in a continuum of constant viscosity (η_0). The Stokes–Einstein equation for the diffusion coefficient (D) can then be used for the whole chain motion of a polymeric chain as introduced in **Section 7.1**,

$$D = \frac{kT}{6\pi\eta_0 a} \tag{10.90}$$

where kT is the thermal energy. The structural relaxation time (τ) of a liquid that consists of such spherical particles is defined as the time for them to diffuse their molecular size (a),

$$\tau = \frac{a^2}{D} = \frac{6\pi\eta_0 a^3}{kT} \tag{10.91}$$

However, polymer chains can have internal relaxation modes due to their flexibility and they can have topological entanglements with their neighbours. These two effects significantly complicate the relaxation spectrum of polymers at short times compared with the case of simple rigid spherical colloids.

The *Rouse model* describes the spectrum of relaxation times of a flexible phantom chain in an immobile solvent (**Figure 10.17**). It is the simplest successful theory of polymer chain dynamics, and is formulated for a chain with a Gaussian conformation. The Rouse model assumes the chain statistics are ideal random walks (equation (10.46)). The mathematical description of the Rouse model is based on coupled equations of motion of the links (beads connected with idealised elastic springs), that incorporate random thermalised forces that tend to randomise their motions.

Once the Fourier transform of the basic equations of the Rouse model has been performed, the motion of a polymer chain can be represented as a superposition of independent Rouse modes, much like a harmonic spectrum used to describe the motion of a plucked guitar string. In the Rouse model, the maximum relaxation time of the polymer coil and the diffusion coefficient of the coil as a whole, vary with the number of chain links (N) as N^{-2} and N^{-1}, respectively. The first slow intermolecular relaxation mode and diffusive motion of the coil as a whole, conform to the first and fundamental Rouse modes respectively.

The root mean square displacement of a link on a Rouse chain varies with time (t) as $\sim t^{1/4}$, over time intervals less than the maximum relaxation time of the chain (τ), and only for times greater than the maximum relaxation time ($t > \tau$) does it become proportional to $t^{1/2}$, as in the case of ordinary diffusion of a Brownian

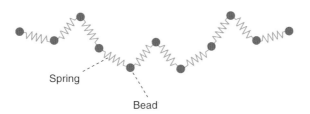

Spring

Bead

Figure 10.17 *Schematic diagram of a Rouse chain, a bead–spring model for the dynamics of flexible chains with no hydrodynamic interactions between the monomers.*

particle, equation (7.8). This behaviour has been well demonstrated experimentally with fluorescent microscopy and quasielastic neutron scattering experiments.

The Rouse model is found to be useful for dense polymeric melts in which the hydrodynamic interactions are screened, e.g. molten polystyrene or polyethylene samples. Similarly in semidilute solutions of flexible polymers combined with solvents, Rouse dynamics describe the motion of the polymer chains beyond the mesh size (ξ) of the network, i.e. the long length-scale dynamics (which have screened hydrodynamics) are well described.

The Rouse model yields results that differ from experimental observation for chains in dilute solution; one problem is the neglect of the hydrodynamic interaction caused by solvent entrainment between different sections of the polymer chains. The Zimm model includes a coarse-grained hydrodynamic interaction and is often in reasonable agreement with experimental data for isolated flexible chains in solution. The root mean square displacement of a link of a Zimm chain varies with time as $\sim t^{1/3}$ and only for times greater than the longest relaxation time ($t > \tau$) does it become proportional to $t^{1/2}$ as expected for diffusion, equation (7.8).

Both the Rouse and Zimm models are therefore useful for the description of the dynamics of flexible chains. For *semiflexible* chains the hydrodynamic beam model is required to explain their dynamics. Such a model can be used to describe relatively rigid polymers such as actin filaments, cochlea, keratin filaments, and microtubules (**Figure 10.18**). The drag force per unit length ($f_\perp(x)$) of the semiflexible filament for transverse fluctuations is given by

$$f_\perp(x) = -c_\perp v_\perp(x) = -c_\perp \frac{\partial y(x)}{\partial t} \tag{10.92}$$

where $c_\perp$ is the mass density of the filament, $v_\perp$ is the velocity, and y is the vertical displacement. The drag force can be balanced with the elastic restoring force of the filament and leads to the *hydrodynamic beam equation*,

$$\frac{\partial^4 y}{\partial x^4} = -\frac{c_\perp}{EI} \frac{\partial y}{\partial t} \tag{10.93}$$

where E is Young's modulus and I is the moment of inertia of the fibre. The relaxation time of a semiflexible chain can then be calculated as the time for a bent rod to relax back to its original straight conformation. The

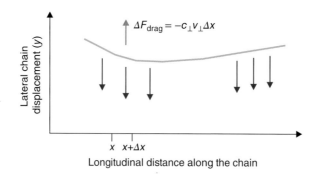

Figure 10.18 The transverse displacement (y) as a function of the longitudinal distance (x) along the semiflexible chain (blue). The lateral drag force (ΔF_{drag}) on a semiflexible fibre is a function of the perpendicular velocity ($v_\perp$), the size of the element (Δx) and the mass density ($c_\perp$).

amplitude of the bend is found to decrease exponentially with time, with a time constant (τ_n) that depends on the mode number (n) considered (**Figure 10.19**),

$$\tau_n \approx \frac{c_\perp}{EI}\left(\frac{L}{\pi(n+1/2)}\right)^4 \tag{10.94}$$

where $n = 1,2,3,...$, and L is the length of the rod. Such predictions for the dynamic modes of semiflexible chains are in good agreement with experiment.

The dynamics of both flexible and semiflexible polymers in *concentrated solutions* have a simple explanation in terms of the one-dimensional diffusion of each chain in a tube created by its neighbours (the Doi–Edwards chop-stick model is required to explain rigid-rod collective dynamics). The model is called *reptation* and the name refers to the snake-like motion of the chains in their tubes (**Figure 10.20**). The utility of the reptation model can be shown with a scaling calculation of the dynamics of a flexible polymer in an

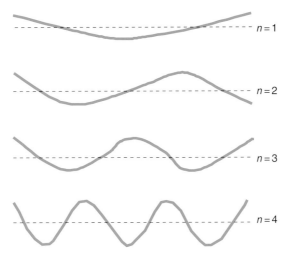

Figure 10.19 *The transverse displacement spectrum of a semiflexible chain (green). The hydrodynamics modes (n = 1,2,3,4) of a semiflexible rod in solution are shown. The free ends act as antinodes for the chain motion.*

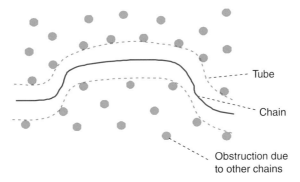

Figure 10.20 *Schematic diagram of a polymer chain (red) that reptates in a concentrated solution of other polymer chains. The chain diffuses in a tube (orange dashed line) formed by the steric constraints of its neighbours (green circles).*

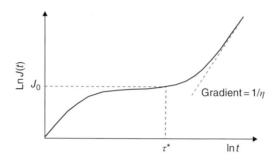

Figure 10.21 *Schematic diagram of the natural logarithm of the compliance (J = extension/stress) of a concentrated solution of entangled polymers as a function of the natural logarithm of the time. There is a plateau (J_0) at intermediate times (τ^*) and at long times the gradient is 1/η.*

entangled solution of other similar chains. Consider the *longest relaxation time* of a concentrated polymer solution. Initially ($t = 0$) a constant elongational stress (σ) is applied and the resultant relative deformation ($\Delta l/l$) is measured. If the stress (σ) is small, a linear approximation holds, and from the definition of the compliance ($J(t)$ = strain/stress),

$$\frac{\Delta l(t)}{l} = \sigma J(t) \tag{10.95}$$

For a typical polymer solution the compliance resembles that shown in **Figure 10.21**. After a sharp rise the compliance reaches a plateau value (J_0). For times greater than the longest relaxation time (τ^*) the stress is no longer proportional to the strain, but to the rate at which the strain increases, i.e. the material becomes liquid-like and Newton's law of viscosity (equation (7.10), **Figure 10.21**) holds,

$$\sigma = \eta \frac{d\gamma}{dt} \tag{10.96}$$

where γ is the strain. In the microscopic picture the longest relaxation time of the system (τ^*) is the time for entangled crosslinks to decay as the chains slither out of their tubes, i.e. the time for tube renewal. The Young's modulus (E) for the entangled solution is given by an analogous equation as was derived for the rubber network, equation (10.77),

$$E \sim \frac{kT}{N_e a^3} \tag{10.97}$$

where N_e is the average number of monomer units along the chain between two effective cross links, a is the Kuhn segment length and kT is the thermal energy.

For the snake-like reptative motion, the chain moves through a cylindrical tube (**Figure 10.22**), and for Gaussian chain statistics the diameter (d) is equal to the blob size (**Section 10.2**, equation (10.56)) given by

$$d \sim a N_e^{1/2} \tag{10.98}$$

where N_e is the number of monomers in a blob and a is the step size. Similar to the calculation of the size of a flexible chain in a capillary, the total contour length (Λ) of a tube is just the number of blobs in a chain (N/N_e) multiplied by the size of a blob

$$\Lambda \sim \frac{N}{N_e} d \tag{10.99}$$

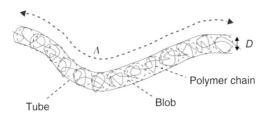

Tube   Blob

Figure 10.22 *Schematic diagram of the conformation of blobs of a flexible polymer in a tube used in a scaling calculation of the reptation model i.e. the dynamics of flexible chains in entangled solutions. Λ is the length of the tube and D is the blob diameter.*

where N/N_e is the number of blobs in a chain. Therefore, since $d = aN_e^{1/2}$ the contour length can be expressed as

$$\Lambda \sim aNN_e^{-1/2} \qquad (10.100)$$

For one-dimensional diffusion the mean square fluctuations in the one-dimensional displacement of the chain along the tube are related to the diffusion coefficient (D) by the standard relationship (equation (7.7), **Section 7.1**),

$$\langle x^2 \rangle = 2Dt \qquad (10.101)$$

The diffusion coefficient can be calculated for the whole chain (D_t) from the fluctuation-dissipation theorem (equation (7.9)) as

$$D_t = \frac{kT}{\mu_t} \qquad (10.102)$$

where kT is the thermal energy and μ_t is the friction coefficient of a chain. The total friction coefficient experienced by the chain can be approximated by the summation of the frictional terms due to each subunit (μ),

$$\mu_t = N\mu \qquad (10.103)$$

where N is the number of chain subunits. The longest relaxation time of the chain from the definition of the diffusion coefficient is

$$\tau^* \sim \frac{\Lambda^2}{D_t} \sim \frac{N^3 a^2 \mu}{N_e kT} \qquad (10.104)$$

The microscopic relaxation time (τ_m) of a molecular liquid is defined as $\tau_m \sim a^2/D$ (equation (10.91)), which gives a slightly simpler form for the relaxation time.

$$\tau^* \sim \frac{N^3}{N_e}\tau_m \qquad (10.105)$$

The viscosity and the diffusion coefficient can then be calculated from this reptation model. They both help to characterise the dynamics of entangled polymer solutions, and are found to be in good agreement with experiment. The viscosity of the solution is approximately equal to the modulus (E, equation (10.97)) multiplied by the longest relaxation time of the fluid (equation (10.104); a standard trick from rheology, which is exactly true for a Maxwell fluid, **Chapter 15**),

$$\eta \sim E\tau^* \sim \frac{\mu N^3}{aN_e^2} \qquad (10.106)$$

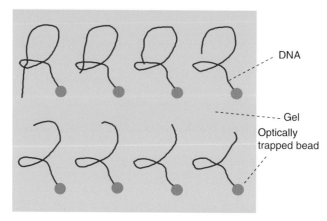

Figure 10.23 *Reptation in entangled polymer chains can be clearly visualised using fluorescence microscopy and optical tweezers. A time sequence of images is shown for a DNA chain that moves in a polyacrylamide gel. The optical tweezers cause driven reptation of the DNA chain. [From T.T. Perkins, D.E. Smith, S. Chu, Science, 1994, 264, 819. Reprinted with permission from AAAS.]*

The self-diffusion coefficient of the polymer chains is the rate at which the chain fluctuates the square of its size (R), using equation (10.104) again with $R^2 = a^2 N$,

$$D_s \sim \frac{R^2}{\tau^*} \sim \frac{N_e T}{N^2 \mu} \tag{10.107}$$

Experimental evidence for reptation is now very good. More sophisticated elaboration of the model provides a quantitative theory for the motion of DNA through polyacrylamide gels in electrophoresis gels (**Section 19.14**). Quasielastic neutron scattering experiments can explore the motion of labelled sections of polymers in a melt and are found to be in agreement with the reptation model. Fluorescent recovery after photobleaching measurements of concentrated solutions provides data in agreement with the predicted diffusion coefficients. Reptation theories for the rheological response of polymer melts allow quantitative predictions for a range of mechanical spectroscopy experiments (**Chapter 15**). However, perhaps the strongest evidence to date are the microscopy images of a fluorescent DNA molecule in a concentrated solution of other untagged DNA chains. The DNA chain is shown to experience forced reptation when pulled with optical tweezers. The chain clearly has a memory of the position of its tube as it moves (**Figure 10.23**) through a process of reptation.

10.6 Topology of Polymer Chains – Supercoiling

Many biologically important properties of semiflexible polymer chains are of a topological origin. The important characteristic of these chains is that they resist torsional distortion and can thus have a memory of their torsional state. The duplex DNA chains of bacteria are circular and typically occur in a compact supercoiled state, which provides an important motivation for the study of chain topology.

It was discovered that DNA chains with the same molecular weight, but different values of the number of super twists (τ) separate in a well-defined manner during gel electrophoresis experiments. The superhelical state moves across a gel faster due to its more compact conformation. The state of a closed circular DNA chain is now known to be characterised by two topological invariants; the type of knot formed by the double helix as a whole (W) and the *linking number* (L) of one strand with the other. The minimum of the energy of a closed-ring DNA corresponds to the superhelical state; the number of twists in a superhelix depends on the order of strand linking (*L*). Naturally occurring circular DNA chains are always negatively supercoiled. The axial *twisting* of strands around each other can differ from the magnitude of their linking number by the amount of *writhing*, which depends on the spatial form of the axis of the double helix. Viruses and bacteria can change the topology of their DNA during replication using topoisomerases. Topoisomerase I changes the linking number by +1 and topoisomerase II changes it by +2. Further, enzymes that introduce supercoils are called the gyrases. Negative supercoils favour unwinding of DNA and the subsequent processes of replication, transcription and recombination. Therefore, bacterial DNA is stored in a highly supercoiled, plectonomic structure (compact cylinders of DNA) ready for use, but with small space requirements, and topoisomerases/gyrases are employed when the information needs to be accessed.

To be mathematically more precise, consider the topology of closed ribbons. Three numbers (topological invariants) characterise the closed ribbon formed by circular DNA: the *linking number* (*L*), the *twist number* (*T*) and the *writhing number* (*W*). The *linking number* (*L*) is the number of times the two edges are linked in space, and it is an integer. To find the linking number (*L*), a projection of the ribbon on to a plane is required and all points where a segment of one of the curves passes above the other need to be counted. This process is demonstrated in **Figure 10.24**. The *twist* (*T*) is the extent to which *u* (a vector perpendicular to the chain) rotates around *s* (the contour vector along the chain). It can be calculated from the line integral along the path that describes a complete circuit around the chain (**Figure 10.25**),

$$T = \frac{1}{2\pi} \oint \omega ds \qquad (10.108)$$

where ω is the angular rate *u* rotates per unit length, and *ds* is the increment in arc length. The *writhing number* is defined as

$$W = L - T \qquad (10.109)$$

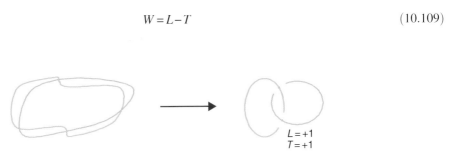

$$L = +1$$
$$T = +1$$

Figure 10.24 *The linking number (L) is equal to the number of times one topologically connected chain passes over the other. The diagram illustrates the process by which the topology of the chain can be clearly identified by inspection. T is the twist number.*

Figure 10.25 *The twist number can be calculated from the extent that the normal vector to the ribbon (u), rotates around the tangent vector (s) that defines the ribbon direction.*

$$T = 0 \qquad\qquad T = +1$$
$$L = 0 \qquad\qquad L = +3$$
$$W = 0 \qquad\qquad W = +2$$

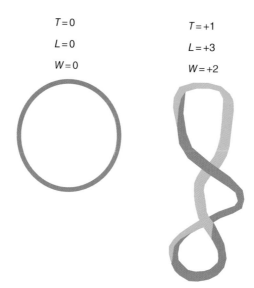

Figure 10.26 *A rich variety of topological states are possible with duplex DNA and two alternatives are shown. Similar to a telephone cable, duplex DNA chains have a torsional memory of their topology. [for more information the reader should see C.R. Cantor, P.R. Schimmel, Biophysical Chemistry III, Freemannnn 1980.]*

and is a measure of the number of loops formed taking into account their handedness. Equation (10.109) thus describes how loops can be converted into twists or links, and these topological constraints can only be circumvented if the circular chain is actually broken. **Figure 10.26** shows two possible configurations of a circular duplex DNA chain.

Suggested Reading

If you can only read one book, then try:

Top recommended book: Rubinstein, M. & Colby, R.H. (2004) *Polymer Physics*, Oxford University Press. A very clear account of the physical phenomena involved with synthetic polymers. Follows an elegant scaling approach and there is a great set of tutorial questions.

Cantor, C.R. & Schimmel, P.R. (1980) *Biophysical Chemistry Pt. III. The Behaviour of Biological Macromolecules*, Freeman. Old fashioned, but readable account of the statistical physics of biological molecules.

Doi, M. & Edwards, S.F. (1986) *The Theory of Polymer Dynamics*, Oxford University Press. Mathematically challenging advanced approach to polymer dynamics.

Grosberg, A.Y. & Khoklov, A.R. (1995, 1997) *Statistical Physics of Macromolecules,* AIP, 1995 and *Giant Molecules Here There Everywhere* AIP, 1997. Two informative books on polymer physics.

Howard, J. (2001) *Mechanics of Motor Proteins and the Cytoskeleton*, Sinauer. Useful account of the mechanics of semiflexible biopolymers.

Phillips, R., Kondev, J., Theriot, J. & Garcia, H.G. (2013) *Physical Biology of the Cell*, 2nd edition, Garland. Good discussion of semiflexible polymers.

Tutorial Questions 10

10.1 The Young's moduli of elastin* and collagen are 1 MPa and 1.5 GPa, respectively. Calculate the density of crosslinks that would give rise to these moduli in a random Gaussian network. Assume that the sections of chain are completely random between crosslinks. Explain the lack of agreement for the collagen data if the actual crosslinking density is on the order of 8×10^{25} m^{-3} in the two samples.

10.2 Consider the thermal fluctuation of a polymeric sickle cell haemoglobin aggregate observed with a fluorescence microscope. θ is the angle of the change in direction of a tangent to a filament along its length, s is the length of the aggregate and κ_f is the bending rigidity. The relationship between the bending energy and the average square fluctuation of the angle ($<\theta^2>$) is

$$E_{\text{bend}} = \frac{\kappa_f \left\langle \theta^2 \right\rangle}{2s}$$

For an aggregate of contour length 300 nm, determine the value of $<\theta^2>^{1/2}$ that arises from thermal fluctuations, if the persistence length of the fibrous aggregate is 10^{-3} m.

10.3 A flexible titin molecule occurs in an approximately cylindrical hole inside striated muscle (**Figure 10.22**, adapted from E. di Cola, et al., *Biophysical Journal*, 2005, 88, 4095–4106). If the molecule is unattached at either end, calculate the equilibrium length of the polymer ($R_\parallel$), assuming a Kuhn segment length of 30 nm, a contour length of 750 nm and a capillary diameter of 40 nm. The molecule is attached to the sacromere at either end at a fixed distance of 0.6 μm. Explain whether the loss in entropy of the chain due to its confinement in the capillary will contribute to the elasticity of chain.

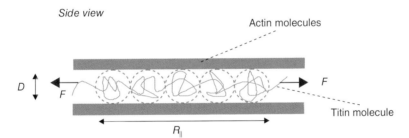

Figure 10.27 *Schematic diagram of a titin chain compressed inside an actin cage in striated muscle. The diameter of the pore is* D, *the force on the chain is* F *and the length of the chain is* R$_{||}$.

10.4 A polymer chain in a good solvent is held between two optical traps. Describe the form of the force/extension curve. If the chain was gradually placed in a 'bad solvent' at constant separation explain how the force exerted on the traps change.

10.5 Calculate the root mean square end-to-end distance $<r^2>^{1/2}$ for phage DNA chain that has a persistence length of 450 Å and a contour length of 60 μm using a Kratky–Porod model.

---

* Note that there is more ordering in the structure of elastin chains than this calculation might imply. The agreement of the calculations with experiment is therefore to some extent fortuitous.

11

Charged Ions and Polymers

There is a large range of biological phenomena in which electrostatics is important. The list includes the entrance of viruses into cells, the binding of drug molecules to DNA, the binding of enzymes to DNA (the lac repressor, ribonuclease, etc.), the compaction of DNA into chromosomes, the swelling of eye balls and the friction in knee joints. As a specific example consider the infection of a cell by the influenza virus. A pH-dependent process occurs by which the nucleocaspid of the influenza virus is delivered into the cytoplasm of a mammalian host cell. After endocytosis, specialised proton pumps are added to the endosomal membrane which transports hydrogen ions from the cytoplasm and causes a drop in the pH. The vesicle bursts and the virus contents are released into the cell cytoplasm; a process predominantly directed by the system's electrostatics.

Most readers will be familiar with Coulombs law, but there are many hard unsolved (perhaps unsolvable) problems in biological electrostatics due to the long-range nature of the charge interactions and the simultaneous interaction of large numbers of molecules. There are some ways to coarse grain the behaviour of all the charges, which can be satisfactory under some circumstances. In molecular biophysics the effects of billions upon billions of salt molecules often need to be averaged in the calculations and there are limits to the accuracy of such averaging procedures.

Charged ions exist in biological systems in a wide range of forms. The ions can be dressed atoms (e.g. Na^+ with a hydration shell, ~ 2 Å), small molecules (~ 1 nm), protein nanocomposites (10 nm) or giant linear aggregates (cm for DNA with millions of charged groups). Examples of small molecular ions found in biology include the carboxylic acid (COO^-) groups of aspartic and glutamic acid in proteins, the polar heads of fatty acids ($COO^- - (CH_2)_n - CH_3$), the positively charged amine groups (NH_2^+) in lysine, arginine, and histdine, and the phosphate group (PO^{4-}) in nucleic acid and phospholipids (**Figures 11.1–11.3**). All these ions are typically surrounded by a shell of associated water molecules and their interaction with this hydration shell has a dramatic effect on their physical properties.

Partial charges due to polarisation of covalent bonds also can exist in biomolecules and provide substantial interaction energies between neighbouring atoms. A particularly strong form of this polar interaction is observed in hydrogen-bonded molecules.

The Physics of Living Processes: A Mesoscopic Approach, First Edition. Thomas Andrew Waigh.
© 2014 John Wiley & Sons, Ltd. Published 2014 by John Wiley & Sons, Ltd.

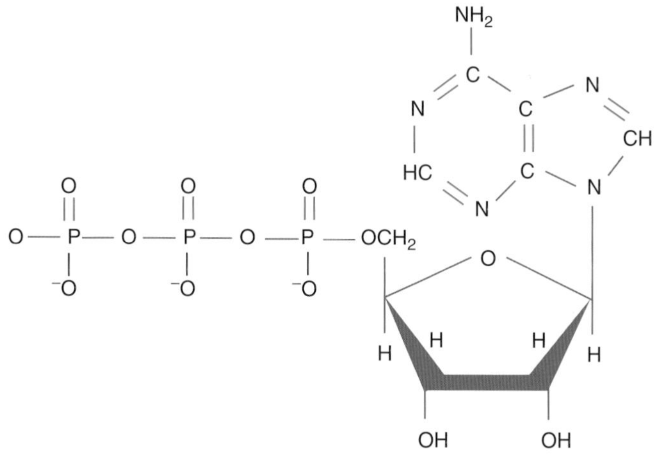

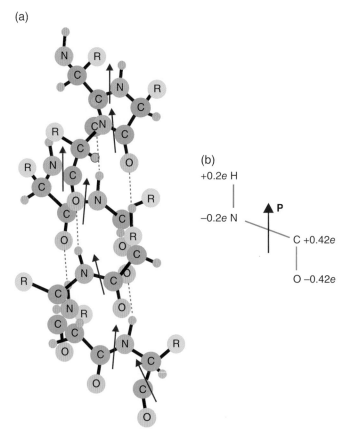

Figure 11.1 *Example of a small charged biological molecule, adenosine triphosphate (ATP); an important molecular source of energy in cells. The phosphate groups are highly negatively charged.*

Figure 11.2 *(a) An alpha helix in a protein can have a considerable dipole moment due to the alignment of the individual peptide dipoles along its backbone. (b) A single peptide dipole (P) from the helix. e is the electronic charge.*

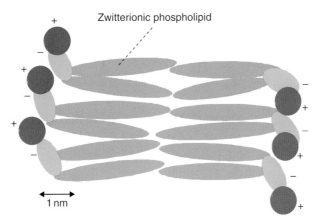

Figure 11.3 *Charged lipid molecules in bilayers can produce one of the highest surface charge densities in Nature with up to one electronic charge per 0.6 nm².*

There are a wide range of uses for ionised molecules. For example charged ions are used in signalling. Minute quantities of calcium (Ca^{2+}) ions control the molecular motors in striated muscle (μM) and a cocktail of Na^+, K^+ and Cl^- ions are required for the action of electrical impulses in nerve cells. There are therefore a wide range of enzymes (ion pumps) specifically designed to move and regulate small ionic species in cells.

Many of the processes that involve charged ions in biological systems involve the mechanism of *acid–base equilibria* (**Section 1.1**). A molecule (AH) is acidic if it can dissociate in water to provide a supply of hydrogen (H^+) ions, and conversely a molecule is basic if it can accept hydrogen ions. For an acid the process of dissociation can be described as (equation (1.4))

$$AH \xrightleftharpoons{} A^- + H^+ \tag{11.1}$$

The study of acid–base equilibria is of central importance to aqueous physical chemistry and reference should be made to specialised texts for more details. Dissociation of water molecules (equation (1.1)) reflects a competition between the energetics of binding and the entropy of charge liberation. In thermal equilibrium the law of mass action (**Chapter 20**) gives

$$\frac{[H^+][OH^-]}{[H_2O]} = \frac{[H^+]_0[OH^-]_0}{[H_2O]_0} e^{-\left(\mu^0_{H+} + \mu^0_{OH-} - \mu^0_{H2O}\right)/kT} \tag{11.2}$$

where μ^0_{H+}, μ^0_{OH-} and $\mu^0_{H_2O}$ are the chemical potentials of H^+, OH^- and H_2O, respectively. $[H^+]_0$, $[OH^-]_0$ and $[H_2O]_0$ are the concentrations of each species in some reference standard state. From simple space filling requirements, the distance between H^+ ions as a function of pH is found to vary from a few nm at pH 3 to 100 nm at pH 8 due to the reduction in their concentration.

The behaviour of charged macromolecules such as DNA, actin, carrageenan (a carbohydrate derived from sea weed), and chrondroitin sulfate (a glycosoamminoglycan found in articulated joints of mammals) will be highlighted. The charged groups are vitally important for both the solubility of these molecules and their correct biological function in a range of architectural, catalytic and information storage roles.

Simple electrostatic forces are a dominant factor in the determination of the structure of a large range of charged biological molecules and these forces will be examined using some simple physical calculations. For example, the dipole moments of the constituent amino acids in a protein tend to line up in an α helix that gives the secondary structure a large cumulative dipole moment (**Figure 11.2**), and these forces help to determine the resultant morphology.

Although many useful theories exist for strongly charged biological systems, it is necessary to emphasise some of their shortfalls. Electrostatic interactions are often long range in biomolecules which leads to an intractable many-body problem. Thus, a series of ingenious schemes and approximations need to be invoked to avoid some of these problems in a variety of different biological scenarios.

11.1   Electrostatics

Some basic electrostatics will first be reviewed. *Coulomb's law*, gives the force (f) between two point charges of magnitude q and q'

$$f = \frac{qq'u}{4\pi\varepsilon r^2} \tag{11.3}$$

where u is the unit vector between the two charges, ε is the permittivity of the material and r is the distance between the charges. This electrostatic force is thus directed along the line that joins two point charges and takes the direction in which a positive test charge would be translated. The local force (f) on a charge (q) is related to the electric field strength (E) it experiences due to the surrounding charge distribution,

$$f = qE \tag{11.4}$$

The energy (W) of an electric dipole (p) in an electric field (E) is the scalar product of the two vectors

$$W = -p.E = -pE\cos\theta \tag{11.5}$$

where θ is the angle between the vectors p and E. A torque (Γ) thus tends to orient dipoles in an applied electric field to minimise the energy (W), and the torque is the vector product of the electric dipole and the electric field,

$$\Gamma = p \times E \tag{11.6}$$

Gauss's theorem relates the electric flux that crosses a closed surface to the amount of charge (q) contained within the surface. The electric flux (ϕ) is defined as the integral of the component of the electric field perpendicular to a surface, over the complete surface,

$$\phi = \oint E.ndS \tag{11.7}$$

where n is the unit vector normal to the surface (dS) and the integral ($\oint$) is taken over the complete closed surface. Gauss's theorem for an arbitrary charge distribution in a vacuum has the simple form

$$\phi = \frac{q}{\varepsilon_0} \tag{11.8}$$

where ε_0 is the permittivity of free space and q is the charge contained within the surface. The total electric flux through a surface is proportional to the charge enclosed.

Gauss's theorem can be applied to a sphere that encloses a charge distribution (total charge q) of surface area $4\pi r^2$ in a vacuum that has a constant normal component of the electric field (E). It is thus easy to calculate the electric field around a *spherically symmetrical* charge distribution. Equation (11.8) applied to the specific geometry of a sphere gives

$$4\pi r^2 E = \frac{q}{\varepsilon_0} \tag{11.9}$$

Therefore, the radial electric field (E) around a point charge is

$$E = \frac{q}{4\pi\varepsilon_0 r^2} \tag{11.10}$$

Similarly, Gauss's theorem can be applied to a *cylindrically symmetrical* charge distribution, which will later be used to model a polyelectrolyte. Consider a cylinder of known radius (r), length (L), with an electric field (E) normal to its surface, that contains a line of charge (charge per unit length λ), and ε_0 is the permittivity of free space. The electric flux that crosses the cylinder's surface is created by the charge contained,

$$2\pi r L E = \frac{\lambda L}{\varepsilon_0} \tag{11.11}$$

The electric field from a line charge (or cylindrically symmetrical object) is therefore

$$E = \frac{\lambda}{2\pi\varepsilon_0 r} \tag{11.12}$$

Comparison with the expression for the electrical field around a sphere (equation 11.10) shows that the electric field for a cylinder decrease more slowly than that of a sphere ($E \sim 1/r^2$ for a sphere compared with $E \sim 1/r$ for a cylinder). These results have direct relevance for the electric forces experienced by spherical colloids (e.g. globular proteins) compared with those of cylindrical polyelectrolytes (e.g. DNA).

For more general charge distributions vector calculus needs to be used to calculate the electric field. The equivalent vector calculus expression of Gauss's law relates the potential (V) to the charge density (ρ)

$$\underline{\nabla}.\left(\underline{\nabla}V\right) = \frac{\rho}{\varepsilon_0} \tag{11.13}$$

The gradient of the potential is equal to the electric field (E) by definition

$$\underline{E} = -\underline{\nabla}V \tag{11.14}$$

A combination of equations (11.13) and (11.14) provides the *Poisson equation* for the potential around an arbitrary charge distribution. This was solved in **Chapter 4** to find the charge density near a plane.

When a material is placed in an electric field all the mobile dipole moments tend to line up due to the torque they experience (equation (11.6)). It is found experimentally that for a polarised material in a vacuum the induced electric moment per unit volume (P) is proportional to the applied electric field (E),

$$\underline{P} = \varepsilon_0 \chi \underline{E} \tag{11.15}$$

where χ is a constant of proportionality called the electric susceptibility (**Figure 11.4**).

It is observed that the effect of a dielectric is to reduce the electric field in proportion to the relative dielectric constant (ε). This observation explains why water is such a good solvent of ionic crystals, such as sodium chloride (Na^+Cl^-). The relative permittivity of water (ε) is 80 at 20 °C and the electrostatic energy of the salt ions is reduced by this factor. The detailed modelling of water as a dielectric can be complicated, since the dielectric constant varies from point to point and depends on the exact state of the molecular polarisation.

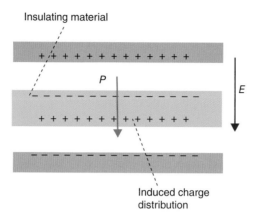

Figure 11.4 *Schematic diagram of the polarisation (P) induced in an insulating material (blue) placed in an electric field (E) between two conductors (gold). A charge distribution is induced in the insulator.*

The polarisation is affected by both the orientation of the water molecules in an electric field and the disorder due to their thermal agitation. Water is such an important solvent that a number of models have been developed to calculate the dielectric constant. One of the most successful of these is due to Lars Onsager, where the effective dielectric (ε) of water molecules is approximated (2–3% error) as

$$\varepsilon = \frac{p'^2 g n}{2\varepsilon_0 kT} \tag{11.16}$$

where g is the correlation parameter of the water molecules, p' is the mean dipole moment per molecule, n is the number of molecules per unit volume, and kT is the thermal energy. Typical values of the correlation parameter and the mean dipole moment are 2.6 (no units) and 8.2×10^{-30} C m, respectively.

An increase of the temperature of an aqueous system increases the rotational Brownian motion of the water molecules and shortens the lifetime of the tetrahedral hydrogen bonds that connect them together. The dielectric of water (ε) is therefore a function of the time scale and the temperature at which the aqueous system is examined. The dielectric of water is assumed to have a single dominant time scale, so the dielectric as a function of frequency (ω) is given by

$$\varepsilon = \varepsilon_\infty + \frac{\varepsilon_0 - \varepsilon_\infty}{\left[1 + (\omega\tau)^2\right]} \tag{11.17}$$

where τ is a dielectric relaxation lifetime of the water dipoles and is on the order of 10^{-11} s. ε_0 and ε_∞ are two characteristic dielectric constants. It is found that the effective dipole ($<p>$) moment of a dielectric induced by an applied electric field (E) is inversely proportional to the temperature, which leads to the useful relationship

$$\langle p \rangle = \frac{p^2 E}{3kT} \tag{11.18}$$

where p is the dipole moment of the constituent molecules.

Another useful electrostatic calculation is provided by a dielectric sphere placed in a uniform electric field. The relative magnitudes of the dielectric of the sphere (ε_2) and the surrounding material (ε_1) (**Figure 11.5**) determines the electric field that the sphere experiences. For $\varepsilon_2 \gg \varepsilon_1$ the electric field inside the sphere is

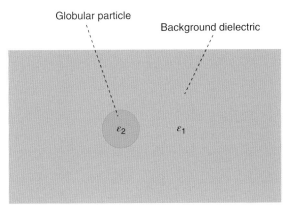

Figure 11.5 *Schematic diagram of a globular particle (green, dielectric ε_2) immersed in a continuous material with a different dielectric constant (ε_1).*

$$E_2 \approx \frac{3\varepsilon_1 E_0}{\varepsilon_2} \tag{11.19}$$

whereas for $\varepsilon_2 << \varepsilon_1$ the electric field in the sphere is now

$$E \approx \frac{3E_0}{2} \tag{11.20}$$

A practical example of this situation is the electric field a protein experiences in water during electrophoresis, where $\varepsilon_2 < \varepsilon_1$, and the second of the two equations, equation (11.20) must be used. An electric dipole at the centre of a sphere is also affected by the ratio of the dielectric of the sphere to that of the surrounding material. The electric dipole potential (V_1) of a sphere in this case is given by

$$V_1 = 3V_0 \left(\frac{\varepsilon_2}{2\varepsilon_1 + \varepsilon_2} \right) \tag{11.21}$$

where V_0 is the value of a dipole potential in a vacuum. With an alpha helix in a protein, this dipolar energy may have a role in specific internal interactions that affect the conformation of the protein.

It is also interesting to calculate the energy to move an ion between the two different dielectric environments, e.g. from oil (ε_1) to water (ε_2). It is given by

$$\Delta W = \frac{q^2}{8\pi\varepsilon_0 r} \left(\frac{1}{\varepsilon_1} - \frac{1}{\varepsilon_2} \right) \tag{11.22}$$

This is, however, only a crude approximation for an aqueous ion, since it does not include the effects of the hydration shell (**Section 11.3**).

The energy cost associated with the assembly of a collection of charges is another instructive calculation. For a general charge distribution the energy (U_{el}) can be calculated for both discrete and continuous charge distributions,

$$U_{el} = \frac{1}{2}\sum_i q_i V_i = \frac{1}{2} \int V(r)\rho(r)d^3 r \tag{11.23}$$

where q_i and V_i are the charge and potential on a discrete charge I, respectively. $V(r)$ and $\rho(r)$ are the continuous potential distributions and density distributions at a point r. Equation (11.23) can be used to compute the electrical energy of a ball of radius R and charge Q. The ball is considered as a series of concentric spherical shells of thickness dr and charge dq. The increase in energy when another shell is added is equal to the potential multiplied by the change in the charge,

$$dU_{el} = V(r)dq \tag{11.24}$$

The electric field outside a charged ball is the same as for a point charge placed at its centre, and by definition the potential is equal to the spatial integral of the electric field $(E(r'))$. Therefore

$$V(r) = \int_r^\infty E(r')dr' = \int_r^\infty \frac{1}{4\pi\varepsilon_0\varepsilon}\frac{q}{r'^2}dr' = \frac{1}{4\pi\varepsilon_0\varepsilon}\frac{q}{r} \tag{11.25}$$

The energy stored in the electric field due to the addition of charge dq is a combination of equations (11.24) and (11.25),

$$dU_{el} = \frac{1}{4\pi\varepsilon_0\varepsilon}\frac{\rho\frac{4}{3}\pi r^3}{r}\rho 4\pi r^2 dr \tag{11.26}$$

Integration over the sphere of radius R gives

$$U_{el} = \int_0^R \frac{1}{4\pi\varepsilon\varepsilon_0}\frac{16\pi^2}{3}\rho^2 r^4 dr = \frac{1}{4\pi\varepsilon\varepsilon_0}\frac{3}{5}\frac{Q^2}{R} \tag{11.27}$$

where Q is the total charge on the sphere.

The energy used by a motor to force DNA to move inside a bacteriophage is found to be around 2×10^5 pN nm using single molecule experiments. The total charge carried by a 20 000 base pair piece of duplex DNA is $Q = \frac{2e}{bp} \times 20000\,bp$ (the factor of 2 accounts for the 2 strands). The DNA is assumed to form an approximate sphere of radius 20 nm and equation (11.27) gives U_{el} to be 10^8 pN nm. The huge difference between the theoretical and experimental energies is due to charge screening and charge condensation. Counterion condensation is required to avoid the huge energy penalty in such highly charged systems and counterions drop on to the charged surface to neutralise some of the charge. This occurs at the expense of the counterion entropy when they are free to move in solution.

11.2   Deybe–Huckel Theory

A useful mean field theory (fluctuations are neglected) of charge interactions is that due to Debye and Huckel. The starting point for the development of the Debye–Huckel theory for an ion in solution is the Poisson equation that relates the potential (V) to the charge density (ρ). Equation (11.13) is taken and adapted for a solution of relative permittivity (ε),

$$\nabla^2 V = -\frac{\rho}{\varepsilon\varepsilon_0} \tag{11.28}$$

The condition of electrical neutrality is imposed on the solution; the quantity of charge on the positively and negatively charged ions must balance,

$$\sum_i z_i n_i = 0 \tag{11.29}$$

where z_i is the valence (size) of the charges, which can be both positive and negative, and n_i is the number of charges. It is then assumed that around any ion the charge distribution is spherically symmetric (the arrangement of the charge depends only on the radius r) and the variation in charge follows the Boltzmann distribution (**Chapter 3**), since it is in thermal equilibrium,

$$\rho_i(r) = \sum_i n_i z_i e = \sum_i n_i^0 z_i \exp\left[-w_{ij}(r)/kT\right] \tag{11.30}$$

where $w_{ij}(r)$ is the potential energy that corresponds to the mean electrostatic force exerted between ion i and j. n_i^0 is the charge density at the origin. In the electrostatic environment of the ions the interaction energy ($w_{ij}(r)$) is to a good approximation equal to the radially averaged Coulombic potential ($V_j(r)$) multiplied by the size of the charge ($z_i e$),

$$w_{ij}(r) \approx z_i e V_j(r) \tag{11.31}$$

In spherical coordinates, equation (11.28) combined with equations (11.30) and (11.31), gives the Poisson–Boltzmann equation,

$$r^{-2} \frac{d}{dr}\left(r^2 \frac{dV}{dr}\right) = (\varepsilon \varepsilon_0)^{-1} \sum_i e z_i n_i^0 e^{-z_i e V(r)/kT} \tag{11.32}$$

This equation is difficult to solve due to the exponential function on the right-hand side, which stops analytic integration of the expression to solve for the potential (V), i.e. it is a nonlinear differential equation. To make the calculation more tractable the exponential is assumed to be small, i.e.

$$\frac{z_i e V}{kT} << 1 \tag{11.33}$$

The Poisson–Boltzmann equation (11.32) can then be expanded in the rescaled potential and only the first-order terms are used to a first approximation, which gives

$$r^{-2} \frac{d}{dr}\left(r^2 \frac{dV}{dr}\right) = \kappa^2 V \tag{11.34}$$

Here, an important constant has been introduced, the *Debye screening length* (κ^{-1}), defined to be

$$\kappa^2 = e^2 \sum_i n_i^0 z_i^2 / \varepsilon \varepsilon_0 kT \tag{11.35}$$

Equations (11.34) and (11.35) constitute the crucial components of the Debye–Huckel theory for simple ions. The Debye screening length (κ^{-1}) is the length scale over which the electrostatic forces decay when screened by salt ions (**Section 4.3**). To find the exact potential and counterion profile, equation (11.34) has to be solved for a particular set of boundary conditions, but the Debye screening length can be used as an order of magnitude estimate for the range of an electrostatic interaction in many biological problems that relate to electrical charge.

The Poisson–Boltzmann equation can also be used to calculate the charge distribution around a membrane in a salty solution. The presence of the charged membrane sets up a nonuniform distribution of positive and

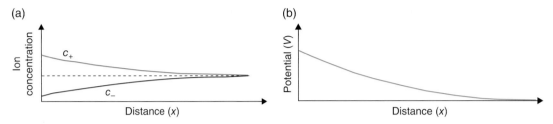

Figure 11.6 *(a) Counterion concentration (positive (c_+) and negatively (c_-) charged) as a function of the perpendicular distance (x) close to a charged planar surface. (b) Potential as a function of perpendicular distance (x) from the same planar surface.*

negative ions in the vicinity of the membrane. The density of counterions is increased in the vicinity of the membrane, while the density of like-charged ions is decreased (**Figure 11.6**).

The number density of positively ($c_+(x)$) and negatively ($c_-(x)$) charged ions in solution are given by the Boltzmann formula (**Chapter 3**),

$$c_+(x) = c_\infty e^{-zeV(x)/kT}, \quad c_-(x) = c_\infty e^{+zeV(x)/kT} \tag{11.36}$$

where c_∞ is the ion concentration in the bulk solution, z is the ion valence and e is the electronic charge. The total charge density at position x is obtained by summation over the contributions from both the positive and negative charges,

$$\rho(x) = zec_+(x) - zec_-(x) \tag{11.37}$$

The charge density is related to the electric potential at the same position by the Poisson equation (11.28), which in Cartesian coordinates is given by

$$\frac{d^2V(x)}{dx^2} = -\frac{\rho(x)}{\varepsilon\varepsilon_0} \tag{11.38}$$

The corresponding Poisson–Boltzmann equation is

$$\frac{d^2V(x)}{dx^2} = \frac{zec_\infty}{\varepsilon\varepsilon_0}\left(e^{zeV(x)/kT} - e^{-zeV(x)/kT}\right) \tag{11.39}$$

The potential is known to decay sharply with distance (exponentially) around membranes in solution. If the potential is 'small' the Poisson–Boltzmann equation can be simplified, similar to equation(11.33), since

$$e^{\pm zeV(x)/kT} \approx 1 \pm zeV(x)/kT \tag{11.40}$$

Therefore, the linearised version of equation (11.39) is

$$\frac{d^2V(x)}{dx^2} = \frac{2z^2e^2c_\infty}{\varepsilon\varepsilon_0 kT}V(x) \tag{11.41}$$

This has a solution of the form

$$V(x) = Ae^{-x/\kappa} + Be^{x/\kappa} \tag{11.42}$$

where A and B are integration constants and the Debye screening length is again defined by equation (11.35). Determination of the final integration constant requires the boundary conditions on the electric field,

$$E_x(0) = \frac{\sigma}{\varepsilon_0 \varepsilon}, \tag{11.43}$$

where $E_x(x) = -\dfrac{dV(x)}{dx}$ and σ is the charge density on the surface. The charge density away from a charged membrane is therefore

$$\rho(x) = -\sigma \kappa e^{-x\kappa} \tag{11.44}$$

A similar story is found for a variety of geometries (e.g. protein surfaces, DNA chains). Thus, the *Debye screening length* is again seen to give the length scale over which charge effects are felt by charged objects in salty solutions.

11.3   Ionic Radius

Stoke's law for the hydrodynamics of small spheres (equation (7.11)) is not valid when the size of the water molecules is similar to that of the particles that move. Indeed, ultrafast femtosecond laser experiments point to the importance of the viscoelasticity of water molecules in the determination of ion dynamics at small time scales, which is mirrored in the frequency dependence of the dielectric constant of water (equation 11.17). The Stokes radius (r_c) of an ion is found in a range of mobility studies (including laser and pulsed NMR) to be much larger than that expected from X-ray scattering experiments in the liquid state. This mismatch is interpreted as a hydration shell that surrounds each of the ions in solution. Femtosecond laser experiments combined with molecular dynamics simulations demonstrate the lifetime of these cages of water is on the order of ~10 ps (**Figures 11.7** and **11.8** show some pulsed laser data and a simulation of the hydration shell that surrounds aqueous iodine ions).

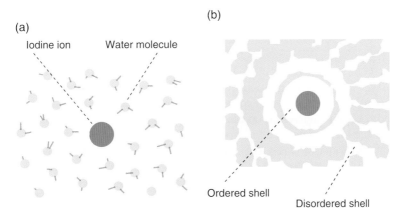

Figure 11.7   *Schematic diagrams that show the results of molecular dynamics simulations of water molecules around an iodine ion at (a) 1 ps and (b) 30 ps. The blurring in (b) indicates the time evolution of the water molecules which create a more ordered shell of water molecules around the iodine molecule. [Reproduced with permission from A.H. Zewail, Chapter 2, p81, Femtosecond Chemistry, eds J. Manz, L. Woste, 1995, Wiley VCH. Copyright © 1995 VCH Verlagsgesellschaft mbH.]*

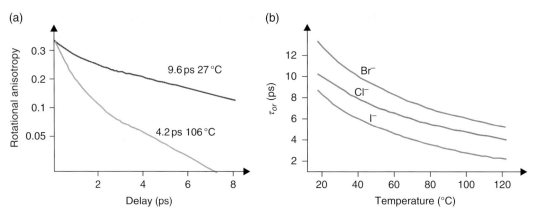

Figure 11.8   *(a) The rotational anisotropy as a function of delay time measured in pulsed laser experiments. Such experiments can probe the rotational motion of halide ions in aqueous solutions at different temperatures (27 °C and 106 °C). (b) The measured orientational diffusion time constants (τ_{or}) of the solvation shells of chloride (gold), bromide (green) and iodide (blue) are shown as a function of temperature. [Reproduced with permission from Femtosecond Chemistry, eds A.H. Zewail, Chapter 2, p81, Femtosecond Chemistry, eds J. Manz, L. Woste, 1995, Wiley VCH. Copyright © 1995 VCH Verlagsgesellschaft mbH.]*

Bjerrum postulated that the distance between ions in a concentrated solution can be small enough that transient associations can be created between ions of opposite charge. An ion pair is created when the distance between two elementary charges (e.g. $+e$ and $-e$) is such that the electrostatic energy of attraction is greater than or equal to the thermal energy (kT). Thermal energy thus tends to disrupt these temporary ion pairs, which constantly associate and dissociate in solution. The separation at which the thermal energy and the electrostatic energy is balanced defines the *Bjerrum length* (l_B), the length at which stable ion pairs begin to form. The Bjerrum length is therefore mathematically defined to be

$$l_B = \frac{e^2}{4\pi\varepsilon\varepsilon_0 kT} \tag{11.45}$$

The dielectric constant of water (ε) is 80 at 20 °C, and l_B is thus 7.12 Å at this temperature (these are useful numbers to remember). The ion pairs are short lived in standard simple electrolyte solutions due to the small Bjerrum length. However, ion pairs are an important effect in polyelectrolyte solutions, particularly with respect to the phenomenon of counterion condensation, and the Bjerrum length is a useful quantity to gauge the range of electrostatic interactions as a function of the dielectric of a solvent with these materials.

To understand the behaviour of charged ions in solution in more detail consider equation (11.5) for the orientational energy (W) of a dipole (p) in an electric field (E). The energy associated with the dipole moment of water is quite strong (10–20 kT) and charged ions are thus surrounded by layers of oriented water, which are not randomised by the thermal energies (**Figure 11.7**). **Figure 11.8** shows data from pulsed femtosecond laser experiments on the rotational motion of water molecules that surround halogen ions in aqueous solution. The simulations are in good agreement with these experiments and show a restricted shell of water molecules around the ions in solution.

Detailed time-averaged X-ray and neutron diffraction studies of liquid water show two regions of ordering around charged ions. In the first hydration shell there is ordered ice-like water and surrounding this there is a second hydration shell of more disordered water.

The free energy of hydration (ΔG_H) for an ion is the work required to move an ion from a vacuum ($\varepsilon_1 = 1$) to a medium of dielectric ε_2. The free energy of hydration (from equation (11.22)) is given by

$$\Delta G_H = \frac{q^2}{8\pi\varepsilon_0 r}\left(1 - \frac{1}{\varepsilon_2}\right) \tag{11.46}$$

It is assumed in this calculation that the ion is initially in a vacuum, the uncharged ion is then transferred from a vacuum to the water and finally the ion is recharged in the water. The ionic radii found from such calculations needs some analysis. Values for the radii are consistently found to be bigger than those from crystallography (**Figure 11.9**). More sophisticated approaches are therefore needed that explain the change in entropy due to the structure breaking properties of the ion on the dipole moments of the surrounding water molecules to account for the shortfall.

It is also possible to quantify the effects of ions on a solution from the change in viscosity (η) of a solution to which the ions are added. The reduced viscosity of the ionic solution is found to take a characteristic form

$$\frac{\eta}{\eta_0} = 1 + Ac^{1/2} + Bc \tag{11.47}$$

where A is a constant related to the electrostatics, B is a constant related to the degree of water structural rearrangement, c is the ion concentration and η_0 is the viscosity of the unperturbed solvent. Salty water thus has a distinct (1 M NaCl ~5%) increase in viscosity compared with a pure dialysed sample and the hydration shells that surround the water ions contributes to this increase.

The ability of an ion to restructure water is related to its exact chemical nature. Hoffmeister classified salts in terms of their ability to precipitate proteins. His series relates to the experimentally observed charge density of

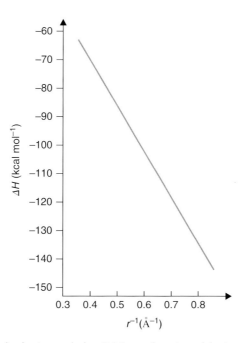

Figure 11.9 *The variation of the hydration enthalpy (ΔH) as a function of the inverse ionic radius (r) for small ions in aqueous solution. [Reprinted with permission from A.A. Rashin and B. Honig, Journal of Physical Chemistry, 1985, 89, 5588–5593. Copyright © 1985, American Chemical Society.]*

Table 11.1 The Hoffmeister series *for anions and cations in terms of their solubility for proteins, protein denaturation, ability to form cavities, surface tension of solutions and protein stability.*

Anions

CO_3^{2-}	SO_4^{2-}	$S_2O_3^{2-}$	$H_2PO_4^-$	F^-	Cl^-	Br^-	NO_3^-	I^-	ClO_4^-	SCN^-

→

Increased solubility of proteins, increased protein denaturation and increased ability to form cavities

←

Increased surface tension of solutions, increased protein stability

Cations

NH_4^+	K^+	Na^+	Li^+	Mg^{2+}	Ca^{2+}	Guanidium$^+$

ions in aqueous solutions when they are surrounded by their hydration shells and can help predict a wide range of biological reactions in water. Some typical anions and cations are listed in **Table 11.1**. This is an approximate series and specific competing reaction schemes can sometimes reverse the order of close neighbours in this series.

The process of salting out that was used by Hoffmeister to classify ions often occurs practically in the purification and crystallisation of biological molecules (**Section 5.4**). Salt ions in solution cause a rearrangement of the water molecules, which compete with the biological molecules for their hydration shell and hence reduce their solubility. Thus, the higher charge density ions in the Hoffmeister series are more effective for salting out proteins.

When an ionic compound is moved between two dielectric environments, it is found that the thermodynamic driving force for the solubility of ionic compounds is proportional to

$$RT \ln f = A\left(\frac{1}{\varepsilon} - \frac{1}{\varepsilon_w}\right) \tag{11.48}$$

where f is the thermodynamic activity coefficient (an effective mole fraction of the ions), R is the ideal gas constant, ε_w is the dielectric constant of water, ε is the dielectric constant of the initial medium and A is a constant of proportionality ($A = \dfrac{e^2}{8\pi\varepsilon_0 b}$, where b is a weighted average of the ion radii). Consider an ion that is transferred from ethanol to water. If a standard sample is taken, b is $\sim 10^{-10}$ m, the temperature is 300 K , e is 1.6×10^{-19} C, the dielectric of ethanol (ε) is 25, the dielectric of water (ε_w) is 80, and it is found that the activity coefficient (f) is 2000. Therefore, salt is two thousand times less soluble in ethanol than water. For a specific protein empirical laws exist for the solubility as a function of the ionic strength (I) of the form

$$\log S = \log S_0 - KI \tag{11.49}$$

where S is the solubility, S_0 is the value of the solubility at zero ionic strength ($I = 0$) and K is a constant proportional to the size of the protein. A fundamental derivation of such a formula in terms of the sticky protein–protein interaction is a current challenge for colloidal science.

11.4 The Behaviour of Polyelectrolytes

The volume that surrounds a polyelectrolyte molecule in solution contains ions of opposite sign to the polyelectrolyte that have dissociated from the molecule, which maintain global charge neutrality, e.g. Na^+, K^+ and Mg^{2+} for nucleic acids or acidic polysaccharides, and OH^-, Cl^- anions for polyamines.

Thus, polyelectrolytes are surrounded by a cloud of counterions and the dissociation of the charged groups along the chain is often the dominant contribution to the solubility of the molecules in water. Fluctuations can occur in the spatial distribution of these counterions consistent with the requirements of thermal equilibrium, but globally charge neutrality must be maintained.

Polyampholytes (e.g. proteins) contain positive and negative charged groups bound to the same polymer chain. These charges can lead to anisotropic charge distributions within the protein. Hydrogen (H^+) and hydroxide (OH^-) are 'special' ions associated with polyacids and polybases that makes the charge fraction on these polymers particularly sensitive to pH changes (**Section 1.3**). Neutralisation of the acidic or basic units with a corresponding alkali or acid removes this effect, i.e. variation of the polymer concentration does not change the charge fraction on each neutralised polyelectrolyte chain, a useful first simplifying step in many experiments on charged polyions.

The conformation of polyelectrolytes is strongly affected by the repulsion of charges along the backbone. Charge repulsion encourages the creation of extended conformations for chains at low ionic strengths. For example, the charged groups are important for the rigidity of semiflexible polymers such as DNA and are a dominant factor with flexible polyelectrolytes such as alginates (a constituent of sea weed). As polyelectrolytes overlap in more concentrated solutions the charge interaction becomes screened and the end-to-end distance of the molecules reduces.

The physics of charged macromolecules in solution is very rich and the chain conformation depends on the fraction of monomers that are charged, the concentration of monomers, the concentration of low molecular weight salt, the intrinsic rigidity of the polymer backbone and the quality of the solvent for the backbone chemistry (e.g. *good*, *bad*, and *theta* solvents, **Section 10.2**). Polyelectrolytes can be classified in terms of strongly and weakly charged behaviour. In strongly charged polyelectrolytes every monomer carries a charge. Therefore, Coulombic monomer–monomer interactions are the dominant forces that determine the conformation of the chains. If e is the charge held on a monomer and ε is the dielectric, the potential energy of the screened Coulomb interaction between charged links i and j, separated by a distance r_{ij}, is given by the Yukawa equation,

$$V\left(r_{ij}\right) = \frac{e^2}{\varepsilon r_{ij}} e^{-r_{ij}\kappa} \tag{11.50}$$

where κ^{-1} defines the Debye length that determines the screening of the electrostatic interaction by other ions in solution (equation (11.35)).

The end-to-end distance of strongly charged polyelectrolyte macromolecules (L) in a dilute salt-free solution is proportional to the number (N) of charged links, since a charged macromolecule is fully stretched ($L \sim Na$, where a is the monomer size). A weakly charged flexible polyelectrolyte macromolecule in dilute solution can be visualised as an extended chain of electrostatic blobs (**Figure 10.7**, **Section 10.2**). The length of the chain is again proportional to the number of monomers, but the chain size needs to be rescaled by the number of monomers in a blob ($L \sim ND/g$; g is the number of monomers in a blob, and D is the blob size).

In a polyelectrolyte solution with overlapping chains, the Coulomb interaction is screened by counterions, so that the chain conformations become random walks on large length scales, while their rod-like conformations are retained at short length scales. The distance between the neighbouring chains in solution is of the order of the Debye screening radius. For flexible polyelectrolytes this charge screening causes a large contraction of the end-to-end distance upon coil overlap.

The *transport phenomena* involved with charged chains is an important, but complicated subject. The counterion clouds associated with a polyelectrolyte chain must be dragged around when the polyelectrolyte moves, and this dissipates energy. The field of polyelectrolyte dynamics relates to the subject of electrophoresis, the driven motion of a charged molecule in an electric field, which will be investigated in **Section 19.14** due to its application in DNA sequencing.

Chemical modification of a polymeric molecule so that it is charged has a dramatic effect on its osmotic pressure in solution. The osmotic pressure of the solution is dominated by the contribution of the counterions. Consider a negatively charged polyelectrolyte in solution. The number of cations (n_+) in the solution are provided by both the salt and the counterions. However, the number of negatively charged anions (n_-) are predominantly due to the salt, since the number of polyelectrolyte chains is negligibly small compared with the number of small salt ions. The number of positive (n_+) and negative (n_-) charged units in the solution can therefore be written as

$$n_+ = n_s + n_p v \tag{11.51}$$

$$n_- = n_s + n_p \approx n_s \tag{11.52}$$

where n_s is the total number of salt cations and salt anions per unit volume (they are equal for a monovalent salt) and n_p is the number of molecules of a polyion per unit volume when negatively charged polyions of valence v are dissolved. Each independent unit in the solution contributes kT to the osmotic pressure, much like the molecules in an ideal gas as they bounce off the walls of their container. The total osmotic pressure of the solution is thus

$$\pi = kT\phi(n_+ + n_-) = kT\phi(n_p v + 2n_s) \tag{11.53}$$

where ϕ is the osmotic coefficient which accounts for the fact that only a fraction of the polyelectrolyte ions are dissociated per unit volume. This gives another method to probe the phenomenon of charge condensation that is described in detail in **Section 11.8**. The osmotic pressure of charged polymers is much larger than their neutral counterparts by an amount $kT\phi n_p v$ and this has a range of associated phenomena, e.g. the swelling behaviour of nappies and the shape of the cornea in the eye.

The calculation that gives equation (11.53) does not assume that all the polyelectrolyte's counterions are dissociated. To decide on the nature of counterion binding in polyelectrolytes the osmotic coefficient (ϕ) is defined experimentally as the ratio of the osmotic pressure (π) to a reference value, π_0 $(\phi = \pi/\pi_0)$, for full dissociation,

$$\pi_0 = (2n_s + n_p v)kT \tag{11.54}$$

Thermodynamically the osmotic pressure can be shown to be the negative rate of change of free energy with volume (V) of the solution $(-\partial F/\partial V)$. The difference between the osmotic pressure and the reference value for full dissociation $(\pi-\pi_0)$ is due to additional electric free energy in the system that causes the counterions to bind to the polyions.

11.5 Donnan Equilibria

The process by which the osmotic pressure of a system is regulated by the affinity of molecules for their counterions is called *Donnan equilibrium*. Donnan equilibrium is an important phenomenon in a wide range of biological systems. Striated muscle fibres (actin and myosin assemblies) are held apart by a Donnan pressure. The cornea in the eye is osmotically stressed and the Donnan pressure is important to maintain the interfibrillar spacing (**Figure 11.10**). In mammalian cells ions pumps are required to regulate the Donnan pressure and thus the working environment of the cell. The malfunction of the Donnan equilibrium in any of these examples would be catastrophic for the organism involved.

Donnan equilibrium describes the extent to which small counterions travel through a partition between two ionic environments and this is a useful simple scenario with which to approach the subject. Consider two compartments A and B that are separated by a membrane through which only small molecules such as water

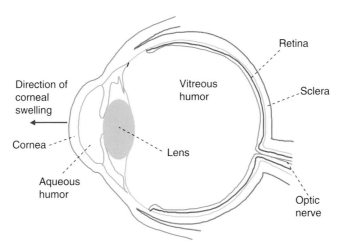

Figure 11.10 *Donnan equilibrium controls the swelling of the human eye and can affect the cornea, the aqueous humor, the lens and the vitreous humor.*

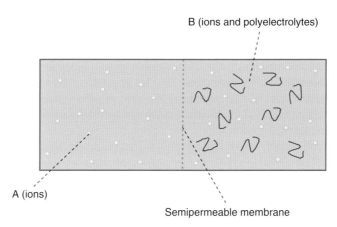

Figure 11.11 *Schematic diagram of the process of Donnan equilibrium for the concentration of ions and polyions in aqueous solution maintained between two compartments (A and B) separated by a semipermeable membrane.*

and ions can pass (**Figure 11.11**). Compartment B contains polyions and salt, whereas compartment A only has the small salt molecules. The chemical potentials of the salt are identical on both sides of the membrane (μ_A and μ_B), since there is no change in the free energy per particle when particles pass through the membrane in thermal equilibrium,

$$\mu_s^A = \mu_s^B \tag{11.55}$$

Electrical neutrality must be maintained in each compartment to avoid an excessive electrostatic penalty. n is the molar concentration of a particular particle species, the salt is monovalent and the polyions carry v charges. The number of charges in each compartment must balance and therefore

$$\text{In compartment A} \quad n_+^A = n_-^A \tag{11.56}$$

$$\text{In compartment B} \quad n_-^B + \phi v n_p = n_+^B \tag{11.57}$$

where n_p is the molar concentration of the polyelectrolyte and ϕ is the fraction of charges that dissociate from the polyelectrolyte. When the system is fully equilibrated the cation concentration is lower in A than in B, whereas the anion concentration is higher in A than in B. This process of Donnan equilibrium is characterised by a Donnan coefficient (Γ) defined as

$$\Gamma = \lim_{n_p \to 0} \frac{\left(n_+^B - n_+^A\right)}{\nu n_p} = \frac{\phi}{2} \tag{11.58}$$

For the case where no counterions are bound to the polyion ($\phi = 1$), the Donnan coefficient (Γ) is 1/2. For DNA in a low ionic strength medium a typical value of the Donnan coefficient (Γ) is -0.1. This implies 80% of the counterions behave as if they are bound to the polyelectrolyte, in agreement with the discussion on counterion condensation in **Section 11.8**. In the intracellular environment there is a complex process of Donnan equilibrium due to the interchange of ions between the mixed cocktail of polyionic species.

11.6 Titration Curves

There is an important difference in the mechanisms through which charges can be placed on a polyion and they are denoted *annealed* and *quenched* mechanisms. When a weakly charged polyelectrolyte is obtained by copolymerisation of neutral and charged monomers the number of charges and their position is fixed. This is called a *quenched polyelectrolyte*.

Polyacids or polybases are polymers in which the monomers can dissociate and acquire a charge that is dependent on the pH of the solution. The dissociation of a hydrogen ion (H^+) ion from an oppositely charged polymer (e.g. COO^- groups) gives a negative charge. This is an *annealed polyelectrolyte*, the total number of charges on a given chain is not fixed, but the chemical potential of the H^+ ion and the chemical potential of the charges is imposed by the pH of the solution.

Acid–base equilibria often apply to biological polyelectrolyte molecules (**Section 1.3**). For example hydrogen ions (protons, H^+) can bind to a basic unit on a polyelectrolyte (A^-) to give an acid AH, as described in equation (11.1). The association constant (K_a) for the acid-base equilibria is defined as

$$K_a = \frac{c_{AH}}{c_{A^-} c_{H^+}} \tag{11.59}$$

where c indicates the concentration of the species in moles. θ is defined to be the fraction of acid monomers (A^-) with a bound proton,

$$\theta = \frac{c_{AH}}{c_{A^-} + c_{AH}} \tag{11.60}$$

This can be rearranged to give

$$\frac{\theta}{1-\theta} = K_a c_{H^+} \tag{11.61}$$

Polyelectrolyte titration is the experimental process in which a well-characterised acid or base is added to a polyion solution to determine its degree of dissociation. The process of titration can be determined through measurement of the conductivity of the solution or by the use of pH-sensitive dyes. A titration curve is shown in **Figure 11.12**. ν is defined as the mean number of bound protons and n is the total number of ions, so the fraction of dissociated ions is $\theta = \nu/n$. Therefore, equation (11.61) can be re-expressed as

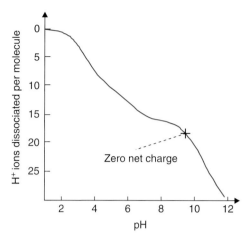

Figure 11.12 *Titration of a polyacid (ribonuclease at 25°C) can be used to calculate the isoelectric point. The concentration of hydrogen ions dissociated per molecule is shown as a function of the solution pH. [Reprinted with permission from C. Tanford, J.D. Hauenstein, J. Am. Chem. Soc, 1956, 77, 5287–5291. Copyright © 1956, American Chemical Society.]*

$$\frac{\nu}{n-\nu} = K_a c_{H^+} \tag{11.62}$$

The degree of dissociation is one minus the degree of association,

$$\alpha = \frac{c_{A^-}}{c_{A^-} + c_{AH}} = 1 - \theta \tag{11.63}$$

The dissociation constant is equal to the inverse of the association constant,

$$K_d = K_a^{-1} \tag{11.64}$$

Equation (11.62) can therefore be rewritten,

$$\frac{(1-\alpha)}{\alpha} = K_d^{-1} c_{H^+} \tag{11.65}$$

It is useful to define two important parameters that relate to the strength of the hydrogen ion concentration (pH) and the degree to which a polyacid can associate (pK_a),

$$pH = -\log c_{H^+} \tag{11.66}$$

$$pK_a = -\log K_a \tag{11.67}$$

The logarithm is introduced to facilitate calculations with concentrations that can vary by many orders of magnitude. The exponential nature of the electrostatic interaction (equation (11.50)) makes solutions sensitive to a vast range of hydrogen ion concentrations from nM to M, and there are a correspondingly vast range of hydrogen ion concentrations and equilibrium constants that are found to be practically important. The logarithm of equation (11.59) gives the Henderson–Hasslebach equation. The equation is an extremely useful relationship, which connects the pH of a solution and the intrinsic pK_a value of an ionisable group to the charge fraction (α) of ionisable groups in solution (**Section 1.3**),

$$pH = pK_a + \log\left(\frac{\alpha}{1-\alpha}\right) \tag{11.68}$$

The Henderson–Hasslebach equation requires the assumption that the values of the association constant (K_a) are independent of the charge on the polyion. This is not a completely reasonable assumption, since the binding of a counterion on a polyion can change the potential experienced by the neighbouring charges on the chain. The change in free energy $(\Delta G_0$ upon the association of the polyions with their counterions) can be related to the association constant (K_a),

$$\Delta G_0 = -RT \ln K_a \tag{11.69}$$

where R is the ideal gas constant. From the definition of the pK_a value, equation (11.67), this gives

$$\Delta G_0 = 2.3RTpK_a \tag{11.70}$$

and the factor of 2.3 occurs due to the change in base from 10 to e, i.e. the log becomes a ln. W_{el} is defined as the work required to bind hydrogen ions onto a polyion and is the work done against the polyion potential provided by all the charged groups. Therefore, an expression for the work of binding is

$$W_{el} = eV(a) \tag{11.71}$$

where $V(a)$ is the potential at the surface of the polyion and e is the electronic charge. The total change in free energy of a polyion with N groups is therefore

$$\Delta G = \Delta G_0 + NW_{el} \tag{11.72}$$

Thus, there is an effective pK_a value (pK_a') measured in an experiment for the association of a polyion with hydrogen ions given by

$$pK_a' = pK_a + 0.43\frac{eV}{k_B T} \tag{11.73}$$

This expression allows the effect of neighbouring charged groups on the pK_a value of a polyion to be quantified to a first approximation.

11.7 Poisson–Boltzmann Theory for Cylindrical Charge Distributions

The Poisson equation for the potential (V) as a function of the charge density is given by equation (11.28). A Boltzmann distribution for the energies of the counterions in a monovalent salt can be substituted in equation (11.28), the Poisson–Boltzmann equation becomes (identical to equation (11.39))

$$\nabla^2 V = 2n_s e \frac{\sinh(eV/kT)}{\varepsilon\varepsilon_0} \tag{11.74}$$

where n_s is the density of salt ions. In a similar manner to that which the equation was solved in spherical and planar geometries in **Section 11.2**, numerical solutions of the PB equation for a line charge with cylindrical symmetry are shown in **Figure 11.13**.

There are three separate regions of charges that surround a polyion that can be classified in terms of their distance from the surface of the polymer. In the *Debye–Huckel* region the ions are treated as point charges and form a double layer around the polymer. The Poisson–Boltzmann equation (11.74) can be linearised for the polyion in a similar spirit to that of a single spherical ion in equation (11.34). However, close to the surface,

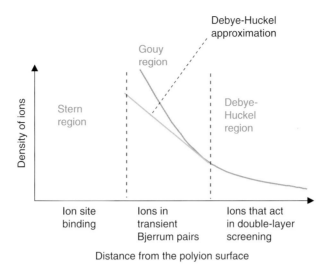

Figure 11.13 *Density of ions as a function of the distance from the surface of a polyion that shows the* Stern *(ion site binding),* Gouy *(ions in transient pairs) and* Debye *(ions that act in double-layer screening) regions.*

in the *Gouy region*, the cylindrical symmetry can still be used, but the PB equation must be solved without any approximation, since the potential is large and this requires analysis with a computer. Numerical solutions show that a condensed phase of counterions occurs near the polyelectrolyte. Even closer to the surface, in the *Stern region*, the cylindrical symmetry of the chain disappears at very close distances to the charged groups along the polyion, and the structural and geometrical factors of the specific chain chemistry have to be considered. The Poisson–Boltzmann equation gives no useful predictions for the electrostatics very close to a charged chain.

11.8 Charge Condensation

In a solution of sufficiently strongly charged polyelectrolytes, a fraction of counterions will stay in the immediate vicinity of the polymer chains, and they effectively neutralise some of the chains' charge. This phenomena is referred to as *counterion condensation* and was encountered previously in the discussion of Donnan equilibrium and in the calculation of the self-energy of a globular protein. Manning devised a simple model for understanding charge condensation with polyelectrolytes and the model is a useful starting point to develop an intuition on the interactions between charged biological polymers. The effective potential that determines the forces between polyions is regulated by the condensation of the counterions that surround the chains and the effective charge fraction is thus often much smaller than expected from the chain chemistry with complete dissociation.

The assumptions required in the Manning model are that the solvent is a continuum with a uniform dielectric constant (ε), the ions are represented by a continuous charge density ($\rho(r)$), and the polyion is modelled by a line charge of infinite length characterised by a charge density parameter (ξ). Using Gauss's theorem for a line charge, the electric field (E) can be calculated around a polyelectrolyte chain in water (compare equation (11.12)),

$$E = \frac{e}{2\pi\varepsilon\varepsilon_0 br} \tag{11.75}$$

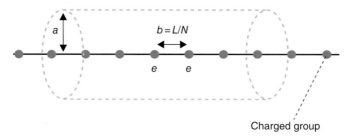

Figure 11.14 *Geometry of a cylindrical polyelectrolyte used in the calculation of Manning charge condensation. b is the distance between two adjacent charged groups (charge e) along the backbone, equal to the length of the chain (L) divided by the total number of charged units (N). a is the radius of a cylinder that surrounds the charged groups along the chain backbone.*

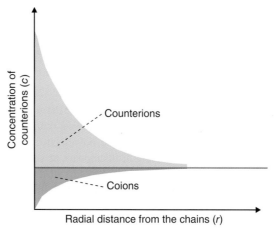

Figure 11.15 *Radial distribution of counterions and coions that surround a polyelectrolyte. The concentration of counterions is shown as a function of the radial distance from the polymer backbone.*

where b is the linear charge density (Coulomb m^{-1}), e is the electronic charge, and r is the equipotential radius for a cylinder around the line of charge (**Figure 11.14**). The electric field (E) is related to the potential (V) by the spatial derivative,

$$E = -\frac{dV}{dr} \tag{11.76}$$

Equation (11.76) can be integrated to provide the potential using equation (11.75),

$$V = A - 2e \ln\left(\frac{r}{4\pi\varepsilon\varepsilon_0 b}\right) \tag{11.77}$$

A monovalent counterion with a unit charge (e) at a certain distance (r) away from the polyion acquires a potential energy $E_p(r)$ (**Figure 11.15**),

$$E_p(r) = eV(r) \tag{11.78}$$

A Boltzmann's distribution for the thermalised energies of the counterions is assumed (equation (3.22)). The Boltzmann equation can be written in terms of the charge parameter (ξ)

$$e^{-E_p/kT} = W_0 r^{-2\xi} \tag{11.79}$$

where $W_0 = e^{eA/kT}$ and the charge parameter is defined as the ratio between the Bjerrum length (l_B) and the linear charge density (b),

$$\xi = \frac{l_B}{b} \tag{11.80}$$

The number of counterions inside a cylinder (radius r_0) of unit length is proportional to the integral of equation (11.79) over the radius of the cylindrical geometry,

$$\int_0^{r_0} W_0 r^{-2\xi} 2\pi r dr = 2\pi W_0 \int_0^{r_0} r^{1-2\xi} dr \tag{11.81}$$

This integral diverges at the origin ($r = 0$) if the charge density parameter is greater than one ($\xi > 1$), which is an unphysical result. A condensed layer of counterions is invoked to avoid this problem, which maintains the charge parameter at a value of one and a finite total energy for the system. The fraction of charge neutralised is simply expressed in terms of the charge density parameter,

$$\frac{(\xi-1)}{\xi} = 1 - \frac{1}{\xi} \tag{11.82}$$

This analysis can be extended to the case of multivalent counterions. For an ion of charge ze the neutralised charge fraction is

$$1 - \frac{1}{z\xi} \tag{11.83}$$

The condensation process is explained physically in terms of the formation of ion pairs (Bjerrum) on the surface of the polymer. **Figure 11.16** illustrates the effect of Manning condensation above a critical charge fraction on a polyelectrolyte chain.

For example, consider a double-stranded chain of DNA at 20 °C. The distance between phosphate groups (b is 1.7 Å) is taken for the B form of DNA and hence the charge parameter (ξ) is 4.2. The fraction of charge neutralised is given by equation (11.83) and on average three quarters of the phosphate groups are neutralised.

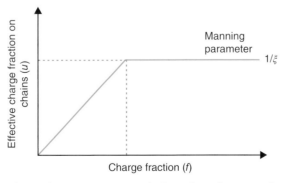

Figure 11.16 *The effective charge fraction (u) on a polyelectrolyte chain as a function of the charge fraction expected from the number of chargeable groups (f). At high charge fractions the effective charge fraction saturates at the Manning value (1/ξ).*

Charge condensation thus clearly has a strong effect on the electrostatic forces experienced by a DNA chain. This model's prediction is in reasonable agreement with experiment.

The *Osmotic coefficient* (ϕ, equation (11.53)) can be measured experimentally by osmometry. The osmotic pressure (π_{exp}) found using osmometry on a polyelectrolyte solution in distilled water is proportional to the concentration of counterions (c_e),

$$\pi_{exp} = kT\phi c_e \tag{11.84}$$

and the osmotic coefficient can be identified with the fraction of neutralised charge,

$$\phi = 1 - \frac{\xi}{2} \tag{11.85}$$

The Manning theory is in reasonable agreement with osmometry experiments. Manning condensation can also be used to explain *association phenomena* in highly charged biological molecules, e.g. drug binding to DNA.

In summary, the continuum Manning calculation predicts there is an excess of counterions in the vicinity of the polyion and the charge variable (ξ) is important for the description of polyelectrolytes. However, there are two limits on the validity of this type of Manning model. The polyelectrolytes are required to be of infinite length and there is assumed to be a zero free-ion concentration. A more satisfactory approach is to solve the full Poisson–Boltzmann (PB) equation for the cylindrical geometry (equation (11.74)) and use this to determine the degree of counterion association (**Figure 11.17**). The PB equation is valid at physiological salt concentrations (0.15 M), which is not true for the Manning theory. The Manning model is, however, conceptually simpler than the PB equation and provides a useful starting point to learn about charge condensation.

Charge condensation is also found with spherical colloids, as discussed in **Section 11.1**, and other intermediate geometries (ellipsoids, polygons, etc.). Solution of the Poisson–Boltzmann equation in dilute solutions with spherical symmetry for spherical colloids (equation (11.34)) can provide an effective charge to renormalise the

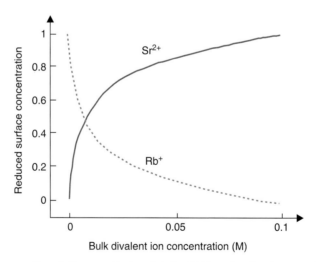

Figure 11.17 *Anomalous small-angle X-ray scattering data for DNA molecules that associates with either rubidium (Rb⁺) or strontium (Sr²⁺) counterions. Rubidium ions are displaced from the surface of the DNA as the bulk divalent counterion concentration is increased in accord with Poisson–Boltzmann theory. [Reprinted with permission from K. Andresen, R. Das, H.Y. Park, H. Smith, L.W. Kwok, J.S. Lamb, E.J. Kirkland, D. Herschlag, K.D. Finkelstein, L. Pollack, PRL, 2004, 93, 248103. Copyright 2004 by the American Physical Society.]*

interaction strength in a DLVO treatment (equation (4.19)) for the interparticle potential. Renormalisation is required for highly charged colloids (large Q) with small radii of curvature (R), and they follow an inequality for monovalent counterions,

$$\frac{Q_{geom}l_B}{4R} \geq 1 \tag{11.86}$$

where l_B is the Bjerrum length. For these highly charged and curved colloids the charge is pinned at a value (the effective colloid charge, Q_{eff}) given by

$$Q_{eff} = \frac{4R}{l_B} \tag{11.87}$$

11.9 Other Polyelectrolyte Phenomena

For weakly charged polyelectrolytes pronounced counterion condensation only occurs in poor solvents (where the blobs are globular) and in this case it constitutes an avalanche-like process that results in the condensation of nearly all the counterions on to the macromolecule. This is thought to be an important contribution to the process of folding of charged globular proteins.

The Coulombic interactions of a strongly charged polyelectrolyte tend to stiffen the chain and lead to an increase in its persistence length (l_p). The contribution to the total persistence length due to the electrostatics (l_e) is called the *electrostatic persistence length*. A useful theory that predicts the persistence length of charged semiflexible chains is due to Odjik, Skolnick and Fixman. The OSF theory is applicable to semiflexible biopolymer chains such as actin and DNA. The electrostatic component (l_e) is simply added on to the intrinsic rigidity due to the backbone chemistry,

$$l_T = l_P + \left(4l_B\kappa^2\right)^{-1} = l_p + l_e \tag{11.88}$$

where l_B is the Bjerrum length, κ^{-1} is the Debye screening length, and l_p is the intrinsic persistence length. A similar behaviour is expected for flexible polyelectrolytes whose conformation consists of a rigid cylinder of charged blobs, but there continues to be some confusion in the literature as to how the blob size renormalises the effective length of the charged chain (**Figure 11.18**). Both κ^{-1} and κ^{-2} dependences of l_e are predicted theoretically for flexible polyelectrolytes. Some caution is therefore required when applying equation (11.88) for chains with flexible backbones.

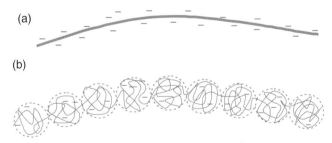

Figure 11.18 *The OSF approach for the calculation of the electrostatic contribution to the persistence length is possible in both (a) semiflexible and (b) flexible polyelectrolytes, although there continues to be some debate about the exact calculation involved in case (b).*

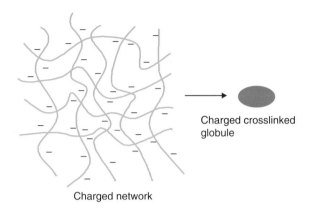

Figure 11.19   *A charged crosslinked polymeric network contracts into a crosslinked globule with a first order transition as the quality of the solvent for the polymer chain deteriorates. The positively charged counterions are not shown for simplicity.*

When a small fraction of the links of a polymer network are charged, the collapse of the network in a poor solvent follows (as the solvent quality deteriorates) a discrete first-order phase transition. The abrupt change in the size of the network is associated with a large reduction in the osmotic pressure of the gas of counterions in the charged network (**Figure 11.19**). This process of collapse is analogous to the globule–coil transition of a single chain (**Section 5.3**) and has been used to produce biomimetic muscular contraction of polyelectrolyte gels, since the phase transition is reversible.

The compatibility of a mixture of two polymers improves substantially after one of the components is weakly charged. Many of the extreme liquid-phase separation phenomena observed in **Section 5.5**, such as polymer/polymer phase separation, can thus be avoided in the intracellular environment using charge effects.

An area of hot debate over many years is that of an attractive force between like-charged polyelectrolytes. The current consensus is that an attractive force does exist due to counterion sharing effects between the polyions, but an accurate conclusive quantitative model for the associated experimental phenomena is still elusive.

Suggested Reading

If you can only read one book, then try:

Top recommended article: Dobrynin, A.V. & Rubinstein, M. (2005) Theory of polyelectrolytes in solutions and at surfaces, *Progress in Polymer Science*, 30, 1049–1118. Good introduction to the physics of flexible polyelectrolytes.

Benedek, G.B. & Villars, F. (2000) *Physics with Illustrative Examples from Medicine and Biology*, Springer. Well-written expansive account of biophysical subjects.
Daune, M. (1999) *Molecular Biophysics*, Oxford University Press. Useful introduction to ions and polions.
Grosberg, A.Y. & Khoklov, A.R. (1994) *Statistical Physics of Macromolecules*, AIP Press. Clear compact account of the properties of polyelectrolytes.
Phillips, R., Kondev, J., Theriot, J. & Garcia, H.G. (2013) *Physical Biology of the Cell*, 2nd edition, Garland. Contains a chapter with a clear discussion of charged polymers.

Tutorial Questions 11

11.1 The dissociation pK value for lysine is 10 and the pK_a value for polylysine is 9.5. Qualitatively explain the shift in value.

11.2 Calculate the effective charge fraction on a polymer of hyaluronic acid according to the Manning model, given that the distance between the acids groups along the polymer backbone is 5 Å. Each charged group is assumed to have a single unscreened electronic charge before condensation.

11.3 An amyloid fibre associated with a pathogenic misfolded protein has a charge fraction (f) of 0.5, and a repeat unit of 1 nm. Estimate using the Odjik, Skolnick, and Fixman model the electrostatic contribution to the persistence length of the fibre in a buffer solution with a Debye screening length (κ^{-1}) of 4 nm.

11.4 A very big industrial application for polyelectrolytes is in the gels that fill disposable diapers. Describe some physical properties of polyelectrolytes that makes them especially suitable for this critical technology. The osmotic pressure of a charged polymer is dominated by counterions. Compare the osmotic pressure of a charged PAMPS (poly acryl amide methyl propane sulfonic acid) gel with that of its neutral counterpart at a polymer monomer concentration of 1 mM. State any assumptions used.

11.5 Explain how the size of the blobs in a uncharged polylysine chain (a polypeptide) varies if the chain is charged by a change in pH. Describe what happens to the conformation of the chain during the process of charging.

12

Membranes

Every living cell is surrounded by an outer membrane. The membrane acts as a dividing partition between the cell's interior and exterior environments. It is the interface through which a cell communicates with the external world. Biological membranes are involved in a wide range of cellular activities. Simple mechanical functions occur using the membrane such as motility, food entrapment and transport. Also, highly specific biochemical processes are made possible by the membrane's structure, which include energy transduction, nerve conduction and biosynthesis (**Figure 12.1**). Adhesion between cellular membranes is thought to be a critical factor for the determination of the morphology and the development of organisms (morphogenesis) from the initial ball of dividing cells (the blastula, **Section 22.9**).

Biological lipids in solution self-assemble into thin bilayer membranes that can compartmentalise different regions within a cell and protect the inside of the cell from the external environment. The membrane remains intact even when the bathing medium is extremely depleted of lipids due to the lipids' extremely low critical micelle concentration (**Section 8.1**). Also, due to unsaturation and branching of the constituent lipids, the membranes are in a fluid state at physiological temperature, with rapid two-dimensional rearrangements possible of the neighbouring lipid molecules.

Long-chained polypeptides are often embedded in membranes and consist of long strings of amino acid residues (~500 000). The polypeptides are rigid compared with the lipids in the surrounding cell membrane and they are amphiphilic, with their surface exposed to both hydrophobic and hydrophilic regions. These membrane proteins induce stress in the surrounding lipids, change the membrane's curvature, and determine a range of functions that include adhesion and signalling.

The basic structure of all cells is the same (**Chapter 2**). Fluid sheets surround the cell and its internal compartments, whilst semiflexible filaments form rigid internal scaffolding within the cell, and contribute to its mechanical integrity. Deformations of the fluid sheets at the cell boundary are due to a number of processes; compositional inhomogeneity of the bilayers (phase separation of membrane components), anisotropic structuration of the cell walls under lateral stress and pressure from internal structural elements, such as microtubules, can occur. Membranes are also important in many intracellular organelles, such as the endoplasmic reticulum, the golgi apparatus, the nucleus and vesicles.

The Physics of Living Processes: A Mesoscopic Approach, First Edition. Thomas Andrew Waigh.
© 2014 John Wiley & Sons, Ltd. Published 2014 by John Wiley & Sons, Ltd.

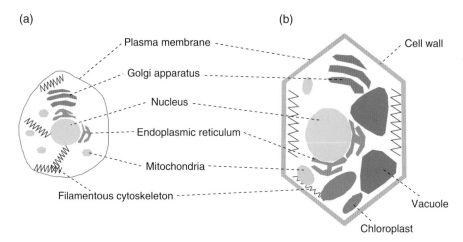

Figure 12.1 *Schematic diagrams of (a) an animal and (b) a plant cell that show the compartmentalised structures and some individual pieces of intracellular machinery in two different types of eukaryotic cell. [Copyright 2008 from Molecular Biology of the Cell, Fifth Edition by Alberts et al. Reproduced by permission of Garland Science/Taylor & Francis LLC.]*

Much of the soft-matter research on membrane surfactants has been driven by the soap industry. Soap molecules are amphiphilic, similar to the phospholipids in membranes, and the role of soap in the disintegration of cell membrane structure is crucial for part of their cleaning action, e.g. their antibacterial activity. There are many well-developed membrane models that trace their roots back to studies of soap.

Membranes live in flat land, i.e. in two dimensions. Synthetic examples of membranes range from graphene, which has an extremely large solid-like shear modulus, to surfactant films that are much more fluid-like. Polymers can be modelled as one-dimensional objects embedded in three dimensional space, whereas membranes require models for two-dimensional objects embedded in three-dimensional space.

Both eukaryotic and prokaryotic cells have a large range of specialised membrane structures. For E.coli bacteria (prokaryotes) the cell membrane consists of two fluid-like bilayers separated by a rigid cell wall.

Lots of motor proteins are associated with membranes. Furthermore, membrane proteins can act as specialist enzymes, transporters/channels and receptors. There is a rich variety of membrane structures that achieve such tasks. They can be integral membrane proteins that provide a bridge across the bilayer structure or peripheral proteins that associate onto only one of the sides of the bilayer using a molecular anchor or nonspecific electrostatic interactions.

Historically, the first major paradigm for membrane structure studies was the *fluid mosaic model*. In this model the bilayer structure is seen as stable and long lived, whereas within the bilayer the lipids and proteins are fluid-like, and constantly change positions due to the available thermal energy and weak cohesive interactions. This model has subsequently been refined to include the *lipid raft hypothesis*, in which different regions of the membrane are enriched in different components, with internal membrane phase separation a possible physical mechanism for their creation (**Section 5.5**). The rafts are long-lived structures that have different thicknesses and bending rigidities compared with other sections of the membrane.

12.1 Undulations

Membranes are two-dimensional objects and the fluctuations of their shape (undulations) are specific to their dimensionality and are of primary importance to their physical properties. There is an important difference in

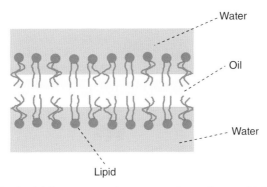

Figure 12.2 *The structure of a fluid bilayer at an oil/water interface. The amphiphilic molecules are confined to the interface.*

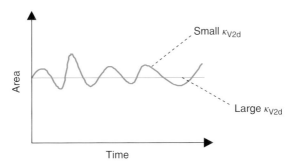

Figure 12.3 *Schematic diagram that shows the fluctuations in area as a function of time for a membrane. A large areal compressibility (κ_{V2d}) suppresses volume fluctuations and vice versa.*

the undulations of fluid bilayers, in which there is no shear resistance (**Figure 12.2**) and those that can sustain an inplane shear stress due to ionic or covalent bonds between neighbouring atoms or molecules, e.g. a fluid phospholipid bilayer versus a rigid sheet of graphene (a two-dimensional sheet of carbon). Membranes with and without substantial shear resistance can occur naturally in living systems, although membranes with shear resistance are very common due to the association of bilayers with cytoskeletal proteins.

From standard continuum mechanics the compression modulus (κ_V) in three dimensions is defined to be the rate of change in volume (V) with pressure (P) (**Section 13.1**) via

$$\kappa_V^{-1} = -\frac{1}{V}\left(\frac{\partial V}{\partial P}\right)_T \tag{12.1}$$

where the scaling factor $1/V$ makes the expression independent of specimen size and $\partial V/\partial P$ is calculated at constant temperature. For membranes an analogous two-dimensional compressibility (κ_{V2D}) can be related to the fluctuations in the area of the membrane (ΔA),

$$\frac{kT}{K_{V2D}} = \frac{\left\langle(\Delta A)^2\right\rangle}{A_0} \tag{12.2}$$

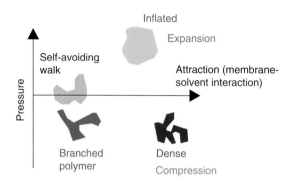

Figure 12.4 *Phase diagram for membranes that shows the pressure as a function of the strength of the attractive membrane-solvent interaction. [Reprinted with permission from D.H. Boal, Phys. Rev. A, 1991, 43, 6771–6777. Copyright 1991 by the American Physical Society.]*

Table 12.1 *Scaling behaviour of closed membrane bags in three dimensions. R_g is the radius of gyration, l_c is the contour length and A is the area.*

Configuration	Scaling Law	
Three dimensions	$\left\langle R_g^2 \right\rangle \propto l_c^{2v}$	$\langle A \rangle \propto l_c^{2\eta}$
Inflated (good solvent)	$v = 1$	$\eta = 3/2$
Flory type (theta solvent)	$v = 4/5$	
Branched polymer	$v = 1$	$\eta = 1$
Dense (poor solvent)	$v = 2/3$	$\eta = 1$

where A_0 is the area at zero temperature, k is Boltzmann's constant and T is the temperature. A membrane with a large areal compressibility (K_{v2D}) only experiences small fluctuations in its area at fixed pressure and *vice versa* (**Figure 12.3**).

The mechanism through which undulations affect the size of membranes is still an open area of research (**Figure 12.4**). In a similar way to which the size of a polymer is dependent on the quality of the solvent (**Section 10.2**), the average size of a membrane is related to the interplay between solvent/membrane interactions and the excluded volume of the two dimensional surface. The scaling behaviour of the radius of gyration and the area of a range of model closed bags is shown in **Table 12.1**. A visualisation of the different scenarios for membrane/solvent interaction is shown in **Figure 12.4**.

12.2 Bending Resistance

At zero temperature a membrane minimises its bending energy and adopts a shape that is flat or uniformly curved. At finite temperature, the spatial decorrelation of the normals to the membrane surface provides a measure of the fluctuations of the membrane's structure. This spatial decorrelation of the normals is characterised by a persistence length ξ_p (**Figure 12.5**); the two-dimensional analogue of that found for a semiflexible

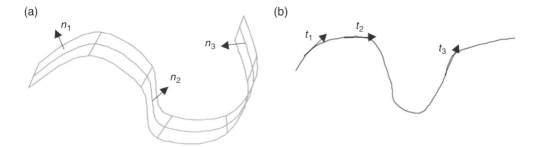

Figure 12.5   *(a) The correlation of the normals ($\underline{n}_i$) to a membrane provides a definition for the persistence length of a membrane and (b) the correlation of tangent vectors ($\underline{t}_i$) to a polymer is related to the persistence length of a polymer.*

polymer (l_p, **Section 10.1**). The normal to a membrane surface at a point in space ($\underline{r}$) is defined as $\underline{n}(\underline{r})$ and the correlation of the normals at an average separation ($\underline{\Delta r}$) is found to decay exponentially,

$$\left\langle \underline{n}\left(\underline{r}_1\right) . \underline{n}\left(\underline{r}_2\right) \right\rangle = e^{-|\underline{\Delta r}|/\xi_p} \tag{12.3}$$

$$\underline{\Delta r} \equiv \underline{r}_1 - \underline{r}_2 \tag{12.4}$$

where r_1 and r_2 are two points on the membrane. In contrast to the case of a polymer where the persistence length (l_p) was found to be proportional to the bending modulus divided by the thermal energy ($l_p \sim \kappa/kT$, **Section 10.1**), the persistence length of a membrane (ξ_p) depends exponentially on the bending modulus (κ_b) divided by the thermal energy,

$$\xi_p \sim b e^{2\pi\kappa_b/kT} \tag{12.5}$$

where b is a characteristic step length (with units of metres) along the membrane. The membrane persistence length is thus a much more sensitive function of the modulus and temperature than in the case of a polymer. A flat zero-temperature membrane obeys $R_g^2 \sim L^2$; the area (R_g^2) of a membrane in three dimensions is approximately equal to its contour area (L^2, where L is the contour length of its sides). When the membrane is subject to thermal fluctuations, the size of the membrane without self-avoidance grows very slowly with contour area, $R_g^2 \sim \ln L$, which can be shown by simulation and analytically using Fourier decomposition of the surface profile.

The scattering of X-rays and light from a membrane can be used as a sensitive measure of their undulations (**Figure 12.6**). Consider an X-ray scattering experiment with a stack of membranes. Typically, elastic scattering experiments measure the structure factor ($S(q)$) as a function of momentum transfer (q, the inverse length scale, **Section 19.10**),

$$S\left(\underline{q}\right) = N^{-2} \left\langle \sum_{m,n} e^{i\underline{q} . \left(\underline{r}_m - \underline{r}_n\right)} \right\rangle \tag{12.6}$$

where N is the number of membranes in a stack and the summation is carried out over all the separate molecules in the stack with coordinates $\underline{r}_m$ and $\underline{r}_n$. For a frozen zero-temperature stack of membranes, an experiment would measure delta function Bragg peaks corresponding to the lamellar periodicity. However, there are

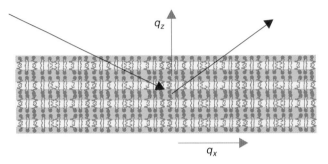

Figure 12.6 *Schematic diagram of the scattering geometry of radiation from a stack of membranes.* q_z *is the momentum transfer perpendicular to the membrane and* q_x *is that lateral to the membrane.*

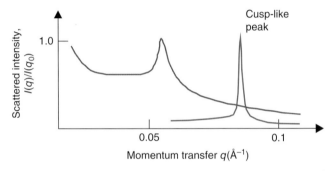

Figure 12.7 *Cusp-like peaks are found in X-ray experiments due to the constructive interference of X-rays scattered from stacks of lipid membranes. The scattered intensity is shown as a function of the perpendicular momentum transfer of the X-rays* (q_z).

no true Bragg peaks at finite temperatures in such scattering experiments due to the undulations of the membrane stack, and the scattering profiles exhibit power law singularities (**Figure 12.7**). In the direction perpendicular to the surface (z) the scattered intensity ($I(0,0,q_z)$) of some radiation on a detector (e.g. X-ray) is given by

$$I(0,0,q_z) \propto (q_z - q_m)^{-2 + \eta_m} \tag{12.7}$$

where q_m is the value of the momentum transfer centred on the mth-order peak ($q_m = 2\pi/md$, d is the membrane spacing). Parallel to the surface the scattered intensity has the form

$$I(q_{||}, 0, q_m) \propto q_{||}^{-4 + 2\eta_m} \tag{12.8}$$

The exponent η_m for equations (12.7) and (12.8) is related to the elastic modulus of the membrane,

$$\eta_m = \frac{m^2 \eta_1^2 kT}{8\pi\sqrt{BK}} \tag{12.9}$$

where $K = K_c/d$, d is the interlamellar spacing, K_c is the elasticity modulus, η_1 has different forms that depend on the dominant force of interaction between the layers (e.g. undulation forces or electrostatics), and kT is the thermal energy. It is also possible to examine single membranes using the reflectivity geometry with the total internal reflection of X-rays, neutrons and light radiation. Calculation of the structure-factor equations (12.7)

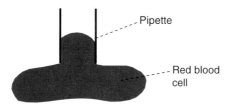

Figure 12.8   *Micropipette apparatus can be used to examine the elasticity of red blood cells using digital analysis of microscopy images of the resultant membrane profile and knowledge of the applied pressure.*

and (12.8) are slightly more complicated in this case, but quantitative measurements of membrane fluctuations can be made from the profile of the reflected radiation. Image analysis in light-microscopy experiments of the curvature of membranes aspirated with a micropipette also provides a method to calculate the membrane bending elasticity (**Figure 12.8**).

12.3   Elasticity

The Poisson ratio (σ_p) is a measure of how a material contracts in the transverse direction when it is stretched longitudinally (**Section 13.1**). It is defined in two dimensions as the ratio of the strains parallel (u_{xx}) and perpendicular (u_{yy}) to the direction of extension,

$$\sigma_\text{p} = -\frac{u_{yy}}{u_{xx}} \tag{12.10}$$

where the stress is applied along the x-axis. The negative sign is included so 'normal materials' have positive values for the Poisson ratio. However, unusually, crumpled membranes have negative Poisson ratios. Another unusual example of negative Poisson ratios will be found in the **Section 13.3** with foams in three dimensions, which have openly connected membraneous structures, e.g. corks in wine bottles.

The shear modulus (μ) for a two-dimensional elastic fibrous network is given by the expression

$$\mu \approx \rho k T \tag{12.11}$$

where ρ is the density of crosslinks and kT is the thermal energy. This is a very similar result to the three dimensional case of crosslinked polymers (equation (10.77)) and has a similar derivation. The exact prefactor in equation (12.11) depends on the coordination number of the network, i.e. the number of chains that meet at a crosslink.

There is a useful expression that relates the stress and internal pressure in an elastic spherical membrane to the curvature (**Figure 12.9**). The resultant tensile force (F_t) due to the wall stress is equal to the average area of the wall times the average stress ($<\sigma>$),

$$F_\text{t} = \pi \left(r_0^2 - r_i^2 \right) \langle \sigma \rangle \tag{12.12}$$

where r_0 is the external radius, and r_i is the internal radius of the membrane. In equilibrium, the stresses in the wall of the membrane are balanced by the internal pressure (p_i),

$$\pi \left(r_0^2 - r_i^2 \right) \langle \sigma \rangle = \pi r_i^2 p_i \tag{12.13}$$

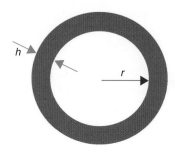

Figure 12.9 *Cross section through a spherical membrane of thickness (h) and internal radius (r) used to calculate the shear stress in the wall.*

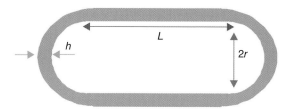

Figure 12.10 *Cross section through a cylindrical membrane of length (L), thickness (h) and end cap radius of curvature (r) used to calculate the shear stress in the wall.*

Equation (12.13) can be rearranged and the shell thickness (h) is defined as r_0-r_i, which gives

$$\langle \sigma \rangle = p_i \frac{r_i^2}{r_0^2 - r_i^2} = p_i \frac{r_i^2}{h(r_0 + r_i)} \tag{12.14}$$

If the elastic membrane on the sphere is thin, the internal radius is approximately equal to the external radius ($r_0 \approx r_i = r$) and equation (12.14) can be further simplified,

$$\langle \sigma \rangle = \frac{\text{force}}{\text{area}} = \frac{rp}{2h} \tag{12.15}$$

where p is the interior pressure, h is the shell thickness and r is the radius of the sphere. This explains why balloons are initially difficult to expand, since the applied shear stress ($\langle \sigma \rangle$) is small for small balloon radii (r).

The stress in the cell wall of a *cylindrical membrane* (e.g. a cylindrical bacterium) is a little different from that of a sphere and has important medical applications (**Figure 12.10**), e.g. the mechanical properties of a range of tubular organs, such as the intestines. For a cylinder, $<\sigma_\theta> (r_0 - r_i)L$ is the resultant force on the cross section, balanced by the pressure that acts on the inside of the cylinder $2r_iLp_i$,

$$2\langle \sigma_\theta \rangle (r_0 - r_i)L = 2r_iLp_i \tag{12.16}$$

where $<\sigma_\theta>$ is the average value of the angular varying stress (σ_θ, cylindrical coordinates) over the cross section and L is the length of the cylinder. Therefore, with $h = r_0 - r_i$ the stress in the hoop direction is

$$\langle \sigma_\theta \rangle = \frac{rP}{h} \tag{12.17}$$

The stress in the axial direction (σ_z) is similar to the case of a sphere,

$$\langle \sigma_z \rangle = \frac{rp}{2h} \tag{12.18}$$

where p is the pressure and r is the cylinder radius. The largest hoop stress ($\langle \sigma_\theta \rangle$) occurs along the line of the shell. This explains why sausages in frying pans and internal organs often burst in the direction of the hoop, since the hoop stress is two times the axial stress.

More generally, it can be shown for two principal curvatures ($1/R_1$ and $1/R_2$), the pressure (P) and the two line tensions τ_1 and τ_2 are related by

$$\frac{\tau_1}{R_1} + \frac{\tau_2}{R_2} = p \tag{12.19}$$

For fluid sheets the tension is isotropic and equation (12.19) becomes the Young–Laplace equation that was encountered earlier (**Section 9.4**),

$$\tau \left(\frac{1}{R_1} + \frac{1}{R_2} \right) = p \tag{12.20}$$

Mathematically, a unit tangent vector ($\underline{t}$) to the membrane can be defined,

$$\underline{t} = \frac{\partial \underline{r}}{\partial s} \tag{12.21}$$

where $\underline{r}$ is the position on the membrane and s is a distance along the membrane surface. And the curvature (c) is given by

$$c = \underline{n} \cdot \frac{\partial \underline{t}}{\partial s} = \underline{n} \cdot \left(\frac{\partial^2 \underline{r}}{\partial s^2} \right) \tag{12.22}$$

where $\underline{n}$ is the normal to the membrane curvature at position $\underline{r}$.

The simplest model for the free energy density (F) of a membrane is

$$F = \frac{\kappa_b}{2} (c_1 + c_2 - c_0)^2 + \kappa_G c_1 c_2 \tag{12.23}$$

where c_0 represents the spontaneous curvature, c_1 and c_2 are quadratic terms in the Taylor expansion of the surface around a point (principal curvatures), κ_b is the bending modulus introduced earlier, and κ_G is the saddle splay modulus. For a stable film κ_b in the free energy equation (12.23) must be positive. The product $c_1 c_2$ is called the *Gaussian curvature* and has a large impact on the morphology that a membrane adopts (**Figure 12.11**).

The integral of the Gaussian curvature over a closed surface is found to be an invariant,

$$\oint c_1 c_2 dA = 4\pi (n_c - n_h) \tag{12.24}$$

where ($n_c - n_h$) is the difference between the number of connected components and the number of saddles, and $\oint$ is an integral over the complete surface. For a bilayer membrane with no spontaneous curvature and topology conserving fluctuations, the bending energy is simpler than equation (12.23),

$$E_{el} = \oint \left[\frac{\kappa_b}{2} (c_1 + c_2)^2 \right] dA \tag{12.25}$$

where κ_b is the bending modulus.

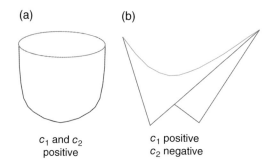

Figure 12.11 *Schematic diagram of surfaces with curvature (C_1 and C_2) of (a) the same and (b) opposite signs.*

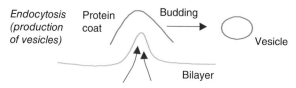

Figure 12.12 *The formation of a vesicle used for transportation from a membrane bilayer occurs through a process called endocytosis. For example it could be used to signal to another cell.*

Biologically, vesicles are used for the transportation of biochemical products from their point of manufacture to the sites of usage; a postal service for the cell. The products must be packaged to prevent their loss during transport and the process involves small membrane-bound entities (the vesicles). The packages are labelled so they can be recognised at their point of destination. The packages need to be shipped along an efficient transportation route and they often use motor proteins (**Figure 12.12**). Equation (12.25) for the bending energy of a membrane can be used to calculate the membrane size of simple spherical vesicles (they are normally less complicated in their construction than a complete cell) and to understand their mechanism of formation.

For real biological membranes often more detailed models of membrane elasticity are required to explain the observed phenomena. Thus, in addition to the process of bending considered in equation (12.25), stretches (increases in membrane area), thickness changes, and shear deformations also need to be considered (**Figure 12.13**). The free-energy penalty associated with changing the area (a) of a lipid bilayer is

$$G_{\text{stretch}} = \frac{K_a}{2} \int \left(\frac{\Delta a}{a_0} \right)^2 da \tag{12.26}$$

where K_a is the area stretch modulus, a_0 is the initial area of the membrane, and Δa is the change in area of the membrane. If the areal strain is assumed to be uniform, the equation can be simplified,

$$G_{\text{stretch}} = \frac{K_a}{2} \frac{\Delta a^2}{a_0} \tag{12.27}$$

There is also a free energy penalty for a change in the thickness of a lipid bilayer,

$$G_{\text{thickness}}[w(x,y)] = \frac{K_t}{2} \int \left(\frac{w(x,y) - w_0}{w_0} \right)^2 da \tag{12.28}$$

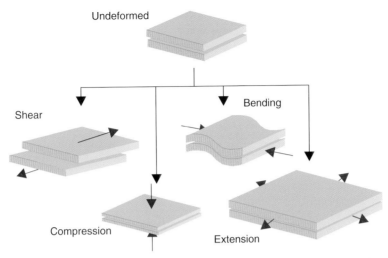

Figure 12.13 *Deformation mechanisms on an elastic membrane include shear, compression, bending and extension.*

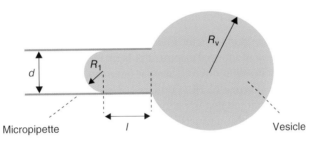

Figure 12.14 *Geometry of a micropipette aspiration experiment to measure the mechanical properties of a vesicle. The micropipette is of diameter* d, *the entrance length of the vesicle is* l *and the radii of curvature of the vesicle are* R_1 *and* R_v.

where $w(x,y)$ is the thickness of the membrane at point (x,y), w_o is the original membrane thickness, and K_t is the membrane stiffness to a perpendicular compression or extension.

The micropipette aspiration technique can be used to measure the mechanical properties of a vesicle's membrane (**Figure 12.14**). The pressure difference between the interior pressure and the surrounding material (Δp_{out}) is given by the Laplace–Young equation,

$$\Delta p_{out} = \frac{2\tau}{R_v} \qquad (12.29)$$

where τ is the surface tension of the vesicle's membrane, and R_v is the radius of the vesicle. The pressure difference between the inside of the vesicle and the inside of the micropipette is controlled experimentally,

$$\Delta p_{in} = \frac{2\tau}{R_1} \qquad (12.30)$$

The difference between equations (12.29) and (12.30) can be taken to give $\Delta p_{in} - \Delta p_{out} = \Delta p$ and therefore

$$\tau = \frac{\Delta p}{2} \frac{R_1}{1 - (R_1/R_v)} \qquad (12.31)$$

Since the geometry and tension are well defined, the mechanical response of the membrane can be calculated. To calculate the area stretch modulus (K_a) the equation

$$\tau = K_a \frac{\Delta a}{a_0}$$

(12.32)

can be used. From optical images the area changes of the capped cylindrical membrane that is attached to the micropipette can be calculated using

$$\Delta a = 2\pi R_1 l + 2\pi R_1^2$$

(12.33)

The reference area is a spherical vesicle and therefore

$$\frac{\Delta a}{a_0} = \frac{R_1^2 (1 + l/R_1)}{2R_v^2}$$

(12.34)

where $a_0 = 4\pi R_v^2$. Then using equation (12.31) and (12.32) with knowledge of the pressure and information from image analysis, the area stretch modulus (K_a) can be calculated.

12.4 Intermembrane Forces

A range of mesoscopic forces are important in the interactions of membranes. At small length scales the membranes attract due to the interaction of induced electric dipole moments (the van der Waals force). The balance between van der Waals and electrostatic forces occurs at intermediate length scales, which often defines an average intermembrane distance (**Figure 12.15**). Furthermore, there is often a furry coat of polymers attached to the exterior of the plasma membrane that impedes cell adhesion (**Figure 12.16**) due to the induced steric repulsive force (**Section 4.4**). Water also interacts with the structure of membranes and can give rise to a long-range hydrophobic interaction which sensitively depends on the ionic environment. Noncovalent binding is possible between specific molecules attached to membranes at very close distances of approach and this promotes adhesion. There are also a range of proteins that are designed to connect cells together with ionic bonds, and they are a major factor that determines the cohesion of cells into tissues.

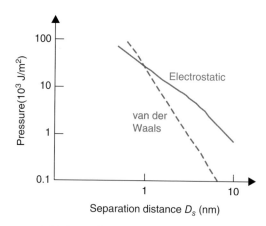

Figure 12.15 *Pressure between two rigid charged plates as a function of their degree of separation. The equilibrium distance is at around 1 nm. [Reproduced from D. Boal, Mechanics of the Cell, 2012. With permission from Cambridge University Press.]*

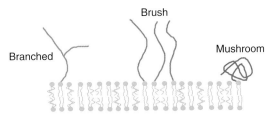

Figure 12.16   *The conformations of the polysaccharide component of the glycoproteins in cell membranes can demonstrate branched, parallel brush and mushroom morphologies.*

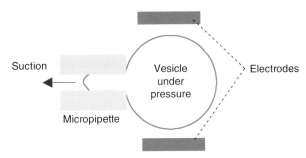

Figure 12.17   *The electroporation technique for the study of the surface tension of cells. Holes are formed by the action of an electric field and their size is measured with a microscope.*

The measurement of line tensions in membranes is possible by the creation of holes in membranes (**Figure 12.17**) and observation of the critical hole radius. At zero temperature a membraneous system in which a hole has been introduced (energy U) acts to minimise its enthalpy (H, there is a negligible entropic contribution in this problem),

$$H = U - \tau A \tag{12.35}$$

where τ is the tension of the membrane in two dimensions, and A is the area. The energy of a circular hole (U) formed in a bilayer is given by

$$U = 2\pi R \lambda \tag{12.36}$$

where λ is the line tension and R is the hole radius. The difference in area between the sheet and the hole system with respect to the intact sheet is πR^2. From equation (12.35) the change in enthalpy (ΔH) on the production of a hole is therefore given by

$$\Delta H = 2\pi R \lambda - \tau \pi R^2 \tag{12.37}$$

The extrema of this expression can be calculated and the maximum value of the enthalpy (ΔH) occurs at a critical hole radius (R^*) of

$$R^* = \frac{\tau}{\lambda} \tag{12.38}$$

and the line tension can be calculated experimentally as

$$\lambda = R^* / \tau \tag{12.39}$$

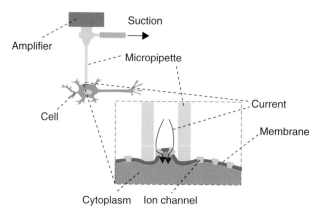

Figure 12.18 *The* patch clamp *method can be used to examine the electrical properties of single membrane proteins. A pipette is attached to a cell and the electrical activity from an ion channel is detected with a differential amplifier.*

Proteins embedded in a cell membrane can provide it with electrochemical activity. Patch clamps are used to measure the electrical properties of membranes provided by such proteins. A very narrow clean capillary pipette is taken and a small degree of suction is applied to a membrane. The tiny electrical potentials formed by cells can be measured with respect to a reference voltage (**Figure 12.18**, also **Section 23.3**). A very high degree of amplification is needed to measure the voltage produced by a single membrane protein and the trick is to separate the current due to the cellular events from the background noise. Such electrical data provides information on the conduction of membranes and has even enabled the kinetics of individual channel opening and closing events to be followed (**Chapter 23**).

Anomalous X-ray experiments have provided detailed information on the distribution of counterion clouds near membranes (**Figure 12.19**). These clouds determine the electrostatic potential experienced by the membranes.

12.5   Passive/Active Transport

Selective transport of molecules through membranes is a crucial process to regulate the intracellular environment and for the cell to communicate with the outside world. This is achieved passively, when no additional energy is required, and actively when ATP molecules are used up. Passive transport depends on diffusion (introduced in **Chapter 7**) in which molecules move from a region of high concentration to a region of low concentration across the membrane.

Passive transport can be considered with a diffusive model in which the flux of molecules (j_s) is related to the concentration drop across the membrane (Δc). A linearised version of Fick's first law (equation (7.15)) can be used,

$$j_s = -P_s \Delta c \tag{12.40}$$

where P_s is the permeability of the membrane to the molecule studied. The permeability roughly reflects the width of the pores, the thickness of the membrane (the length of the pores) and the diffusion constant of the solute molecules. Δc is the change in concentration of the molecules across the membrane.

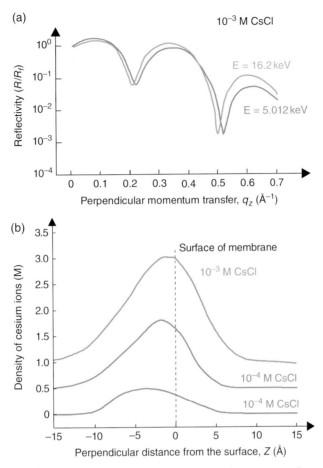

Figure 12.19 *Anomalous X-ray reflectivity data from phospolipid bilayers. (a) Reflectivity curve that shows the reflected intensity as a function of momentum transfer (q_z), and (b) is the calculated Poisson–Boltzmann counterion profile that shows the density of cesium counterions as a function of the perpendicular distance from the bilayer surface. [Reprinted with permission from W. Bu, D. Vaknin, A. Travesset, PRE, 2005, 72, 060501. Copyright 2005 by the American Physical Society.]*

For example, a cell can be modelled as a spherical bag of radius 10 μm bounded by a membrane that passes alcohol with a permeability of 10 μm s^{-1}. Rearrangement of equation (12.40) gives

$$-\frac{d(\Delta c)}{dt} = \left(\frac{AP_s}{V}\right)\Delta c \tag{12.41}$$

This equation can be solved for a jump in the internal alcohol concentration,

$$\Delta c(t) = \Delta c(0)e^{-t/\tau} \tag{12.42}$$

where $\tau = \dfrac{V}{AP_s}$. The model is in good agreement with measurements for the decay time of the concentration. Detailed measurements show that $P_s \propto BD$. The ratio of the concentration in the two phases (in the membrane

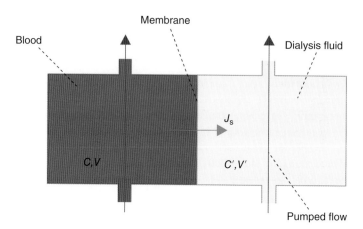

Figure 12.20   *Geometry of a simple dialysis experiment. Blood is separated from an artificial dialysis fluid by a membrane and there is diffusive flux (J_s) of small molecule pollutants (urea, etc.) that need to be removed from the blood.*

and in solution) at equilibrium is the partition coefficient B. The permeability of a pure bilayer to a molecule is roughly BD multiplied by a constant that is independent of the solute, and D is its diffusion coefficient of the molecules in the lipid.

Kidney dialysis is a partially successful solution to chronic renal failure when cellular homeostasis breaks down and unwanted biochemicals accumulate in the patient's blood. The body's fluid and the dialysis fluid are separated by a membrane that is porous to the small molecules to be removed and impermeable to the larger molecules. Dialysis fluid is prepared with the desired composition of small molecules such as Na, K and glucose. A static dialysis system can encounter problems with bacterial growth, so the solutions are pumped into the vessel. A similar model to equation (12.40) can be exploited to analyse the behaviour of an artificial kidney used in dialysis, i.e. to remove unwanted small molecules from bodily fluids (**Figure 12.20**). All the body fluid is treated as one compartment and the transport across the membrane is assumed to take a longer time than the transport from various body compartments to the blood. The solution for the kinetics of the artificial kidney is the same as equation (12.42). Typically, the relaxation time (τ) is on the order of 1.1 h and thus dialysis requires several hours to complete.

Real kidneys are now known to contain aquaphorins that selectively transport water (Nobel Prize for chemistry in 2003), i.e. extremely efficient pores exist that only allow the passage of water molecules and do this at rates of billions of molecules per second. It is hoped that in the future developments in dialysis machines could be helped by research into such naturally occurring membrane channels.

In general, the passive motion of water across a membrane is particularly important, since it is the principle solvent for biological molecules and is thus given the specific name of *osmosis* (**Chapter 19**). There are many clinical examples of passive transport through membranes. As blood flows through capillaries in humans, oxygen and nutrients leave the blood and pass to the cells. Waste products leave the cells and enter the blood. Diffusion is the main process that accomplishes this transfer.

There are typically three specific varieties of passive transport through membranes (**Figure 12.21**) and each process is powered by the concentration gradient of the molecules that are transported. In the cell the three main scenarios are diffusion through lipid layers, diffusion through protein channels, and carrier proteins in membrane that assist the movement of molecules across the membrane, down the concentration gradient, without expending energy.

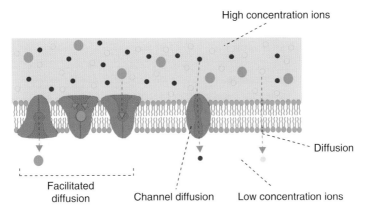

Figure 12.21 *The mechanisms of passive diffusion across a membrane include simple diffusion, diffusion through a channel and facilitated diffusion by a channel.*

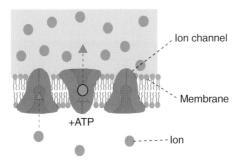

Figure 12.22 *Active transport can be used to move ions against their concentration gradient if ATP is consumed by the ion channels.*

The balance of osmotic pressure in the human circulation system can be disturbed in human disease and gives rise to an edema, a collection of fluid in the tissue. Higher than average pressure along the capillary can be a contributing factor. Also, reduction in osmotic pressure due to lower protein concentrations in the blood can cause similar problems. Another possibility is that increased permeability of capillary walls to large molecules can reduce the osmotic pressure and again can result in an edema. Specific examples of edemas include those due to heart failure, nephrotic syndrome, liver disease and ascites, leakage of protein into the urine due to kidney damage, and inflammatory reactions (accumulation of fluid beneath the skin). Associated phenomena include headaches in renal dialysis, osmotic diuresis that causes the increased production of urine and osmotic fragility of red cells.

Active transport *requires energy* to move substances from an area of lower concentration to an area of higher concentration (**Figure 12.22**). As an example of active transport consider the sodium/potassium pump (**Figure 12.23**). The sodium/potassium pump can expel unwanted ions and keep the needed ones. It also helps to maintain the cell volume through modulation of the osmotic pressure. ATP is used to expel three sodium ions for every two potassium ions brought into the cell. Increased cell volume due to increased water in the cytoplasm is produced by a reduction in the pump activity, which allows more sodium to exist inside the cell.

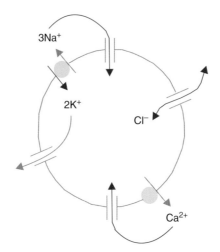

Figure 12.23 *Schematic diagram of a range of different ion channels that occurs in a living cell, e.g. sodium/ potassium, chloride and calcium ion channels.*

A decrease in cell volume through a reduction of water in the cytoplasm is produced by an increase in the pump activity, which expels more sodium ions.

Variations in tonicity of the extracellular environment have a large effect on cellular volume due to osmotic pressure differences. In *isotonic* cells the extracellular and intracellular ionic concentrations are equal. *Hypotonic* describes when the extracellular ionic concentration is less than intracellular concentrations. Cells tend to become inflated compared to their native state. With *hypertonic* cells the converse is true, the extracellular ionic concentration is higher than the intracellular concentration. Cells become deflated compared with their native state.

Cells tend to carefully regulate their intracellular ionic concentrations to ensure that no osmotic pressure changes occur. As a consequence, the major ions Na^+, K^+, Cl^- and Ca^{2+} often have different concentrations in the extracellular and intracellular environments, and thus a voltage difference arises across the cell membrane. Essentially, two different kinds of cell occur in Nature; *excitable* and *nonexcitable*. All cells have a resting membrane potential, but only excitable cells modulate it actively (**Part V**).

Models for the behavior of pumps, ionic currents, and osmotic forces are needed to understand the quantitative behavior of the steady-state cell. Homeostasis describes the steady-state properties of a cell, which is the result of lots of feedback processes on the microscale. Small ion species are very important for both electrically excitable and nonexcitable cells, e.g. poisons that affect sodium channels such as tetrodotoxin from puffer fish or batrachotoxin from poison dart frogs can kill organisms very quickly by the disruption of the homeostatic balance. Cells dynamically regulate their environment to keep themselves alive. There are some extreme examples of regulatory behaviour in cellular survival, e.g. bacteria can live at temperatures close to 100 °C and much of their cellular machinery is adapted to these high temperatures. Furthermore, plant seeds can survive thousands of years without water, e.g. seeds taken from inside Egyptian pyramids can germinate and give rise to viable plants.

Active pumping is used by cells for regulation when they cannot receive what they need passively. A huge amount of energy is required to create concentration gradients that are opposed by diffusion, and often such energies are indeed expended, since the concentration gradients are critical for cellular survival.

Typical cellular ionic concentrations are shown in **Table 12.2**. Clearly the action of pumps is crucial for the maintenance of ionic concentration differences in many different types of cell. There are many different kinds of ion pumps involved. Some use ATP as an energy source to pump against a gradient (active), and others use a

Table 12.2 *Typical ion concentrations (mM) found in giant squid axons, frog Sartorius muscle and human red blood cells.*

	Squid Giant Axon (mM)	Frog Sartorius Muscle (mM)	Human Red Blood Cell (mM)
Intracellular			
Na^+	50	13	19
K^+	397	138	136
Cl^-	40	3	78
Extracellular			
Na^+	437	110	155
K^+	20	2.5	5
Cl^-	556	90	112

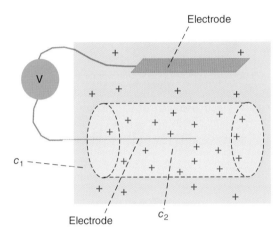

Figure 12.24 *Membrane potentials can be measured for a cell by comparison of the voltage of an electrode inside and outside the cell. c_1 and c_2 are the ion concentrations outside and inside the cell, respectively.*

gradient of one ion to pump another ion against its gradient (passive). A huge proportion of the energy intake of a human is devoted to the operation of such ionic pumps.

Measurement of the membrane potential (ΔV) is shown in **Figure 12.24**. The bulk concentration c_2 of interior cations is greater than the exterior concentration c_1. A schematic diagram of the ion concentration giving rise to the electrostatic potential across a membrane is shown in **Figure 12.25**.

Consider the energy requirement of the Na^+–K^+ ATPase pump (**Figure 12.26**). The NaK pump is a specific molecular machine embedded in cell membranes that hydrolyses ATP, and uses some of the resultant free energy to pump sodium ions out of the cell. At the same time the pump imports potassium. This inward motion of potassium partially offsets the loss of electric charge from the exported sodium. In fact the potassium current *in* is two thirds the sodium current *out*. There are three Na^+ binding sites and two K^+ sites per protein. The total free energy cost to pump one sodium ion is

$$-e\left(\Delta V - V_{Na^+}{}^{\text{Nernst}}\right) = e(60mV + 54mV) = 114emV \tag{12.43}$$

where ΔV is the change in voltage across the membrane and $V_{Na^+}{}^{\text{Nernst}}$ is the Nernst potential of the Na^+ ions introduced in **Section 23.5**.

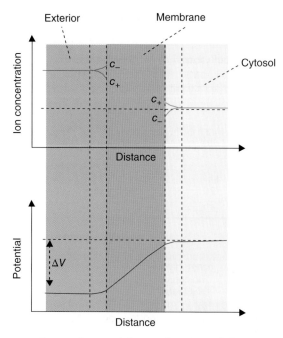

Figure 12.25 *The ion concentration and potential as a function of the perpendicular distance across the membrane. The variation in concentration of ions across a cell's membrane creates the membrane potential of the cell due to the summation of the Nernst potentials of the individual ions (**Section 23.5**).*

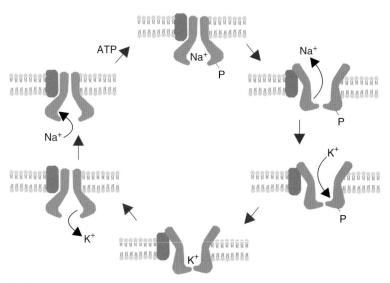

Figure 12.26 *Individual molecular stages during the action of the Na–K ion channel embedded in a cell membrane. Na⁺ ions are pumped out of the cell and K⁺ ions are pumped in.*

The total free energy cost for inward pumping of one potassium ion is

$$+e\left(\Delta V - V_{K^+}^{\text{Nernst}}\right) = e(-60mV - (-75mV)) = 15emV \tag{12.44}$$

where $V_{K^+}^{\text{Nernst}}$ is the Nernst potential of the K^+ ions.

The total cost of the exchange of three sodium ions for two potassium ions across the membrane is

$$3(114emV) + 2(15emV) \approx 15kT \tag{12.45}$$

The hydrolysis of one ATP molecule liberates ~19 kT of free energy (there is a small dependence on the environment of the reaction). It therefore can be deduced that the pump is efficient, only 4 kT is lost as thermal energy. It is found that the ion species with the greatest conductance per unit area provide the biggest vote in the determination of the steady-state membrane potential (equation (23.39)). The resting membrane potential ΔV is closer to the Nernst potential of the most permeant pumped species ($V_{K^+}^{\text{Nernst}}$) than that of the less permeant one ($V_{Na^+}^{\text{Nernst}}$), equation (23.39).

12.6 Vesicles

Information-carrying molecules are carried around the internal environment of cells within vesicles. The vesicles have a very specialised sorting, signalling and transportation system associated with them. The biochemistry is very rich and complicated.

Endocytosis of vesicles moves material into cells (**Figure 12.27a**), whereas *exocytosis* moves material out of cell (**Figure 12.27**). Membrane-bound proteins serve to determine the destinations of the vesicular packages. Examples of vesicle transportation events include clathrin-coated pits that occur in endocytosis, phagocytosis when large particles are engulfed, vesicle transport in a neuron axon, synaptic signalling between a muscle cell and a nerve cell, goblet cells that release mucus, and the internal motions of the Golgi apparatus.

There is a fixed free energy cost associated with spherical vesicles of all sizes. Consider the bending energy (equation (12.25)),

$$G_{\text{vesicle}} = \frac{\kappa_b}{2} \oint \left(\frac{2}{R}\right)^2 da \tag{12.46}$$

(a)

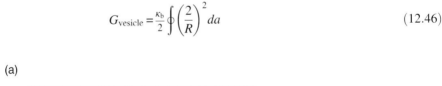

(b)

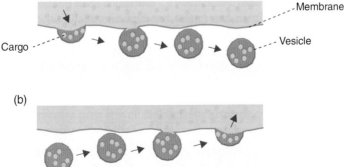

Figure 12.27 *Process of vesicle formation in (a) endocytosis (cargoes move into the cell) and (b) exocytosis (cargoes move out of the cell). The inside of the cell is yellow and the outside of the cell is blue.*

Since the curvature is constant on the sphere the expression is easy to integrate to give

$$G_{vesicle} = \frac{\kappa_b}{2} \left(\frac{2}{R}\right)^2 4\pi R^2 = 8\pi\kappa_b \qquad (12.47)$$

The energy to make a vesicle is independent of its size. $\kappa_b \approx 10\text{--}20\,kT$, so the free-energy cost to make a vesicle is in the range $250\text{--}500\,kT$. The energy cost can be reduced by the introduction of proteins and lipids that favour the curved state. Membrane proteins can induce elastic deformations in the surrounding membrane, which leads to changes in their structure. Vesicle fusion is also catalysed by specialised proteins.

Suggested Reading

If you can only read one book, then try:

Top recommended book: Phillips, R., Kondev, J., Theriot, J. & Garcia, H.G. (2013) *Physical Biology of the Cell*, 2nd edition, Garland. Good introduction to statistical mechanics models of membranes.

Boal, D. (20011) *Mechanics of the Cell*, 2nd edition, Cambridge University Press. Thorough well explained account of the physics of membranes.

Forgar, G. & Newman, S.A. (2005) *The Biological Physics of Developing Embryos*, Cambridge University Press. Interesting discussion of the role of membrane adhesion in cellular morphogenesis.

Hille, B. (2001) *Ionic Channels of Excitable Membranes*, 3rd edition, Sinauer. Classic account of membrane ion channels, although it is now becoming a little dated.

Luckey, M. (2008) *Membrane Structural Biology*, Wiley. Detailed modern account of membrane biochemistry combined with some physical chemistry.

Mouritsen, O.G. (2004) *Life – as a Matter of Fat: The Emerging Science of Lipidomics*, Springer. Popular account of lipid biochemistry.

Tutorial Questions 12

12.1 Calculate the line tension of a membrane if the critical hole radius in its surface is 2.6 nm and its surface tension is 0.03 J m^{-2}.

12.2 Calculate the axial and hoop stresses on a cylindrical bacterium if its internal pressure is 1×10^5 Pa, the membrane thickness is 1 nm and the radius is 1 μm.

12.3 Calculate the factor by which the persistence length of a membrane ($\kappa/kT = 5$) will change if its bending modulus is doubled. Also, calculate the factor by which the exponent would change, measured in an ideal X-ray scattering experiment, for the power-law cusp from a stack of such membranes. Assume all of the parameters of the membrane remain unchanged when it is incorporated in the stack. Explain whether this is a reasonable assumption.

12.4 A membrane is moved from a good solvent to a poor solvent and its contour length is L_c. Calculate the factor by which the radius of gyration and area change according to the scaling laws for closed bags.

13

Continuum Mechanics

Two architectural themes occur again and again in the continuum mechanics of naturally occurring biomaterials; *fibrous composites* and *cellular solids* (and combinations of the two). *Fibrous composites* consist of stiff rigid rods (e.g. collagen or cellulose) combined with a highly viscous dissipative filler (e.g. proteoglycans or lignins). The stiff rods resist extension and compression to provide the composite with its strength, whereas the dissipative filler increases the materials toughness by many orders of magnitude. Composite materials are now widely used in a range of synthetic products (e.g. skis, the shafts of golf clubs, the fuselages of planes, etc.), but biological composites continue to outperform many of their synthetic counterparts, due to their well-optimised structures on the nanometre length scale.

Cellular materials are widespread in biology. Cellular solids have reasonable strengths and toughnesses, but they only use a fraction of the structural component to achieve these properties. Thus, cellular materials provide a mechanical solution with greatly improved stiffness/weight ratios and consequently occur in a wide range of biological tissues that include bones and woods.

Table 13.1 shows a range of characteristic parameters commonly used to quantify the mechanics of materials. All of these properties can be important in biological situations and their extension to anisotropic nanostructured materials needs to be carefully considered.

One of the simplest experiments that can be performed on a biomaterial, concerns the application of a force (stress) to a material that causes it to extend (become strained). **Figure 13.1** shows the wide range of stress–strain properties observed with elastic proteins. Typically the stress–strain curves are linear for small deformations and Hooke's law is obeyed, with the force (and corresponding stress) proportional to the extension (and corresponding strain). At larger stresses nonlinear elasticity is observed. Numerical values of the material properties of some proteins and standard synthetic materials are shown in **Table 13.2**.

The Physics of Living Processes: A Mesoscopic Approach, First Edition. Thomas Andrew Waigh.
© 2014 John Wiley & Sons, Ltd. Published 2014 by John Wiley & Sons, Ltd.

Table 13.1 *Some important parameters that describe the mechanical properties of solid biomaterials.*

Functional attribute	Material property	Units
Stiffness	Modulus of elasticity, E_{init}	$N\,m^{-2}$
Strength	Stress at fracture, σ_{max}	$N\,m^{-2}$
Toughness	Energy to break at fracture	$J\,m^{-3}$
Extensibility	Strain at failure, ε_{max}	No units
Spring efficiency	Resilience	%
Durability	Fatigue lifetime	s to failure
Spring capacity	Energy storage capacity, W_{out}	$J\,kg^{-1}$

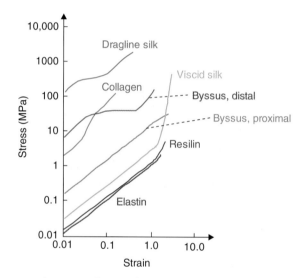

Figure 13.1 *Tensile stress as a function of strain for a range of structural proteins. Hooke's law, with stress proportional to strain, is observed with all these materials at small degrees of strain. [Adapted from J. Gosline, M. Lillie, E. Carrington, P. Guerette, C. Ortlepp, K. Savage, Philos. Trans. R. Soc. Lond. B, 2002, 357, 121–132. With permission 2002, The Royal Society.]*

13.1 Structural Mechanics

For an isotropic material there is a simple relationship between the stress ($\sigma = F/A$, the force (F) divided by the area (A)) and the strain ($\varepsilon = \Delta l/l$, the change in length (Δl) divided by the original length (l)). For a linear material the stress–strain relationship is

$$\sigma = E\varepsilon \tag{13.1}$$

where the constant of proportionality is Young's modulus (E, units of Pascals).

The Poisson ratio (ν) for an isotropic material is a dimensionless number given by the ratio of the perpendicular strain (ε_{perp}) to the longitudinal strain (ε) (**Figure 13.2**),

Table 13.2 *Mechanical properties of a range of structural proteins commonly encountered in biology and some synthetic equivalents. [Adapted from J. Gosline, M. Lillie, E. Carrington, P. Guerette, C. Ortlepp, K. Savage, Philos. Trans. R. Soc. Lond. B, 2002, 357, 121–132. With permission 2002, The Royal Society.]*

Material	Modulus (GPa)	Strength (GPa)	Extensibility	Toughness (MJ m^{-3})	Resilience
Elastin	0.0011	0.002	1.5	1.6	90%
Resilin	0.002	0.004	1.9	4	92%
Collagen	1.2	0.12	0.13	6	90%
Synthetic rubber	0.0016	0.0021	5	10	90%
Mussel byssus proximal	0.016	0.035	2.0	35	53%
Dragline silk	10	1.1	0.3	160	35%
Viscid silk	0.003	0.5	2.7	150	35%
Kevlar (e.g. in bullet-proof jackets)	130	3.6	0.027	50	
Carbon fibre	300	4	0.013	25	
High tensile steel	200	1.5	0.008	6	

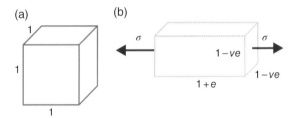

Figure 13.2 *The geometry of a stressed cube of an elastic material is used to illustrate the Poisson ratio (ν), which quantifies the reduction in size perpendicular to the direction of applied stress. (a) Unstressed cube of material and (b) the material with an extension (e) in the direction of a uniaxial tensile stress (σ), equations (13.1) and (13.2).*

$$v = -\frac{\varepsilon_{\text{perp}}}{\varepsilon} \tag{13.2}$$

It describes how the change in morphology of a material under stress is coupled in the directions parallel and perpendicular to the direction of the stress. The Poisson ratio is one half for an incompressible material in uniaxial extension.

The shear modulus (G) describes the resistance of a material to a shearing motion, e.g. the motion of two parallel faces of a material relative to one another (**Figure 13.3**). For small angles of deformation the shear modulus is equal to the stress (σ) divided by the angle of deformation (θ in radians) and it therefore has the units of Pascals,

$$G = \frac{\sigma}{\theta} \tag{13.3}$$

The bulk modulus (K, units of Pascals) quantifies the variation in volume (dV) of a material when the pressure is changed (dp) (**Figure 13.4**). It is defined through the equation

$$\frac{1}{K} = -\left(\frac{1}{V}\right)\frac{dV}{dp} \tag{13.4}$$

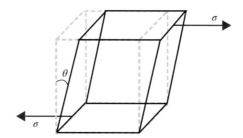

Figure 13.3   *Geometry used to define the shear modulus (G) of a cube of an elastic material in terms of the shear stress (σ) and the angle of deformation (θ), equation (13.3).*

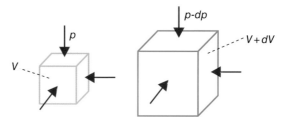

Figure 13.4   *The bulk modulus of an elastic solid (K), defines the change in volume (dV) of a material in response to a change in pressure (dp), equation (13.4).*

The factor $1/V$ makes the bulk modulus independent of the volume of material considered and the negative sign allows positive compressibilities to describe the typical case where an increase in pressure causes a decrease in volume.

For an isotropic elastic material the bulk modulus (K), Young's modulus (E), shear modulus (G) and Poisson ratio (ν) are interrelated by two simple relationships,

$$K = \frac{E}{3(1-2v)} \tag{13.5}$$

$$G = \frac{E}{2(1+v)} \tag{13.6}$$

Thus, if two of the characteristic elastic constants (from E, K, v, and G) are known for an isotropic material, the other two can be calculated.

For anisotropic materials (the predominant moiety among biological structures) a full tensorial analysis is required to describe the stress and strain of a material (**Figure 13.5**). In three dimensions the stress tensor (σ_{ij}) defines the extensive, compressive, dilatational and shear forces on a small volume element. In Cartesian coordinates (x,y,z) the stress tensor is given by

$$\sigma_{ij} = \begin{bmatrix} \sigma_{xx} & \sigma_{xy} & \sigma_{xz} \\ \sigma_{xy} & \sigma_{yy} & \sigma_{yz} \\ \sigma_{xz} & \sigma_{yz} & \sigma_{zz} \end{bmatrix} \tag{13.7}$$

There are also nine constants that describe the strain tensor (e_{ij}) of the material in response to the tensorial stress in three-dimensional Cartesian coordinates. For linear elastic materials the stress is related to the strain by a tensorial equation,

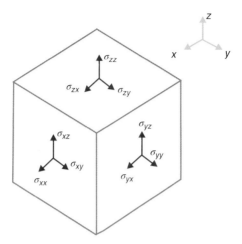

Figure 13.5   *General tensorial nature of the stress (σ_{ij}) on a cubic volume element of an elastic material, equation (13.7).*

$$e_{ij} = s_{ijkl}\sigma_{kl} \tag{13.8}$$

where s_{ijkl} are the compliance constants that characterise the elastic response of a particular material. Thus, eighty one ($3^4 = 81$) compliance constants are required in general to describe an anisotropic material that has no symmetry in its morphology. Fortunately, in most practical situations there is a reduction in the number of independent constants due to molecular symmetry and the morphology of the specimen of a material that is tested.

13.2   Composites

Composite materials are constructed from a mixture of discrete rigid units combined with a dissipative matrix. A range of composite morphologies commonly occur in Nature. These include two-dimensional laminates (the rigid units are planar) and one-dimensional fibres (the rigid units are rods, **Figure 13.6** and **Figure 13.7**). Furthermore, a wide range of tessellations are possible for the rigid embedded units, that provide important consequences for the mechanical properties. For fibrous composites the length of the rigid units embedded is another critically important parameter for the resultant mechanical properties and separate models have been developed to describe the stress in *short* and *long composites*. Examples of biocomposites include elastin/collagen in heart walls (fibrous composites) and nacre/protein in sea shells (laminar composites, **Section 17.6**). There is a direct link between the physical properties of fibrous composites and solid liquid crystals, and often the two descriptions are used interchangeably. For example, with polymeric liquid crystals the amorphous material in the flexible linkers can act as a dissipative filler and the nematic inserts provide resistance to extension.

Inorganic/organic biocomposites such as dentine, the hard surface layer in teeth, often have only small amounts of organic material ($\sim$1%), but this dissipative filler has an extremely large impact on the resultant physical properties. In particular, the toughness of the material (e.g. a combination of hydroxyapatite and protein in teeth), the amount of energy the material can dissipate before breakage, is many orders of magnitude larger than the toughness of a single crystal.

To a reasonably good first approximation it is possible to calculate the effective Young's modulus for a layered composite material. For a force applied *parallel* to the composite (**Figure 13.6a**) the force applied

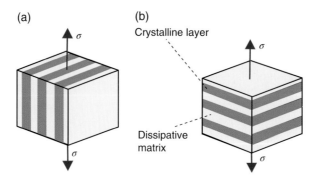

Figure 13.6 *The effective Young's modulus of laminated composites depends on the direction of the applied stress (σ), which is shown (a) parallel to the layers, and (b) perpendicular to the layers.*

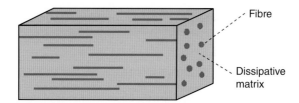

Figure 13.7 *The arrangement of fibres in a composite material with uniaxial orientation.*

to the mixed material (F_m) is equal to the sum of the forces on the crystal (F_c) and the amorphous phases (F_a), e.g. in an adhesive protein,

$$F_m = F_c + F_a \tag{13.9}$$

These forces can be re-expressed in terms of average stresses and areas,

$$\sigma_m A_m = \sigma_c A_c + \sigma_a A_a \tag{13.10}$$

where σ_m, σ_c and σ_a are the stresses on the mixture, the crystal and the protein, respectively. A_m, A_c and A_a are the corresponding areas. Equation (13.10) can be divided by the total volume of the mixture that allows the effective Young's modulus of the mixture (E_m) to be calculated in terms of the Young's modulus of the two constituent phases (E_c and E_a) and the volume fraction of the crystalline phase (ϕ_c),

$$E_m = E_c \phi_c + E_a (1 - \phi_c) \tag{13.11}$$

For a force applied *perpendicular* to a layered composite (**Figure 13.6b**) the total extension (δl_m) is the sum of the extension of each individual phase (δl_c and δl_a for the crystal and amorphous phase, respectively),

$$\delta l_m = \delta l_c + \delta l_a \tag{13.12}$$

The extension can be converted to strains using the definition

$$\varepsilon = \frac{\delta l}{l} \tag{13.13}$$

where δl is the change in length and l is the initial length. Therefore, if l_m, l_c, and l_a are the lengths of the mixture, crystal and amorphous components respectively; ε_m, ε_p and ε_c are the corresponding strains. Thus, equation (13.12) can be written as

$$\varepsilon_m l_m = \varepsilon_c l_c + \varepsilon_a l_a \tag{13.14}$$

The strains can be expressed in terms of the Young's modulus of the individual components, since the stresses (σ) on each component are identical,

$$\frac{\sigma l_m}{E_m} = \frac{\sigma l_c}{E_c} + \frac{\sigma l_a}{E_a} \tag{13.15}$$

The stresses cancel from this expression and the lengths can be written in terms of the volume fraction of the components. The final result for the Young's modulus of a layered composite strained in a direction *perpendicular* to the layers is

$$E_m = \frac{E_c}{E_a(1-\phi) + E_c\phi} \tag{13.16}$$

For a fibre composite the modulus *parallel* to the direction of stress is

$$E = \eta E_f \phi_f + E_m(1-\phi_f) \tag{13.17}$$

where E_f and E_a are the Young's modulus of the fibres and amorphous matrix, respectively. ϕ_f is the volume fraction of the fibres. This expression looks similar to that for a layered composite (equation (13.11)) with the inclusion of the scale factor (η). The scale factor is given by

$$\eta = 1 - \frac{\tanh ax}{ax} \tag{13.18}$$

where a is the aspect ratio (length (l) of the fibres divided by their diameter ($2r$), $l/2r$) and x is given by the expression

$$x = \left[\frac{2G_a}{E_f \ln(R/r)}\right] \tag{13.19}$$

where G_a is the shear modulus of the matrix, E_f is Young's modulus of the fibre, R is the distance of separation between the fibres and r is the fibre radius.

Anisotropic composites also occur in Nature. For example, helicoidal materials are often observed in Nature (elastin/collagen composites are a major component of mammalian heart walls) and this orientation provides additional resistance against torsional rotation and tearing.

13.3 Foams

Foams (or equivalently cellular solids) are a morphology optimised for high strength and low weight. A range of mechanical properties for some standard biological cellular solids are included in **Table 13.3**.

It is possible to motivate the elasticity of foams using a simple scaling theory. For the compression of a foam, the stress is resisted by strut-like sections of the foam structure (**Figure 13.8**). The deformation of a single strut (δ) at its midpoint by a force (F) is related to the Young's modulus of the strut (E), the length of the strut (a) and the thickness (t),

$$\delta \sim \frac{Fa^3}{Et^4} \tag{13.20}$$

This result is from the solution to the beam equation, i.e. an angular equivalent of Hooke's law, where the bending moment is proportional to the curvature. The structure is considered to be supported at either end

Table 13.3 *Mechanical properties of some biological cellular solids.*

Material	Young's Modulus (E, MPa)	Fracture stress σ_f (MPa)	Volume fraction of solid (%)	Density (ρ, kg m^{-3})	Poisson ratio (v)	Toughness (kJ m^{-2})
Soft wood	10 000	180	0.96	10^2–10^3	0.3–0.68	12
Hard wood	17 500	240	0.73	10^2–10^3	0.01–0.78	11
Cork	20	15	0.15	170	0–0.1	0.060–0.130
Bone	12 000	105	0.05–0.7	10^2–10^3	0.36	0.6–3.0
Carrot	7	1	0.03–0.38	10^3	0.21–0.49	0.2

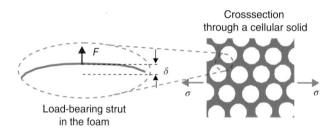

Figure 13.8 *The mechanical strength of a foam is related to the elasticity of the struts in the material. The average force on the stuts is F and their deformation is δ. σ is the average stress applied to the foam.*

by its connections to other struts. The force required to deform the strut (F) is related to the compressive stress (σ) on the strut ($F \sim \sigma a^2$), so

$$\sigma \sim \frac{\delta t^4 E}{a^5} \tag{13.21}$$

The Young's modulus of an unfilled foam (E_f, i.e. a foam that contains air) is approximately the stress (σ) divided by the strain (γ). The strain (γ) for a given displacement is inversely proportional to the length of a fibre ($\gamma \sim a^{-1}$) and therefore equation (13.21) can be rewritten as

$$E_f \sim \frac{\sigma}{\gamma} \approx \left(\frac{t}{a}\right)^4 E \tag{13.22}$$

This Young's modulus of an unfilled foam can be rewritten in terms of the relative density of the material

$$\frac{E_f}{E} = C_1 \left(\frac{\rho^*}{\rho_s}\right)^2 \tag{13.23}$$

where C_1 is a constant of proportionality, ρ^* is the density of the foam and ρ_s is the density of the strut. This equation is in reasonable agreement with experiment for a wide range of foam morphologies. This calculation is also close to the behaviour expected with crosslinked semiflexible polymers such as actin and fibrin, and there are many mechanical features held in common between polymer networks and open cell foams.

Filled foams that predominantly contain sealed water-filled compartments, such as biological tissues, are incompressible, and thus have markedly different mechanical properties from an unfilled foam. Failure in this

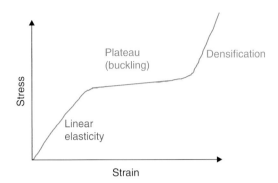

Figure 13.9 *The compressional behaviour of an open-celled foam (figure 13.8). The stress is shown as a function of strain. As the strain increases there is a linear Hookean region, followed by a plateau due to buckling of the struts and finally the structure densifies before failure.*

case corresponds to the cells bursting and strain is experienced by the stretching of the cellular walls not bending of these materials. The prediction in this case is a lower dependence of the thickness and length of the struts, $E_f \sim (t/a)^2 E$ compared with $(t/a)^4 E$ for the unfilled case, equation (13.22).

Figure 13.9 shows a schematic diagram of the compressional behaviour expected for an open-celled foam, e.g. cork or bone. Compaction of the foam occurs under compressive stress and its densification at medium and high strains is characteristic of a cellular solid. An Euler buckling transition is associated with the collapse of struts at high strain (equation (10.45)).

The Poisson ratio for unfilled cellular solids can have an unusual behaviour. It can be negative, with the classic example of cork stoppers used to seal wine bottles. Cork stoppers expand as they are stretched in the neck of a bottle, which seals the contents.

13.4 Fracture

The mechanism through which biological materials break and fracture is a vital concern in the lives of many organisms. Mechanical materials are often nanostructured in order to terminate cracks as they form, which greatly increases the energy they can absorb before failure. Such termination of cracks in soft viscoelastic fillers contributes to the toughness of nacre and bone.

The energy released by the propagation of a crack in a one component solid has two competing terms, the energy released by an advance of the crack and the energy absorbed to make the crack surface (**Figure 13.10**). There is thus a critical length (L_c) above which a crack begins to propagate. The critical crack length is related to the work of failure per unit area (W_f) and the strain energy stored per unit volume (W_s),

$$L_c = \frac{W_f}{\pi W_s} \tag{13.24}$$

Typically, for Hookean materials the strain energy stored (W_s) by an elastic material is half the stress times the strain ($\sigma\varepsilon/2$) and therefore

$$L_c = \frac{2W_f}{\pi\sigma\varepsilon} \tag{13.25}$$

To avoid the propagation of cracks biological materials use a series of different mechanisms; a high work of fracture, low strain energy at fracture extension, limitation of the material to low values of stress and strain,

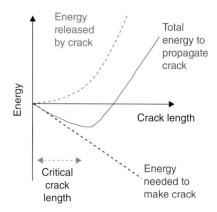

Figure 13.10 *The energy released by the propagation of a crack in a one-component solid as a function of the crack length. There is a critical crack length due to the interplay between the energy needed to nucleate the crack and the energy released by the crack growth.*

division of the material transversely to the direction of load, and use of materials that develop blunt cracks, so stresses are minimised at their tips. Cellular solids profit from their ability to terminate cracks in voids in this respect, and the effect greatly increases their toughness.

13.5 Morphology

As any architect of a macroscopic building will testify, the geometry of a structure has a significant impact on its mechanical properties. *Tubes*, *struts*, *braces* and *helicoids* are standard motifs that occur in biology that optimises the mechanical performance of the structures for a required role, e.g. the modification of flexural stiffness to provide additional torsional stiffness.

Tubes have high flexural and torsional stiffness with a minimum of structural material. *Struts* and *braces* have the same bending resistance as a single continuous piece of material, with a large reduction in weight (foams have a random array of struts and braces). Active truces (arrays of motile struts) occur in the spines of many organisms, and soft ligaments provide active modifications in their bending rigidity. *Helicoidal* structures in materials can improve fracture and tear energies by an order of magnitude, as linear fractures must occur across the strong fibre axis for at least part of their path length.

Suggested Reading

If you can only read one book, then try:

Top recommended book: Vincent, J. (1990) *Structural Biomaterials*, Princeton University Press. Very good introductory course to biological materials.

Ahlborn, B.K. (2004) *Zoological Physics*, Springer. Some simple zoological examples of the physics of materials.
Denny, M. & McFadzean, A. (2011) *Engineering Animals: How Life Works*, Harvard. Interesting popular account of biomechanics.

Fung, Y.C. (1993) *A First Course in Continuum Mechanics*, Prentice Hall. Classic introduction to continuum mechanics from a biomechanics perspective.

Gibson, L.J. & Ashby, M.F. (1997) *Cellular Solids*, Cambridge University Press. Classic text on the mechanics of foams.

Gosline, J., Lillie, M., Carrington, E., Guerette, P., Ortlepp, C. & Savage, K. (2002) Elastic proteins; biological roles and mechanical properties, *Philosophical Transactions of the Royal Society of London B*, **357**, 121–132. Review of the mechanical properties of proteins.

Hull, D. & Clyne, T.W. (1996) *An Introduction to Composite Materials*, Cambridge University Press. Classic account of synthetic composite materials.

Humphrey, J.D. & Delange, S.L. (2004) *An Introduction to Biomechanics: Solids and Fluids, Analysis and Design*, Springer-Verlag. Detailed engineering approach to biomechanics.

Jacobs, C.J., Huang, H. & Kwon, R.Y. (2013) *Introduction to Cell Mechanics and Mechanobiology*, Garland Science. Modern approach to biomechanics at the cellular level.

Landau, L.D. & Lifshitz, E.M. (1986) *Theory of Elasticity*, volume 7, Butterworth-Heinemann. Mathematically advanced approach to continuum mechanics.

Lautrup, B. (2011) *Physics of Continuous Matter: Exotic and Everyday Phenomena in the Macroscopic World*, 2nd edition, CRC Press. Good modern review of physical phenomena in continuous matter. Includes both mechanics and fluids mechanics.

Vogel, S. (2003) *Comparative Biomechanics: Life's Physical Worlds*, Princeton University Press. Detailed introductory text to biomaterials.

Ward, I.M. & Sweeney, J. (2012) *Mechanical Properties of Solid Polymers*, 3rd edition, Wiley. Classic text on solid synthetic polymer materials.

Tutorial Questions 13

13.1 Calculate the ratio of the Young's modulus parallel and perpendicular to a layered biocomposite if the volume fraction of the crystalline material is 0.9 and the Young's moduli of the crystalline and amorphous fractions are 50 MPa and 50 GPa, respectively.

13.2 Estimate the Young's modulus of a dried cellular solid if the Young's modulus of the walls is 9 GPa, the diameter of the struts is 1 μm and the length of the struts is 20 μm. Also estimate the change in the Young's modulus if water now fills the pores. Assume that the mechanical properties of the struts are unchanged in the hydrated environment.

14

Fluid Mechanics

A huge range of physical phenomena are observed in the motion of fluids such as turbulence, vorticity, surface waves, viscosity, non-Newtonian flow, the Rayleigh instability, Bernard convection, drag and lift. Many of these effects are important in biology. However, within the space requirements of the current chapter only a few basic problems in pipe flow, vascular networks and circulatory systems will be considered.

The Navier–Stokes equation provides a very good continuum description of Newtonian liquids. This equation (or equations if you combine it with the continuity equation) is nonlinear, which means that its solution presents some mathematical challenges (the general solution of the Navier–Stokes equation is often found on lists of the top 10 unsolved mathematical problems). Furthermore, the equation only describes idealised Newtonian fluids. The majority of biological fluids are viscoelastic (particularly at higher frequencies >0.01 Hz) and a number of subtleties are involved in the construction of a description of such fluids to correctly modify the Navier–Stokes equations. Fluid mechanics is a subsection of the field of rheology and care must thus be taken with the treatment of biological flow problems if viscoelastic aspects are neglected. A more widespread presentation of rheological phenomena is made in **Chapter 15**.

A large number of sophisticated plumbing architectures are found in living organisms for both gases and liquids. Evolution has used many fluid mechanics effects to move fluids within and without (around) organisms (in lungs, veins, arteries, and their networks in organs). For example, blood is transported in a self-healing high-pressure circulatory system in mammals and its performance has important medical implications.

Experimental techniques for *in vivo* fluid mechanics studies include Doppler optical coherence tomography, Doppler ultrasound, magnetic resonance imaging (MRI), particle tracking velocimetry, laser Doppler velocimetry and positron emission tomography (PET). Some of these methods are discussed in more detail in **Chapter 19**.

The Physics of Living Processes: A Mesoscopic Approach, First Edition. Thomas Andrew Waigh.
© 2014 John Wiley & Sons, Ltd. Published 2014 by John Wiley & Sons, Ltd.

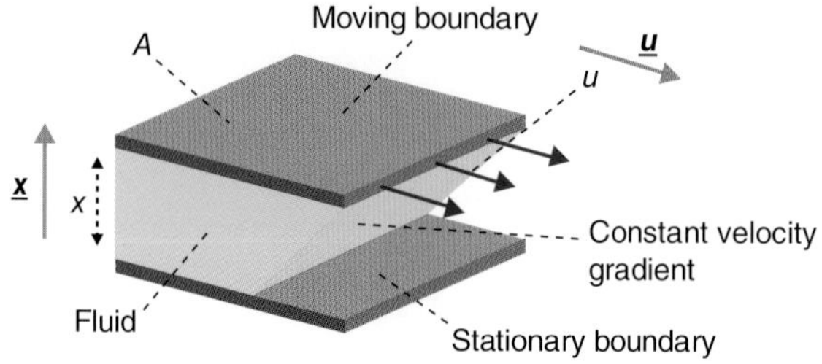

Figure 14.1 *Viscous forces determine the velocity profile of a fluid confined between a stationary boundary and a moving boundary. For a Newtonian fluid the velocity gradient is a constant.*

14.1 Newton's Law of Viscosity

Viscosity provides the resistance to flow when an object is forced to move through a fluid, and the energy is dissipated as heat (**Figure 14.1**). Newton's law of viscosity states that a constant velocity gradient is set up across a sandwich of fluid if the upper and lower plate move relative to one another,

$$\eta = \frac{F/A}{u/x} \tag{14.1}$$

where F is the magnitude of the force, A is the area of the plates, u is the component of the fluid velocity perpendicular to the direction of x and x is the distance between between the plates. The units of viscosity are Pa s (1 Pa s = 10 Poise). Materials that obey equation (14.1) are called *Newtonian fluids*.

14.2 Navier–Stokes Equations

The classical fluid mechanics of incompressible liquids often reduces to the solution of Navier–Stokes equation,

$$\rho\left(\frac{\partial}{\partial t} + \underline{u}.\nabla\right)\underline{u} = -\nabla p + \eta\nabla^2\underline{u} = 0 \tag{14.2}$$

where u is the fluid velocity, p is the pressure, ρ is the density and t is time. This partial differential equation is fairly formidable and cannot be analytically solved in general, but numerical solutions often exist in stable flows. For incompressible fluids (often the case with biological liquids) an additional continuity equation can be used to supplement equation (14.2),

$$\nabla.\underline{u} = 0 \tag{14.3}$$

This equation is a simple result of mass conservation; the mass of fluid that enters a volume element must be equal to the mass that leaves it.

The flow properties of real biological fluids are often much more complicated than expected for Newtonian fluids, since they are predominantly viscoelastic and extensions of the Navier–Stokes equations are needed, e.g. the Oldroyd B model, which adapts the Navier–Stokes equation for a Maxwell viscoelastic fluid (see **Section 15.1**).

Under low Reynold's number conditions (**Section 7.2**) the Navier–Stokes equation simplifies to

$$-\nabla p + \eta\nabla^2\underline{u} = 0 \tag{14.4}$$

and the continuity equation (14.3) is still valid. The Stokes equation (14.4) is linear and linear superposition principles can be used to solve it for the pressure and velocity fields with a range of boundary conditions, e.g. the flow pattern around different shapes of micro-organism or in different flow geometries. The solution (the Green's function) for the velocity field ($u(x)$) of a Stokes flow with a delta function driving force, $\delta(\underline{x}-\underline{x}')\underline{F}$, is given by

$$\underline{u}(\underline{x}) = \underline{G}(\underline{x}-\underline{x}') . \underline{F} \tag{14.5}$$

$$\underline{G}(\underline{r}) = \frac{1}{8\pi\eta}\left(\frac{1}{r} + \frac{\underline{r}\,\underline{r}}{r^3}\right) \tag{14.6}$$

where $\underline{F}$ is the driving force, $r = |\underline{r}|$, $\underline{G}$ is called the Oseen tensor and equation (14.6) is called a *Stokeslet*. The pressure field ($p(x)$) of the Stokeslet can be calculated using,

$$p(\underline{x}) = \underline{H}(\underline{x}-\underline{x}') . \underline{F} \tag{14.7}$$

$$\underline{H}(\underline{r}) = \frac{\underline{r}}{4\pi r^3} \tag{14.8}$$

The motility of microscopic particles in solution (e.g. micro-organisms) can often be modelled through an analysis of the Stokeslet contributions.

In addition to translational motion, rigid particles can also rotate. Translational and rotational motions are in general coupled. The linearity of the Stoke's equation (14.4) motivates a relationship for solid rigid particles,

$$\begin{pmatrix} \underline{U} \\ \underline{\Omega} \end{pmatrix} = \begin{pmatrix} \underline{M} & \underline{N} \\ \underline{N}^T & \underline{O} \end{pmatrix} . \begin{pmatrix} \underline{F} \\ \underline{L} \end{pmatrix} \tag{14.9}$$

where $\underline{F}$ is the external force, $\underline{L}$ is the external moment, $\underline{U}$ is the velocity and $\underline{\Omega}$ is the rotational rate. For a solid sphere $\underline{M} = (6\pi\eta R)^{-1}\underline{1}$ and $\underline{O} = (8\pi\eta R^3)^{-1}\underline{1}$ and $\underline{N} = \underline{N}^T = 0$. However, in general $\underline{N}$ and $\underline{N}^T$ do not vanish for asymmetric objects. The drag anisotropy on slender filaments allows then to be used for the propulsion of micro-organisms.

14.3   Pipe Flow

Consider viscous fluid flow in a pipe and neglect any viscoelastic effects. A cylindrical volume element (radius r) in the pipe is shown in **Figure 14.2**. The force due to the pressure (p_1) on one side of the pipe is $p_1\pi r^2$ and that on the other side (p_2) is $p_2\pi r^2$. The resultant force (F) due to the pressure difference is therefore

$$F = (p_1 - p_2)\pi r^2 \tag{14.10}$$

A viscous force described by equation (14.1) opposes the fluid's motion driven by the difference in pressures. It is possible to replace u/x by du/dr for a cylindrical geometry in equation (14.1) and the cross-sectional area is $A = 2\pi rL$, where L is the length of the pipe. The frictional force on the cylindrical fluid element is therefore

$$F = -\eta 2\pi rL\frac{dv}{dr} \tag{14.11}$$

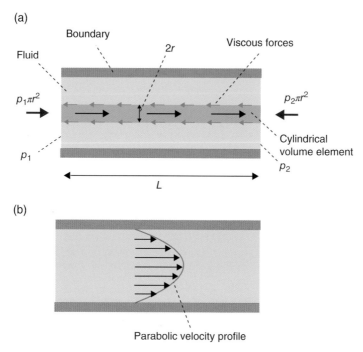

Figure 14.2 *Flow of a liquid in a cylindrical pipes. (a) Frictional viscous forces on a cylindrical volume element. (b) Resultant parabolic velocity profile observed in the pipe.*

Equations (14.10) and (14.11) can be equated to give

$$\frac{dv}{dr} = -\frac{(p_1 - p_2)r}{2\eta L} \tag{14.12}$$

This equation can be integrated from the centre of the pipe to the boundary, since the fluid velocity is zero at the walls of the pipe (the boundary conditions are $v = 0$ when $r = R$),

$$-\int_v^0 dv = \frac{p_1 - p_2}{2\eta L} \int_r^R r\,dr \tag{14.13}$$

and the solution is

$$v(r) = \frac{\Delta p}{4\eta L}\left(R^2 - r^2\right) \tag{14.14}$$

where $\Delta p = p_1 - p_2$. The average flow velocity ($\bar{v}$) is therefore

$$\bar{v} = \frac{\displaystyle\int_0^R v(r)2\pi r\,dr}{\pi R^2} = \frac{\Delta p R^2}{8\eta L} \tag{14.15}$$

The total flow rate is the average velocity multiplied by the cross-sectional area and is called the *Hagen–Pouseille* equation,

$$Q = \frac{\pi \Delta p R^4}{8\eta L} \tag{14.16}$$

where Q is the total volume of flow per unit time. The *Hagen–Pouseille* (HP) equation is very useful in biology, e.g. to estimate the blood flow through capillaries. The Hagen–Pouseille equation is a special case of the more general Darcy's law, which will be encountered again in **Chapter 17** with the discussion of fluid flow through cartilage, a porous solid. Darcy's law simply relates the total volume flow rate (Q) through a porous material to the pressure drop (Δp) and the hydrodynamic resistance (Z),

$$Q = \frac{\Delta p}{Z} \tag{14.17}$$

Therefore, for the Hagen–Pouseille equation (14.16) the hydrodynamic resistance in Darcy's law is given by

$$Z = \frac{8\eta L}{\pi R^4} \tag{14.18}$$

Another example of Darcy's law is given by fluid flow across a membrane.

$$Z = \frac{1}{A L_{\mathrm{p}}} \tag{14.19}$$

where A is the membrane area and L_{p} is its hydraulic permeability.

Question: Find the change in radius needed to increase the hydrodynamic resistance of a cylindrical blood vessel by 30%.

Answer: From equation (14.16) $\Delta p/Q$ increases by a factor of 1.3, which means $\frac{(R')^4}{R^4} = 1.3$ from the Hagen–Pouseille equation. Thus, the radius only needs to be reduced by 6%, which has important biomedical implications, e.g. atherosclerosis plaques in heart disease.

14.4   Vascular Networks

Bacteria rely on diffusion to feed themselves, but larger organisms need a more elaborate mechanism for delivery of nutrients and the removal of waste. Typically, large organisms employ vascular networks. Virtually every macroscopic organism imaginable has one or more vascular networks that carry blood, sap, air, lymph, etc. around their bodies. These networks normally have a hierarchically branched structure, e.g. the human aorta splits into iliac arteries that again splits into smaller branches and so on, until the capillary beds that nourish the tissue are finally reached.

14.5   Haemodynamics

Blood flow is very important in the human body. The key tasks of blood include the transport of oxygen and nutrients to active tissues, the return of carbon dioxide to the lungs, the delivery of metabolic end products to the kidneys, buffering of the pH of bodily fluids and the transport of heat. An equivalent compact form of equation (14.1) for a Newtonian fluid is

$$\tau = \eta \dot{\gamma} \tag{14.20}$$

where τ is the shear stress and $\dot{\gamma}$ is the shear rate. However, blood is *non-Newtonian*, since it obeys the equation

$$\tau = \eta_{\mathrm{eff}}\,\dot{\gamma} \qquad (14.21)$$

where the effective viscosity (η_{eff}) is not a constant and can depend on both the stress and the shear rate.

Figure 14.3 shows the velocity profile near a solid surface and shows velocity gradients from Newtonian and non-Newtonian fluids.

The percentage composition of human blood is shown in **Figure 14.4**. The water, ions, proteins, and minority components collectively make up the plasma. In common with the rest of the human body, water is the majority species in blood. The composition of the cellular components is also shown and it is predominantly composed of red blood cells.

Blood is a shear-thinning fluid due to the properties of the red blood cells. At low shear rates, red blood cells can stack into columns called Rouleaux structures. Imposition of a shear tends to break up the Rouleaux structures and causes a decrease in the solution viscosity. Furthermore, shear thinning also results from shear

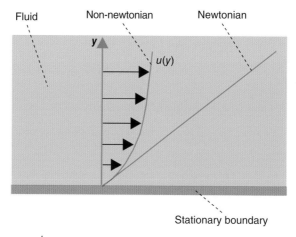

Figure 14.3   *The shear rate ($\dot{\gamma} = \dfrac{du}{dy}$) can depend on position in a* non-Newtonian fluid. *With a Newtonian fluid the shear rate is constant at low shear rates (turbulent nonlinear phenomena can create complex flow patterns at high shear rates).*

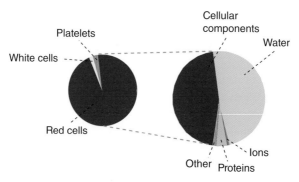

Figure 14.4   *Percentage composition of blood, which has cellular components (red blood cells, white blood cells and platelets), water, proteins, ions and some other minority components.*

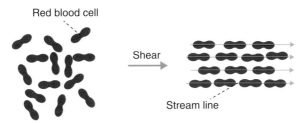

Figure 14.5 *Red blood cells become aligned along stream lines when they are sheared in blood, which causes a reduction in their viscosity.*

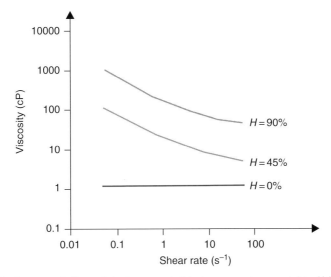

Figure 14.6 *Schematic diagram of effect of the haemocrit (H, the percentage weight of blood cells) concentraton of blood on its viscosity as a function of shear rate. [Reproduced from S. Chien et al, Journal of Applied Physiology, 1966, 21, 81–82. With permission from The American Physiological Society.]*

alignment (**Figure 14.5**). At low shear rates Brownian forces dominate and the red cells are randomly oriented, whereas at higher shear rates the red cells become oriented along streamlines.

Figure 14.6 shows the effective viscosity versus shear rate for blood with different red blood cell concentration (hematocrits, H). Newtonian behavior occurs at zero hematocrit, but at higher red blood cell fractions clear shear thinning behavior occurs.

Blood also exhibits a yield stress, which is another clear non-Newtonian phenomenon (**Figure 14.7**). The yield stress is a critical applied stress (τ_y) below which blood will not flow. Blood acts like a solid when $0 < \tau < \tau_y$ and acts like a fluid when $\tau \geq \tau_y$. The Casson relationship can be used to describe both the yield stress and shear thinning behaviour of blood (**Figure 14.7**),

$$\sqrt{\tau} = \sqrt{\tau_y} + \sqrt{\eta \dot{\gamma}}, \quad \tau > \tau_y, \quad \dot{\gamma} = 0, \tau \leq \tau_y \tag{14.22}$$

Further complexities occur for blood flow patterns *in vivo*. Arteries have complex three-dimensional geometries with bifurcations, changes in diameter, etc. In large arteries flow is classified as moderate to highly

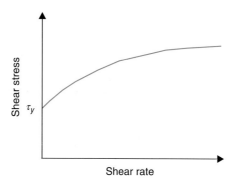

Figure 14.7 *Schematic diagram of the relationship between the shear stress and shear rate for a Casson viscoelastic fluid, e.g. blood. A yield stress (τ_y) is observed for this viscoelastic fluid.*

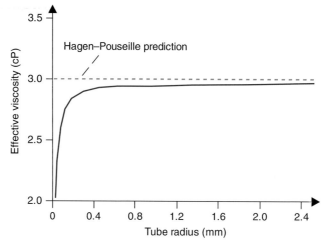

Figure 14.8 *The effective viscosity of blood as it flows through a capillary as a function of its radius. The Hagen–Pouseille model prediction is shown as a straight dotted green line, whereas the experimental data follows the red curve (the* Fahreus–Lindvist *effect). [Reproduced from R.H. Haynes, American Journal of Physiology, 1960, 198, 1193–1200. With permission from The American Physiological Society.]*

unsteady and can occur at high Reynolds numbers with turbulent phenomena. Artery walls are distensible and some arteries undergo large motions. Also, blood flow is pulsatile, which makes accurate *in vivo* modelling very challenging.

Fahreus and Lindvist forced blood through fine capillary tubes that connected two reservoirs. **Figure 14.8** shows the effective viscosity of the flow of blood as a function of the capillary radius (red). The dashed green line shows the effective viscosity in a tube of very large diameter. The volume flow rate of blood in the capillaries was expected to follow the HP form (equation (14.16)),

$$Q = \frac{\pi R^4 \Delta p}{8\eta_{\text{eff}} L} \tag{14.23}$$

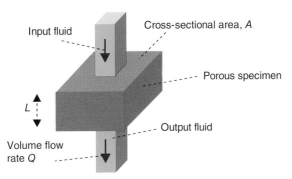

Figure 14.9 *Darcy's law relates the fluid volume flow rate (Q) through a porous specimen to its length (L) and hydrodynamic resistance (Z), equation (14.17).*

where the effective viscosity (η_{eff}) has been introduced to describe small deviations from the HP predictions. It was found experimentally that the effective viscosity was not constant and decreases with the capillary radius (**Figure 14.8**). An explanation is that when deformable particles flow in a tube, there is a net hydrodynamic force that tends to force the particles towards the centre of the tube, which reduces their effective viscosity.

14.6 Circulatory Systems

In humans, the circulation system consists of two varieties of pulsatile pump (the heart, driven by striated muscle) and the vasculature (arteries, arterioles, capillaries, venules and veins, driven by smooth muscle). Unsteady pulsatile flow propagates throughout these circulatory systems.

The interstitum is the tissue that surrounds the capillaries. It consists of cells embedded in extracellular biomacromolecules. Darcy's law, equation (14.17) can be used to explain the clearance time of swollen tissues and the motion of interstitial fluid, e.g. the motion of fluid through bruises. A schematic diagram of flow through the interstitum is shown in **Figure 14.9**.

14.7 Lungs

The main function of the lungs is to exchange oxygen for carbon dioxide in the blood. The process of mass transfer in alveolar sacs is shown in **Figure 14.10**, where blood flows through the capillary that is gas permeable. The blood gas concentration changes with axial position along the capillary as the oxygen concentration increases.

Lungs have a huge surface area to volume ratio to increase the rate of gas transfer and thus have convoluted quasifractal geometries (**Figure 14.11**). A mucus layer coats the surface of the lung to protect it from pathogens such as dust and bacteria. Mucus often occurs as a boundary lubricant in biological flow systems and its primary role is to move the plane of shear away from fragile biological surfaces, e.g. the surface of the eye. The mucus enables the cilia in the lung (small motile hairs) to act as a dust escalator to move harmful particles away from the fragile alveolar sacs. The cilial motion forms a Mexican wave, so that coherent motion is possible for the dust particles as they ride the metachronal waves (the technical name for the Mexican waves). The mucin layers sit above the cilia on epithelial tethered mucin brushes, i.e. there are bound (epithelial) and unbound gelling mucins in the lung that are expressed by different genes and play different roles (**Figure 1.27**).

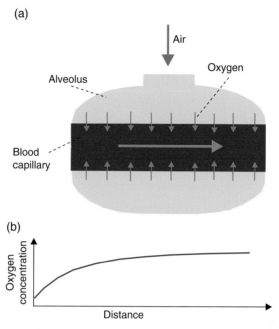

Figure 14.10 *(a) Air enters the alveolar sac (***Figure 14.11***) inside a mammalian lung and oxygen enters the blood of an adjacent capillary. The level of oxygen increases along the length of the capillary. (b) The oxygen concentration in the capillary as a function of distance along the capillary.*

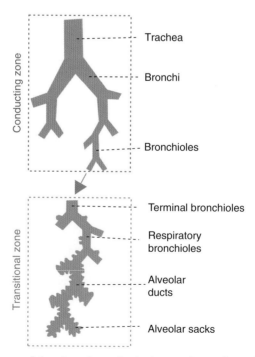

Figure 14.11 *Schematic diagram of the air pathway in the human lungs, from the scale of the trachea (largest structure) down to the alveolar sacks (smallest structure).*

Otherwise, the gelling mucins would obstruct the cilial motion and such processes of obstruction are thought to be a contributing factor in diseases such as cystic fibrosis and chronic obstructive pulmonary disorder.

Suggested Reading

If you can only read one book, then try:

Top recommended book: Ethier, E. & Simmons, C. (2007) *Introductory Biomechanics: From Cells to Organisms*, Cambridge University Press. Well balanced account of bioengineering fluid mechanics.

Denny, M.W. (1993) *Air and Water: The Biology and Physics of Life's Media*, Princeton University Press. Simple introduction to biological fluid mechanics.

Faber, T.E. (1995) *Fluid Dynamics for Physicists*, Cambridge University Press. One of the best texts on physical fluid mechanics, although it is a shame there are no tutorial questions.

Herman, I.P. (2007) *Physics of the Human Body: A Physical View of Physiology*, Springer. Interesting discussion of the medical applications of fluid mechanics.

Pozrikidis, C. (2011) *Introduction to Theoretical and Computational Fluid Mechanics*, Oxford University Press. Excellent guide to the applied mathematics of fluid dynamics calculations, which includes a particularly good introduction to low Reynolds number calculations.

Tritton, D.J. (1988) *Physical Fluid Mechanics*, Oxford University Press. Classic undergraduate physics textbook with a good introduction to turbulence.

Vogel, S. (1993) *Vital Circuits: On Pumps, Pipes and the Wondrous Workings of Circulatory Systems*, Princeton University Press. Another useful popular account from Vogel, this time on biomedical plumbing.

Vogel, S. (1994) *Life in Moving Fluids: The Physical Biology of Flow*, Princeton University Press. Classic popular account of fluid mechanics in biology.

Tutorial Questions 14

14.1 Research some phenomena in biology in which vorticity (related to the rotational motion of fluids, $\underline{\omega} = \nabla \times \underline{u}$, where u is the velocity and ω is the vorticity) is important, e.g. with motile micro-organisms. Explain the conditions under which vorticity can be safely ignored.

14.2 Describe some physical effects that limit the top speed of an organism swimming on the surface of water, e.g. a duck or a whirligig beetle swimming on a pond.

14.3 Calculate the pressure difference needed to force water up through the vascular system of a tree. The height of the tree is 50 m, the capillaries have a radius of 20 μm and the average velocity required is 1 mm s^{-1}.

15

Rheology

Traditionally, experimental rheology has consisted of placing a sample in the scientific equivalent of a food blender (a *rheometer*). The response of the material in the blender provides a probe of the material's *viscoelasticity*. Rheology thus considers the measurement of the viscoelasticity of materials. A classic example of a viscoelastic substance is the child's toy, silly putty. This material can flow when it is drawn in the hand and bounces when subjected to a rapid collision with the floor. Silly putty therefore acts as a viscous fluid (a liquid) at long times and an elastic solid at short times; thus the viscoelasticity of the material is seen to be a time-dependent phenomenon. All materials are viscoelastic to some degree (water and glycerol become viscoelastic at frequencies of ~10 GHz), but many biological materials have very carefully tailored mechanical behaviour, simultaneously exhibiting finely tuned viscosities and elasticities over physiologically important time windows.

In terms of a fundamental understanding of condensed matter the field of rheology introduces the key concept of irreversible dissipative behaviour. Fluid mechanics can be considered as a subfield of rheology. Dissipative behaviour needs to be included in the development of realistic statistical models of biological processes. Thus, experimental rheology provides a quantitative method to test models from theories of irreversible thermodynamics.

The mechanisms by which biological materials store and dissipate energy are of prime importance in many biological functions, e.g. shock absorbers formed from cartilage (**Section 17.1**), the contraction of striated muscle in the heart (**Section 16.2**), the resilience of skin to impacts and how bloods cells are pumped through arteries (**Figure 15.1**). Rheological measurements are necessary to determine the behaviour of biomaterials in such bioengineering problems.

The rheological behaviour of biological materials can be exceedingly complicated. However, as a first step, two *dimensionless numbers* are introduced to qualitatively understand the flow behaviour of biological specimens; the *Peclet number* and the *Deborah number*.

The *Peclet number* determines when the stress applied to a material will substantially deform the microstructure. For example, consider a colloidal material that consists of spherical particles (radius, *a*).

The Physics of Living Processes: A Mesoscopic Approach, First Edition. Thomas Andrew Waigh.
© 2014 John Wiley & Sons, Ltd. Published 2014 by John Wiley & Sons, Ltd.

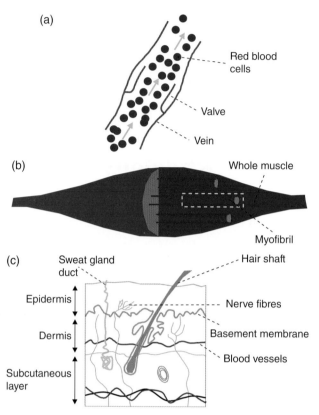

Figure 15.1   *Examples of biological materials in which the viscoelasticity is vital to their function (a) blood pumped through veins (microrheology, thixotropy, etc.), (b) the action of striated muscle (active stresses) and (c) the complex viscoelastic composite that is skin (the viscoelasticity of mixtures).*

The Stokes–Einstein relationship (equation (7.9)) gives an expression for the diffusion coefficient (D) of the constituent particles as

$$D = \frac{kT}{6\pi\eta a} \tag{15.1}$$

where η is the viscosity and kT is the thermal energy. From equation (7.37), the relaxation time associated with the diffusive motion (t_a) of the particles (the time for the colloids to diffuse their own radius) is then

$$t_a = \frac{a^2}{D} = \frac{6\pi\eta a^3}{kT} \tag{15.2}$$

where for simplicity the prefactors on the diffusion equation have been neglected. These spherical particles experience an applied stress (σ), and the characteristic time for shear flow (t_s) is the reciprocal of the shear rate ($t_s = \dot{\gamma}^{-1}$). The Peclet number (Pe) is simply the ratio of the time for diffusive rearrangement (t_a) to the time associated with the rate of shear flow (t_s),

$$\text{Pe} = \frac{t_a}{t_s} = \frac{6\pi\eta a^3 \dot{\gamma}}{kT} \tag{15.3}$$

If the microstructure is undisturbed the fluid has a low Peclet number and therefore Pe < 1. For a predominantly viscous liquid the stress is directly proportional to the shear rate (Newton's law of viscosity),

$$\sigma = \eta \dot{\gamma} \tag{15.4}$$

Therefore, the expression for the Peclet number (equation (15.3)) for a colloidal solution can be re-expressed as

$$Pe = \frac{6\pi a^3 \sigma}{kT} < 1 \tag{15.5}$$

where $kT/6\pi a^3$ is the thermal stress and σ is the mechanical stress. The convective motion due to the applied stress must be less than that due to Brownian motion or the microstructure will be significantly disturbed.

A further condition in order for an experiment to measure the linear viscoelasticity of a sample is that the structural relaxation by diffusion must occur on a time scale (τ) less than the measurement time (t). The De*borah number* must be less than 1 (De < 1) for linear rheology experiments and by definition

$$De = \frac{\tau}{t} \tag{15.6}$$

When the Deborah number is much greater than one (De>> 1) the material is solid-like and for De of order one or below the material is liquid-like. Both the conditions on the Peclet number and the Deborah number need to be satisfied to observe linear viscoelasticity experimentally. However, these conditions are necessary, but not always sufficient for steady linear flows. For example, elastic turbulence can occur in low Reynolds number flows at high shear rates with strongly elastic solutions and nonlinear instabilities are also possible due to Marangoni-type surface effects (**Section 9.8**).

Drag flow rheometers can normally function in both oscillatory and unidirectional shear modes. Steady-state shear experiments typically measure nonlinear rheology, whereas oscillatory measurements are sensitive to the linear viscoelasticity in the limit of small-amplitude oscillations.

15.1   Storage and Loss Moduli

A fundamental aim of models for the rheology of a material is to relate the applied force (stress) to the resultant deformation (strain) as a function of time; this is provided by a *constitutive equation*. A simple general constitutive equation for a viscoelastic material will first be constructed to illustrate the key concepts involved.

For an elastic Hookean solid the stress (σ) is proportional to the strain (γ) with an elastic constant (μ),

$$\sigma = \mu \gamma \tag{15.7}$$

This expression can then be extended into three dimensions using tensorial notation for an isotropic material (with a single elastic constant, μ) as

$$\sigma_{xy} = \mu \gamma_{xy} \tag{15.8}$$

For a viscoelastic solid the mathematical relationship that connects the stress to the strain requires a more complicated expression, since the stress slowly reduces to zero with time as the material relaxes. The solid will be modelled in one dimension for simplicity and it is assumed that the viscoelastic response of the system is additive in time. The stress can then be expressed as an integral over the relaxation modulus,

$$\sigma = \int_{-\infty}^{t} G(t-t') \left(\frac{d\gamma}{dt'} \right) dt' \tag{15.9}$$

where $G(t-t')$ is the relaxation modulus (the stress per unit applied strain), and $d\gamma/dt'$ is the shear rate. Now, consider the system subject to an oscillatory strain (e.g. $\gamma(t)$ due to mechanical vibration of the sample) of the form

$$\gamma(t) = \gamma_0 \sin \omega t \tag{15.10}$$

where γ_0 is the amplitude of the strain, t is the time and ω is the frequency of the oscillation. The corresponding strain rate is simply the differential with respect to time of equation (15.10) and is therefore

$$\frac{d\gamma}{dt} = \omega\gamma_0 \cos \omega t \tag{15.11}$$

Through substitution of this expression for the strain rate in equation (15.9) it can be shown that an alternative functional form for the stress as a function of time (t) is

$$\sigma = G'(\omega)\gamma(t) + \frac{G''(\omega)}{\omega}\frac{d\gamma}{dt} \tag{15.12}$$

where G' is a frequency-dependent shear modulus, a measure of the elastic energy stored in the network (the *elastic modulus*), and G'' is the component of the shear modulus that corresponds to the energy dissipated in the material (the *loss modulus*). G''/ω can be defined as the frequency-dependent *dynamic viscosity*. Newton's law of viscosity relates the shear stress (σ) to the rate of shear ($\dot{\gamma}$),

$$\sigma = \eta\frac{d\gamma}{dt} \tag{15.13}$$

Therefore, it can be seen that the conventional viscosity of the material is given by the limit of the dynamic viscosity ($\eta = G''/\omega$) as the frequency tends to zero ($\omega \to 0$) for a material with no elasticity ($G' = 0$).

As an example of the response of a viscoelastic material to an oscillatory strain the storage modulus of some polymer solutions is shown as a function of frequency in **Figure 15.2**. The expectation is that the shear modulus of a viscoelastic polymer solution will increase with the polymer concentration, since there are more large structure-forming molecules that resist the shear. However, the expectation for the frequency dependence of the shear modulus needs much more careful thought and it will be considered in detail in **Section 15.3**.

A useful function for the characterisation of the linear viscoelasticity of a material is the ratio of the elastic modulus to the dissipative modulus and it is defined to be equal to the tangent of a phase angle (δ),

$$\tan\delta = \frac{G''}{G'} \tag{15.14}$$

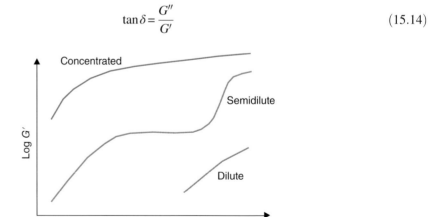

Figure 15.2   *Schematic diagram of the elastic shear modulus (G') as a function of frequency (ω) for solutions of flexible polymers. The frequency dependence of the three dominant regimes of polymer concentration is illustrated (dilute, semi-dilute and concentrated).*

It is then found that the resultant stress on a linear viscoelastic material in response to an applied sinusoidal strain (equation (15.10)) is also sinusoidal, but has a phase lag (δ),

$$\sigma = \sigma_0 \sin(\omega t + \delta) \tag{15.15}$$

This phase lag helps quantify the linear viscoelasticity of the material. Evidently a material that is predominantly elastic will have a small value for tanδ in equation (15.14) (e.g. elastin), whereas a fluid (e.g. blood plasma) will have a very large value.

To begin to understand the time response of biological materials simple mechanical models can be useful. The simplest models are those due to *Kelvin* and *Maxwell*, which are constructed from a dashpot and a spring in parallel and series, respectively (**Figure 15.3**). The solution of these models provides some simple constitutive equations that can then be used to model the viscoelastic response of real biological materials. Ideally, the connection between the viscoelasticity and dynamics of the constituent biological molecules is required (what is the molecular origin of the spring and dashpot?), but simple constitutive equations are a useful first step towards this goal.

For an *elastic spring* (an ideal elastic solid) the shear modulus (G) is given by equation (15.7), which can be written in the form

$$\sigma = G\gamma \tag{15.16}$$

The stress is proportional to the strain (Hooke's law). For a *viscous dashpot* (an ideal viscous fluid) in contrast, a totally dissipative structure, the stress is proportional to the rate of strain as per Newton's law, equation (15.13)

$$\sigma = \eta\dot{\gamma} \tag{15.17}$$

where η is the viscosity. For the *Kelvin model* (a dashpot and an elastic spring in parallel, **Figure 15.3a**) the stresses in both elements can be added, since they resist the imposed stress in concert,

$$\sigma = G\gamma + \eta\dot{\gamma} \tag{15.18}$$

For the *Maxwell model* (a dashpot and an elastic spring in series, **Figure 15.3b**) the strain rates in both elements add linearly, since the total strain is just the sum of that in each component. Differentiation of equation (15.16) followed by addition of the result to equation (15.17) gives

$$\dot{\gamma} = \frac{\dot{\sigma}}{G} + \frac{\sigma}{\eta} \tag{15.19}$$

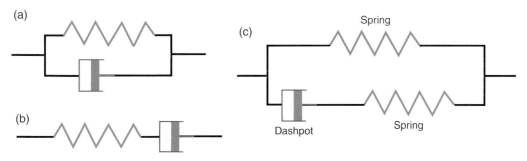

Figure 15.3   *Standard mechanical constitutive equations for viscoelasticity can be derived for (a) the Kelvin model (spring and dashpot in parallel), (b) the Maxwell model (spring and dashpot in series) and (c) the standard linear solid (spring and dashpot in parallel with a spring).*

The decay of stress with time when a rapid strain is applied to the Maxwell model can be calculated from equation (15.19) and is given by

$$\sigma(t) = \sigma_0 e^{-t/\tau_m} \tag{15.20}$$

where the time constant is given by $\tau_m = \eta/G$ and σ_0 is the initial stress. The time-dependent relaxation modulus is given by $G(t) = \sigma(t)/\gamma$ and a simple expression for it can then be calculated for a Maxwell material,

$$G(t) = \frac{\sigma_0}{\gamma} e^{-t/\tau_m} \tag{15.21}$$

Similarly the decay of strain (γ) with time if a stress (σ_0) is applied can be found using the Kelvin model through solution of equation (15.18). It is

$$\gamma = \frac{\sigma_0}{G} \left[1 - e^{-Gt/\eta} \right] \tag{15.22}$$

More realistic mechanical models can be created that capture both the stress and strain relaxation simultaneously, such as the standard linear solid (**Figure 15.3c**). However, these models are still very much the domain of bioengineering, a molecular biophysicist hopes to make the connection between the viscoelasticity and the molecular structure for a detailed reductionist understanding of the phenomena (**Section 15.3**).

15.2   Rheological Functions

Experiments that study linear viscoelasticity are a form of mechanical spectroscopy. The sample is struck with a mechanical perturbation (either a stress or a strain) and its response is measured as a function of time. A series of equivalent rheological measures can be used to quantify such experiments (**Figure 15.4**); these include the

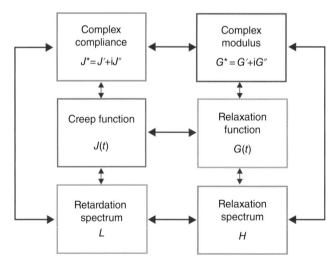

Figure 15.4 *Interrelationship between linear measures of viscoelasticity; complex compliance (J*), complex modulus (G*), creep function (J(t)), relaxation function (G(t)), retardation spectrum (L) and the relaxation spectrum (H). [Reprinted with permission from J.W. Goodby, R.W. Hughes, 'Rheology for Chemists', RSC, 2000. Copyright © 1990, American Chemical Society.]*

complex compliance ($J^* = J' + iJ''$, $J^* = \gamma^*/\sigma^*$), the complex modulus ($G^* = G' + iG''$, $G^* = \sigma^*/\gamma^*$), the creep function ($J(t)$, $J(t) = \gamma(t)/\sigma$), the relaxation function ($G(t)$, $G(t) = \sigma(t)/\gamma$), the retardation spectrum (L) and the relaxation spectrum (H). This range of measures is convenient for a number of reasons. Often, an experimentalist needs to impose a low stress or strain to obtain a linear response and this can be facilitated by the measurement of a particular rheological function. The data can then be mathematically translated to a different function for comparison with an actual process at a different stress or strain. The use of linear transformations also allows the prediction of viscoelastic behaviour over a wide range of time scales, e.g. outside the experimental time window for a single rheological function.

A flavour of some of the mathematical interrelationships between the rheological functions are shown in equations (15.23)–(15.26). All of these transformations are linear and well determined. The data can in principle be transformed back and forth between these functions with no loss of information,

$$G(t) = \int_{-\infty}^{\infty} He^{-t/\tau} d\ln\tau = \frac{2}{\pi} \int_{-\infty}^{\infty} \frac{G''(\omega)}{\omega} \cos\omega t d\omega \tag{15.23}$$

$$G(t) = \frac{2}{\pi} \int_{-\infty}^{\infty} \frac{G'(\omega)}{\omega} \sin\omega t d\omega, \quad G'(\omega) = \frac{J'(\omega)}{J'(\omega)^2 + J''(\omega)^2} \tag{15.24}$$

$$G''(\omega) = \frac{J''(\omega)}{J'(\omega)^2 + J''(\omega)^2}, \quad J'(\omega) = \omega \int_0^{\infty} J(t) \sin\omega t dt \tag{15.25}$$

$$J''(\omega) = -\omega \int_0^{\infty} J(t) \cos\omega t dt, \quad J(t) = \int_{-\infty}^{\infty} L\left[1 - e^{-t/\tau}\right] d\ln\tau \tag{15.26}$$

These numerical transformations are very standard and a number of software packages are available that provide robustly implemented algorithms for their calculation.

15.3   Examples from Biology: Neutral Polymer Solutions, Polyelectrolytes, Gels, Colloids, Liquid Crystalline Polymers, Glasses, Microfluidics

15.3.1   Neutral Polymer Solutions

Many polymeric solutions exist in biological systems. Even solid polymers such as collagen and spider silk pass through a concentrated lyotropic phase during their synthesis. The dynamics of flexible neutral polymers is much simpler than the scenario with charged polymers (polyelectrolytes) and this will be the initial focus of the discussion. There are three basic concentration regimes that affect the dynamics of neutral flexible polymeric molecules; *dilute* (the chains do not overlap), *semidilute* (the chains form an overlapping mesh) and *concentrated* (the thermal blob size is equal to the correlation length and the chains are Gaussian on all length scales). **Figure 15.5** shows the range of dynamic behaviour exhibited by a neutral flexible polymer chain as a function of the number of monomers in a chain (N) and the concentration of monomers (c). In the dilute regime the solutions are predominantly viscous fluids. Above the overlap concentration (c^*) the chain dynamics are strongly coupled with an interplay of Zimm and Rouse modes due to screening of the hydrodynamics of one chain by another. The elastic component of the viscoelasticity becomes important at lower and lower frequencies as the polymer concentration is increased. Polymers can *reptate* above a critical polymer length and

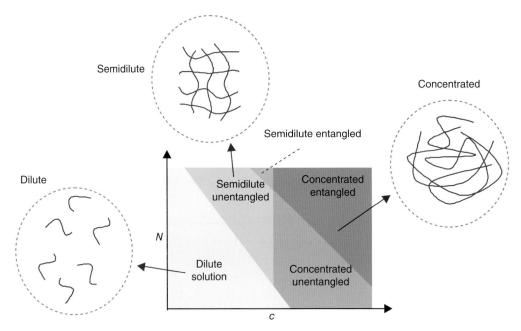

Figure 15.5 *The phase diagram for the dynamics of neutral flexible polymer solutions. The number of monomers in the polymer chains (N) is shown versus the polymer concentration (c).*

monomer concentration (**Section 10.5**), which has a dramatic effect on their viscoelasticity; they demonstrate *entangled* rheology. The concentration at which the dynamic transition to reptation occurs is called the *entanglement concentration* (c_e).

The *overlap concentration* (c^*) is defined for the transition between the dilute and semidilute concentration regimes (**Figure 15.5**). This concentration, in units of number of chains per unit volume, is given by

$$c^* = \frac{1}{4\pi R^3/3} \tag{15.27}$$

where R is the end-to-end distance of a chain and the chains are assumed to inhabit a spherical volume. In the case of rigid molecules the end-to-end distance is equal to the long axis of the molecular rod (the contour length, L).

The *Rouse and Zimm models*, were encountered previously (**Section 10.5**), for polymer dynamics and can be used to describe the viscoelasticity of flexible polymer solutions (**Figures 15.6 and 15.7**). At low frequencies both models predict power laws of 1 and 2 for the frequency dependence of G' and G'' respectively (which corresponds to a Maxwell model). At high frequencies a $\omega^{2/3}$ dependence for both components of the complex shear modulus on frequency is observed for the Zimm model ($G' \sim G''/1.73$), whereas there is a $\omega^{1/2}$ dependence for Rouse modes ($G' \sim G''$). The detailed analysis of the linear viscoelastic spectrum of flexible polymers using the whole range of dynamic modes can become quite involved, so only a few key results will be quoted here. The relaxation modulus ($G(t)$) of semidilute polymer solutions of relatively short chains can be calculated with the Rouse model. For long chains the reptation concept is also required. Stresses in the Rouse model are determined by the orientation of the chains under the influence of an external force. The steady-state shear viscosity (η) and steady-state compliance ($J(t)$) of Rouse macromolecules are proportional to the number of links (N) in the chain. Solutions of long polymer chains exhibit unusual viscoelastic properties due to their

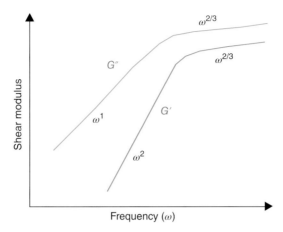

Figure 15.6 *The complex shear modulus (G', G'' as a function of frequency ω) for polymer chains that obey the Zimm model for their dynamics (dilute polymer solutions and semidilute chains at length scales below the mesh size). Shear moduli follow G''∼ω¹, G''∼ω² at low frequencies (identical to the prediction of a Maxwell model) and G' ∼ G''/1.73 ∼ ω²/³ at high frequencies.*

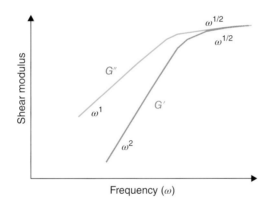

Figure 15.7 *The complex shear modulus as a function of frequency (ω) for polymer chains that obey the Rouse model for their dynamics (concentrated polymer solutions, and semidilute solutions at length scales above the mesh size). The shear moduli follow G'' ∼ ω¹, G' ∼ ω² at low frequencies (identical to the prediction of a Maxwell model) and G' ∼ G'' ∼ ω¹/² at high frequencies.*

reptative motion. The reptation model presented in **Section 10.5** provides a calculation of the viscosity (η) of polymeric solutions and the viscosity was found to be a strong function of the degree of polymerisation ($\eta \sim N^3$). However, experiments indicate that the viscosity follows a 3.4 power law dependence on the degree of polymerisation ($\eta \sim N^{3.4}$) in real polymeric solutions (**Figure 15.8**). This deviation from the prediction of the reptation model is thought to be associated with fluctuations of the contour length of the tubes and the tube-renewal process. The simple picture of a stationary tube of constraints provided in **Section 10.5**, thus needs to be modified. Reptative dynamics in polymeric solutions can be induced by a range of factors; an increase in the degree of polymerisation (the viscosity varies from a N^1 to $N^{3.4}$ dependence as the material starts to reptate), an increase in the persistence length of the chain, addition of bulky side groups to the polymeric molecules and by an increase of the polymer concentration (**Figure 15.8**).

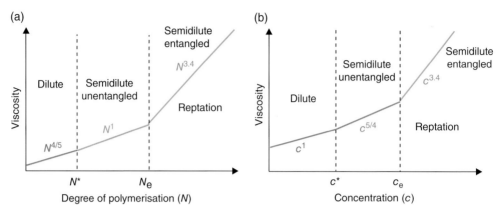

Figure 15.8 *The dependence of the viscosity of a* neutral flexible polymer *solution on (a) the degree of polymerisation of the chains (N) and (b) the monomer concentration (c). In (a) the transition between the degree of polymerisations for overlap (N*) and entanglement (N_e) are shown. In (b) the semidilute overlap concentration (c*) and the entanglement concentration (c_e) are shown. [Reprinted with permission from R. Colby, M. Rubinstein, Macromolecules, 1990, 23, 2753–2757. Copyright © 1990, American Chemical Society.]*

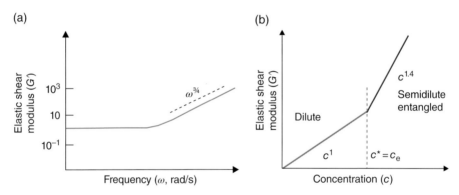

Figure 15.9 *The elastic shear modulus (G′) of long semiflexible polymeric networks has a number of unique experimental signatures. (a) The viscoelasticity as a function of frequency (ω) follows $G′ \sim ω^{3/4}$ at high frequencies and (b) the modulus as a function of polymer concentration (c) follows a $G′ \sim c^{1.4}$ scaling above the overlap concentration. [Reprinted with permission from D. Morse, Macromolecules, 1998, 31, 7044–7067. Copyright © 1998, American Chemical Society.]*

When the persistence length of a polymer chain is significant compared to its contour length *semiflexible* models are required to describe the contributions of the internal modes of the polymer chain to the viscoelasticity (**Chapter 10**). There are a wide range of biological materials that consist of networks of such semiflexible rods and ropes. These materials conform to the predictions of relatively new theoretical models for semiflexible chain dynamics, e.g. intermediate filaments, actin, peptide fibrils and microtubule solutions. Two experimental effects that are well predicted by the semiflexible models are the high-frequency viscoelasticity and the polymer concentration dependence of the shear modulus. Shear storage moduli of semidilute solutions of semiflexible actin filaments determined using diffusing wave spectroscopy are shown in **Figure 15.9**. At very high frequencies the elastic modulus for semiflexible chains differs from that predicted by Rousse and Zimm models. The shear modulus follows a power-law dependence on the frequency (*ω*) with an exponent of 0.75 ($G' \sim \omega^{3/4}$), which is thought to be characteristic of the transverse fluctuations of the semiflexible chains.

15.3.2 Polyelectrolytes

The viscoelasticity of polyelectrolytes is a tricky field to approach both experimentally and theoretically, and is still under development. Examples of biological polyelectrolytes include nucleic acids, sea-weed extracts, hyaluronic acid, proteoglycans, glycoproteins and muscle proteins. Hyaluronic acid is contained in articulated joints and the rheology of this polyelectrolyte in synovial fluid is important for the mobility of the joint. Proteoglycans and glycoproteins (both are comb polyelectrolytes) fulfil a wide range of biological roles, as shock absorbers, for the reduction of friction, and as barrier materials, e.g. the protection of the stomach from self-digestion and the minimisation of the adhesion of eye balls to eyelids. The rheology of DNA has a range of biotechnological applications, e.g. to provide a detailed understanding of the results of electrophoresis experiments or to alleviate respiratory problems in cystic fibrosis sufferers that have DNA from bacterial infections in their lungs (patients are treated with DNAase to reduce the sputum viscosity). Due to the relatively large size of DNA molecules they can be examined on a single-molecule basis, which has allowed single-biomolecule polyelectrolyte rheology experiments to be performed.

Polyelectrolytes can be subdivided into flexible, semiflexible and rigid classifications that depend on the persistence length of the backbone in a similar manner to neutral polymers. However, the inclusion of charges along the backbone of a polyelectrolyte tends to increase the chains' persistence length when compared with their neutral counterparts (**Section 11.9**). Semiflexible and rigid polyelectrolyte chains can be modelled by direct adaptation of the neutral results. Indeed, the predictions are identical to those of neutral polymers in the case of high salt, where most of the charged interactions are screened. To predict the viscoelasticity of flexible polyelectrolytes, blob models can be used that adjust the results for neutral flexible polymers through the inclusion of the statistics of charged blob chain conformations and examples of the predictions of scaling models are shown in **Figure 15.10**. The entangled regime is highlighted by a change of the power-law dependence on the concentration of the viscosity from ½ to 3/2 ($\eta \sim c^{1/2}$ to $\eta \sim c^{3/2}$) (**Figure 15.10**) and the unentangled semidilute regime (with a $\eta \sim c^{1/2}$ dependence of the viscosity) is over

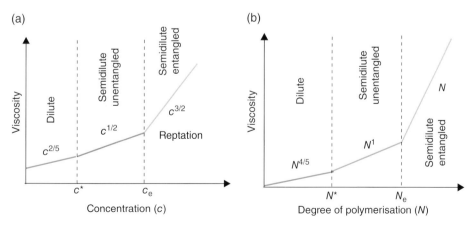

Figure 15.10 *Predictions for the scaling dependence of the viscosity of a solution of flexible polyelectrolytes on (a) the polymer concentration (c) and (b) the degree of polymerisation (N). [Reprinted with permission from A.V. Dobrynin, R.H. Colby and M. Rubinstein, Macromolecules, 1995, 28, 1859–1871. Copyright © 1995, American Chemical Society.]*

an anomalously large range of polymer concentrations. This behaviour is well established experimentally for flexible polyelectrolytes. Polyelectrolytes shear thin at very low shear rates, which makes measurement of intrinsic viscosities more challenging with these materials, but particle-tracking microrheology experiments and low shear rate bulk rheometers appear to give reliable results. The shear-thinning property of polyelectrolytes is practically useful in articulated joints, since the resistive forces decrease as the shear rate is increased (**Figure 15.11**).

A complete understanding of the linear viscoelasticity of flexible polyelectrolytes is still being developed. An example of experimental results from an entangled flexible polyelectrolyte solution is shown in **Figure 15.12**, which shows the complex shear moduli ($G^*(\omega) = G'(\omega) + iG''(\omega)$) of synovial fluids from umbilical cord and articulated joints.

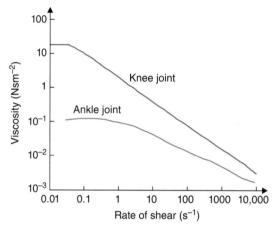

Figure 15.11 *Nonlinear shear thinning rheology of synovial fluids (flexible polyelectrolytes) from the ankle and knee joints. The viscosity is shown as a function of shear rate. [With kind permission from Springer Science + Business Media: R.G. King, Rheol. Acta, 1966, 5, 41–44.]*

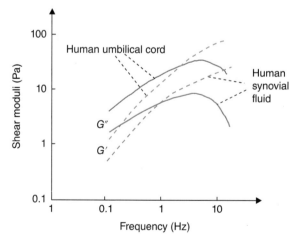

Figure 15.12 *The shear moduli of fluids from the human umbilical cord (blue) and synovial fluids (green) as a function of frequency. Both materials contain large quantities of hyaluronic acid, a flexible polyelectrolyte. G″ is indicated by the continuous lines and G′ by the dashed lines. [Reprinted with permission from D.A. Gibbs, Biopolymers, 1968, 6, 777–791. Copyright © 1968 John Wiley & Sons, Inc.]*

15.3.3 Gels

A gel is a material in which the polymeric components are crosslinked to form a network (**Figure 15.13**). The crucial difference between a rubber (**Section 10.3**) and a gel is that there is a large amount of solvent associated with the components of a gel, which swells its microstructure and alters the mobility of the chains. Gelled phases of matter are realised by a range of biological molecules, e.g. many foods are gelled biopolymers such as starches (custard, and Turkish delight), pectins (jams) and denatured collagen (table jelly).

Crosslinks in biopolymers are separately categorised as *physical* (of a weak nature, e.g. electrostatic or hydrogen bonds) or *chemical* (strong in nature, e.g. covalent/ionic bonds). With physical gels the moduli become low when the crosslinks (e.g. helical, or egg-box connections between chains) are melted at high temperatures, but the moduli increase back to their previous levels upon a subsequent decrease in temperature. Chemical gels in contrast cannot be reformed, since permanent chemical crosslinks exist between the subunits (often disulfide links exist in collagenous systems). Heating chemical gels to high temperatures would result in a complete irreversible breakdown of the chemical structure and a dramatic reduction in their viscoelastic moduli.

In many physical biopolymer networks the degree of crosslinking can be switched on or off by variation of the temperature, solvent quality, electrostatics or the introduction of specialised biomolecular crosslinkers. At the point at which the degree of crosslinking is sufficiently large for one complete aggregate to span the sample volume (called the *percolation threshold* or the *gel point*), the relaxation modulus ($G(t)$) is often well described by a simple power law (the gel equation),

$$G(t) = St^{-n} \tag{15.28}$$

where S is a scaling constant called the strength of the gel, n is the scaling exponent, and t is the time. Using equations (15.23)–(15.26) to transform between the different measures of linear viscoelasticity, the complex shear moduli (G^*) can be constructed from this simple power law for the relaxation function ($G(t)$) and then compared with an oscillatory drag flow rheometry experiment. Above the percolation threshold and at low frequencies where the internal modes cannot be observed, the shear moduli of a gel has a weak frequency dependence (**Figure 15.14**). Furthermore, characteristic rubbery-type behaviour occurs at low frequencies with the elastic modulus much higher than the dissipative modulus ($G' >> G''$). To a first approximation, the elastic shear modulus is proportional to the density of crosslinks in a gel, similar to the behaviour of crosslinked rubbery networks of flexible polymers (**Section 15.3**).

The model of *sticky reptation* has been developed to describe the rheology of associating physical gels (**Figure 15.15**). The temporary stickers that form the crosslinks between the polymers introduce a second time

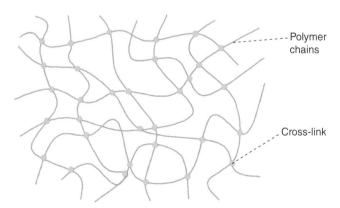

Figure 15.13 *Schematic diagram of a polymeric gel showing the chemical crosslinks (blue circles). The solvent molecules that swell the network and help determine the chain conformations are not shown for simplicity.*

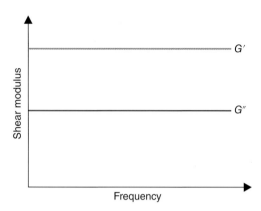

Figure 15.14 *Schematic diagram of the complex shear moduli of a model gelled flexible polymeric network at low frequencies;* G' > G" *and the frequency dependence is fairly weak. Note the similarity with* **Figure 15.9a** *for a semiflexible network at low frequencies for* G'.

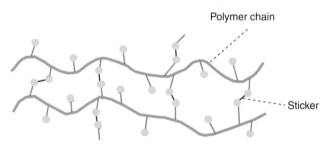

Figure 15.15 *The entangled rheology of associating physical gels can be described using the theory of sticky reptation. An additional time scale is introduced into the chain dynamics due to the sticker lifetime, which radically slows down the motion of the chains. Chains move like many-legged millipedes through their reptation tubes.*

scale to the dynamics of the polymer chains; the time for a sticker to dissociate. Above the entanglement concentration the chains move like centipedes through their reptation tubes with stickers for legs. The viscosity therefore becomes a very strong function of the polymer concentration (**Figure 15.16**) compared with that for unassociating polymer solutions, due to the dramatic slowing down of the chain motion with the introduction of more stickers. Other unusual phenomena can also occur with sticky polymers such as *shear thickening* (polymer solutions normally shear thin), where the resistance of the network to flow (the shear modulus) increases with increasing shear rate. The increased shear rate increases the number of sticky links between chains.

Many naturally occurring gels are formed from polyelectrolytes. The osmotic pressure of the counterions associated with the polymer chains in these materials has a series of dramatic consequences for the physical properties of charged gels. The free counterions increase the swelling of the gel by a large degree, and modify the elastic moduli. The mechanical properties of charged gels are an important consideration in a range of biological problems, such as the viscoelasticity of cartilage, the cornea and striated muscle.

Many common biological polymers are semiflexible and their crosslinked networks have a range of unique properties that relate to the increased rigidity of the chains (**Figure 15.17**). One interesting phenomena is that of

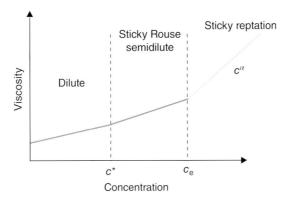

Figure 15.16 *The viscosity of a solution of associating neutral flexible polymers as a function of polymer concentration (*c*). The viscosity of an associating polymeric solutions is a very strong function of the polymer concentration. The dynamic behaviour can be very rich. A simplified phase diagram is shown and the sticker lifetimes modify the dynamics in both the semidilute (Rouse* $c^* < c < c_e$*) and concentrated (reputation* $c > c_e$*) regimes. The exponent* α *in the sticky reputation regime (*$\eta \sim c^\alpha$*) can be much larger than the value for nonassociating polymers, i.e.* $\alpha > 3.4$ *(**Figure 15.8b**). [Reprinted with permission from M. Rubinstein, A.N. Semenov, Macromolecules, 1998, 31, 1386–1397. Copyright © 1998, American Chemical Society.]*

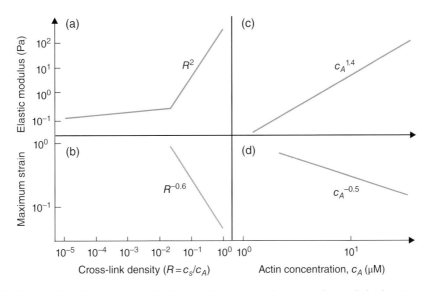

Figure 15.17 *Trends found experimentally in the linear viscoelasticity of crosslinked actin, a well-defined semiflexible polymeric gel. (a) The elastic modulus as a function of crosslinking density (R), (b) the maximum strain resisted by the gels as a function of crosslinking density (R), (c) the elastic modulus as a function of actin concentration (*c_A*) at a fixed crosslinking density (R = 0.13), and (d) the maximum strain as a function of actin concentration (*c_A*) at fixed crosslinking density (R = 0.13). [From M.L. Gardel, J.H. Shin, F.C. MacKintosh, L. Mahadevan, P. Matsudaira, D.A. Weitz, Science, 2004, 304, 1301–1305. Reprinted with permission from AAAS.]*

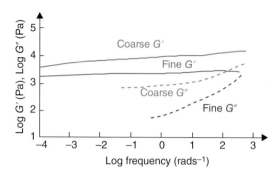

Figure 15.18 *Linear viscoelasticity of fibrin clots responsible for the clotting process in blood (both fine and coarse clots are shown). The elastic (G′) and the loss (G″) shear moduli are shown as a function of frequency. Semiflexible fibres occur in the clots that contain a large number of crosslinks and demonstrate almost perfect frequency independent elasticity over six orders of magnitude in frequency (compare with* **Figure 15.14**). [Reprinted from W.W. Roberts, O. Kramer, R.W. Rosser, F.H.M. Nestler, J.D. Ferry, Biophy. Chemist, 1974, 152–160. With permission from Elsevier.]*

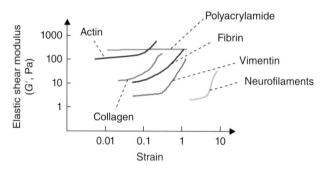

Figure 15.19 *Elastic shear modulus (G′) as a function of strain for polymeric networks. Strain hardening of semiflexible networks is shown by the increase in the storage modulus (G′) with strain for a range of semiflexible polymers. A polyacrylamide gel is shown for comparison as it is a synthetic flexible polymer network that does not strain harden. [Reprinted by permission from Macmillan Publishers Ltd: C. Storm, J.J. Pastore, F.C. MacKintosh, T.C. Lubensky, P.A. Janmey, Nature, 2005, 435, 191–194.]*

strain hardening, the elastic modulus increases as the samples are strained. The semiflexible networks of fibrin in blood clots have a virtually perfect signature of elasticity in their linear shear moduli over a wide range of frequencies (seven orders of magnitude, **Figure 15.18**). However, the semiflexibility of the fibrin chains is clearly observed in the strain-hardening phenomena at high strains (the shear moduli increase with strain). Flexible chain elastomeric gels (e.g. polyacrylamide gels) do not show such behaviour (**Figure 15.19**)*. Actin chains that form the cellular cytoskeleton are semiflexible and there are a wide range of specialised crosslinking proteins that hold the chains either parallel (bundling proteins) or at an angle (gel-forming proteins). Thus, evolution has created a wide variety of methods to fine tune the viscoelasticity of the cytoskeleton for a particular biological role.

---

*Recent concerns have been raised that the profiles in figure 15.19 may be very sensitive to the rate of strain used. See C. Semmrich, R.J. Larsen, A.R. Bausch, Soft Matter, 2008, 4, 1675–1680.

15.3.4 Colloids

Colloidal science encompasses the large field of research that relates to particulate dispersions. One of the simplest examples of a colloidal system is a monodisperse suspension of identical spherical particles, and important biological examples of this scenario are provided by globular proteins and icosohedral viruses in solution.

For dilute dispersions of spheres the flow field is shown in **Figure 15.20**. From solution of the Navier–Stokes equations at low Reynold's number, Einstein found a surprisingly simple expression that relates the viscosity (η) to the volume fraction (φ) of colloids,

$$\eta = \eta_0 \left(1 + \frac{5\varphi}{2} + O\left(\varphi^2\right) \right) \tag{15.29}$$

where $O(\varphi^2)$ is a term of order φ^2 that is neglected, and η_0 is the viscosity of the solvent. This Einstein equation can only model the behaviour of fairly dilute solutions (**Figure 15.21**) at low shear rates. At higher shear rates dilute colloidal solutions typically shear thin and the effects of a number of competing hydrodynamic effects then need to be calculated.

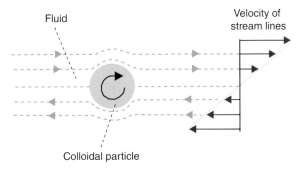

Figure 15.20 *Steady flow field in a fluid around a spherical colloidal particle. The shear field has a vorticity and the colloidal particle rotates with a constant angular velocity.*

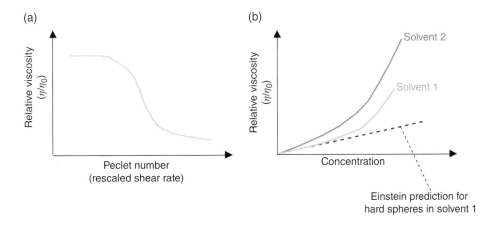

Figure 15.21 *The relative viscosity (viscosity (η)/viscosity of solvent (η_0)) of colloidal suspensions as a function of (a) Peclet number (a rescaled shear rate) and (b) the concentration of the colloids. Two different solvents 1 and 2 are shown in (b).*

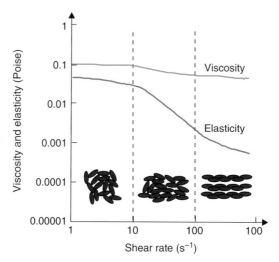

Figure 15.22 *The viscosity and elasticity of blood cells as a function of shear rate. The nonlinear rheology of blood cells and colloidal gels can demonstrate shear-thinning behaviour (and the related phenomenon of thixotropy due to transients in the structural rearrangement of the colloids).*

The phase behaviour of a colloidal dispersion has a strong effect on the viscosity of the solutions. An increase of the volume fraction of colloids above 2–3% can also cause a break down in equation (15.29) due to the change in the colloidal microstructure and a wide range of gel, fluid, or jammed phases are possible, e.g. concentrated suspensions of corn starch can shear thicken due to a jamming transition (people can walk across swimming pools filled with corn starch). Other novel phenomena occur when high density colloidal solutions are sheared. **Figure 15.22** indicates a shear-thinning phenomenon in suspensions of blood cells, also discussed in **Section 14.5**. At low shear rates the blood cells aggregate into a *rouleaux* structure. An increase in the shear rate causes a break up of these structures, which decreases the viscosity. At even higher shear rates, there is a further reduction in viscosity associated with the formation of strings of blood cells, as they slide more easily past one another in a shear field. This formation of chain like aggregates has also been observed with synthetic colloids and it is thought to be due to a coupled hydrodynamic interaction.

15.3.5 Liquid-Crystalline Polymers

Many biological polymers have liquid-crystalline phases, e.g. DNA, cellulose, carrageenan, and α-helical peptides. These liquid-crystalline polymers strongly shear thin in ways that depend on their defect structures (**Section 6.3**). It is thus important to measure the orientation of the nematic director (with chiral and smectic order parameters if necessary) with respect to the direction of shear to understand the viscoelastic properties of these materials. An example of the nonlinear flow properties of nematic liquid-crystalline polymers is shown in **Figure 15.23**. Different degrees of shear thinning occur with shear rate. Often, three power-law behaviours are observed experimentally with nematic liquid crystals, a type I Newtonian plateau, followed by type II shear thinning $\eta \sim \dot{\gamma}^{-1/2}$ and finally a further type III Newtonian plateau occurs.

15.3.6 Glassy Materials

Many amorphous biological materials exhibit glassy phenomena (e.g. resilin at frequencies below 10 Hz). Ergodicity (the exploration of all the molecules' energy states) is lost in these materials. Glasses result

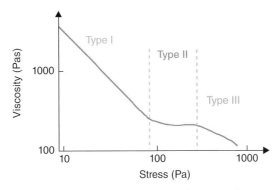

Figure 15.23 *The viscosity of a liquid-crystalline polymer as a function of stress. The flow behaviour of liquid-crystalline polymers as a function of stress (or shear rate) often has three characteristic regimes; Type I shear thinning, the Type II Newtonian plateau and Type III shear thinning. (Data from a cellulose derivative). [Reproduced from S. Onogi, T. Asada, in Rheology, Proceedings of the Eighth International Congress on Rheology, Naples, Italy, 1980, Plenum, 126–136. With permission from Springer ScienceSpringer Science + Business Media.]*

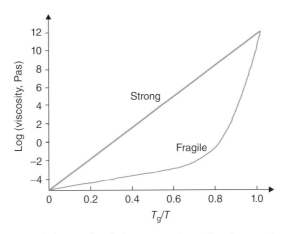

Figure 15.24 *The viscoelasticity of glasses (fragile/strong, etc.) can be characterised by plots of the viscosity as a function of the reciprocal of the temperature (rescaled by the glass transition temperature* T_g*). [Reprinted from C.A. Angell, J. Non-Cryst. Sol, 1988, 102, 205. With permission from Elsevier.]*

from, in the simplest picture, undercooled liquids in which the viscosity increases so rapidly that it prevents the formation of a crystalline phase. Everyday examples of biological glasses are boiled sweet (sugars) and a wide range of biopolymers in lowly plasticised states, e.g. amorphous starches in dehydrated foods.

Glasses are further categorised as *strong* and *weak*. In strong glasses the short-range order tends to persist above the glass transition, whereas fragile glasses have no such memory. The viscosity of glasses has a characteristic dependence on the temperature (**Figure 15.24**).

The linear viscoelasticity of the cytoskeleton of many cells is found to be scale invariant. The cytoskeletal dynamics conform to a model of *soft glassy* relaxation behaviour. Nonergodicity is therefore a common phenomena in a range of *in -vivo* biological processes. The globularisation of polymer chains below their glass transition can lead to morphologies with hysteresis in their structure and sometimes an analogy is made to describe the behaviour of globular proteins in terms of glassy dynamics.

15.3.7 Microfluidics in Channels

There is a large current research effort in the area of microfluidics and microrheology. These studies consider the rheology of viscoelastic fluids as a function of length scale. Examples include how blood cells squeeze through the micrometre-sized capillaries in arteries (the effective viscosity is much smaller than expected from a simple Pouseille flow calculation, **Figure 15.1a**, the Fahreus–Lindvist effect, **Section 14.5**), the interaction between mucins and cilia in the lungs, and the motion of DNA on templated surfaces for separation technologies.

15.4 Viscoelasticity of the Cell

Studies of the viscoelasticity of single cells have led to many new physical insights into physiology and disease. Often, the viscoelasticity of cells changes in diseased states (e.g. cancerous cell lines) due to large-scale remodelling of the cytoskeleton, osmotic pressure deregulation, and cell-membrane modification. Cell viscoelasticity sensitively depends on the type of cell considered and plays an important role in differentiation, morphogenesis and the cell's activity during its life cycle.

Consider a Maxwell model for the creep response of a cell to a step change in the applied force, (F), from zero to F_0 at a time $t = 0$. The solution for the compliance $(J(t)$, **Figure 15.25**$)$ is

$$J(t) = \frac{x(t)}{F_0} = \frac{1}{G} + \frac{t}{\eta} \tag{15.30}$$

A standard linear solid model for the creep response of a cell in response to a step change in force F from zero to F_0 at time $t = 0$ is shown in **Figure 15.26**. The solution is

$$J(t) = \frac{x(t)}{F_0} = \frac{1}{G_1} \left(1 - \frac{G_0}{G_0 + G_1} e^{-t/\tau} \right) \tag{15.31}$$

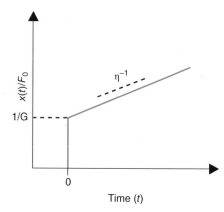

Figure 15.25 *Rescaled extension $(x(t)/F_0 = J(t)$, the compliance) as a function of time for a Maxwell model subject to a constant force (F_0) at time $t = 0$.*

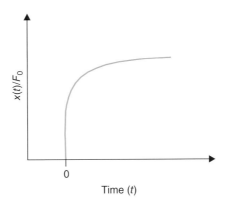

Figure 15.26 *Rescaled extension (x(t)/F$_0$ = J(t), the compliance) as a function of time for a standard linear solid model subject to a constant force (F$_0$) at time* t = 0.

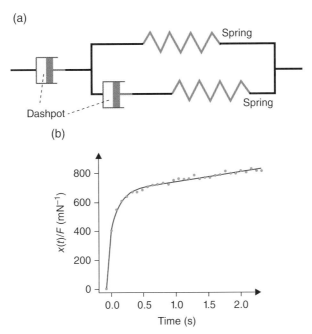

Figure 15.27 *(a) Lumped parameter model for the viscoelasticity of a cell is constructed from two dashpots and two springs. (b) Creep compliance (J(t) = x(t)/F) as a function of time for a cell measured in a magnetic-bead microrheology experiment with a fit of the lumped-parameter model. [Reprinted from A.R. Bausch et al, Biophysical Journal, 1998, 75, 2038–2049. With permission from Elsevier.]*

where τ is the relaxation time. This model does a much better job of the representation of the creep behaviour of a cell subjected to magnetic-bead microrheology experiment than that of the Maxwell model. However, a more sophisticated lumped parameter model (additional springs and dashpots are involved) does even better and is shown in **Figure 15.27**. The lumped-parameter model for the viscoelastic creep response of a cell subjected to a sudden force by magnetic-bead rheometry is compared with the model on the figure.

An important factor that drives the research on single-cell mechanics has been the development of new sensitive experimental techniques (that can measure fN–pN forces) to measure the viscoelasticity of these fragile, soft materials (microrheology, **Chapter 19**). Cell stretchers, optical/magnetic tweezers, shear flow devices, micropillars, biaxial membrane stretchers, and particle tracking image analysis have all proven to be important tools for cellular biomechanics.

Suggested Reading

If you can only read one book, then try:

Goodwin, J.W. & Hughes, R.W. (2008) *Rheology for Chemists*, 2nd edition, Royal Society of Chemistry. A very clear simple introduction to the key concepts of rheology.

Larson, R.G. (1999) *The Structure and Rheology of Complex Fluids*, Oxford University Press. Advanced level text, an extensive range of material is covered.

Macosko, C.W. (1994) *Rheology: Principles, Measurements and Applications*, Wiley. Good discussion of rheometer geometries and instrumentation.

Mewis, J. & Wagner, N.J. (2011) *Colloidal Suspension Rheology*, Cambridge University Press. Excellent introduction to the dynamics of colloidal materials.

Morrison, F.A. (2001) *Understanding Rheology*, Oxford University Press. Good introduction to three-dimensional flow analysis with viscoelastic materials.

Rubinstein, M. & Colby, R.H. (2003) *Polymer Physics*, Oxford University Press. Excellent introduction to polymeric models of viscoelasticity.

Tutorial Questions 15

15.1 An athlete is in a marathon. During their run, the cartilage surfaces in their synovial knee joints move relative to one another at a velocity of $1 \, \text{cm s}^{-1}$. The distance between the two cartilage surfaces can be of the order of $10 \, \mu\text{m}$. Calculate the shear rate experienced by the synovial fluid. Estimate the Peclet number of a hyaluronic acid chain in the fluid if its radius is $1 \, \mu\text{m}$ and the viscosity of the surrounding fluid is $0.001 \, \text{Pa s}$.

15.2 The longest time scale important in the viscoelasticity of a boiled sweet (glassy sugar molecules) can be described using a Maxwell model. Calculate the time scale for stress relaxation if the viscosity is $1 \, \text{MPa s}$ and the elastic modulus is $10^{-3} \, \text{Pa}$.

15.3 It is assumed that there is no interaction between blood cells in a suspension. Estimate the viscosity of the dilute blood suspension if the viscosity of the surrounding buffer is $10^{-3} \, \text{Pa s}$ and the volume fraction of the blood cells is 2%.

15.4 Explain what is meant by the term thixotropy. Give some examples of thixotropic materials.

16

Motors

A current challenge for the synthetic nanotechnology industry is how to transport chemical cargoes at the molecular scale in order to construct new materials, remove waste products and catalyse reactions. Nature has already evolved a wide range of efficient nanomotors that are used in a huge number of biological processes. Cells actively change their shape and move with respect to their environment, e.g. the contraction of muscle cells in the heart, replication, transcription and translation of DNA, movement of macrophages to capture and remove hostile cells, division of cells during mitosis and the rotation of flagella to propel bacteria. As a common theme, chemical energy derived from the hydrolysis of ATP (or GTP with microtubules) or stored in a proton gradient (with bacteria), is transformed into mechanical work to drive the cell's motility. There are currently thought to be five major mechanisms for molecular motility that occur naturally; *self-assembling motors*, *linear stepper motors*, *rotatory motors*, *extrusion nozzles* and *prestressed springs* (**Figure 16.1**).

Adenosine triphosphate (ATP) is the central currency in energy transduction in biological systems and it is useful to examine the reaction of the molecules in more detail (**Section 1.10**). The dissociation of ATP into ADP and a free phosphate ion liberates a reasonable amount of energy ($\sim 20\,kT$) and is used to power a wide range of biochemical reactions,

$$\text{ATP} \underset{\longleftarrow}{\longrightarrow} \text{ADP} + \text{P}_i \tag{16.1}$$

where ATP signifies a range of species with different degrees of ionisation, e.g. $MgATP^{2-}$, ATP^{4-}, etc., P_i is the free phosphate ion and ADP is adenosine diphosphate. The equilibrium constant (K) for energy transduction from ATP has the same units as concentration (M),

$$K = \frac{c_{ADP} c_{Pi}}{c_{ATP}} = 4.9 \times 10^5 \text{M} \tag{16.2}$$

where c denotes the concentration of the chemical species.

The Physics of Living Processes: A Mesoscopic Approach, First Edition. Thomas Andrew Waigh.
© 2014 John Wiley & Sons, Ltd. Published 2014 by John Wiley & Sons, Ltd.

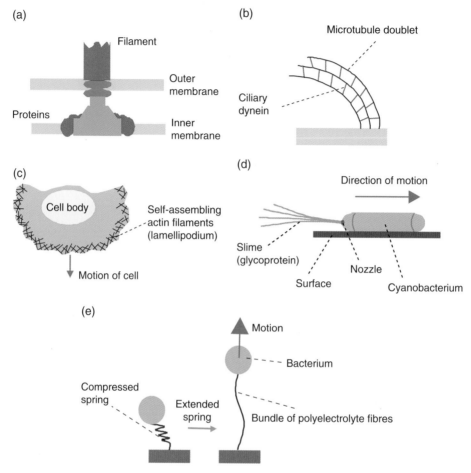

Figure 16.1 *Examples of the five separate strategies for biological motors; (a) rotatory motors in bacterial locomotion, (b) linear stepper motors in cilia, (c) self-assembling motors formed from actin filaments in lamellipodium, (d) extrusion nozzles in cyanobacterium, and (e) prestressed springs in bacterial locomotion.*

The equilibrium constant depends on several factors such as the free magnesium concentration, the pH and the ionic strength. The value given for the equilibrium constant is for the standard conditions found in the cytoplasm of a vertebrate cell. The amount of energy liberated (ΔG) by the ATP reaction is given by

$$\Delta G = \Delta G_0 - kT \ln \frac{c_{ATP}}{c_{ADP}c_{Pi}} \tag{16.3}$$

where ΔG is the standard free energy and $\Delta G_0 = -54 \times 10^{-21}$ J. Thus, the free energy of the ATP hydrolysis reaction depends on both the standard free energy and the concentration of ATP, ADP and P. The $\sim 20\ kT$ per ATP hydrolysis event quoted is therefore a useful rule of thumb, but more accurate calculations are possible if the specific conditions are known. Motor protein enzymes can thus liberate energy from ATP during conformational (mechanical) changes in their low Reynold's number aqueous environments.

The standard speed for many biological processes driven by simple molecular motors is on the order of 1 μm s^{-1}. The actin filament speed of growth is around 10^{-2}–1 μm s^{-1} and is dependent on the concentration of the actin filaments. Actin-filament-based cell crawling is also in the range 10^{-2}–1 µm s^{-1} and this involves the processes of growth and disassembly of actin filaments at the leading edge of a lamellipodium (**Figure 16.1c**). Myosin interacts with actin at a rate of 10^{-2}–1 µm s^{-1}. Striated muscle parallelises the myosin/actin interactions and provides much larger forces than available from individual molecules, but with a similar time response to that of the individual molecules, e.g. in human heart muscle. Microtubule growth and shrinkage is on the order of 0.1–0.6 µm s^{-1} which is similar to the rate of motion of self-assembling actin fibres. Fast and slow axonal transport in neurons occurs at rates in the range 10^{-3}–10^{-1} µm s^{-1} as the motor proteins kinesin and dynein walk towards the plus and minus ends of a microtubule, although these motor proteins can walk at rates as fast as 1–10 µm s^{-1} in other cell types.

Nanomotors perform under low Reynolds number conditions (**Chapter 7**) and their small size means that thermal fluctuations have a large impact on their motion (Brownian perturbations). Thus, nanomotors act as if they are swimming through syrup in a stochastic hurricane and a challenge is to build an intuitive picture of the unusual physical environment experienced by a nanomotor. Nanomotors are important in physiology for network structures and for signalling. An example is the movement of neurotransmitters around nerve cells, in which loss of the processivity of the motors can result in motor neuron disease.

16.1 Self-Assembling Motility – Polymerisation of Actin and Tubulin

Both actin and microtubules together form the cytoskeleton of cells. Actins are found around the periphery of most mammalian cells, whereas microtubules act as a transport network from the periphery of the cell to its centre (the centrosome, next to the nucleus). The polymerisation of actin and tubulin are examples of one-dimensional aggregating self-assembly (**Section 8.4**). The rate of addition of subunits is found to be proportional to the concentration of free monomers in solution (c_m) and there is a rate constant of proportionality for the addition of monomers (k_{on}). The number of monomers captured per unit time (dn/dt) is proportional to the number of monomers available for capture (first order reaction kinetics, **Chapter 20**),

$$\frac{dn}{dt} = k_{on} c_m \tag{16.4}$$

In contrast, it is found that the release rate of monomers does not depend upon the free monomer concentration. k_{off} is a rate constant for subtraction of monomers that is independent of the monomer concentration (zeroth order reaction kinetics),

$$\frac{dn}{dt} = -k_{off} \tag{16.5}$$

The total elongation rate of the filament is the sum of the processes for addition (equation (16.4)) and release (equation (16.5)) of the monomers, provided a nucleation site for filament growth has been made available,

$$\frac{dn}{dt} = k_{on} c_m - k_{off} \tag{16.6}$$

The critical concentration (c_{mcrit}) for self-assembly occurs when the elongation rate (dn/dt) vanishes, i.e. when dn/dt is equal to zero, and equation (16.6) then gives

$$c_{mcrit} = \frac{k_{off}}{k_{on}} \tag{16.7}$$

Figure 16.2 shows a graphical solution of equation (16.6) for one-dimensional aggregating self-assembly. Above the critical monomer concentration the fibres expand, whereas below this concentration they shrink.

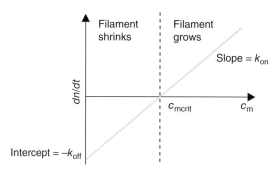

Figure 16.2 *The rate of polymerisation of actin filaments (dn/dt) as a function of monomer concentration (c_m). c_{mcrit} is the critical monomer concentration for self-assembly. Below c_{mcrit} the filaments shrink and above c_{mcrit} they grow. The gradient of the figure gives the association rate constant (k_{on}) and the dn/dt intercept gives the dissociation rate constant ($-k_{off}$).*

Figure 16.3 *The self-assembly of tubulin (and actin) is anisotropic, due to the anisotropy of the constituent subunits. Fast addition occurs at the positive end (+) and slow addition on the negative end (–).*

Similar processes of self-assembly are observed experimentally for both actin and tubulin filaments. In principle, it is easy to extract the rate constants for addition and subtraction of the monomer subunits (k_{on} and k_{off}) from in vitro experiments from a plot of the elongation rate as a function of the monomer concentration. There are, however, some additional complications with real self-assembling biological motors. Subunits are not symmetrical and add to each other with a preferred orientation, which gives rise to oriented filaments (**Figure 16.3**). The two ends of the polymer are not chemically equivalent. The faster-growing end is referred to as the plus end and the slower growing end is labelled with a minus sign. Thus, experimentally the two ends (+ and –) of the self-assembling filament need to be considered separately to extract the two sets of rate constants for addition and subtraction. It is found that the rate constants depend on both the solvent and salt concentration, so the aqueous environment that surrounds the filaments needs to be carefully monitored.

This situation of anisotropic self-assembly can be analysed through an extension of the Oosawa model described by equation (16.6) (**Figures 16.4** and **16.5**). Since the two ends of the filament are not equivalent, two equations are needed for the rate of elongation of each of the two ends,

$$\frac{dn^+}{dt} = k_{on}^+ c_m - k_{off}^+ \tag{16.8}$$

$$\frac{dn^-}{dt} = k_{on}^- c_m - k_{off}^- \tag{16.9}$$

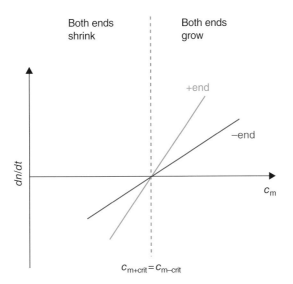

Figure 16.4 *Model for the dynamics of actin self-assembly that considers the different rate constants for both ends of the anisotropic filament. The rate of assembly (dn/dt) is shown as a function of the monomer concentration (c_m). In the case illustrated $c_{m+crit} = c_{m-crit}$.*

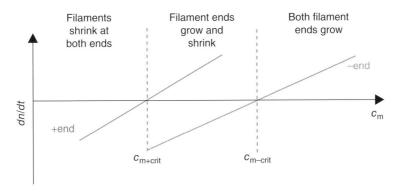

Figure 16.5 *Model for the dynamics of actin self-assembly in which $c_{m+crit} \neq c_{m-crit}$ during the assembly of the anisotropic filaments. The rate of assembly (dn/dt) is shown as a function of the monomer concentration (c_m).*

Each of these equations has a separate critical monomer concentration for the process of self-assembly

$$\left(\frac{dn^+}{dt} = 0, \frac{dn^-}{dt} = 0\right),$$

$$c_{m+\,crit} = \frac{k_{off}^+}{k_{on}^+} \qquad (16.10)$$

$$c_{m-\,crit} = \frac{k_{off}^-}{k_{on}^-} \qquad (16.11)$$

In the special case that the critical concentration of both ends are equal ($c_{m+crit} = c_{m-crit}$) both ends grow or shrink simultaneously, although the rates of assembly may be different. For steady-state conditions

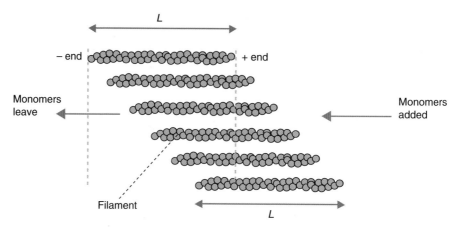

Figure 16.6 *The treadmilling process involved in actin self-assembly. Monomers leave from the negative end and are added to the positive end. The filament length (L) is conserved during the process, as the centre of mass of the filament moves to the right.*

(treadmilling) the rate of growth and shrinkage of the two ends must be equal. This can be expressed mathematically as

$$\frac{dn^+}{dt} = -\frac{dn^-}{dt} \tag{16.12}$$

And there is therefore a single critical concentration (c_{mtm}) for this process of treadmilling self-assembly,

$$c_{mtm} = \frac{\left(k_{off}^+ + k_{off}^-\right)}{k_{on}^+ + k_{on}^-} \tag{16.13}$$

The process of treadmilling is schematically shown in **Figure 16.6**, the length of the filament is invariant during the process, but its centre of mass is displaced to the right. Typical values for the rate constants and the critical concentrations during the self-assembly of actin and microtubules are 0.2–12 (μM s)$^{-1}$ (k_{on}^+, k_{off}^+, k_{on}^-, k_{off}^-), 0.1–2 μM (c_{m+crit}, c_{m-crit}) and 4–920 (μM s)$^{-1}$(k_{on}^+, k_{off}^+, k_{on}^-, k_{off}^-), 0.1–2 μM (c_{m+crit}, c_{m-crit}), respectively. Tread milling often occurs *in vivo*, since it is highly efficient in the reuse of subunits.

The varieties of interactions and their pattern formation in active self-assembling motor protein networks can be very complex. An example of a dynamic morphology created during cell division is shown in **Figure 16.7**. Here, an animal cell is shown in the final stages of cell division (cytokinesis, **Section 2.9**) where an actin/myosin ring contracts to pinch off the two divided cells. Also shown is the remains of the mitotic spindle formed from microtubules that drives the movement of the chromosome in the initial stages of cell division.

The amoeboid motility of many cells is achieved by self-assembly of actin filaments at the leading edge of the cell (**Figure 16.8**). However, *in vivo* there are a complex collection of actin associating proteins (e.g. branching or capping proteins) that direct the process of self-assembly.

16.2 Parallelised Linear Stepper Motors – Striated Muscle

The basic constituents of striated muscle are actin and myosin that are arranged in a parallel array (**Figure 16.9**). These motors are perhaps the most important for human health, since heart disease provides the largest contribution to annual human mortality rates and the majority of the muscle in hearts is striated.

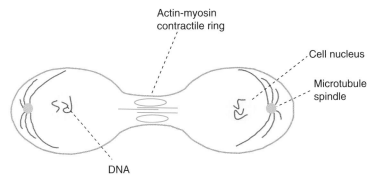

Figure 16.7 *The motor proteins involved during the process of cytokinesis in cell division (**Section 2.9**). The actin-myosin ring pinches off the cell in the final stages of replication. The microtubule spindle is used in a prior step in the process of chromosomal division.*

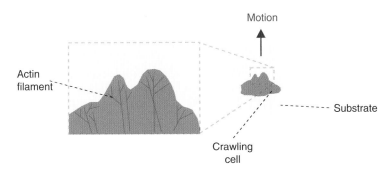

Figure 16.8 *Branched self-assembly of actin filaments drives the amoeboid crawling motion of cells in vivo.*

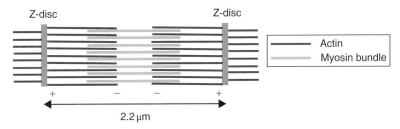

Figure 16.9 *The arrangement of actin and myosin that are parallelised into arrays in striated muscle, e.g. heart muscle or the muscle in a bicep. The distance between the Z-discs decreases during muscular contraction as the myosin molecules walk along the actin filaments.*

A scheme for the chemomechanical transduction process that use ATP to provide motility in striated muscle is provided by the rotating crossbridge model (**Figure 16.10**). This involves two key ideas; the myosin motors cycle between attached and detached states, and the motor undergoes a conformational change (the working stroke) that moves the load-bearing region of the motor in a specific direction along the filament. The rotating crossbridge model incorporates the Lymn–Taylor scheme that describes chemically how nucleotides (ATP,

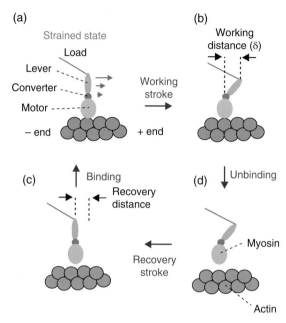

Figure 16.10 *The rotating crossbridge model for myosin/actin association consists of four distinct steps. (a) The myosin attaches to the actin filament, (b) the myosin molecule does work as it stresses the binding site, (c) the myosin unbinds from the actin filament and (d) the stress in the myosin molecule is dissipated as it moves one step along the actin filament.*

ADP) regulate the attachment and detachment of myosin from the filament, the swinging lever arm hypothesis that provides a mechanism for the amplification of small structural changes around the nucleotide-binding pocket into much larger conformational changes of the crossbridge, and the power-stroke model that accounts for how the motor generates force through the use of an elastic element within the crossbridge that is strained during the power stroke.

During the action of striated muscle there is a sequence of transitions between different chemical states of the myosin molecules; ATP binding, ATP hydrolysis and ADP release. These transitions alter the association between the motor domain and the filament, which leads to the alternation between attached and detached states.

There are three distances required to understand the inch worm motion of myosin along the actin filaments. The *working distance* (δ) is the distance a crossbridge moves during the attached phase of its hydrolysis cycle. The *distance per ATP*, is the distance that each motor domain moves during the time it takes to complete a cycle, which is also equal to the speed of movement divided by the ATPase (the part of the myosin that acts as an ATP enzyme) rate per head. The *path distance* is the distance between consecutive myosin binding sites (or stepping stones) along the actin fibre.

A series of single-molecule techniques have been used to measure the forces and characteristic distances used by myosins that associate with single fibres of actin. Force transducers that are typically used are cantilevered glass rods, atomic force microscopes (AFM) and dual-trap optical tweezers (**Section 19.7**). A particularly elegant experiment uses an actin filament attached at either end to two optically trapped spheres and the actin interacts with single myosin II molecules (**Figure 16.11**). Single working strokes of the myosin

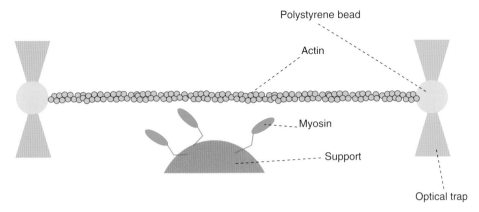

Figure 16.11   *Double trap optical tweezers can be used to measure the step size of myosin II motors when they interact with actin filaments. The actin filament is attached at either end to optically trapped colloidal beads (polystyrene).*

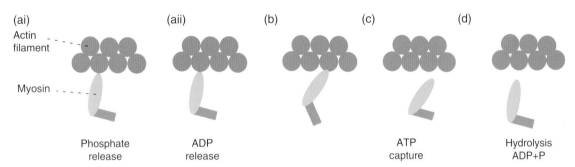

Figure 16.12   *Chemical steps in the cyclic attachment of myosin to actin filaments that corresponds to the rotating crossbridge model (**Figure 16.10**).*

molecules are resolvable with this method. Single-molecule fluorescence is another powerful technique to follow the pathway of a motor protein reaction and microrheology techniques can resolve the changes in viscoelasticity due to the motion of active molecular motors.

It is useful to consider the exact nature of the molecular steps of myosin II that travel along an actin filament. The myosin has five structural configurations during its interaction with actin in muscular motion (**Figure 16.12**). Initially, there is tight binding of the myosin head to the actin filament (b), called the rigor position (as in rigor mortis where the additional crosslinks account for the rigidity of dead muscle). Next, the myosin filament is released when it captures ATP (c), which provides the energy for the force on the actin fibre. There is then a configurational change to the cocked position during hydrolysis of ATP (d). Subsequently there is weak binding of the head to the myosin filament in a new position and finally a phosphate group is released (a).

The speed and processivity of the crossbridge motion can be understood using the concept of the duty ratio, which is the fraction of the time each motor domain spends attached to its filament. There is a cyclic process (**Figure 16.13**) in which the motor repeatedly binds to and unbinds from the filament. During each cross bridge

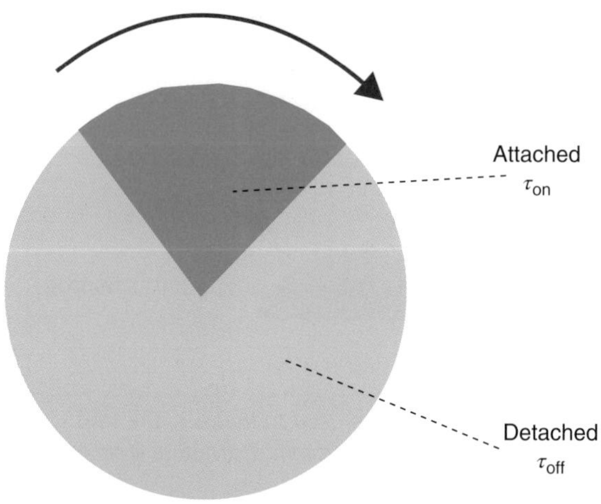

Crossbridge cycle

Attached
τ_{on}

Detached
τ_{off}

Figure 16.13 *Crossbridge cycle with myosin binding to actin filaments. τ_{on} is the attached time and τ_{off} is the detached time. The cycle rotates through alternating periods of attachment and detachment.*

cycle, a motor domain spends an average time attached to the filament (τ_{on}) when it makes it working stroke and an average time detached from the filament (τ_{off}) when it makes its recovery stroke. The duty ratio is the fraction of time that each head spends in its attached phase,

$$r = \frac{\tau_{on}}{\tau_{on} + \tau_{off}} \tag{16.14}$$

The minimum number of heads (N_{min}) required for continuous movement (N_{min}) is thus related to the duty ratio,

$$r \approx \frac{1}{N_{min}} \tag{16.15}$$

Single stepper motors are parallelised to provide much larger forces and continuous movement, but with a reasonably fast time response in striated muscle.

For the striated muscle in the human bicep it is possible to make a quick calculation for the force exerted by each myosin molecule. The number of myosin chains is equal to the cross-sectional area of the muscle divided by the cross-sectional area of a single thick filament multiplied by the number of myosins per thick filament,

$N_{myosin} \approx$ (Cross-sectional area of muscle/Cross-sectional area of thick filament) $\times N_{myosin}$ per thick filament
$$\approx \pi(3\,\text{cm})^2 / \pi(60\,\text{cm})^2 \times 300 \approx 10^{14} \tag{16.16}$$

A 10-kg weight can be comfortably lifted by a human bicep, so an order of magnitude estimate gives the force per myosin as

$$F_{myosin} \approx 10\,\text{kg} \times 10\,\text{m s}^{-2} / 10^{14} = 1\,\text{pN} \tag{16.17}$$

This value of the force per myson is in reasonable agreement with that measured in a single-molecule experiment.

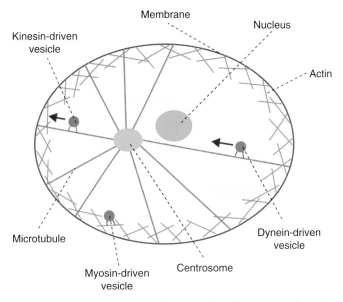

Figure 16.14 *The stepper motors dynein, kinesin and myosin drive the motion of vesicles inside eukaryotic cells. Myosins travel along actins at the cell periphery. Dyneins drive transport towards the cell nucleus along microtubules and kinesins drive transport in the opposite direction.*

Stepper motors also occur in a wide range of roles other than in striated muscle. Inside eukaryotic cells motor proteins are required to drive the motion of vesicles. Dynein motor proteins drive cargoes towards the nucleus, whereas kinesins move vesicles to the cell periphery and myosins move cargoes around the actin cytoskeleton at the cell periphery (**Figure 16.14**). For single stepper motors another simple estimate of the force exerted during a single motor step can be made. Consider a dynein molecule that moves 8 nm along a microtubule per *ATP* hydrolysis event (8 nm is the repeat distance of monomers along the microtubule). The force (F_{max}) exerted by the dynein walking on a microtubule (or kinesin using an equivalent approximation) is the energy divided by the distance moved,

$$F_{max} = \text{free energy of ATP hydrolysis/step size} \approx 220\,kT/8\,\text{nm} \approx 10\,\text{pN} \qquad (16.18)$$

Therefore, 10 pN is an upper limit on the force that dynein can exert, since it assumes that the motor is 100% efficient.

A current challenge is to understand the regulation of the different modes of transport of vesicles inside cells. Stochastic stepping motion of vesicles is observed in live cells with well-regulated changes in speed and direction. How this relates to the activity of multiple motor proteins attached to a single vesicle is not well understood, e.g. does a tug of war take place between opposing polarity microtubule motors (opposing teams of dynein molecules versus kinesin molecules)?

16.3 Rotatory Motors

Following on from the discussion of Poisson motility processes of bacteria in **Section 7.3** the molecular biophysics of the rotatory flagellar motor for the propulsion of bacteria will be considered (**Figure 16.15**). The ATP synthase in the mitochondria of eukaryotic cells converts ADP into ATP. It is also a rotary motor and is

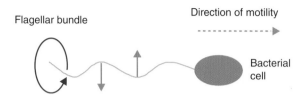

Figure 16.15   *Helical flagellar filaments provide a bacterium with motility in its low Reynold's number environment. E.coli cells typically have multiple flagellae (6–8) attached to their surface, so forward motility is driven by the action of a coherent bundle of flagellae.*

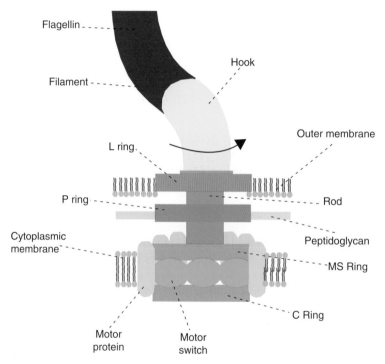

Figure 16.16   *The rotatory motor attached to the flagellae of bacteria is constructed from over twenty different types of protein (also seen in **Figure 16.15**). The motor is driven by a gradient of hydrogen ions.*

driven by a pH gradient (in common with the bacterial motor, but the process occurs in reverse, ADP → ATP for the synthase rather than ATP → ADP in the flagellar motor). Nature thus invented the wheel billions of years before man and it has a number of different roles.

A curved segment separates the motor from the main length of the filament in bacteria (**Figure 16.16**). The filament is bent away perpendicularly from the surface of the membrane for several nanometres. This filament executes a helical motion as it is rotated by the motor and acts like a propeller that provides a source of motility for the bacteria.

A series of proteins form the flagellum of the bacterium and each has a specific function; the bushings seal the cell membrane, the circular stator is attached to the cell and the rotor is connected to the flagellum (**Figure 16.1a**). The flagellar propeller is not run directly by ATP. Instead protons run down a pH gradient across the membrane and produce an electric potential. Variation of the pH drop across a membrane can change the direction of the motor. Sodium ions also can fulfil the same function as hydrogen ions in some marine bacteria. As bacteria move through a solution their flagella can rotate at up to 100 rev s^{-1}, which is comparable to the rate at which an automobile petrol engine (30 rev s^{-1}) functions. The flagellar motor works equally well in both clockwise and counterclockwise modes. These bacterial motors are relatively complicated devices and consist of over twenty separate protein components that only together can provide motility. The evolutionary history of the creation of such a finely orchestrated engine is a fascinating story.

Steps in the rotatory motion of a bacterial motor can be observed by attachment of a fluorescent actin filament to the hook subunit in fluorescence microscopy experiments. Discrete quantised angular movements of the motor are observed. The rate of rotatory motion is found to be proportional to the potential difference across the motor under physiological conditions.

16.4   Ratchet Models

An interesting, but inefficient (and thus inaccurate) model of molecular motility is provided by the thermal ratchet. The model demonstrates how directed motility of a muscle protein can be derived from rectified Brownian motion (**Figure 16.17**), i.e. a constant probability bias is superposed on the thermal fluctuations of displacement of a particle in a particular direction. Widely differing processes of motor driven motility such as stepper motors and rotary motors can be described in terms of rectified Brownian motion, as a unifying principle, and ratchet models have thus been widely used to analyse their motion.

The thermal ratchet is a simple mechanism to produce motion in a low Reynold's number environment. It uses a spatially asymmetric potential that oscillates with time (**Figure 16.18**). The probabibity distribution of motor proteins $(P(x))$ as a function of distance (x) evolves due to diffusion in the standard Brownian fashion due to thermal motion when the potential is switched off (**Section 7.1**). The asymmetry of an oscillating sawtooth potential, when supperposed on the thermal fluctuating force, causes a net motion of the proteins in a given direction. The net probability of directed motion (P_{net}) is the difference between the probability to move right (P_R) and that to move left (P_L),

$$P_{net} = P_R - P_L \qquad (16.19)$$

A simple mathematical form for the probability distribution of the motor proteins results from the action of the sawtooth potential. If the proteins do not diffuse a sufficient distance there is no net flux,

$$P_{net} = 0 \qquad 0 < \sigma < \alpha x \qquad (16.20)$$

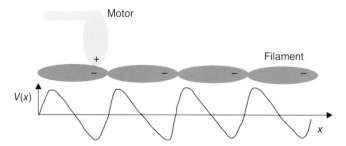

Figure 16.17   *The interaction of motor proteins (e.g. myosin) with actin can be modelled with a single one-dimensional potential (V(x)) as a function of distance (x). The monomers of a biofilament have a dipolar charge distribution and the myosin motors experience a sawtooth interaction potential (V(x)) as they interact with the filament.*

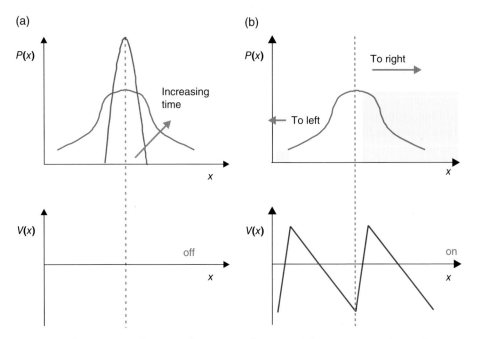

Figure 16.18 *An oscillating sawtooth potential (V(x)) as a function of distance (x) can be used to move a series of particles (probability density P(x)). Within this realisation of the ratchet model particles are moved to the right by the asymmetry of the sawtooth potential. (a) The particle probability distribution and potential when the ratchet is switched off. (b) The particle probability distribution and potential when the ratchet is switched on. [Reprinted with permission from J. Prost, J.F. Chauwin, L. Peliti, A. Ajdari, PRL, 1994, 72, 16, 2652–2655. Copyright 1994 by the American Physical Society.]*

where x is the wavelength of the sawtooth potential, σ is a measure of the spread of the particle distribution and αx is the peak-to-trough separation of the potential. If the probability distribution created by the thermal motion is sufficiently broad, a net flux occurs,

$$P_{\text{net}} = (1 - \alpha x/\sigma)^2/2 \quad \alpha x < \sigma < (1-\alpha)x \tag{16.21}$$

The probability distribution initially broadens when the potential is not applied due to thermal diffusion of the motor proteins. When the potential is switched on again there is a higher probability that the particles are drawn to the right than to the left, due to the asymmetric nature of the potential. This allows the dipolar nature of the motion of the monomers along the biofilament to be modelled.

The major problem with such a simple Brownian ratchet model is its efficiency. A thermal ratchet can take the hydrolysis of up to 10 ATP molecules for 1 step ($P_{\text{net}} = 0.1$). In real biological systems the efficiency is typically 5 times better than that found for the model. More sophisticated extensions of such models have recently been proposed that resolve this shortfall.

Ratchet models have also been applied to rotatory motors (**Figure 16.6**). An elastic link is invoked between the stator unit and the cell wall that rectifies the angular thermal fluctuations in a certain sense (anticlockwise or clockwise), which creates directed motion.

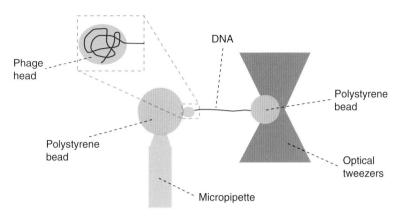

Figure 16.19 *A motor on the head of a bacteriophage can be used to compact a DNA chain into the interior of the phage head. The forces involved can be measured using optical tweezers.*

16.5 Other Systems

Other less-common mechanisms for biological motility have been discovered. *Extrusion nozzles* are present in the myxbacteria, cyanobacteria and flexibacteria. Slow uniform gliding motion is achieved for these organisms by a continuous secretion of a glycoprotein slime (**Figure 16.1d**).

Supramolecular springs store conformational energy in chemical bonds that can then act as latches for its release (**Figure 16.1e**). The specific power of such motors can be very high. One example is the scruin/actin system in which scruin captures actin in a slightly overtwisted state. Calcium-dependent changes in the scruin are then used to release the conformational energy of the actin and provide a force for motility.

In bacteriophage the DNA is stored at very high osmotic pressures. A translocation motor is thus needed to accomplish this process of packaging when the DNA is pushed through the pore in the bacteriophage's coat proteins (**Figure 16.19**). Binding of proteins inside the bacteriophage are thought to facilitate the transport of the DNA chain into the bacteriophage.

It is expected that many more nanomotor systems will continue to be discovered and the list above is by no means exhaustive, e.g. gas-filled bubbles in aquatic micro-organisms allow them to adjust their feeding depth.

Suggested Reading

If you can only read one book, then try:

Top recommended book: Howard, J. (2001) *Mechanics of Motor Proteins and the Cytoskeleton*, Sinauer. Very good introductory text on motor proteins.

Boal, D. (2011) *Mechanics of the Cell*, 2nd edition, Cambridge University Press. Contains a useful section on active molecular networks.
Bray, D. (2000) *Cell Movements: from Molecules to Motility*, Garland. Classic text on both biological motors and micro-organism motility.

Hoffmann, P.M. (2012) *Life's Ratchet*, Perseus. Excellent popular account of the biophysics of nanomotors.

Phillips, R., Londev, J., Theriot, J. & Garcia, H. (2012) *Physical Biology of the Cell*, Garland. Useful description of statistical models for motor protein activity.

Vogel, S. (2002) *Prime Mover: A Natural History of Muscle*, W.W. Norton. Popular introduction to the biology of muscle.

Tutorial Questions 16

16.1 Make a list of the molecular motors that occur in biology. Name some diseases associated with malfunctioning motor proteins.

16.2 Describe the factors that affect the maximum rate of the motion and the force that can be exerted by an actin filament.

17

Structural Biomaterials

A wide range of biomaterials have evolved naturally and many are optimally matched to their structural roles. Examples include cartilage in synovial joints, spider silk for web building, resilin in the hinge joints of dragon fly wings, mollusc glue for adhesion and cancellous bone in skeletons. These examples are chosen to illustrate the rich variety of physical phenomena involved and the exquisite nature of the design principles that evolution has implemented to solve structural problems.

17.1 Cartilage – Tough Shock Absorbers in Human Joints

Normal healthy human joints have friction coefficients (μ) in the range 0.001–0.03, which is lower than that found with the materials that coat nonstick frying pans ($\mu \approx 0.01$ for teflon on teflon). These values are also remarkably low when compared with hydrodynamically lubricated journal bearings that occur in efficient mechanical engines such as those in cars. However, hydrodynamic lubrication is not in effect (it occurs in car engines, aircraft turbines, etc. at high speeds), since the bone surfaces in synovial joints never move relative to one another at more than a few cm s^{-1} (**Section 9.6**). Boundary lubrication is in effect in synovial joints.

A schematic diagram of an articulated joint is shown in **Figure 17.1**. It consists of three main mechanical components: *bone* (a living mineral foam composite), a *viscoelastic fluid* (a semidilute solution predominantly composed of the polyelectrolyte hyaluronic acid and water) and *cartilage* (an elastic protein/proteoglycan composite).

Cartilage acts as a shock absorber in a series of applications throughout the human body, which includes articulated joints. Cartilage is a living tissue and specialised cells (chondrocytes) contained within the tissue play a role in the repair of damage from wear due to the motion of its surfaces and protect it against bacterial attack (**Figure 17.2**).

In human knee joints the average pore size of the cartilage between collagen fibres is approximately 6 nm. The surface of cartilage has ripples (amplitude 3 µm, and wavelength 40 µm) superimposed on a micro-roughness (amplitude 0.3 µm, and wavelength 0.5 µm). The synovial fluid contained between the sections

---

The Physics of Living Processes: A Mesoscopic Approach, First Edition. Thomas Andrew Waigh.
© 2014 John Wiley & Sons, Ltd. Published 2014 by John Wiley & Sons, Ltd.

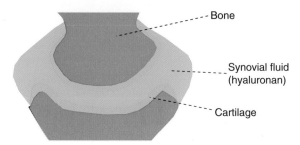

Figure 17.1 *Schematic diagram of an articulated joint that shows the two sections of bone, the synovial fluid and the cartilage shock absorbers e.g. a knee joint.*

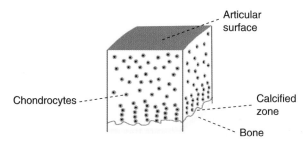

Figure 17.2 *Cartilage is a living shock absorber, chondrocyte cells are arranged throughout its structure. The chondrocytes replenish the extracellular material that constitutes the cartilage.*

of cartilage is a non-Newtonian liquid that has the property of shear thinning; its viscosity decreases almost linearly with shear rate. This rheological behaviour is typical of flexible polyelectrolyte solutions (**Section 15.3**).

Cartilage presents one of the biggest challenges in tissue engineering for replacement materials to treat arthritic conditions. The sections of the bone in articulated joints can be successfully replaced with polyethylene and new hyaluronic acid can be injected in to the knee cavity to replace damaged boundary lubricants. However, osteoarthritis involves the breakdown of cartilage on the bones' surface that causes cracks to form in the articular surface, and currently no effective replacement exists for these low-friction shock absorbers. The lifetime of polyethylene replacement joints is seriously compromised by the high friction wear mechanism that results from the absence of a cartilage covering and on average they need to be replaced after ten years. New replacement materials are required and this necessitates an improved physical understanding of synovial joints.

The collagen in cartilage exists with an anisotropic distribution of fibre orientations (**Figure 17.3**) and thus it has anisotropic mechanical properties. The shear modulus of cartilage is higher perpendicular to the collagen chain orientation than in the parallel direction.

In general, the function of cartilage in articulated joints is to increase the area of load distribution and to provide a smooth wear-resistant surface optimised for low friction. Biomechanically articular cartilage can be viewed as a two phase (solid–fluid) material; the collagen/proteoglycan solid matrix (25% contribution to the wet weight) is surrounded by freely moving interstitial fluid (75% by wet weight). Important biomechanical properties of articular cartilage are the resistance of the solid matrix to deformation and the frictional resistance to the flow of the interstitial fluid through the porous permeable solid matrix. Articular cartilage has the ability to provide joints with a self-lubrication behaviour that operates under normal physiological

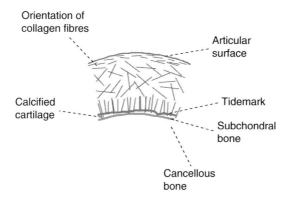

Figure 17.3 *Cartilage has an anisotropic fibrous structure that leads to anisotropic mechanical properties. Collagen fibres are attached perpendicularly to the surface of the bone and the arrangement shifts to a parallel alignment at the surface of the articular cartilage.*

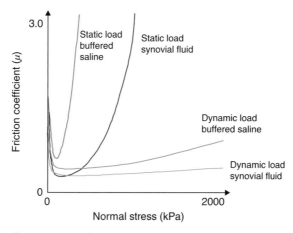

Figure 17.4 *The friction coefficient (μ) as a function of normal stress between two cartilage surfaces. Synovial fluid is seen to cause an important reduction in the friction coefficients of both static (red) and dynamic (blue) tests compared with the saline control (purple and green respectively). [Ref. L.L. Malcom et al, 1978, Biorheology, 15, 5–6, 482.]*

conditions. Pressure on the surface of the cartilage forces water through the porous matrix. Water moves out through the surface, and provides a lubricating fluid film on the surface of the cartilage (**Figure 17.4**). Damage to articular cartilage can disrupt the normal load carrying ability of the tissue and thus the normal lubrication process that operates in the joint. Insufficient boundary lubrication is thought to be a primary factor in the development of osteoarthritis, which causes acute damage to the cartilaginous surfaces and extreme pain for the sufferers.

Cartilage is a charged crosslinked elastomeric composite material and can be compared with resilin and elastin, two uncharged bioelastomers examined separately in **Section 17.3**. Although cartilage is reasonably elastic, unlike resilin and elastin it is not resilient; energy dissipation is maximised. The crosslinks that form the elastic matrix are provided by the collagen in cartilage and the dissipative properties are provided by giant polyelectrolyte combs (the aggrecans, **Figure 17.5**, **Section 1.8**).

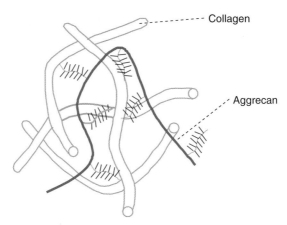

Figure 17.5 *Cartilage is a composite mixture of collagen and aggrecan molecules. The collagen molecules provide strength and elasticity to the network, whereas aggrecan is used to dissipate energy.*

The physical phenomena that contribute to the behaviour of cartilage are very rich. Principle questions include the origin of the repulsive forces between the two charged cartilaginous plates, the friction coefficient of polymeric cartilage gels, the modulus of the cartilage gels with their rigid molecular nematic inserts (collagen) and the time effects observed with longitudinal stress relaxation after the cartilage has been loaded. A simple model for the extremely low friction coefficients found in articular cartilage will first be examined. The model still requires development, and is only at the qualitative level of understanding. However, it does demonstrate the bottom-up approach of biophysics to explain the molecular basis of some sophisticated material properties.

First, consider *the forces between two charged plates*. The Poisson–Boltzmann equation can be used for the potential ($V(r)$) due to the surface charges at a perpendicular distance (r) from the plates (**Section 4.3**),

$$\nabla^2 V(r) = -\left(\frac{e\rho_0}{\varepsilon}\right) e^{-eV(r)/kT} \tag{17.1}$$

where ρ_0 is the ion density profile at the point of zero potential ($V = 0$), i.e. the midpoint between the plates. e is the electronic charge, ε is the dielectric constant, and kT is the thermal energy. The solution for the charge density as a function of the distance from a single-charged surface is shown in **Figure 17.6**. For a charged homopolymer gel carrying one charge on each monomer unit, the surface charge density (σ, units of electrons per m^3) is

$$\sigma = (cN_A)^{2/3} \tag{17.2}$$

where c is the gel concentration and N_A is Avogadro's number. For positively charged counterions in one dimension the Poisson–Boltzmann equation (17.1) can be written

$$\frac{d^2 V}{dz^2} = -\frac{e\rho_0}{\varepsilon} e^{-eV/kT} \tag{17.3}$$

where z is the perpendicular distance from the plates. The Poisson–Boltzmann equation is subject to two sets of boundary conditions; one on the cartilage gel surfaces and the other on the symmetric plane between the two sections of cartilage. From the electrostatic Poisson equation the boundary conditions can be written mathematically. On the gel surface the gradient of the potential equals the surface charge density,

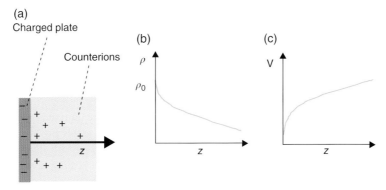

Figure 17.6   *Schematic diagram of the charge density near a charged plate that shows (a) arrangement of the counterions, (b) counterion density (ρ) and (c) the potential (V) near the surface as a function of the perpendicular distance from the plate (z).*

$$\left(\frac{dV}{dz}\right)_{z=\pm l} = \mp\frac{e\sigma}{\varepsilon} \tag{17.4}$$

On the symmetric plane (the half-way distance between the two plates) the gradient of the potential is zero,

$$\left(\frac{dV}{dz}\right)_{z=0} = 0 \tag{17.5}$$

Electrical neutrality requires that the surface charge density (σ) is equal to the total charge of the oppositely charged counterions associated with the polyelectrolyte chains,

$$\sigma = \rho_0 \int_0^l e^{-eV/kT} dz \tag{17.6}$$

where l is the solvent layer thickness. The Poisson–Boltzmann equation can be solved subject to the three requirements of equations (17.4), (17.5) and (17.6) to give

$$\sigma = \sqrt{\frac{2\rho_0}{l_b}} \tanh\left(l\sqrt{\frac{\rho_0 l_b}{2}}\right) \tag{17.7}$$

Here, $l_b = e^2/\varepsilon kT$ is the Bjerrum length (a constant). The repulsive osmotic pressure (π) between two charged surfaces is determined by the ion charge distribution (density ρ_0) at the symmetry plane from the contact value theorem (**Section 4.4**),

$$\pi = \rho_0 kT \tag{17.8}$$

where kT is the thermal energy. In the equilibrium state, with a constant pressure on the cartilage, the osmotic pressure (π), that is predominantly due to the counterions in a charged gel (**Section 11.4**), is counterbalanced by the applied pressure (P), e.g. the weight of a person's upper body distributed across the area of their knees,

$$P = \pi \tag{17.9}$$

The solvent layer thickness (2*l*) that remains between the two sections of cartilaginous gels can then be calculated by rearrangement of equation (17.7) in terms of the solvent layer thickness combined with equations (17.8) and (17.9),

$$2l = 2\sqrt{\frac{2kT}{Pl_b}}\tanh^{-1}\left(\sigma\sqrt{\frac{kTl_b}{2P}}\right)$$ (17.10)

It is concluded that highly charged surfaces are able to sustain more pressure (at fixed pressure the equilibrium distance is larger) than the equivalent neutral surface (**Figure 17.7**). Cartilage has a relatively rigid crosslinked network, so swelling of the chains by the osmotic pressure at equilibrium is neglected in this model.

The friction coefficient of a polymer gel can also be analysed using a scaling approach (**Section 10.2**). Amonton's law for the friction of conventional materials states that the frictional coefficient is independent of load (**Section 9.6**). The frictional force (*F*) of many solids on solids is therefore related to the normal force (*W*) by the universal law

$$F = \mu W$$ (17.11)

where μ is the frictional coefficient. Experimentally, in the case of solids with repulsive interfacial interactions (e.g. like-charged polyelectrolyte gels) it is found that the velocity dependence of the friction is strongly dependent on the normal compressive strain. The smaller the strain, the weaker the velocity dependence of the friction. Indeed, with gels it is found that the frictional force is no longer linearly proportional to the normal force and equation (17.11) is replaced by

$$F \propto AP^\alpha$$ (17.12)

where *P* is the average normal pressure, equal to the weight (*W*) divided by the contact area (*A*), and α is a constant in the range 0 to 1. Furthermore, when two pieces of negatively charged gels are allowed to slide

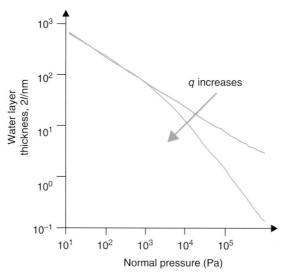

Figure 17.7   *Thickness of water between two charged plates as a function of the normal pressure. As the swelling of the gels (q) increases the two plates move closer together due to the corresponding decrease in surface charge density. [Reprinted with permission from J. Gong, Y. Iwasaki, Y. Osada, K. Kurihara, Y. Hamai, J. Phys. Chem. B, 1999, 103, 6001–6006. Copyright © 1999, American Chemical Society.]*

past each other the frictional force is found to be proportional to a power law of the velocity (v) with a constant exponent (β),

$$F \propto v^{\beta} \tag{17.13}$$

where β depends on the normal compressive strain. This power-law dependence of the force on the velocity indicates that the viscoelasticity of polymer networks plays an important role in the frictional properties in addition to hydrodynamic lubrication.

A qualitative *molecular theory* for polymer friction at the surface of a gel will be motivated. For an uncharged semidilute polymer solution the mesh size (ξ) (equation (10.58)) depends only on the polymer concentration (c), assuming the polymer chains are in a good solvent,

$$\xi \sim ac^{-3/4} \tag{17.14}$$

where a is the effective monomer length, and ξ can be considered a measure of the size of the pores in the polymer mesh. It is found experimentally that polymer gels have the same scaling with regard to their correlation length (ξ) and osmotic pressure (π) as the equivalent semidilute solution prepared at the same polymer concentration (this is called the c^* *theorem*). The change in interfacial energy ($A-A_0$) between a polymer gel and a solid surface is

$$A - A_0 \approx \pi_0 \xi \tag{17.15}$$

where π_0 is the osmotic pressure of the bulk solution, A is the interfacial energy between the solid and the gel, and A_0 is the interfacial energy between the substrate and the pure solvent. From a polymer scaling theory the osmotic pressure of the polymer gel is known to be related to the correlation length

$$\pi_0 \approx kT\xi^{-3} \tag{17.16}$$

where T is the temperature and k is Boltzmann's constant. The work done by the solid surface to repel the polymer from the surface a distance ξ_g against the osmotic pressure is equal to the increase in surface free energy

$$P\xi_g \approx A - A_0 \approx \pi_0 \xi \tag{17.17}$$

where P is the average normal pressure. When no surface adsorption of the polymer occurs the frictional force is due to the viscous flow of the solvent at the interface. Viscous solvent flow obeys Newton's second law and hydrodynamic lubrication theory can be applied between two particles separated by a solvent layer to obtain the frictional force (f) using nonslip boundary conditions,

$$f = \frac{\eta v}{\xi_g + D} \tag{17.18}$$

where ξ_g is the thickness of the solvent layer, D is the thickness of the polymer film (the layer thickness that is sheared), η is the viscosity of the solvent, and v is the relative velocity of the surfaces. It is possible to show that the frictional coefficient depends on the temperature (T), Young's modulus of the gel (E) and applied pressure (P),

$$f = \frac{\eta v P}{E^{2/3} T^{1/3}} \tag{17.19}$$

And a similar calculation for a charged polyelectrolyte gel surface gives

$$f = \frac{\eta v}{2\left(D + \sqrt{k_{gel}}\right)} \tag{17.20}$$

where k_{gel} is the hydraulic permeability of the gel (see equation (17.22)). The frictional coefficient (μ) can be shown to depend on the pressure (P) in a nonlinear manner with charged gels,

$$\mu \sim P^{-3/5} \tag{17.21}$$

This is in agreement with the experiment results in (**Figure 17.8**).

   Simply put, these results show for the friction of polyelectrolyte gels, the higher the charge on the gels (for constant swelling), the lower the resultant frictional coefficient (**Figure 17.9**). However, charge effects are only one important factor for the determination of the frictional properties. Practically, in synovial joints the situation is often more complicated. Elastohydrodynamic fluid films of both the sliding and squeeze type play an important role in the lubrication of the joints. There is thus a mixed method of frictional reduction in cartilage with contributions from both the repulsion of charged polymers at the surface and the water exuded from inside the cartilage. Microcontacts (asperities) between the surfaces can also play a role in the determination of frictional coefficients in regions where the double-layer forces are insufficient to withstand the local pressure increases (**Figure 17.10**).

   The time dependence of the *relaxation modulus* of cartilage can also be considered. When cartilage is sheared or longitudinally compressed the material properties become time dependent due to the motion of fluid through the pores of the gel. Fluid motion through porous materials occurs in a range of biological situations, e.g. when blood plasma moves through blood clots. The problem has been solved by D'arcy in the case of a Newtonian fluid that moves through an ideally porous material (**Section 14.3**). Cartilage is a composite material (**Figure 17.5**); a rigid collagen scaffold combined with a dissipative proteoglycan matrix. The modulus of

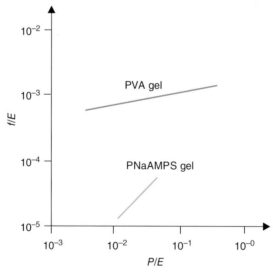

Figure 17.8 *Experimentally determined dependence of the frictional force (f) on the normal pressure (P) for a flexible neutral polymer (PVA) gel and a flexible polyelectrolyte (PNaAMPs) gel. The frictional force and pressure are renormalised by the Young's modulus (E) for comparison. The frictional force is much lower for the charged gels (PNaAMPS) in water. [Reprinted with permission from Journal of Chemical Physics, 1998, 109, 8062. Copyright 1998, AIP Publishing LLC.]*

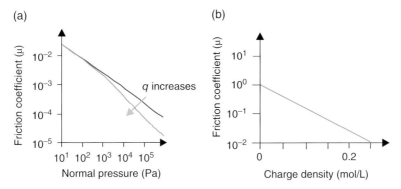

Figure 17.9 *(a) Dependence of the frictional coefficient on the normal pressure (q is the swelling of the gel). The frictional coefficient decreases as the gel swelling increases. (b) Dependence of the frictional coefficient on the charge density for a polyelectrolyte gel. The frictional coefficient of the charged gel decreases with increasing charge density of the gel. [Reprinted with permission from J. Gong, Y. Iwasaki, Y. Osada, K. Kurihara, Y. Hamai, J. Phys. Chem. B, 1999, 103, 6001–6006. Copyright © 1999, American Chemical Society.]*

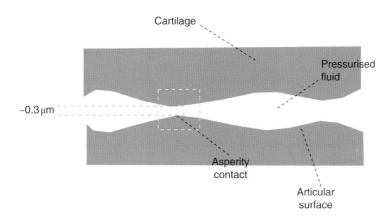

Figure 17.10 *Microscopic contacts (asperities) provide an important contribution to the frictional properties of cartilage.*

the composite is 10^5 times that of a concentrated proteoglycan solution, which implies the proteoglycans do not contribute to the shear stiffness for articular cartilage (**Figure 17.11**). The two major contributions to the shear stiffness are thus the crosslinked anisotropic collagen molecules and the flow of fluid through the network.

The stress-relaxation curve as a function of time after a step change in shear strain has a relaxation time of about 30 min. Thus, shearing the cartilage in the knee (when a heavy object is lifted) has a long time effect on the elasticity. It can be shown that the relaxation time for a crosslinked gel is proportional to the mutual diffusion coefficient of the polymer gel. D'Arcy's law states that the average fluid velocity through a porous system is linearly related to the pressure gradient (also see **Section 14.3**),

$$U = -b\frac{\partial P}{\partial z} \tag{17.22}$$

where b is the hydraulic permeability. Using *D'Arcy's law* (equation (14.17)) it is possible to show that the slowest relaxation time in a polymer gel is given by

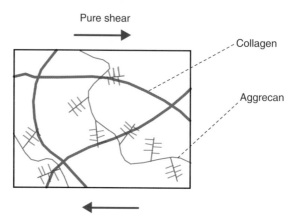

Figure 17.11 *Schematic diagram of collagen and proteoglycans (aggrecan) molecules in cartilage that experience a shear deformation.*

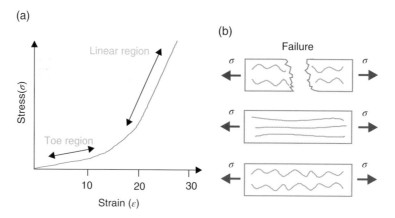

Figure 17.12 *(a) Tensile stress (σ) as a function of strain (ε) for cartilage. The toe region corresponds to the Hookean elasticity of the collagen fibres. The linear region is due to straightening of the fibres that causes the Young's modulus to be increased. At high stresses/strains the cartilage fails and the collagen network is broken (failure). (b) The schematic behaviour of the collagen fibres under extension.*

$$\tau_1 = \frac{\delta \delta_{eq}}{\pi^2 E b} \tag{17.23}$$

where δ is the compressed sample thickness, δ_{eq} in the freely swollen sample thickness and E is Young's modulus.

In tension, the mechanical properties of cartilage are strongly *anisotropic*. Cartilage is stiffer and stronger in the direction of load. It exhibits viscoelastic behaviour in tension that is attributed to both internal friction associated with polymer motion and the flow of the interstitial fluid. A typical equilibrium tensile stress–strain curve for articular cartilage is shown in **Figure 17.12**. For small amounts of strain, in the toe region, the collagen molecules are slowly extended and reoriented. Further extension straightens the collagen molecules that leads to nonlinear strain hardening and finally the molecules break and fail.

17.2   Spider Silk

Spider silk is a classic example of a nanostructured polymer composite. Evolution has created a silk for the spider that is both tough and strong. The silk can be rapidly produced by the spider to an external stimulus (e.g. to escape a predator) and is fabricated inside the spinneret (**Figure 17.13**). The protein is produced in a nematic liquid-crystalline state inside the spider and extruded into an oriented solid polymer with remarkable structural properties. Orientation in polymers is directly related to their tensile strength, as observed with synthetic analogues such as Kevlar in bullet-proof jackets (a synthetic liquid-crystalline polymer). Spider silk is highly oriented. The extreme toughness of spider silk can be observed in the large area contained under the stress–strain curve (**Figure 17.14**). This toughness is a key feature of the silk. It is five times greater than that of Kevlar. The spider can produce a wide range of silk materials (up to eight) whose mechanical properties are optimised for their different roles. **Figure 17.15** shows the different mechanical properties found for dragline silk, which is optimised for the maximum stress before fracture and catching silk, which has a high strain

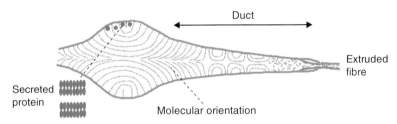

Figure 17.13   *Schematic diagram of the structure of spider silk inside a spider's spinneret. The proteins adopt a nematic liquid crystalline phase as they are extruded through the spinneret and solidify to form the spider silk. [Reprinted from D.P. Knight, F. Vollrath, Philos. Trans. R. Soc. Lond. B, 2002, 357, 155–163. With permission from The Royal Society.]*

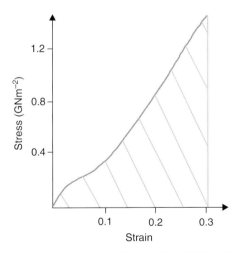

Figure 17.14   *Tensile stress as a function of strain curve for a spider silk fibre. The high toughness of the spider silk is indicated by the large area under the stress–strain curve. [Reprinted from J. Gosline, DeMont and Denny, Endeavour, 1986, 10, 1, 37–43. With permission from Elsevier.]*

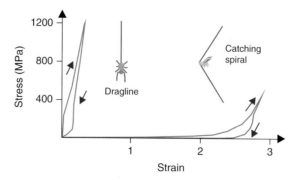

Figure 17.15 *Comparison of the mechanical properties of dragline and catching spiral silk. The tensile stress is plotted as a function of strain. Dragline silk has a high stiffness, but is relatively inextensible. The opposite is true with the catching spiral silk. [Reprinted from J. Gosline, M. Lillie, E. Carrington et al, Philos. Trans. R. Soc. Lond. B, 2002, 357, 121–132. With permission from The Royal Society.]*

before fracture. The range of mechanical properties offered by spider silks encompasses both rubber-like and extremely rigid behaviour. Viscid silk is used in the glue that covers the spiral of orb webs and rigid silk is used as a safety line when the spider moves around.

17.3   Elastin and Resilin

The key function of *elastin* and *resilin* is to provide a low stiffness, highly extensible, efficient elastic energy storage mechanism in animals. Elastin is a major component of arteries, and allows them to adjust to pressure differences in blood flow. Elastin is also frequently used as a shock absorbing material, e.g. in the neck joints of cows. Resilin is an analogous material that is used for elastic energy storage in a series of different animals, such as the jumping mechanism in fleas and the hinge in the wings of dragon flies. In such mechanical roles a predominantly elastic response is required over a wide range of frequencies, and both elastin and resilin are extremely well-optimised elastomers (rubbers). Structurally this implies flexible protein strands exist between crosslinks to provide entropic elasticity in these rubbery proteins.

The stress–strain curve for elastin and resilin are shown in **Figure 17.16**. The *resilence* (R) is the fraction of work that is stored in a mechanically stressed system and can be calculated as

$$R = e^{-2\pi\delta} \tag{17.24}$$

where δ is the damping factor equal to the ration, $\delta = E'/E''$, E' is the storage Young's modulus and E'' is the dissipative Young's modulus. Both resilin and elastin are extremely resilient materials according to this measure.

A number of structural proteins such as resilin and spider silk can now be expressed in different organisms (e.g. bacteria) using recombinant DNA technology and these materials could have a range of biomedical applications, e.g. for replacement arteries. The synthetic processing of the genetically expressed proteins presents a number of challenges to provide well-defined mechanical properties and is a bottleneck that restricts the application of the technology. For example, the creation of correctly oriented spider silk fibres is a challenge, although recent advances have seen carpets and violin strings woven from transgenic spider silks.

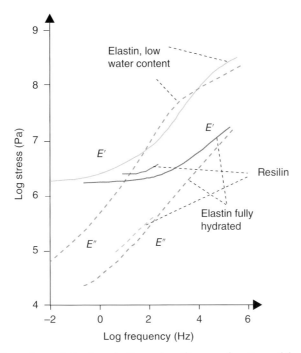

Figure 17.16 *Complex Young's modulus for elastin and resilin as a function of frequency. Both samples are predominantly elastic with E' > E''. Addition of water to elastin causes a decrease in its moduli. [Reprinted from J. Gosline, M. Lillie, E. Carrington et al, Philos. Trans. R. Soc. Lond. B, 2002, 357, 121–132. With permission from The Royal Society.]*

17.4 Bone

Bone is a protein/inorganic crystalline composite material. Compact bone is similar in structure to nacre and is discussed in **Section 17.6**. Cancellous bone has a more porous structure. It is a cellular solid (a foam, **Section 13.3**) and is well optimised for strength and weight. At small strains the linear elastic response of isotropic cancellous bone is due to the elastic bending of the cell walls (**Figure 17.17**). At higher strains the cell walls fail by elastic buckling. This buckling plateau continues until the cell walls meet and touch, which causes a large increase in the stress. The modulus of the material is very sensitive to its degree of hydration, since the plasticisation of the adhesive proteins attached to the hydroxyappatite crystallites radically alters their mechanical properties.

17.5 Adhesive Proteins

Surface coatings of proteins play a crucial role in a number of biological scenarios. The proteins that attach molluscs onto rocks are important in the ship-building industry, since molluscs adhere equally well to the hulls of boats as to rocks (**Figure 17.18**) and layers of molluscs compromise the performance of the hulls. Nature has produced a well-optimised adhesive with molluscs that acts in a harsh hydrated environment. The primary natural function of the protein is to act as a sealant, so the muscular foot of the organism can hold itself on to the rock with suction in a variety of weather conditions. Molluscs appear to adhere preferentially to

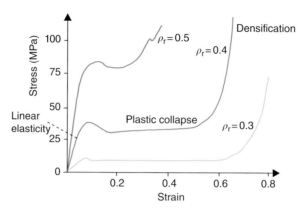

Figure 17.17 *Compressive stress as a function of strain for cancellous bone at a series of relative densities (ρ_r) of the open-celled foams. At small strains the elastic response is linear, the cells then collapse by elastic buckling (plastic collapse), and then densification causes a further increase in the Young's modulus. [Reproduced from L.J. Gibson, M.F. Ashby, Cellular Solids, 1997 With permission from Cambridge University Press.]*

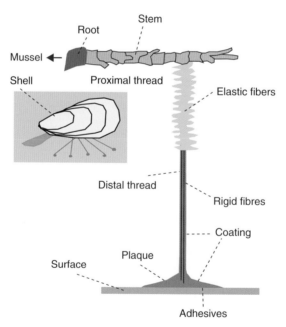

Figure 17.18 *Schematic diagram of the process of adhesion of a mussel onto a surface, e.g. a rock or the hull of a boat. Adhesive proteins displace water on the surface plaques and are attached by rigid fibres (distal threads) which connect the mussel securely to the surface. [Reproduced from S.W. Taylor, J.H. Waite, Protein Based Materials, ed K. McGrath, D. Kaplan, 1997, Birkhauser Boston Inc. With permission from Springer ScienceSpringer Science + Business Media.]*

high-energy surfaces (e.g. metals) and the adhesive proteins are optimised for the displacement of water at these surfaces. Glues are also thought to occur with the feet of starfish, but with most organisms that attach to surfaces mixed adhesive mechanisms are in effect. van der Waals forces (e.g. Geckos), capillary forces (e.g. frogs) and microhooks (e.g. plant burrs) are often found to be important.

17.6 Nacre and Mineral Composites

Nacre and bone are both fibrous composite materials with hard nanocrystallites embedded in a protein matrix (**Figure 17.19**). The stress distribution along the length of mineral crystals is assumed to be linear, so the maximum tensile stress (σ_m) and the average tensile stress ($\overline{\sigma_m}$) in the mineral component can be written as

$$\sigma_m = \rho\tau_p \tag{17.25}$$

and

$$\overline{\sigma_m} = \rho\tau_p/2 \tag{17.26}$$

where $\rho = L/h$ is the aspect ratio of mineral platelets (length L and width h) and τ_p is the shear stress of the proteins. It is assumed that the protein does not carry a tensile load in the composite model and the effective tensile stress (σ) in the nanocrystalline composite is given by

$$\sigma = \phi\overline{\sigma_m} \tag{17.27}$$

where ϕ is the volume fraction of the mineral and $\overline{\sigma_m}$ is the average stress in the mineral component. The effective strain (ε) in the composite when the mixture is stressed is the sum of both the protein and mineral components,

$$\varepsilon = \frac{\Delta_m + 2\varepsilon_p h(1-\phi)/\phi}{L} \tag{17.28}$$

where h is the thickness of the platelets, and L is the length of the platelets. Δ_m and ε_p are the elongation of the mineral platelets and shear strain of the proteins respectively. The elongation of the mineral platelets (Δ_m) is given by

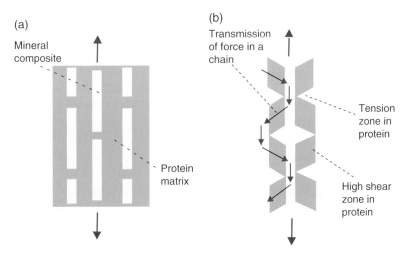

Figure 17.19 *Schematic diagram of a generic biocomposite that can be used to model materials such as cortical bone and nacre. (a) The mineral/protein composite is placed under stress. (b) The tensile stress is predominantly experienced by the mineral component and the protein shears under the stress. [Reprinted from B. Ji, H. Gao, Journal of Mechanics and Physics of Solids, 2004, 52, 1963–1990. With permission from Elsevier.]*

$$\Delta_\mathrm{m} = \frac{\sigma_\mathrm{m} L}{2E_\mathrm{m}} \qquad (17.29)$$

where E_m is the Young's modulus of the mineral component. The shear strain in the protein (ε_p) is related to the shear modulus of the protein (G_p) and the shear stress (τ_p) the protein experiences,

$$\varepsilon_\mathrm{p} = \frac{\tau_\mathrm{p}}{G_\mathrm{p}} \qquad (17.30)$$

The total effective Young's modulus (E) of the biocomposite is therefore

$$\frac{1}{E} = \frac{4(1-\phi)}{G_\mathrm{p}\phi^2\rho^2} + \frac{1}{\phi E_\mathrm{m}} \qquad (17.31)$$

and reasonably high Young's moduli are possible with such a model of a biocomposite, as observed in Nature.

The *toughness* of biocomposites is very well optimised. Nanoscale mineral inclusions have fewer flaws in their crystalline lattices than the macroscopic equivalents and their strength approaches that of the atomic bonds between the crystalline atoms. The viscoelasticity of the proteins that adhere to the crystallites helps the material dissipate fracture energy and ensures that large cracks do not occur. Biocomposites achieve a high stiffness using a large aspect ratio for the crystallite inclusions (they are long and thin) and a nanotextured, staggered alignment of these nanocrystallites.

Suggested Reading

If you can only read one book, then try:

Top recommended book: Vincent, J. (1990) *Structural Biomaterials*, Princeton. Extremely well written, compact introduction to biological materials.

Denny, M. & McFadzean, A. (2011) *Engineering Animals: How Life Works*, Harvard. Engaging popular account of bioengineering.

Gibson, L.J. & Ashby, M.F. (1997) *Cellular Solids*, Cambridge University Press. Classic text on solid foams.

Gong, J.P. (2002) Surface friction of polymer gels, *Progress in Polymer Science*, **27**, 3–38. Discussion of some elegant experiments on the friction of polymer gels.

Gosline, J., Lillie, M., Carrington, E., Guerette, P., Ortlepp, C. & Savage, K. (2002) Elastic proteins; biological roles and mechanical properties, *Philosophical Transactions of the Royal Society of London B*, **357**, 121–132. Useful review on the structural roles of proteins.

Vogel, S. (2003) *Comparative Biomechanics: Life's Physical Worlds*, Princeton. Contains an exhaustive range of biomechanical examples.

Tutorial Questions 17

17.1 Estimate the shear modulus of an adhesive protein in nacre if the Young's modulus of a shell is 12.4 MPa, the crystalline volume fraction is 0.95, the aspect ratio of the crystallites is 10 and the shear modulus of the crystallites is 3 GPa.

17.2 A man picks up a piano and the sections of cartilage in his knees are compressed from 1 cm thickness to 0.95 cm. Calculate the relaxation time of the stress in the cartilage gels once the weight is removed if the Young's modulus of cartilage is 0.78×10^6 Pa and the hydraulic permeability is $6 \times 10^{-13}\,\mathrm{m^4\,N^{-1}\,s^{-1}}$.

18

Phase Behaviour of DNA

The *in vivo* behaviour of DNA presents a wide range of fascinating phenomena with respect to the molecule's structure, dynamics and phase transitions. The full impact of DNA's properties can only be appreciated after the mechanisms of gene expression are properly understood (**Chapter 22**). In this chapter some of the molecular biophysics of DNA chains will be considered, which contribute to this understanding (also see **Part IV**).

18.1 Chromatin – Naturally Packaged DNA Chains

The method through which DNA is packaged into the nucleus of a cell has posed evolution an interesting problem. In a human cell all the 46 pieces of DNA arranged in a line constitute a long narrow thread of 1.5 m in length and 2 nm diameter, which needs to be accommodated into a box (the nucleus) whose volume is only a few micrometres cubed ($\sim(5 \ \mu m)^3$). The solution that Nature has evolved is to have the DNA stored in a compacted form with the chains wrapped around proteins spools (histones); much like cotton is wound around bobbins in needle work. The DNA spools are then assembled into fibrous aggregates that are called chromosomes.

The first strong experimental evidence for histones was presented by Hewish and Burgoyne in 1973. They found that the majority of chromosomal DNA, when digested by a DNA-cutting enzyme, formed small fragments of regular size 200, 400, and 600 base pairs (using gel electrophoresis). The explanation for this phenomenon was that the binding proteins (histones) are arranged in a regular manner along the DNA chains and only DNA between the histones could be cut by the enzymes, which sets the fundamental length of the fragments.

The method by which the nucleosomes associate with DNA chains poses many questions and the physicochemical processes that drives the self-assembly of the histones on to a specific sequence of base pairs are still incompletely understood. A constant length of ~150 base pairs of DNA is thought to associate with a single histone (**Figure 18.1**).

Wide-angle X-ray and neutron diffraction experiments have examined the specific interactions at the molecular level between small fragments of DNA and histone proteins (eight histone protein subunits are

The Physics of Living Processes: A Mesoscopic Approach, First Edition. Thomas Andrew Waigh.
© 2014 John Wiley & Sons, Ltd. Published 2014 by John Wiley & Sons, Ltd.

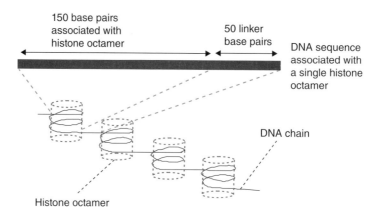

150 base pairs
associated with
histone octamer

50 linker
base pairs

DNA sequence
associated with
a single histone
octamer

DNA chain

Histone octamer

Figure 18.1 *Schematic diagram of the specific association of histone octamers with sections of DNA with well defined lengths. The discrete DNA lengths formed by enzymatic digestion of DNA were a preliminary indication of the existence of histones.*

DNA chain

Histone
subunits

Figure 18.2 *Complexation of a histone octamer (eight subunits) with a small DNA chain fragment. The structure is based on wide angle X-ray and neutron scattering experiments on crystalline chromosomal fragments.*

required to form a single histone bobbin). Accurate molecular models have then been made of small crystalline sections of DNA with histone octamers (**Figure 18.2**). Unfortunately, these techniques are not feasible with noncrystalline samples such as complete chromosomal fibres. Larger lengths of DNA chain could have markedly different conformations with the histones due to their altered elasticity, torsional resistance and counterion condensation effects. Separate experimental evidence is required.

Molecular models of small-angle neutron scattering and small angle X-ray scattering data have extended the resolution of histone structures when combined with longer DNA chains in solution to much larger length scales (nm), **Figure 18.3**. Careful modelling of the liquid-state scattering data is required, but good evidence for compact and extended forms of the complexes has been found dependent on the amount of salt in the solutions. This indicates a process of electrostatic binding between the negatively charged DNA and the positively charged histone proteins. However, these scattering techniques only provide general features of the chromosomal structure and little detailed information is available.

There still exist a number of questions that relate to the ambient large-scale structure (10 nm) of chromatin and the self-assembly of this morphology, due to the noncrystallinity and aperiodicity of the samples. Valuable information has been provided by electron microscopy, a technique that under suitable conditions can provide Angstrom resolution of aperiodic materials, although they need to be frozen and held under high-vacuum conditions. Tomographic reconstruction of transmission electron micrographs led Aron Klug and coworkers to propose the 'beads on a string' model in 1977 (**Figure 18.4**). This model is the result of tomographic reconstruction of a stack of transmission electron microscopy images of freeze-fractured chromosomal fibres

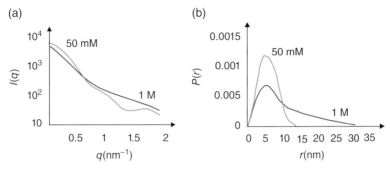

Figure 18.3 *(a) Small-angle neutron scattering data (scattered intensity I(q) versus momentum transfer q) from histone/DNA complexes allows (b) the solution state structure of histone/DNA complexes to be modelled with high salt (1 M) and with low salt (50 mM) concentrations. The radial distribution function (P(r)) is calculated for the DNA/histone complex at two salt concentrations. A much more expanded structure is observed at high salt concentrations as the DNA chains start to unbind from the histone octamers (the electrostatic binding force decreases due to the increased screening). [Reprinted from S. Mangenot, A. Leforestier, P. Vachette et al, Biophysical Journal, 2002, 82, 345–356 With permission from Elsevier.]*

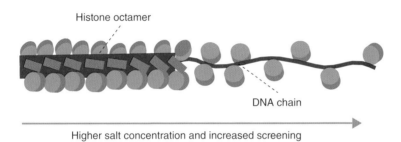

Figure 18.4 *Model for chromosomal fibres based on tomographic reconstruction of transmission electron microscopy images. The chromosomes consist of histone octamers (green) assembled onto a long thread of DNA (red), in a beads on a string manner.*

that contain a staining agent that provides strong contrast for electron scattering. The mathematical reconstruction technique implemented to analyse the images requires careful handling to produce dependable results; the inclusion of a staining agent and the process of freeze fracture during sample preparation are cause for concern, since both could have radically affected the morphology of the chromosomal fibres. Thus, there continues to be a degree of controversy in the field of the chromosomal structure with respect to the orientation of the histones along the axis of the chromosomal fibre, but all the current evidence points to the existence of tightly bundled fibres formed from beads on a string and that these fibres occur *in vivo* in the nucleus.

In contrast to human DNA, the length of bacterial DNA is much smaller and is not associated with histones[*]. However, to make the chains fit inside the bacterial cell compaction is still required. Overtwists are introduced into the circular DNA chains by enzymes and the chains form compact plectonomic structures (**Section 10.6**). Circular DNA associated with topology preserving proteins are said to be 'restrained'. Nicks in 'restrained' DNA do not cause the supercoil to unwind into the relaxed state, since they are still bound to a protein scaffold, and can be repaired with no loss in the degree of winding. The torsional energy stored in a restrained duplex DNA is therefore conserved.

---

[*] Some Archaea (the third general classification of life, in addition to standard prokaryote and eukaryote classifications) are now known to have histone analogues that compact their DNA, emphasising their separate identity from standard bacteria.

18.2 DNA Compaction – An Example of Polyelectrolyte Complexation

Many diseases have a genetic origin, the most important example being cancer in its many different forms. A possible strategy to treat these conditions is to replace the malfunctioning DNA in malignant cells with a benign substitute. An obstacle to this strategy of *gene therapy* is how to transfer the material to the nucleus of a cell without it being destroyed by the cell's defence mechanisms. One reasonably successful strategy that has been used to transfer the DNA, is to combine it with an oppositely charged polyelectrolyte or a virus (a drawback is the virus can itself prove to be pathogenic) to allow transfer through the cell wall. These questions on DNA transfection, combined with more fundamental problems concerning the natural functioning of chromosomes, provide motivation for the understanding of DNA compaction with oppositely charged colloidal spheres, which can act as transfer vehicles for DNA as well as for chain compaction.

The persistence length of a DNA chain is about 50 nm under standard physiological conditions. The persistence length of the uncomplexed DNA can be modelled and is thought to be a combination of the effects of counterion condensation, the electrostatic repulsion of like-charged segments and the intrinsic rigidity of the polymer backbone (the helix acts as an elastic rod). From **Section 11.9** the total persistence length (l_T) of a charged polymer is given by equation (11.90). The charge density (Q) along the chain is limited by charge condensation. The distance between the charged phosphate groups along the backbone of a DNA chain is 1.7 Å, the Bjerrum length is 7 Å at 20 °C and the effective charge fraction (ξ) in the Manning charge condensation model (equation (11.84)) is therefore $1.7/7 = 0.24$. The OSF calculation given by equation (11.90) provides the correct order of magnitude for the persistence length when compared with experiment. The total persistence length can be separated in to the contribution of the intrinsic persistence length ($l_p = 30$ nm) and the charge repulsion of the phosphate groups. The charged contribution to the persistence length is therefore 20 nm at 0.1 M salt.

The effect of chirality on the resultant chromosomal morphology is a further question with twist-storing polymers such as DNA. An additional term can be introduced into the free energy for semiflexible chains (equation (10.19)) which corresponds to the propensity for torsional rotation. Theoretical studies imply that the chirality of nucleosome fibres is due to the specific histone/DNA potential and not due to the intrinsic twist–bend interaction of the DNA fibres, although base-pair chirality does provide a small contribution to the fluctuations and elasticity of the naked uncomplexed DNA chains.

Chromosomal DNA is wrapped around the cylindrical histone core on a helical path of a diameter (D) of 11 nm (**Figure 18.2**). This is smaller than the intrinsic persistence length of the DNA chain, so substantial elastic energy is stored upon complexation. The origin of the attraction between DNA and the histone is electrostatic, but can be considered short range at physiological salt concentrations. The Debye screening length for the electrostatic interaction at physiological salt concentration is around 1 nm and is a first approximation for the length scale of the DNA–histone electrostatic interaction. The binding energy per unit length of the histone can be estimated from the assumption that the wrapping state represents a dynamical equilibrium in which a wrapped portion of the DNA strand spends part of the time in the dissociated state. With this assumption, the binding energy (λ) of a DNA duplex is found to be $1-2\,kT$ per 10 base pairs. *In vitro* experiments on the association between DNA and histones find an athermal first-order phase transition as a function of the DNA histone interaction strength from a wrapped state to a dissociated state consistent with the idea of an all-or-none wrapping transition. This implies that the powerful thermodynamics ideas that determine the behaviour of phase transitions (**Chapter 5**) can be applied to the complexation of DNA.

Figure 18.5 shows a phase diagram for the complexation between an idealised positively charged sphere and a DNA chain in aqueous solution. DNA chains can be touching, have point contacts, or be wrapped onto the charged spheres, dependent on the electrostatic screening length and the amount of charge on the sphere. Molecular dynamics simulations observe similar phenomena to those observed in the figure and the effect of the curvature of the spheres can be probed. A *wrapping phase transition* is predicted as the diameter of the positive sphere is increased using both analytic theory and Monte Carlo simulation.

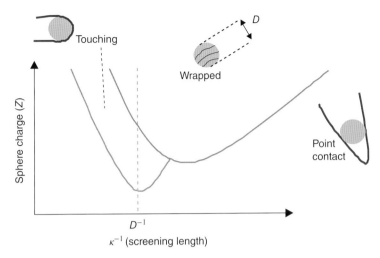

Figure 18.5 *Theoretical phase diagram for the statistics of DNA chains in contact with oppositely charged spheres. The diagram shows the state of the DNA–sphere complexes as a function of the charge per sphere (Z) and the Debye screening length (κ^{-1}) of the solution. It is assumed that the process of binding is purely electrostatic and there is no chemical specificity. First-order phase transitions are predicted between point contact, touching and wrapped states. D is the diameter of the colloid. [Reprinted with permission from R.R. Netz, J.F. Joanny, Macromolecules, 1999, 32, 9026–9040. Copyright © 1999, American Chemical Society.]*

A novel counterintuitive physical phenomenon that occurs in polyelectrolyte complexation is that of overcharging. *Overcharging* can occur with both the complexation of colloids (e.g. histones, amine derivatised spheres, etc.) with oppositely charged polymers (e.g. DNA, polystyrene sulfonate, etc.) and polyelectrolytes with charged surfaces. More polyelectrolyte is adsorbed to a surface than required for simple charge neutralisation; the charge on the adsorbing surface is reversed. With DNA/histone complexes the charge on the DNA greatly outweighs that of the histone. This results in chromosomal complexes being strongly negatively charged (histones are positively charged). The thermodynamic origin of the effect is thought to be related to counterion condensation. The additional contribution to the electrostatic energy of the overcharged complex in the free energy of the system is compensated by the additional entropy of the released counterions.

18.3 Facilitated Diffusion

There are a range of enzymes that bind to DNA. These include enzymes that initiate transcription, RNA polymerase and endonucleases that chemically modify the DNA sequence. The rates of reaction for the DNA-binding proteins are significantly faster than would be expected from calculations that assume three-dimensional diffusion to reaction (**Section 7.5**). Indeed with the classic example of the lac repressor (**Chapter 22**) the degree of association of the repressor protein for the DNA chain is underestimated by a factor of a hundred. A number of models of facilitated diffusion have been proposed to account for this short fall. The three primary scenarios for the interaction between the DNA and the lac protein are thought to be sliding of the protein along the DNA, hopping of the protein from site to site along the DNA chain and intersegmental transfer of the proteins between multiple binding sites (**Figure 18.6**). An important clue that accounts for the increased binding rates is that the efficiency of a collision as a result of diffusion is increased by a reduction in the number of dimensions in which the diffusive process occurs. Compelling the protein to execute a

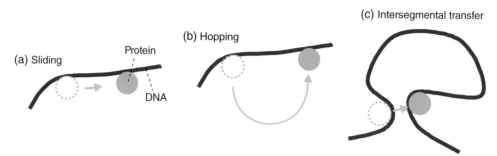

Figure 18.6   *Schematic diagram of the three principle models for facilitated diffusion of a DNA chain with a protein. The possible mechanism are (a) sliding, (b) hopping and (c) intersegmental transfer for proteins with multiple binding sites.*

one-dimensional random walk along the DNA chain greatly decreases the time to search the whole sequence for the correct binding site. Furthermore, the probability of a collision between the protein and the DNA chain is much higher when the protein is confined inside the DNA coil.

The Stokes–Einstein equation (7.10) gives the diffusion coefficient for a small globular enzyme (diameter 5 nm) in three dimensions in water at room temperature to be approximately $10^8\,\mathrm{nm^2\,s^{-1}}$. It is assumed that a DNA-associating enzyme diffuses in three dimensions before the reaction with the DNA and the DNA chain moves much more slowly that the associating protein due to its size, i.e. the DNA chain is effectively stationary. The association rate constant (k, equation (7.57)) is given by

$$k = 4\pi Da \tag{18.1}$$

where a is the size of the binding site on the protein and D is the diffusion coefficient for the protein. Typically, the size of the binding site for the enzyme (a) is much smaller than the size of the protein, $a/d \sim 0.1$, where d is the diameter of the protein. From equation (18.1) the value of the rate constant (k) is found to be $10^8\,\mathrm{M\,s^{-1}}$. However, experimentally the lac repressor is found to have a rate constant $\geq 1 \times 10^{10}\,\mathrm{M\,s^{-1}}$ and the mechanism of facilitated diffusion is needed to explain the shortfall by a factor of 100.

Both sliding of the protein along the DNA and hopping through three dimensions would increase the association rate in models for facilitated diffusion and is used to explain the shortfall in association constants (**Figure 18.6**). The probability that a protein sliding on a DNA chain stays on the chain after N steps is

$$(1-P)^N = e^{N\ln(1-P)} \tag{18.2}$$

where P is the probability of dissociation during a step and the equality is from simple mathematical manipulation. The process of dissociation is another example of a Poisson decay process (**Chapter 3** and **Section 7.3**) where the probability is proportional to $e^{-\mu}$ and μ is the mean. The expectation value of one decay event is when the mean equals one ($\mu = 1$) and the number of steps over which sliding (N) can occur is therefore given by

$$N = \frac{-1}{\ln(1-P)} \tag{18.3}$$

In the case of a very low probability of dissociation,

$$P \ll 1 \quad \ln(1-P) \approx -P \tag{18.4}$$

And therefore

$$N = 1/P \tag{18.5}$$

The characteristic sliding length (l_{sl}) explored by one dimensional diffusive sliding ($<l_{sl}^2> = 2h^2N$, equation 7.7) is then

$$l_{sl} = \frac{\sqrt{2}h}{\sqrt{P}} \tag{18.6}$$

where h is the base-pair step sliding length (**Figure 18.7**). From the definition of diffusion in one dimension (equation (7.7)) the characteristic time (τ_{sl}) for exploration of the sliding length is

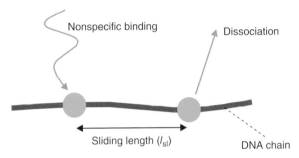

Figure 18.7 *The definition of the sliding length (l_{sl}) used in the process of facilitated diffusion of a protein along a DNA chain. The protein associates with the DNA chain by nonspecific binding, it moves the sliding length and then dissociates from the chain.*

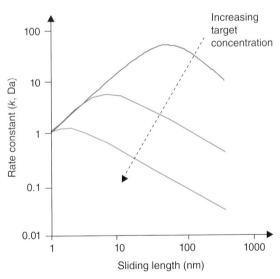

Figure 18.8 *The predicted association rate constants of a protein (k) with a DNA chain as a function of the sliding length (l_{sl}). Three possible solutions are shown corresponding to low, medium and high target concentrations. The optimal sliding length depends on the target concentration. The sliding length is rescaled by the diameter of the binding site (a) in the plot. [Reproduced from S.E. Halford, J.F. Marko, Nucleic Acids Research, 2004, 32, 10, 3040–3052 by permission of Oxford University Press.]*

$$\tau_{sl} = \frac{l_{sl}^2}{D_1} \qquad (18.7)$$

Therefore, it is concluded that very small dissociation constants (P) are required for long l_{sl} (equation (18.6)) which consequently allow significant increases to occur in the reaction rates (equation (18.7)). Inclusion of the combined effects of both hopping and sliding motions (**Figure 18.8**) shows that the reaction rate per unit length (k) is given by

$$k = Da\left(\frac{a}{l_{sl}} + \frac{D}{D_1}aLl_{sl}c\right)^{-1} \qquad (18.8)$$

where L is the total contour length of the DNA chain, a is the size of the binding site, D is the diffusion coefficient in three dimensions of the binding proteins, D_1 is the diffusion coefficient related to sliding of the proteins along the DNA chain and c is the concentration of target DNA. The nonmonotonic behaviour of the rate constant on the ionic strength predicted theoretically by such models has been observed experimentally and is a major success of the theory. This picture of facilitated diffusion provides a molecular model for the control of gene expression in DNA chains (e.g. the lac repressor) which is discussed in more detail in **Chapter 22**.

Suggested Reading

If you can only read one book, then try:

Calladine, C.R., Drew, H.R., Luisi, B. & Travers, A. (2004) *Understanding DNA: The Molecule and How it Works*, Academic Press. Extremely useful introductory account of the behaviour of DNA from a structural engineering point of view, which requires a minimum of mathematical ability.

Halford, S.E. & Marko, J.F. (2004) How do site specific DNA binding proteins find their targets?, *Nucleic Acid Research*, **32** (10), 3040–3052. A clear explanation of the processes involved in DNA binding.

Crick, F. (1990) *What Mad Pursuit: A Personal View of Scientific Discovery*, Basic Books. Entertaining popular account of the discovery of the structure of DNA.

Part III

Experimental Techniques

The largest impact of physics in the life sciences has arguably been the creation of new experimental characterisation techniques. The word 'cell' was coined by the Oxford physicist, Robert Hooke, after he constructed an early compound optical microscope, examined a cork sample, and needed a word to describe what he saw. Francis Crick helped develop the technique of X-ray fibre diffraction modelling in the Bragg laboratory at Cambridge, which resulted in the discovery of the helical structure of DNA with James D. Watson. More recently, the development of functional magnetic resonance imaging has allowed conscious mental processes to be followed in living individuals. The impact of such physical experiments in biology has therefore unarguably been colossal.

Thus, if someone manages to discover (or uncover) a new physical technique, their second thought should always be, can they also apply it to biology? There are a huge number of physical methods that have proven useful in biology and only the physical basis of a few of the major techniques for the examination of molecules and cells will thus be considered here. Other sources in the literature should be consulted for more extensive accounts of specialised methods, e.g. medical physics methods used in hospitals.

Particular emphasis will be given to *photonics techniques*, and these have had a long illustrious history in biology. Highlights include visible wavelength photons in light microscopes, X-ray photons for crystallography and imaging, radio frequency photons used in nuclear magnetic resonance experiments, and infrared photons in optical tweezer experiments with single molecules. The area of photonics does not encompass all of the standard biophysical experimental techniques, but it does cover many of them (particularly if you include phenomena that involve virtual photons such as magnetism). The subfield of visible/infrared photonics techniques is very appealing; since it is possible to build new instruments relatively cheaply in one's own laboratory that provide novel world-class functionalities.

Missing from this discussion is the amazing photonics materials that have been created naturally through evolution, e.g. the iridescent colours of butterflies' wings, the polarisation properties of beetles' exoskeletons, and the luminescence of fireflies. Again this is a matter of space, not a judgement on the intrinsic merit of the field.

Imaging forms a large component of this Parts's contents. Indeed, high-resolution imaging was recently highlighted as one of a series of grand challenges for the twenty-first century by the Institute of Physics.

The Physics of Living Processes: A Mesoscopic Approach, First Edition. Thomas Andrew Waigh.
© 2014 John Wiley & Sons, Ltd. Published 2014 by John Wiley & Sons, Ltd.

Countless questions in cellular biology, molecular biology and materials science could be solved unambiguously if noninvasive imaging could be routinely performed at the nanometre length scale. The huge area of *spectroscopy* is also included, since it provides detailed information on molecular dynamics, chemical structure and reactivity on the atomic scale. A third clear area, rich with modern developments, is *single-molecule* imaging and characterisation. Examination of molecules on an individual basis provides a rich amount of additional physical information, that is often lost during ensemble averaging, and can be crucial for an understanding of live cell activity, e.g. each cell typically only has a single DNA chain for each chromosome and these single molecules play a crucial role in the determination of the cell's fate. A large section is included on *scattering*, which was historically of central importance in the creation of the field of molecular biology (e.g. X-ray crystallography of proteins) and continues to provide a rich range of useful tools in the twenty-first century.

The mechanical form of spectroscopy (*rheology*) will be examined in detail; the sample is hit with a mechanical perturbation and its response in time is observed. Study of this field often provides refreshing new perspectives on biological problems, because viscoelastic effects have a large influence on dynamic phenomena and their effects are felt all the way from the length scale of individual molecules to that of whole organisms.

In addition, the methods of *osmometry, chromatography, sedimentation, tribology, solid mechanics, thermodynamics, hydrodynamics* and *electrophoresis* will be presented. These useful tools have diverse origins in materials science and biochemistry. Each can provide unique insights into biological problems.

The coverage of the experimental techniques section is by necessity a little lopsided due to the huge number of important topics vying for space. However, hopefully it will encourage biophysicists to explore some of the vast range of physical techniques now available and perhaps create (or adapt) some new ones themselves to solve a specific biological problem.

Suggested Reading

Serdynk, I.N. & Zaccai, N.R. (2007) *Methods of Molecular Biophysics*, Cambridge University Press. Extremely detailed review of modern experimental techniques in molecular biophysics.

Lipson, A., Lipson, S.G. & Lipson, H. (2011) *Optical Physics*, 4th edition, Cambridge University Press. Many modern biophysics techniques are based around photonics methods. The book by the Lipsons gives a detailed introduction to some of the relevant underlying physical phenomena and is a good place to start.

19

Experimental Techniques

A vast range of experimental techniques are currently used to analyse the structure and dynamics of biomolecules. A subset of methods that emphasise the physical behaviour of biological molecules and cells will be examined here, and reference should be made to more specialised texts for detailed descriptions of medical and analytical biochemical methods.

19.1 Mass Spectroscopy

Mass spectroscopy allows the charge/mass ratio of atoms and molecules to be measured with high accuracy. Modern developments mean that even large proteins (200 kDa) can be studied. The physical basis of the technique is very simple. Charged ions are created, and accelerated by an electric field, whilst travelling through a transverse magnetic field. The magnetic field creates a lateral force and each ion's trajectory becomes curved. A detailed analysis of this trajectory then allows the charge/mass ratio of the ion to be calculated, which provides valuable structural information on the biological molecule that was used to create the ion. Indeed, mass spectroscopy is now a standard initial step to analyse the structure of unknown biological molecules.

A useful reference calculation is for an ion that moves with constant speed in a circle, where the magnetic force creates the centripetal acceleration. The Lorentz force on the ion provides the acceleration towards the centre of the circle and thus obeys the equation

$$zvB = \frac{mv^2}{r} \tag{19.1}$$

where z is the ionic charge, v is the ion speed, B is the magnetic field, m is the ion mass and r is the radius of curvature of its path. Thus, knowledge of v, B and r for a particular ion allows the charged/mass ratio (z/m) to be calculated and this is the parameter typically output in a mass spectroscopy experiment, i.e. z/m is plotted as a function of signal strength that is proportional to the number of ions. Ions are produced with quantised numbers

The Physics of Living Processes: A Mesoscopic Approach, First Edition. Thomas Andrew Waigh.
© 2014 John Wiley & Sons, Ltd. Published 2014 by John Wiley & Sons, Ltd.

of electronic charge (+1, +2, +3,…), so it is often a relatively simple job to deduce the exact mass of a specific fragment by the consideration of integer multiples of z/m. The sensitivity of the technique is so great that it can also resolve the differences in mass between molecules that contain different isotopes, e.g. carbon ^{12}C and ^{13}C give distinct signals.

Modern advances in mass spectroscopy include a series of sophisticated methods for the creation of charged ions in the gaseous phase using laser desorption and electrosprays. These are currently the most important methods for studies of large biological macromolecules. Different pieces of equipment can be compared based on their mass resolving power and the absolute accuracy of mass determination.

A simple experimental realisation of a mass spectrometer is shown in **Figure 19.1**. Ion trajectories are bent by a magnet and their position is then detected. Quadrupole magnets can be used for additional filtering and magnetic focusing of the ions. The MALDI TOF (matrix-assisted laser desorption/ionisation with time of flight mass analyser) technique resulted in a Nobel Prize in 2002. Ions are desorbed from a surface using a laser (**Figure 19.2**), and it functions well for the creation of suitable ions from fragile

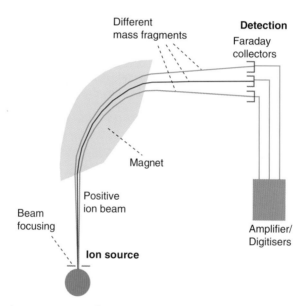

Figure 19.1 *Experimental arrangement of a simple mass spectrometer. Ionised biological molecules are accelerated by an electric field, whilst their trajectory is bent in a magnet. Faraday collectors detect arrival times and trajectory curvatures.*

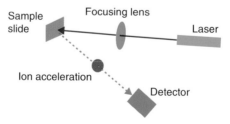

Figure 19.2 *Schematic diagram of matrix-assisted laser desorption ionisation time of flight mass spectrometry (MALDI-TOF).*

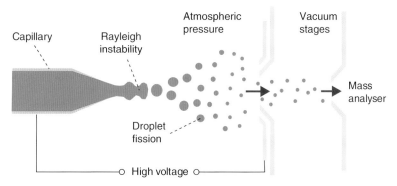

Figure 19.3   *The electrospray method uses the Rayleigh instability to form ions for mass spectroscopy.*

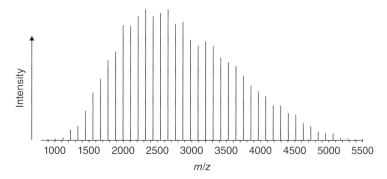

Figure 19.4   *MALDI-TOF mass spectroscopy data from a synthetic polymer, polyvinylpyridine, which shows the detected intensity as a function of the mass/charge ratio (m/z). [Reproduced from A. Brock et al, Anal. Chem. 2003, 75, 3419–3428. Copyright © 2003, American Chemical Society.]*

biomolecules. Electrospray is another method to form the gaseous phase ions (**Figure 19.3**), which is also useful in biology.

A typical *mass spectrum* from a macromolecule is shown in **Figure 19.4**. The m/z ratio of the fragments from polyvinylpyridine seen in the figure provide fundamental compositional information about the molecules. Mass spectra from heterogeneous macromolecules (e.g. proteins) are typically much more complicated, but can be still used for effective structural elucidation when combined with the correct software for computational analysis.

Mass spectroscopy can also be adapted for high-throughput Ohmics-type research, e.g. proteohmics (**Chapter 22**). It is possible to measure microheterogeneity among proteins using the technique, e.g. variations in glycosylation or phosphorylation. It is also possible to sequence proteins by systematic degradation techniques combined with mass-spectroscopy experiments.

19.2   Thermodynamics

Thermodynamics is a fundamental area of physics that determines the colligative properties of materials, i.e. how collections of molecules interact. However, here thermodynamics is used as a specific heading to describe a family of techniques that can be used to probe the thermal or energetic properties of biomolecules.

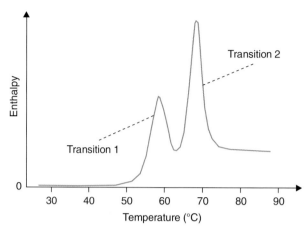

Figure 19.5 *Data from a differential scanning calorimetry experiment, which shows the enthalpy as a function of the temperature. Two well-defined first-order phase transitions are observed, e.g. transition 1 could be a solid to nematic liquid crystal transition and transition 2 could be a helix–coil transition.*

19.2.1 Differential Scanning Calorimetry

Differential scanning calorimetry (DSC) is a well-developed method to study the thermal properties of materials. Heat is applied to a sample and simultaneously to a reference sample. The difference in heats between the sample and the reference (which gives the word 'differential' in the title) is scanned as a function of temperature. It is an important method for the study of thermally induced phase transitions in condensed matter, e.g. helix–coil, liquid crystalline, globule–coil, etc. In **Figure 19.5** a melting endotherm from a DSC is shown with two first-order phase transitions, e.g. the melting transitions of two different globular structures.

Processes measured in a DSC can be endothermic (heat moves in) or exothermic (heat moves out). Statistical models can be made that relate the structure of a biomolecule to its thermodynamics. It is hard to prepare large quantities of many biological molecules with reasonable purity and this has led to the introduction of microcalorimetry methods, i.e. calorimetry for microlitres of material.

19.2.2 Isothermal Titration Calorimetry

The technique of isothermal titration calorimetry can most readily be understood by reference to its name. 'Isothermal' means that the equipment is held at a constant temperature. 'Titration' describes when one substance is slowly added to another and 'calorimetry' is the measurement of changes in heat (**Figure 19.6**). The use of such a method allows enthalpy and entropy changes of molecular binding to be measured, e.g. enzyme substrate binding.

19.2.3 Surface Plasmon Resonance and Interferometry-Based Biosensors

The adsorption of molecules to surfaces loosely fits under the thermodynamics heading, since the processes are driven by the surface-interaction energies. Modern photonics technologies offer a series of methods for an extremely detailed characterisation of surface morphology. Evanescent optical waves can be used as an accurate probe of surface adsorbed biomolecules. This forms the heart of the OWLS (*optical waveguide light mode spectroscopy*) technique. Evanescent waves are enhanced using the phenomena of plasmon resonance,

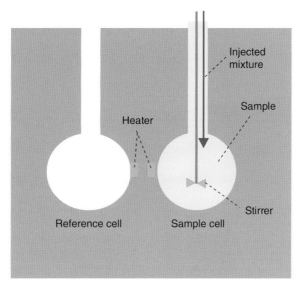

Figure 19.6 *Schematic diagram of an isothermal titration calorimeter. A sample is injected into a stirred vessel and the difference in heat input between it and a reference cell is measured.*

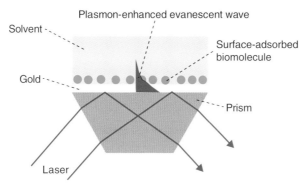

Figure 19.7 *Optical waveguide light mode spectroscopy (OWLS) is a surface plasmon resonance device for the measurement of films of adsorbed biological molecules. The reflected laser light is perturbed by the interaction of the evanescent wave with the sample.*

i.e. photons tunnel through highly conductive thin films attached to the surface, e.g. a gold film (**Figure 19.7**). Combination of the gold film with an optical waveguide allows convenient measurement of the interaction of biomolecules with the surface. Surface sensitivities down to one molecule per mm^2 have recently been demonstrated using plasmon resonance.

Dual polarisation interferometry (DPI) can provide Angstrom level resolution of the thickness of thin films, simultaneously with mass measurements at millisecond time scales. In DPI, rectangular waveguides are used, combined with the switching of the polarisation of a laser light source (using a liquid-crystalline polariser, **Figure 19.8**) and the output of the waveguide forms a Young's two-slit interference pattern. Evanescent waves interact with surface-adsorbed samples and Maxwell's equations can be solved for the light as it propagates in

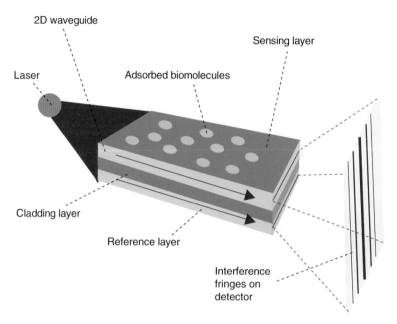

Figure 19.8 *Schematic diagram of a dual-polarisation interferometer used to measure the adsorption of ultrathin layers (0.1–10 nm) of biomolecules on to a surface. The apparatus acts as a Young's double-slit interferometer. The mass and film thickness is extracted from the fringe pattern using a Maxwell equation based model of the two-dimensional waveguide The liquid crystalline polariser which switched the laser polarisation is not shown for simplicity. [Adapted with permission from B. Cowsill, PhD thesis, University of Manchester, 2012.]*

the waveguide. Interference patterns from two different laser polarisation states allow both the mass and thickness of surface adsorbed films to be measured unambiguously. DPI data for the adsorption of globular proteins is shown in **Section 9.7**.

A wide range of other chip-based systems exist that depend on nano- and microphotonics. In addition to interferometry (DPI) and standard surface plasmon resonance (SPR) devices, fluorescence and plasmon resonance Raman scattering are sometimes used to provide high-sensitivity measurements of minute quantities of biological molecules in biosensors.

19.3 Hydrodynamics

A wide range of techniques depend on the motion of biomolecules through fluids. The majority will be described in more detail later in the chapter. *Analytic ultracentrifuges* separate molecular mixtures and allow their sedimentation coefficients to be calculated and thus their hydrodynamic radii. Gel *electrophoresis* allows the length of DNA molecules to be measured. It depends on the phenomenon of reptation, snake-like motion of the DNA chains through the links of the gel. *Electric birefringence* of a solution can be monitored as a function of the applied electric field and this allows molecular sizes to be measured as charged molecules rotate in response to the field. *Flow birefringence* of a solution can be monitored as a function of flow rate and the alignment of molecules can be related to their viscosity. *Fluorescence depolarisation* can be calculated for a fluorescent molecule that is excited by a pulsed laser and the change in polarisation of the emitted photon allows the rotational motions of the molecule to be measured. The method can function at very fast picosecond time scales. *Viscosity* can be related to molecular structure when measured with viscometry or rheometry apparatus. *Dynamic light scattering* uses the Doppler shift of laser light scattered from a biomolecule to measure its dynamics. *Fluorescence correlation spectroscopy* measures the fluctuations of fluorescence molecules into and out of a well focused laser beam (a small volume is excited) and again allows dynamics to be probed.

19.4   Optical Spectroscopy

In general, the full range of the electromagnetic spectrum is useful in biology (**Figure 19.9**). However, optical spectroscopy experiments across visible, infrared and microwave energies have been particularly important. Generally, when a light beam interacts with matter (**Figure 19.10**) it can be elastically scattered (*Rayleigh* scattering), inelastically scattered by sound waves (*Brillouin* scattering) or inelastically scattered by phonons (*Raman* scattering). Fluorescence requires absorption of a photon followed by re-emission, and since it is not simply a scattering process, it will be considered separately.

19.4.1   Rayleigh Scattering

Rayleigh scattering describes the elastic scattering of visible wavelength light of particles whose size is of the order of the wavelength of the light. Rayleigh used this model to explain why the sky is blue; it is due to scattering from small atmospheric dust particles. Larger scattering particles, whose size is larger than

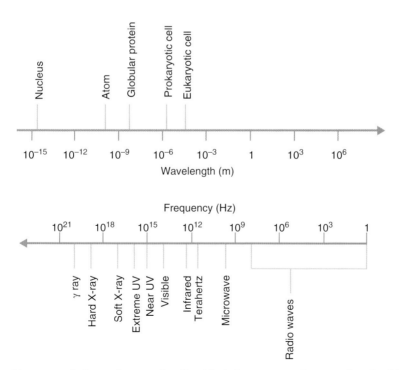

Figure 19.9   *A wide range of photon frequencies (f) with their corresponding wavelengths (λ) can be used in biological physics experiments (c = fλ, where c is the velocity of light).*

(a)

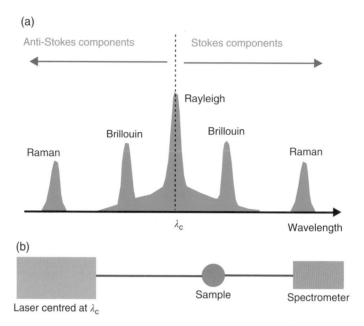

Anti-Stokes components

Stokes components

Rayleigh

Brillouin

Brillouin

Raman

Raman

λ_c

Wavelength

(b)

Laser centred at λ_c

Sample

Spectrometer

Figure 19.10 *Scattering of photons from a material typically involves three separate processes:* Rayleigh scattering *(elastic),* Brillouin scattering *(inelastic) and* Raman scattering *(inelastic). Scattering of light by biological molecules can thus cause a shift in the photons' energies. (a) Intensity of scattered light as a function of the photon wavelength. (b) Simple schematic diagram of a laser spectroscopy experiment.*

the wavelength of the scattered light, require a different model due to Mie to explain their behaviour. The Mie model is based on geometrical optics and is mathematically more challenging than that due to Rayleigh, but the numerical predictions can now be calculated fairly routinely for both Mie and Rayleigh models for standard particle geometries (spheres, ellipsoids, etc.). Thus, good models exist to describe light scattering from a wide range of biomolecules. Laser scattering is thus a standard technique to measure biomolecular size and it is often combined with chromatography to quantify the radii of mixed polydisperse systems.

19.4.2   Brillouin Scattering

Brillouin scattering has been less widely used in biological physics, but it does provide some unique information about the acoustic response of a sample and can be effectively combined with ultrasound imaging. It is possible to make spatially resolved measurements of high-frequency elasticity using Brillouin scattering. Brillouin scattering is also the basis of some standard photonics components, e.g. acousto-optical modulators used as fast shutters for lasers.

19.4.3   Terahertz/Microwave Spectroscopy

Terahertz radiation has been a recent addition to the suite of available biological spectroscopies. A previous barrier to the use of such radiation was the availability of convenient low-cost terahertz sources. Terahertz spectra can be used to fingerprint biological molecules, similar to infrared spectroscopy. Terahertz imaging is also possible, and since it is noninvasive and penetrating, it has wide applications in the creation of whole organism images, albeit at low resolution, e.g. in airport scanners.

19.4.4   Infrared Spectroscopy

The absorption of light at different infrared (IR) wavelengths provides valuable information on the molecules present; it offers another form of molecular fingerprinting. Modern infrared spectrometers are commonly built around a Michelson interferometer (**Figure 19.11**), i.e. they are Fourier transform IR spectrometers. Infrared spectra from polylysine with different secondary structures are shown in **Figure 19.12**.

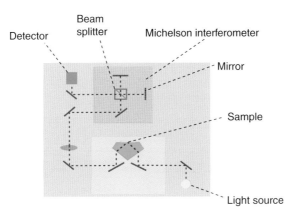

Figure 19.11   *Fourier transform infrared spectscoscopy based on a Michelson interferometer. The technique can be used to probe the secondary structure of biological molecules.*

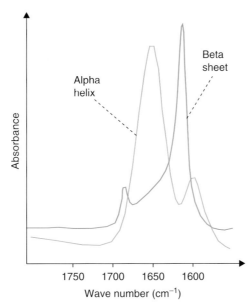

Figure 19.12   *Fourier transform infrared spectra from alpha helix and beta sheet conformations of polylysine. The absorbance is plotted as a function of the wave number ($k = 2\pi/\lambda$, λ is the wavelength of the infrared light). [Reproduced with permission from P.I. Haris, D. Chapman, Biopolymers, 1995, 37, 251–263. Copyright © 1995 John Wiley & Sons, Inc.]*

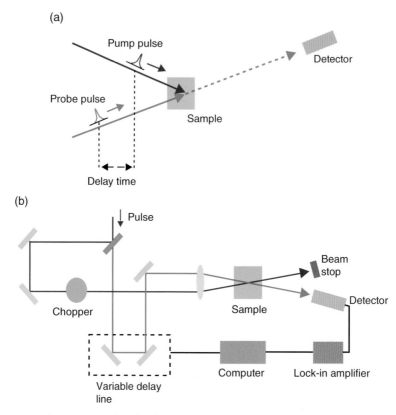

Figure 19.13   *Pump-probe apparatus for ultrafast infrared spectroscopy of biological molecules. (a) The signal is detected as a function of the delay time. (b) The pump and probe beam can be derived from the same pulsed laser source using a variable delay line. The lockin amplifier is locked to the chopper to improve the sensitivity of the signal detection (it helps isolate the signal from the background).*

Pulsed femtosecond IR lasers allow IR spectra to be probed as a function of the minute lag times between pulses, which is used in the pump-probe infrared spectroscopy technique. The sample is excited (pumped) with a first femtosecond pulse and its evolution is measured (probed) with the second femtosecond pulse. Both pulses can be created by the same laser using a delay line. This method can be used to provide femtosecond 'snap shots' of the motion of biomolecules. It is particularly effective in biology since hydrogen bonds strongly absorb IR photons.

A pulse sequence for an ultrafast pump-probe infrared spectroscopy experiment and some simple apparatus used in an experiment are shown in **Figure 19.13**. Transient hydrogen-bonded cages of liquid water have been shown to have lifetimes of ~10 ps using such techniques.

19.4.5   Raman Spectroscopy

Raman spectroscopy is an inelastic scattering process. Changes in the energies of the photons scattered by a biomolecule allow quantised vibrations and rotations (phonons) in the molecule to be probed.

Figure 19.14 shows an imaging Raman spectrometer built around an optical microscope and a diffraction grating. **Figure 19.15** shows a typical Raman spectrum measured from the globular protein lysozyme

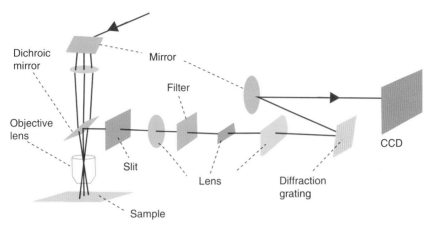

Figure 19.14 *Diffraction grating Raman spectrometer built around an optical microscope. Laser light is used to illuminate a sample and then a diffraction grating acts as a spectrometer to observe the changes in energy of the light due to quantised vibrations in the specimen, e.g. phonons in the molecules of the sample.*

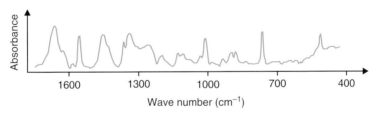

Figure 19.15 *Laser Raman spectra from lysozyme in solution. The absorbance is show as a function of the wave number ($k = 2\pi/\lambda$, λ is the wavelength). [Reproduced from A.H. Clark et al, Int. J. Peptide Res., 1981, 17, 353–364. © 1981 Munksgaard, Copenhagen.]*

in solution. Often, Raman spectra are much richer than fluorescence spectra for biomolecules in solution (more peaks are observed), although they are much weaker in magnitude, which makes intracellular spectroscopy challenging.

19.4.6   Nonlinear Spectroscopy

Developments in pulsed laser spectroscopy have provided intense high-power laser sources that allow non-linear optical processes in materials to be explored. There is a rich variety of possibilities (a few are shown in **Figure 19.16**) and the radiation damage in biological materials is mitigated by the short durations of the laser pulses. Two photon techniques are now well established in optical microscopy, since they allow better penetration and localisation in tissue samples, e.g. two infrared photons can excite a visible wavelength fluorescence transition in a biomolecule and the resultant images have improved contrast.

Nonlinear techniques have also been used in Raman spectroscopy to improve the weak signal strengths detected from biological molecules that restrict many areas of research, such as Raman studies of live cells. *Coherent anti-Stokes Raman scattering* (CARS) microscopy provides a method to circumvent the problems of the weak Raman signal. It uses a multiple photon process to probe the Raman scattering when two high power laser pulses are combined (**Figure 19.17**).

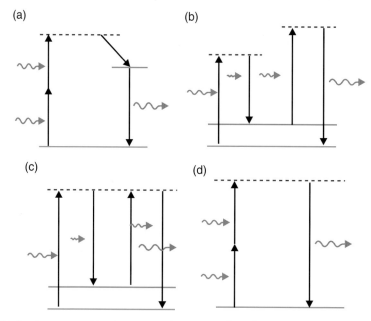

Figure 19.16 *Multiple-photon transitions can occur in a range of non linear processes that include (a) two-photon fluorescence, (b) coherent anti-Stokes Raman scattering, (c) stimulated Raman scattering and (d) second-harmonic generation. The light blue horizontal line is the ground state, the dashed red horizontal line is the excited state and the green horizontal line is an intermediate excited state. Absorbed photons are shown by blue curly lines, whereas emitted photons have orange curly lines.*

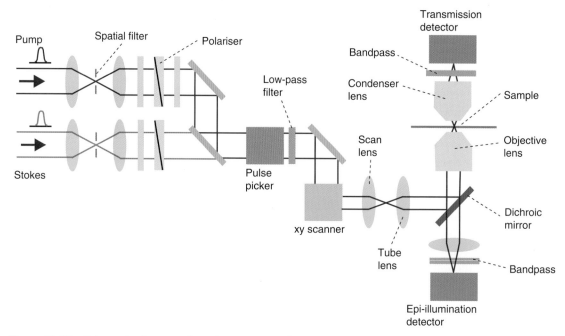

Figure 19.17 *Schematic diagram of a coherent anti-Stokes Raman microscope (CARS). The pump and Stokes beams are optimised for their spatial form and polarisation. The beams then travel into a scanning confocal microscope and an image is built up on a detector in either transmission or epi-illumination modes.*

Stimulated Raman scattering (SRS) microscopy is also used in biology and has less background problems than CARS (**Figure 19.16**). Other, nonlinear spectroscopies are possible, such as *second-harmonic generation*, but this as yet has only a few niche applications in biology, since it requires the molecules to be noncentrosymmetric.

With the advent of high-intensity coherent pulsed X-ray radiation from X-ray free-electron lasers (XFELS), a wide range of nonlinear processes can now be probed in biology at X-ray wavelengths at very small length scales. Interesting developments are thus expected in this area.

19.4.7　Circular Dichroism and UV Spectroscopy

Circular dichroism (CD) measures the difference in the absorption of left-handed versus right-handed polarised light that can arise due to chiral structural asymmetries in samples. Chiral molecules (e.g. helices) preferentially absorb different handedness of light and have a nonzero CD signal. The secondary structure of proteins can be determined by CD spectroscopy in the far-UV spectral region (190–250 nm). At these wavelengths the chromophore is the peptide bond, and a CD signal arises when the peptide bond is located in a regular, folded environment (**Figure 19.18**). Differences can be observed between alpha helical, beta sheet and triple helix secondary structures in proteins using CD UV spectroscopy.

19.5　Optical Microscopy

Light microscopy is a standard laboratory tool, but its conceptual complexity should not be underestimated. At a minimum, a detailed understanding of Fourier optics is required for a quantitative treatment of optical

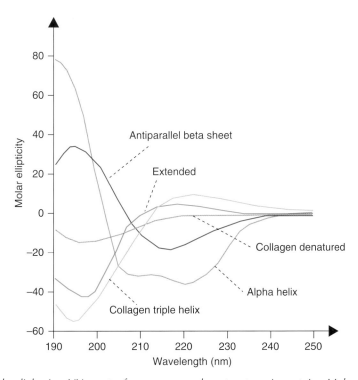

Figure 19.18　*Circular dichroism UV spectra for some secondary structures in proteins. Molar ellipticity is shown as a function of wavelength of the incident UV light.*

microscopy. Zernicke was awarded a Nobel Prize in physics for the creation of the phase-contrast microscope in the 1950s, which is now a routine method in biophysics, since it is frequently used to image cells. Standard light microscopes are quick and convenient tools, but they have a resolution limit of half the wavelength of the visible light used ($\lambda/2$, where λ is the wavelength) and sample opacity is often a problem (thin translucent samples are needed).

Standard optical microscopes use absorption contrast to create images, i.e. the complex part of the refractive index. Phase contrast is also possible and modern coherent digital methods allow both the real and complex parts of the refractive index to be quantified. For example, holographic and ptychographic microscopies allow quantitative phase measurements to be made. Alternatively extrinsic fluorescence, intrinsic fluorescence or intrinsic Raman contrast can be used to create images in modified spectroscopic microscopes.

The simplest quantitative treatment of the physical processes in optical microscopes is due to Abbe, and it allows the microscope's resolution limit to be calculated. The theory considers the imaging process in terms of two separate processes of diffraction (**Figure 19.19**) and the resolution limit of an optical microscope is limited by these diffraction processes. Consider an image formed of a diffraction grating (the object) by a thin lens.

There are two stages for a ray that creates an image on a detector, OF and FI (O is the object plane, F is the back focal plane and I is the image plane), and the two diffraction processes are applied sequentially. Small-angle approximations can be used for the geometry shown in **Figure 19.19**,

$$\theta_j \approx \tan\theta_j = \frac{h_j}{u} \tag{19.2}$$

$$\theta'_j \approx \tan\theta'_j = \frac{h_j}{v} \tag{19.3}$$

where u is the distance from the object to the lens, v is the distance from the lens to the image, h_j is the height of the diffraction order j on the lens and θ_j/θ'_j are the diffraction angles.

Also the diffraction grating equation ($d\sin\theta = j\lambda$, where d is the distance between lines on the grating, j is an integer (=1 for the first order) and θ is the angle of diffraction) can be used to give

$$\theta_j \approx \sin\theta_j = \frac{j\lambda}{d} \tag{19.4}$$

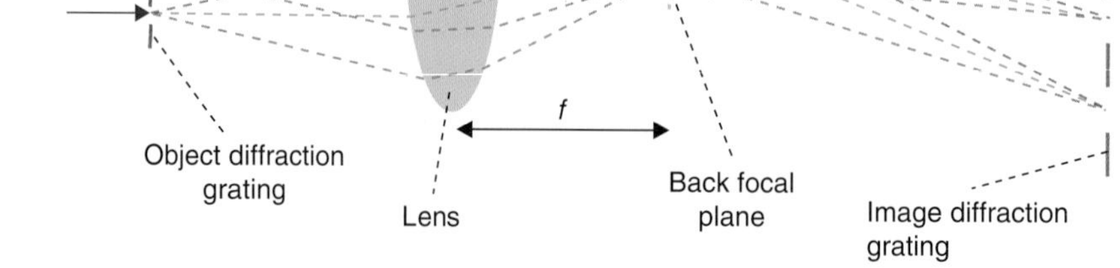

Figure 19.19 *Geometrical arrangement used to explain the formation of an image of a diffraction grating by a lens. The angular semiaperture of the lens is α.*

Therefore, the distances h_j can be cancelled,

$$\theta'_j \approx \frac{u\theta_j}{v} \tag{19.5}$$

The first-order diffraction fringes converge on the image at angles $\pm\theta'_1$ and form periodic fringes (S_1 and S_{-1}). The grating equation can be used again for both S_1 and S_{-1}

$$d' = \frac{\lambda}{\sin\theta'_1} \approx \frac{\lambda v}{\theta_1 u} = \frac{vd}{u} \tag{19.6}$$

Thus, a magnified image is formed in the image plane with magnification $M = v/u$. The finest detail observable in the image is determined by the highest order of diffraction that is actually transmitted by the lens. The spacing is not resolved if

$$\sin\theta_1 = \frac{\lambda}{d} < \sin\alpha \tag{19.7}$$

where α is the acceptance angle of the lens. Thus, the finest distance (d_{min}) that can be resolved is

$$d_{min} = \frac{\lambda}{\sin\alpha} \tag{19.8}$$

The analysis also works for larger angles if equation (19.6) is modified using

$$\frac{\sin\theta_j}{\sin\theta'_j} = M \tag{19.9}$$

Practically, for this to be achieved the lens needs to *aplanatic* and the method forms the basis of *high-power microscope objectives*. The period of fringes in the image created by an aplanatic lens is

$$d'_j = \frac{\lambda}{\sin\theta'_j} = \frac{M\lambda}{\sin\theta_j} = Md_j \tag{19.10}$$

where d_j is the spacing of the jth period in the diffraction grating. Thus, the aplanatic microscope also forms a magnified image with magnification M.

A more quantitative description of image formation in an optical microscope makes use of a Fourier model, which provides a more accurate representation of the propagation of electromagnetic waves through the system (**Figure 19.20**). The amplitude of the wave that reaches point P in the back focal plane of the lens is the Fourier transform of the wave that leaves the object ($f(x)$) with an appropriate phase delay for the path OAP

$$\psi(u) = e^{ik_0\overline{OAP}}F(u) = e^{ik_0\overline{OAP}}\int_{-\infty}^{\infty} f(x)e^{-iux}dx \tag{19.11}$$

where u corresponds to the point P, $k_0 = \frac{2\pi}{\lambda}$ and $u = k_0\sin\theta$. The amplitude $b(y)$ at Q in the image plane also needs to be calculated

$$PQ = \left(PI^2 + y^2 - 2yPI\sin\theta'\right)^{1/2} \approx PI - y\sin\theta' \tag{19.12}$$

If $y \ll PI$ the Abbe sine condition is obeyed so

$$b(y) = \int_{-\infty}^{\infty} \psi(u)e^{ik_0PQ}du \tag{19.13}$$

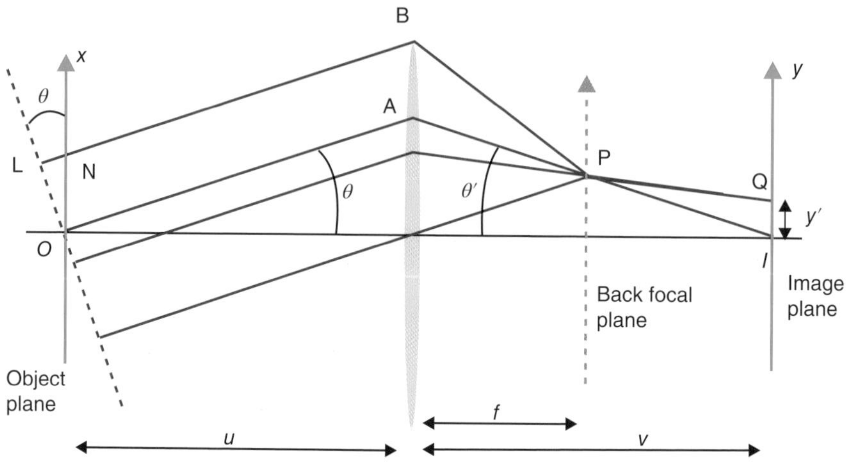

Figure 19.20 *Schematic diagram for the geometry used to calculate the resolution of a single lens microscope using Fourier theory. An image of an object (O) in the object plane (x) is formed at I in the image plane y.*

where $\sin\theta = M\sin\theta'$ and $PQ = PI - yu/Mk_0$. Combination of equation (19.11) and (19.13) gives

$$b(y) = \int_{-\infty}^{\infty}\left\{e^{ik_0\left(\overline{OAP}+PI\right)}\int_{-\infty}^{\infty}f(x)e^{-iux}dx\right\}e^{-iuy/M}du \qquad (19.14)$$

And this can be simplified,

$$b(y) = \int_{-\infty}^{\infty}e^{ik_0 PI}\psi(u)e^{-ix'u/M}du \qquad (19.15)$$

By Fermat's principle the optical path from O to I is independent of the point P,

$$b(y) = e^{ik_0\overline{OI}}\int_{-\infty}^{\infty}\left[\int_{-\infty}^{\infty}f(x)e^{-iux}dx\right]e^{-iuy/M}du \qquad (19.16)$$

Therefore,

$$b(y) = e^{ik_0\overline{OI}}f(-y/M) \qquad (19.17)$$

The image is an inverted copy of the object multiplied by the factor M. Image formation can thus be considered as a process of *double Fourier transformation*. The imaging lens performs the second inverse Fourier transform and the first Fourier transform describes the process of diffraction of the light by the object. It is the inability to satisfactorily perform this second inverse Fourier transform using physical components (focusing optics), that has led to the current resolution limits on X-ray and neutron imaging.

The image of a single point of a distant object in an optical system is called the *point spread function* (**Figure 19.21**). The image of an extended object is the convolution of the object intensity and the point spread function, i.e. it describes the blurring of the image observed in a real microscope with a finite resolution. The Rayleigh resolution criterion is a quantitative measure of resolution and it is related to the ability to separate the

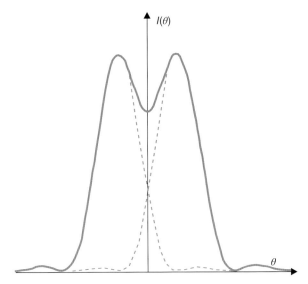

Figure 19.21   *Rayleigh criterion for the point spread function of an optical system. The intensity is plotted as a function of the angle (θ) at which the light is detected.*

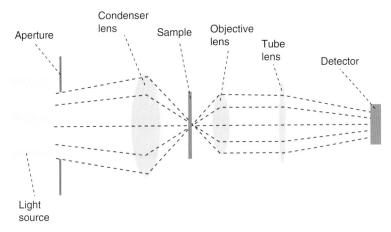

Figure 19.22   *Standard optical arrangement of a visible wavelength optical microscope. A condenser lens focuses a light source onto the sample. An image of the sample is then constructed by a combination of an objective lens and a tube (optional) lens on the detector.*

image of two points convolved with their point spread functions. Rayleigh defined the resolution limit as when the maximum of one point spread function overlaps with the first minimum of the neighbouring point spread function.

Figure 19.22 shows a schematic diagram of a transmission optical microscope. Light leaves the source, it is focused onto the sample with the condenser lens, it scatters off the sample and is imaged by the objective lens onto the detector. From Abbe theory (equation (19.7)),

$$\sin\theta_1 = \frac{\lambda}{d} \tag{19.22}$$

where the object period is d. To image the object the angular aperture (α) of the lens must be greater than θ_1,

$$d_{min} = \frac{\lambda}{\sin \alpha} \tag{19.23}$$

The object is immersed in a medium of refractive index (n), and the wavelength of light in the object is λ/n. The *numerical aperture*, NA (the angular aperture of a lens rescaled by the medium's refractive index) is defined as

$$NA = n \sin \alpha \tag{19.24}$$

Therefore, equation (19.23) can be rewritten as

$$d_{min} = \frac{\lambda}{n \sin \alpha} = \frac{\lambda}{NA} \tag{19.25}$$

This is the coherent resolution limit of a simple optical microscope, without a condenser lens. The illumination has been assumed to be parallel to the axis and both the first-order diffraction and the zeroth-order diffraction are collected. It is possible to improve the resolution by illumination with light that travels at an angle α, using a condenser lens (with the same NA as the objective lens) and the resultant resolution is

$$d_{min} = \frac{\lambda}{2n \sin \alpha} = \frac{\lambda}{2NA} \tag{19.26}$$

This is called the *Abbe resolution limit*; the best resolution that can be achieved with conventional optical microscopy and is defined by the acceptance angle of the two lenses (the condenser and the objective lenses), the refractive index and the wavelength of the light. Thus, a condenser lens is used in most microscope designs, since it can improve the resolution by a factor of two as well as increase the number of photons incident on the sample (an increase in image brightness).

Condenser lens alignment has three standard configurations (**Figure 19.23**). In the first, an image of the source is formed on the specimen, which leads to problems if the source contains imperfections. In the second, an extended source on the specimen is formed (*Köhler illumination*). This makes the illumination incoherent (often this improves the resolution) and more uniform. However, Köhler illumination spreads the light over a wider area and leads to weaker illumination (less-bright images). A third more recent configuration is to place the illumination lens perpendicular to the imaging lens (the objective lens), so the microscope only illuminates the plane that is in focus. A sheet of light is created by a cylindrical condenser lens from the light source to match the plane that is in focus in the objective lens. Sheet-plane illumination reduces the amount of radiation damage (photobleaching of fluorophores, etc.) on the sample and is particularly important for live-cell fluorescence imaging. Sheet illumination forms the basis of the scanning plane imaging microscopy (SPIM) technique and can be adapted to a wide range of other forms of microscopy where radiation damage is important, e.g. Raman SPIM of live cells is possible.

A number of techniques can be used to improve sample contrast with optical microscopy, i.e. improve the quality of images. *Dark-field imaging* eliminates the zeroth-order beam diffracted by the sample through exclusion of the unscattered beam from the image. Dark objects on a bright background become bright objects on a dark background in dark-field images.

Phase-contrast microscopes (**Figure 19.24**) allow *weak phase objects* to be imaged that have little absorption contrast. Cells are weak absorption contrast objects and thus phase-contrast microscopy is widely used in cellular biology. Practically phase-contrast imaging is achieved by illumination of the sample, division of the transmitted light into two components, addition of a phase shift to one of the light paths and then recombination of the light to form a phase-contrast image. The illuminating beam is limited by an aperture below the

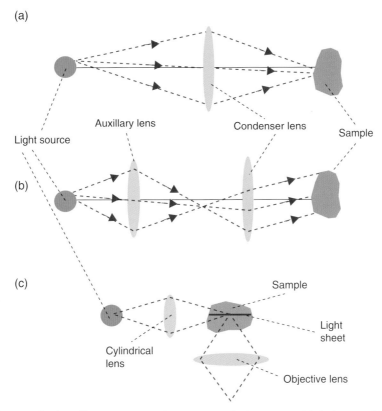

Figure 19.23 *Three methods to illuminate a sample in an optical microscope. (a) An image of the source is formed on the sample. (b) Köhler illumination forms an incoherent uniform disk over the sample. (c) Scanning plane imaging microscopy (SPIM) illuminates the sample with a sheet of light and the objective lens only collects light from this plane.*

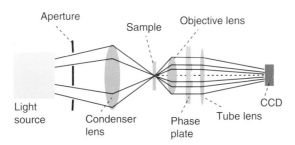

Figure 19.24 *A Zernicke-type phase-contrast microscope uses a phase plate in the optical path. Phase contrast allows very thin specimens to be imaged due to variations in the real part of the refractive index, e.g. biological cells.*

condenser lens and a real image of this opening is formed in the back focal plane of the objective lens. A phase plate is inserted at the back focal plane of the objective lens. The transparent ring (the phase plate) has an optical thickness of a quarter of a wavelength ($\lambda/4$) and dimensions that match the aperture image size. All the undeviated light passes through this plate. A final image is formed by interference between the undeviated light that passes through the phase plate and deviated light that passes by the side of it. This forms a *phase-contrast image* of the object. Phase-contrast microscopes are in principle possible with all photon energies (not just visible) and have proven very useful with both visible and X-ray microscopes.

19.5.1 Fluorescence Microscopy

Fluorescence is a quantum mechanical process, and can be described by a Jabolinski diagram (**Figure 19.25**). A photon strikes a fluorophore and excites it from energy level E_1 to E_3. The excitation process is followed by a fast internal process (usually dissipated as heat) to energy level E_2 (10^{-12} s). A redshifted photon (a photon of lower energy) is then re-emitted as the fluorophore decays to its ground state, E_1 (10^{-9} s). With the simplest fluorophores, E_1, E_2 and E_3 are all in the singlet (spin 1) state, so there is no angular momentum selection rule that forbids the transitions. In fact, this is an oversimplification and a rich range of quantum mechanical phenomena can occur, e.g. phosphorescence involves transitions from the triplet to the singlet states and the angular momentum selection rule leads to long lifetimes. Furthermore, in **Figure 19.25** a Stokes (red) shifted fluorescence process has been assumed and its measurement is often implicit in the filter set used for fluorescence microscopy measurements. Blueshifted fluorescence photons (anti-Stokes fluorescence) also can occur that cause sample cooling (the emitted photon is of a higher energy than the incident photon). Such an effect can be simply measured from a biomolecule with laser excitation and a diffraction grating spectrometer, although the anti-Stokes signal tends to be weaker than the Stokes signal.

The power of fluorescence microscopy stems from the huge range of fluorescent biomolecules available, which allow highly specific labelling. Green fluorescence proteins (GFP) have thus had a big impact in biology. Originally extracted from jelly fish (Nobel Prize chemistry 2008), GFPs can now be genetically expressed in a wide range of cell types with a wide variety of colours and attached to a wide range of proteins, i.e. fluorescent labels can be attached to nearly any conceivable protein in a live cell (**Figure 19.26**). Genetic expression means that the attachment of the fluorescence probe is completely uninvasive, except for minor conformational perturbations due to the interaction of the GFP domains with the rest of the expressed protein. Quantum nanodots are also a promising source of fluorescent labels that have high photon yields and high stabilities. Quantum dots are superior in both these respects compared to GFP, but they do suffer from toxicity problems in live cells. A vast range of fluorescently tagged biomolecules thus now exist and biomolecules can be examined on a single-molecule basis inside living cells with fluorescence microscopy.

Fluorescence microscopes are similar to conventional optical microscopes with a few additional components (**Figure 19.27**). The excitation filter acts as a monochromator, so a single wavelength illuminates the sample, and this component is not required if laser illumination is used. An emission filter is placed before

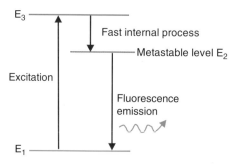

Figure 19.25 Jabolinski diagram for the quantum mechanical processes involved in the vibrational states of a fluorescent molecule during absorption and re-emission of photons. The molecule is excited to a higher energy level ($E_1 \rightarrow E_3$, 10^{-15} s), there is a fast internal relaxation ($E_3 \rightarrow E_2$, 10^{-12} s) followed by fluorescent emission ($E_2 \rightarrow E_1$, 10^{-9} s).

(a)

(b)

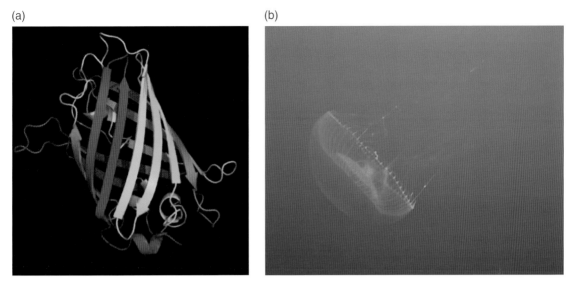

Figure 19.26 *(a) Atomic structure of green fluorescent protein (GFP), (b) jelly fish provided the initial source of GFP. The DNA sequences of a range of naturally occurring fluorescent proteins are now well known and GFPs can be expressed artificially in genetically modified cells.*

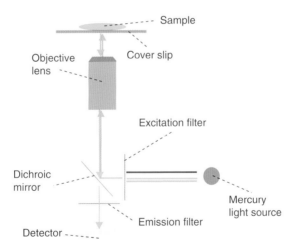

Figure 19.27 *Optical configuration of a fluorescence microscope. A n excitation filter allows a very narrow spectral range from a mercury light source to be transmitted. The filtered light illuminates a sample (say blue light) and the fluorescent light (say green light) is collected by the detector after passing through an emission filter. The emission filter helps remove the background of nonfluorescent photons.*

the detector to block the excitation wavelength, but allow the red/blue shifted photons due to fluorescence to pass. A high-sensitivity cooled CCD camera is also required to count single photons in fluorescence microscopes, since fluorescence signals are much weaker than absorption contrast signals measured in conventional microscopes. The microscope in **Figure 19.27** is constructed in an epi-illumination mode (the objective lens also acts as the condenser lens), but transillumination fluorescence microscopy is also possible.

Important extensions to fluorescence microscopy include confocal microscopy, single molecule measurements (often built around total internal reflection fluorescence (TIRF) microscopes), super-resolution, Forster resonance energy transfer (FRET), and fluorescence correlation spectroscopy (FCS).

19.5.2   Super-Resolution Microscopy

The fundamental resolution limit on a hypothetical photonic microscope will be considered by reference to the 'Heisenberg gedanken γ-ray microscope'. γ-rays are chosen to illuminate the sample, since they in principle offer the highest diffraction-limited resolution due to their small wavelength (**Figure 19.28**). The position of a point particle in the field of the microscope needs to be determined as accurately as possible using a single γ photon. A microscope is chosen with a high-NA objective (good γ-ray lenses do not yet exist, so this is a pedagogic exercise). Photons are scattered through the semiangle α by the sample similar to the Abbe analysis in **Section 19.5.1**. The photon wave number is k_0, and the x component (perpendicular to the direction of travel) of the wave number after scattering lies in the range

$$-k_0 \sin \alpha \le k_x \le k_0 \sin \alpha \tag{19.27}$$

The uncertainty in the x component of the wave number is therefore

$$\delta k_x = 2k_0 \sin \alpha \tag{19.28}$$

From equation (19.26) the Abbe theory gives the resolution limit of the single-photon position measurement of a point particle as

$$\delta x = \frac{\lambda}{2\text{NA}} = \frac{\lambda}{2 \sin \alpha} \tag{19.29}$$

where the refractive index is taken to be 1. Therefore, combination of equations (19.28) and (19.29) gives

$$\delta x \delta k_x = 2\pi \tag{19.30}$$

However, Planck's law gives the momentum (p) of photons in terms of the wave number,

$$p = \frac{hk}{2\pi} \tag{19.31}$$

where h is Planck's constant. Therefore, equation (19.30) is nothing more than Heisenberg's uncertainty principle, in the limit of single-photon resolution

$$\delta x \delta p_x = h \tag{19.32}$$

However, the resolution of a microscope image can be improved through the use of many photons. Every absorbed photon enters the same lens, and the uncertainty in the wave number δk_x is unchanged.

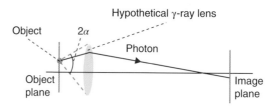

Figure 19.28   *Simple single-lens Heisenberg gedanken γ-ray microscope used to demonstrate the one-photon resolution limit for position measurements of an object that could be used to construct an image.*

A statistical assembly of N photons has total uncertainty in its wave number of ($N^{1/2} \times$ standard deviation, which is given by equation (19.28))

$$2N^{1/2}k_0 \sin\alpha \qquad (19.33)$$

Therefore, the error in a position measurements for an image created with N photons is

$$\delta x \approx \frac{\lambda}{2N^{1/2}\text{NA}} \qquad (19.34)$$

The resolution for multiple photons is improved by a factor of $N^{1/2}$ when compared with that of single photons. This is how some new super-resolution optical microscopy techniques function, i.e. the diffraction limit is broken by the reduction of δx using cumulative measurements with many photons (e.g. with STORM). There are six commonly used strategies to form super-resolution images in optical microscopes, confocal microscopes, NSOM, STED, STORM, SIM and TIRF, but this list is by no mean exhaustive and super-resolution microscopy is a very active area of research.

19.5.2.1   *Confocal Microscope*

Confocal microscopy achieves better resolution than the diffraction limit of standard optical microscopes in three dimensions. The microscope scans one or two apertures (dependent on whether the microscope is used in epi-illumination or transmission, respectively) across the sample. The resultant point spread function of the microscope is sharper than that without the pinholes. Scanning a pinhole in an optical microscope improves the resolution by a factor of $\sqrt{2}$ and reduces the background scatter, i.e. improves the image contrast and enables thicker specimens to be imaged. Many confocal microscopes are constructed using epi-illumination, because it is easier to align a microscope with a single pinhole, and this is thus a standard arrangement for bespoke single-molecule fluorescence microscopes.

Figure 19.29 shows a sample that is illuminated by the diffraction-limited image of a point source in a confocal microscope. Another pinhole is located before the detector. This type of microscope is very good for the creation of sectioned images, since it rejects out-of-focus light. The main disadvantages of confocal microscopy are that it is slow, the resolution is only slightly better than a conventional microscope, and the radiation dose received by the sample is relatively high.

19.5.2.2   *Near-Field Scanning Optical Microscope (NSOM)*

A near-field scanning optical microscope allows surfaces to be imaged using an optical probe that is held very close to the sample. Evanescent waves do not have the same Abbe diffraction limits as far-field waves. The tiny optical probe is placed in close proximity to the sample surface and the sample is scanned in x and y in order to create an image (**Figure 19.30**). The probe can either emit light or collect light from a self-luminous object. Reasonably high resolutions can be achieved, but the technique is slow and the short working distance is often inconvenient (it is hard to create three-dimensional images and difficult to avoid perturbation of the specimen's structure).

19.5.2.3   *Stimulated Emission Depletion Microscopy (STED)*

In a stimulated emission depletion (STED) microscope a fluorescence mechanism is used to make a scanning spot with a point spread function smaller than the diffraction limit. STED microscopes produce a reduced spot size using a quantum mechanical depletion effect and an image is constructed by scanning the spot in a manner similar to a confocal microscope.

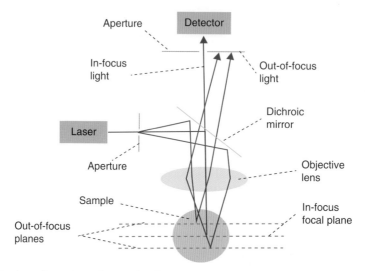

Figure 19.29 *Optical configuration of a confocal microscope with visible-wavelength photons. The confocal microscope is an optical microscope with the addition of two pinholes. One pinhole is placed after the laser and the other before the detector. The holes are used to scan a spot of light through the sample that allows an image to be constructed. The addition of the two pinholes improves the resolution and sectioning ability of the microscope beyond that of a standard optical microscope.*

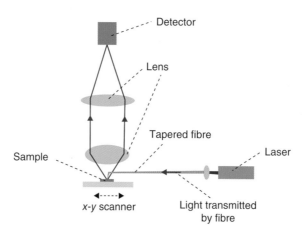

Figure 19.30 *Near-field scanning optical microscopes use light confined in an optical fibre. An image is created as the fibre is scanned in x and y across the sample at very small distances from the sample.*

19.5.2.4 Stochastic Optical Reconstruction Microscopy (STORM)

Stochastic optical reconstruction microscopy (STORM) achieves high resolution by using optically switched fluorescent tags and the multiple photon $N^{1/2}$ effect described by equation (19.34). Switchable fluorescent tags are attached to the sample. The sample is imaged by statistically switching on and off the tags (**Figure 19.31**). The position of the tags can be super-resolved as long as their point spread functions do not overlap, so for a densely labelled sample, a large number of fluorophores need to be shelved in the off state (inactivated) to avoid the overlap of the point spread functions. PALM (photoactivatable light microscopy) is conceptually

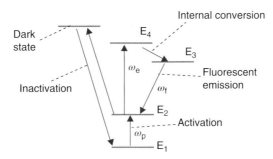

Figure 19.31   *The energy-level scheme for a fluorophore used in a stochastic optical reconstruction microscope. A stochastic subset of the fluorophores are shelved in the dark inactivated state, so that the point spread functions of the fluorophores do not overlap in images. The scheme shown is for CY3–CY5 conjugate synthetic fluorophores. Other successful energy-level schemes for STORM exist for other fluorophores, which in general need both activated and inactivated states.*

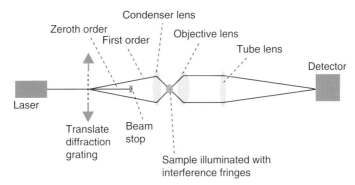

Figure 19.32   *The structured illumination microscope can achieve improved resolution and sectioning compared with a standard optical microscope using the translation of a diffraction grating through the image path. Software reconstructs images of the specimen based on measurements of three or more positions of the diffraction grating.*

a very similar technique to STORM, the main difference is that it uses photoswitchable proteins (GFPs) rather than synthetic photoswitchable fluorophores to image specimens. Synthetic fluorophores tend to be brighter and photobleach slower than GFPs, but they are more biochemically invasive than protein equivalents and cannot be genetically manipulated.

STORM/PALM imaging can be extended to three dimensions using a cylindrical lens after the objective lens, which encodes the height (z) information in the curvature of a noncylindrical point spread function.

19.5.2.5   *Structured Illumination Microscopy*

In a structured illumination microscope the sample is illuminated with patterned light formed from a diffraction grating or a spatial light modulator (**Figure 19.32**). This forms *Moiré fringes* with the light scattered from the sample (interference between the illumination and scattered light) and allows higher spatial frequency (k) information from the scattered radiation to be directed into the detector. Thus, higher-resolution information can be created in images, with a $\sqrt{2}$ improvement in resolution. Improved three-dimensional sectioning also occurs when compared with standard optical microscopy. Extensions of SIM that use pulsed lasers and nonlinear effects have further increased the resolution beyond the $\sqrt{2}$ limit.

19.5.2.6 Total Internal Reflection Microscopy (TIRF)

Total internal reflection fluorescence (TIRF) microscopy uses evanescent waves to improve the resolution of optical images in an epi-illumination microscope. Fluorescent samples must be very close to the lens and are excited with an evanescent wave from specifically designed high-NA TIRF objective lenses. TIRF lenses are made commercially and can be used to adapt standard fluorescence microscopes. STORM/PALM microscopes often use TIRF lenses for an additional improvement in resolution.

19.5.3 Nonlinear Microscopy

Nonlinear microscopy was briefly considered in the section on nonlinear spectroscopy and involves multiphoton processes. Multiphoton processes are more localised in their interaction with a sample and provide better penetration than single-photon methods (infrared light is more penetrating than visible wavelengths). The techniques have therefore found applications in high-resolution *in vivo* imaging, e.g. imaging of active nerve cells in mammalian brain tissue.

19.5.4 Polarisation Microscopy

Polarisation microscopes allow the birefringence of biological materials to be measured, e.g. in chromosomes as cells divide or starch granules as they gelatinise. Quantitative polarisation microscopy thus allows molecular orientation to be calculated, since it is directly related to birefringence. A challenge with such techniques is the need to extend the microscopy apparatus beyond the requirements of the scalar theory of light, i.e. a single value of the electric field at a point in space is insufficient to fully account for polarisation phenomena and a vector theory is needed. Modern polarisation microscopes perturb the polarisation of the imaging light using liquid-crystalline devices, and allow quantitative polarisation contrast images to be constructed. Manipulation of the polarisation of light also finds applications in ellipsometry, dual polarisation interferometers and quantitative phase optical coherence tomography.

19.5.5 Optical Coherence Tomography

A major drawback of biological imaging with visible wavelengths is the occurrence of multiple scattering in a tissue specimen. Single cells can be adequately imaged, but when they are arranged in thicker, *in vivo* structures, the image resolution is degraded to a useless level after transmission through a few cell thicknesses in the sample. Partial solutions include confocal microscopy and two-photon nonlinear imaging. However, a superior solution is to use a coherence gate on the imaging light source to reject photons that are multiply scattered. In its simplest incarnation *optical coherence tomography* backscatters light off a specimen and allows it to interfere with statistically identical light scattered off a mirror, i.e. created from the same partially coherent light source (**Figure 19.33**). Only light within a coherence length of the mirror path length will result in constructive interference on the detector. Furthermore, the apparatus can be built using optical fibres for easy *in vivo* delivery. Three-dimensional images are constructed by scanning the beam across the sample in x and y, and z sectioning is achieved by moving the mirror in the Michelson interferometer. **Figure 19.33** shows a 'time-domain' OCT apparatus with a single point detector. Many modern apparatus use 'frequency-domain' OCT with spectroscopic detection, that simultaneously measures images from across the full sample depth, and provides improved signal-to-noise ratios.

The largest market for optical coherence tomography is currently imaging of the retina at the back of the eye, since the image quality is relatively robust to refraction by the cornea. In the future, OCT is expected to have a wide range of applications in soft matter and biophysics, since many systems are opaque.

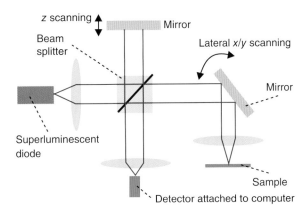

Figure 19.33 *Time-domain optical coherence tomography can be performed using a Michelson interferometer type arrangement. Only light within the longitudinal coherence length of the light source (a super-luminescent diode) forms fringes on the detector and thus a significant signal. A coherence gate is thus formed that suppresses multiple scattering and provides optical sectioning of the sample.*

Lots of components initially created for the telecommunications industry are now used in OCT designs. Infrared wavelengths are favoured for their penetration and noninvasive nature in biological OCT imaging.

It is also possible to make velocity measurements of moving slices of opaque complex fluids using OCT measurements. This has been combined with a fluids rheometer (**Section 19.16**) to form the OCT rheology technique for viscoelastic fluids using the Doppler shift on photons scattered by the motion of samples.

19.5.6   Holographic Microscopy

The field of holography was started by Dennis Gabor in the 1940s for use with electron microscopes. However, it was only with the invention of the laser (a cheap source of coherent light) that holography became a mainstream technology. In its simplest form holography uses the interference between coherent light scattered from a sample and a coherent reference beam, to create an image of the sample. Inclusion of the reference beam allows the phase problem to be solved (if diffraction patterns are measured), three-dimensional information can be easily reconstructed, and a quantitative map of refractive indices in the sample can be created.

A relatively simple holographic microscope is shown in **Figure 19.34**. Light from the laser is split in two. A portion of the light illuminates the sample. The light scattered from the sample is then recombined with the reference wave and an image formed on the CCD detector. Such a setup has proven useful for low resolution (~micrometre) tracking of micro-organisms in three dimensions. However, the arrangement in **Figure 19.34** suffers from stability issues with beam alignment. Modern developments, such as common-path interferometry, have been developed to alleviate some of these problems. Holographic X-ray imaging is also possible and is described in **Section 19.10**.

19.5.7   Other Microscopy Techniques

The photonics industry continues to invent a wide range of new optical technologies, which in turn drive innovations in optical microscopy. The availability of faster computers means more computationally sophisticated image construction tools are possible, such as *synthetic aperture* microscopes. In this technique overlapping images are synthesised together to form composite images with a wider field of view and superior resolution.

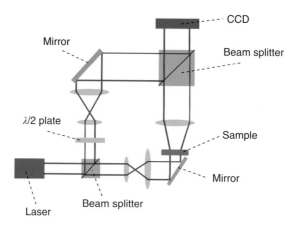

Figure 19.34   *Digital holographic microscope for transmission imaging of biological samples with visible light. Both three-dimensional and quantitative phase information are encoded on the CCD camera. Such apparatus can suffer from optical stability problems due to different mechanical perturbations to the separate paths.*

Spatial light modulators allow the phase, shape and intensity of light beams to be more easily manipulated. Bessel-beam microscopy uses a spatial light modulator to create a beam that is less strongly scattered by a specimen, which results in improvements in image resolution and penetration depth. Spatial light modulators also find applications with optical tweezers to construct multiple traps.

19.6   Single-Molecule Detection

The ability to explore single molecules is a technically challenging, but extremely useful area of research. The standard methods for single-molecule detection in solution depend on fluorescence. These include fluorescence microscopy, near-field scanning optical microscopy (NSOM), confocal microscopy, and total internal reflection fluorescence (TIRF) microscopy. The tricky part of the instrumentation is to increase the signal-to-noise ratio of the molecule of interest, while the background effects from the surrounding molecules are suppressed, e.g. the reduction of the autofluorescent background.

Single molecule detection is a quickly evolving field. Macroscopic quantities (pressure, volume, heat capacity, etc.) are typically averages over billions upon billions of molecules (measurements averaged over Avogadro's number sized ensembles). There are some subtleties when the statistics of single molecules are considered and Bell's equation (4.7) is often required for meaningful calculations. There is frequently much more information in single-molecule measurements (although averaging for a good signal-to-noise ratio can take considerable time) and such experiments indicate that single molecules are crucial to the life cycles of organisms, i.e. cell fates often depend on stochastic processes that require molecules with very low copy numbers.

19.7   Single-Molecule Mechanics and Force Measurements

Once a single molecule has been detected the next challenge is to learn how to manipulate it, e.g. measure its elasticity. There are a wide range of techniques for the measurement of mesoscopic forces and they were briefly examined in **Chapter 4**. Some of the more modern developments in the field of force measurement include *atomic force microscopy* (AFM), *glass microneedles*, *surface-force apparatus* (SFA) and *magnetic/optical*

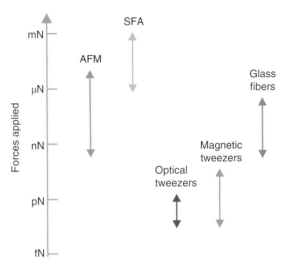

Figure 19.35   *The range of forces that can typically be applied to biomolecular systems with five standard apparatus.*

tweezers. **Figure 19.35** compares the range of forces that can typically be measured with each of these techniques. In general terms, AFM and SFA offer the largest forces and magnetic/optical tweezers offer the greatest sensitivity. All of these methods, except SFA, are routinely used to study single molecules.

Optical and *magnetic tweezers* are both similar in application, a feedback mechanism is used to clamp the position of a colloidal particle in three dimensions under an optical microscope and the force applied to the particle is subsequently used to manipulate biological molecules. However, the physical processes that control the mode of action of the two types of tweezers are very different. Optical tweezers focus laser light to trap a dielectric particle using the pressure of photons from an incident laser beam (**Figure 19.36**). Magnetic tweezers use the magnetic force on superparamagnetic or ferromagnetic beads, exerted by gradients in an applied magnetic field (**Figure 19.37**).

For *optical tweezers*, the electric dipole (p) induced in a particle trapped in a light beam is given by

$$\underline{p} = \underline{\alpha}.\underline{E} \tag{19.35}$$

where $\underline{\alpha}$ is the polarisability of the trapped particle and $\underline{E}$ is the electric field of the incident laser. The induction of the optical dipole moment by the laser beam provides a force ($\underline{F}_{\text{light}}$) on the trapped particle, proportional to the Laplacian of the electric field or equivalently the gradient of the intensity of the incident light (∇I). The gradient force exerted on a trapped particle is (calculated from the Lorentz force combined with Maxwell's equations)

$$\underline{F}_{\text{light}} = \alpha \nabla^2 \underline{E} = \alpha \underline{\nabla} I \tag{19.36}$$

Intensity gradients in a converging light beam draw small objects towards the focus, whereas the radiation pressure of the beam tends to blow them down the optical axis. Only under conditions in which the gradient force dominates can a particle be trapped near the focal point. A single well-focused laser beam can trap dielectric particles (~0.5 μm) in three dimensions using the pN forces induced from the optical dipolar force. For a specific laser, microscope and optical set up, the magnitude of the tweezer force (F) is related to the incident laser power by the equation

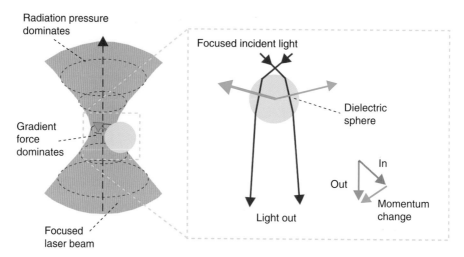

Figure 19.36 *A focused laser beam can be used to trap a dielectic sphere. The momentum transfer due to the change in direction of the refracted beam induces a force on the sphere. The radiation pressure tends to blow the trapped sphere down the optical axis, whereas the gradient force pulls the particle into the focus of the laser beam.*

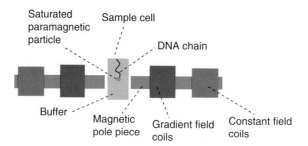

Figure 19.37 *Two pole piece magnetic tweezers for single-molecule extension experiments. The addition of extra pole pieces can provide particle manipulation in three dimensions and force control can be created using feedback from position measurements made with an optical microscope.*

$$F = \frac{Q n_{\mathrm{m}} P}{c} \tag{19.37}$$

where Q is the efficiency of the trap, n_{m} is the index of refraction of the particle, c is the speed of light, and P is the incident laser power. Dual optical traps can be used to extend single molecules that are attached at either end to colloidal probes. The correct choice of laser for a trapping experiment is important to reduce damage to fragile biological molecules. Often, infrared lasers are used with delicate biological materials to minimise this damage. **Figure 19.38** shows some optical components arranged in a standard optical tweezer set up.

Photons exist with well-defined spin states (± 1) directed along their line of travel. This spin has angular momentum associated with it. Photons can thus be used to provide a torque on trapped particles when their angular momentum is absorbed. It is also possible to create laser beams with nonspin angular momentum using Laguerre–Gauss beams, which again can be used to rotate specimens.

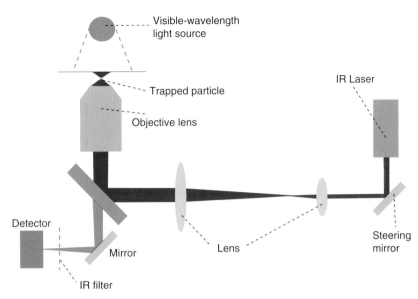

Figure 19.38   *Schematic diagram of an optical trap in an inverted optical microscope. A light beam from an infrared laser is introduced into the microscope and its momentum can be used to manipulate trapped particles.*

Multiple optical traps can be created simultaneously using holography. Typically, a laser beam illuminates a spatial light modulator (similar to a LCD screen, with individually addressable pixels that perform independent phase/amplitude shifts on parts of a reflected beam) and the multiple beams created can trap multiple biological particles or distribute the gradient force more evenly across a single particle.

Standard geometries for optical tweezer experiments in molecular biophysics are shown in **Figure 19.39**. The interaction of trapped beads attached to motor proteins with actin/microtubule fibres can be followed (a), biological molecules can be extended between two trapped beads (b), trapped beads can probe the viscoelasticity of fluids (microrheology, c), crosscorrelation of dual trapped beads can often provide a superior measure of the bulk viscoelasticity of fluids (d), single trapped beads can probe membrane forces or viscoelasticity (e), and the extension of molecules can be explored that have been attached to a surface and a trapped bead (f).

In contrast to optical tweezers, with *magnetic tweezers* the potential energy (U) of a probe's magnetic dipole (m) placed in a magnetic field (B) is given by the scalar product

$$U = -\underline{m}.\underline{B} \tag{19.38}$$

Thus, a free permanent magnetic dipole experiences a torque as it minimises its energy through alignment of the dipole with the applied magnetic field. The magnetic forces experienced by a probe particle depend sensitively on its type of magnetism. The colloidal probes used in magnetic-tweezer experiments are typically either ferromagnetic or superparamagnetic. The corresponding magnetic force (F) is the gradient of the potential (∇U),

$$F = -\nabla\left(\underline{m}.\underline{B}\right) \tag{19.39}$$

Superparamagnetic particles contain magnetic nanoparticles (e.g. Fe_2O_3 of diameter ~10 nm) that are sufficiently small that the thermal energy (kT) can disrupt the alignment of the spins. Thus, the beads have no magnetic moment in a zero magnetic field and medium magnetic dipole moment at higher magnetic fields, but the beads experience much smaller hysteretic effects that with ferromagnetism, so the tweezers are much

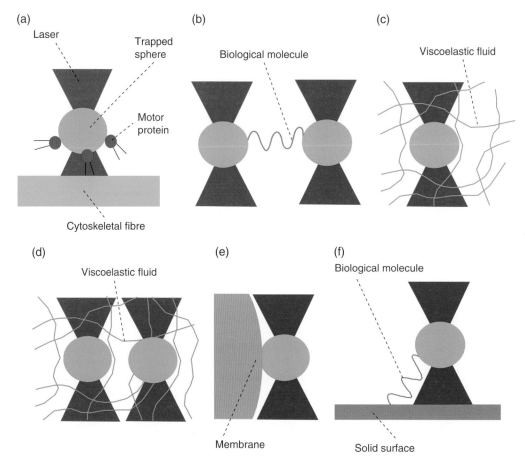

Figure 19.39 *Standard geometries for optical tweezer experiments in biology include (a) the motion of motor proteins along cytoskeletal filaments, (b) the extension of single molecules by dual traps, (c) microrheology of viscoelastic fluids using a single trap, (d) microrheology of viscoelastic fluids using dual traps, (e) membrane dynamics/viscoelasticity using a single trap and (f) the extension of a molecule tethered to a surface.*

easier to calibrate. For superparamagnetic spheres the magnetisation is often approximately equal to the saturated value (M_{max}), $m \cong M_{max}$,

$$F \approx M_{max} V \frac{dB}{dx} \tag{19.40}$$

where V is the particle volume. With superparamagnetic particles the application of a magnetic field gradient provides forces of the order of 10 pN. The torque on a large ferromagnetic particle (4 μm) can be quite considerable (~1000 pN μm) and magnetic-tweezer cytometry that uses ferromagnetism (as opposed to superparamagnetism) has found applications in the determination of the elasticity of cells adhered to magnetic beads. Hysteresis effects in the magnetism curves of the probe particles and pole pieces pose a number of technical challenges for accurate quantitative analysis of magnetic forces with ferromagnetic probes, but calibration is possible. Recent reports also point to a small hysteresis term in the magnetic response of superparamagnetic spheres, although in practice this only leads to a small correction to the applied forces.

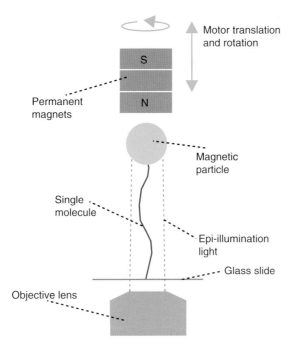

Figure 19.40 *Magnetic tweezers can be used to extend and twist biological molecules. A simple experimental arrangement consists of a stack of permanent magnets attached to an electric motor for translation or rotation. A magnetic particle is attached to the molecules of interest. The position of the magnetic particle is measured with an epi-illumination optical microscope (only the objective lens is shown).*

Magnetic tweezers can be cheaply built with either electromagnets or permanent magnets attached to electronic translation motors (**Figure 19.40**). Such magnetic tweezers can be used to extend/compress single biomolecules attached to superparamagnetic spheres. They can also be used to twist molecules if the magnetic are rotated and thus the torsional properties of the molecules can be measured.

Magnetic tweezers have the advantage over optical tweezers that they do not cause beam damage in living cells, which can be a problem with optical tweezers. However, rapidly oscillating magnetic fields can place lots of heat into magnetic nanoparticles and it is a candidate for *in vivo* treatment to kill cancer cells. Most magnetic tweezer designs do not use such fast switching electromagnets and the heat changes in samples are negligible (although care must be taken to cool the electromagnets themselves to reduce heating effects due to thermal conduction).

Standard geometries for magnetic tweezers experiments with biological physics are shown in **Figure 19.41**. Molecules can be extended or compressed (a), molecules can be twisted to observe torsional properties (b), beads attached to cell membranes can probe membrane viscoelasticity (c), beads inside living cells can probe their viscoelasticity (d), and beads embedded in solutions of biological molecules can also be used to probe their viscoelasticity (e).

The hydrodynamic drag force (F_{drag}) experienced by a trapped particle moved (velocity, v) through its surrounding solvent with optical or magnetic tweezers is given by Stoke's law,

$$F_{\text{drag}} = 6\pi\eta av \tag{19.41}$$

This equation provides a method for the calibration of both optical and magnetic traps. A trapped colloidal probe is held at rest with respect to the laboratory and a solvent is given a constant velocity using a flow cell.

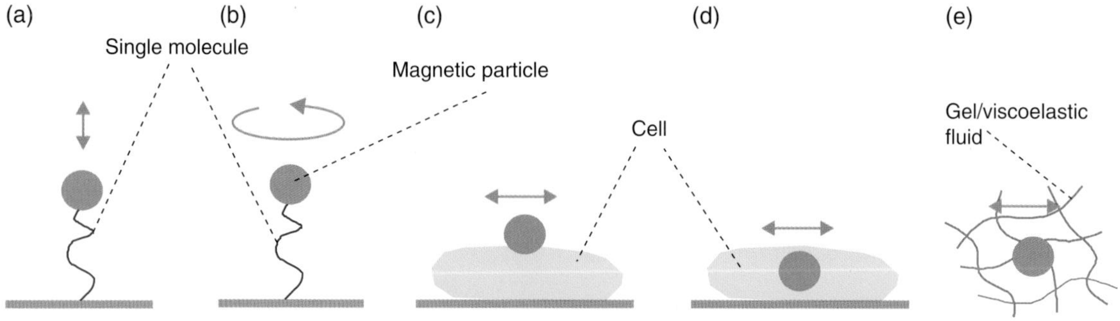

Figure 19.41 *A range of standard geometries for magnetic-tweezer experiments include (a) the extension of single molecules, (b) the rotation of single molecules, (c) the exploration of the external viscoelasticity of cells, (d) the exploration of the internal viscoelasticity of cells and (e) the exploration of the viscoelasticity of complex fluids.*

The critical velocity at which the trapped bead becomes dislodged is measured, which allows the force applied to the tweezers to be calculated using equation (19.41). A more accurate method for trap calibration uses an analysis of the thermal fluctuations of the trapped particle in the flow cell and is based on Langevin's equation for the particle's motion,

$$m\frac{dv}{dt} = F_{\text{thermal}}(t) - \gamma v - \kappa x \tag{19.42}$$

This is just Newton's second law, a balance of the inertial force (mdv/dt), the thermal force $F_{\text{thermal}}(t)$, the drag force γv, and the elastic trap force (κ is the effective lateral trap spring constant, and x is the particle displacement). To solve equation (19.42), the thermal force (F_{thermal}) is assumed to be completely random over time and mathematically this is equivalent to

$$\langle F(t)F(t-\tau)\rangle = 2kT\gamma\delta(\tau) \tag{19.43}$$

where $\delta(\tau)$ is the Dirac delta function, kT is the thermal energy and γ is the frictional coefficient. In the low Reynolds number regime the inertial term can be neglected ($mdv/dt = 0$), which greatly simplifies equation (19.42) and this is typically the case in most tweezer experiments at low frequencies. The Fourier transform of the Langevin equation (19.42) can be taken that provides an expression of the power spectrum ($S(\omega)$, **Section 7.2**) of the fluctuations of the bead displacement,

$$S(\omega) = \frac{kT}{\pi^2\gamma\left(\omega_c^2 - \omega^2\right)} \tag{19.44}$$

where kT is the thermal energy, γ is the drag coefficient, ω_c is the cornering frequency and ω is the frequency. The power spectrum of bead fluctuations can be easily determined experimentally using a fast fourier transform of the mean square displacement of the bead as a function of time (**Figure 19.42**). Equation (19.44) then allows the cornering frequency (ω_c) to be calculated and the spring constant (κ) of the trap can be subsequently found using the equation

$$\omega_c = \frac{\kappa}{2\pi\gamma} \tag{19.45}$$

The technique of *atomic force microscopy* (AFM) allows the force between the tip of a cantilever (with a small radius of curvature) and virtually any kind of surface to be measured (**Figure 19.43**). The cantilever is

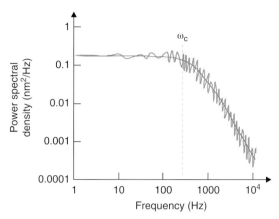

Figure 19.42 *The power spectral density of the position of a particle trapped in the focused laser beam of an optical tweezer set up as a function of frequency. The cornering frequency (ω_c) is shown.*

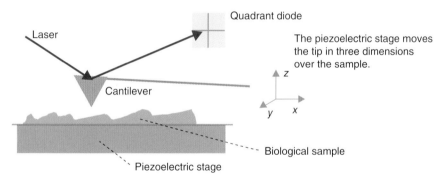

Figure 19.43 *Schematic diagram of an atomic force microscopy (AFM) experiment to study the surface of a biological material. The laser reflects off the back of the cantilever and is then detected by a quadrant diode. A piezoelectric stage attached to the sample moves the tip over the surface to produce a three-dimensional map of the surface topography.*

moved over the sample with a piezoelectric stage and the displacement of the cantilever is measured with a reflected laser beam. The AFM technique also has the significant advantage that the tip can be used to form an image of the surface as it is scanned across the surface. In a typical AFM experiment a small pyramid-shaped tip is mounted on the cantilever that acts as a spring, with a spring constant of ~0.1 N m^{-1}. The sample is arranged on a piezoelectric drive stage. The tip moves in the vertical direction in response to sample movement and the resultant displacement is measured by reflection of a laser beam from the cantilever onto a quadrant photodiode. The photodiode allows the bending motion of the cantilever to be detected to better than 1 Å. A range of feedback methods are used to control the position of the cantilever on the sample, for example, to hold the cantilever at a constant force. The detailed construction of an AFM is shown in the **Figure 19.44**.

AFM allows much larger forces to be applied to a sample than with optical/magnetic tweezers and imaging is also possible. The AFM images in **Figure 19.45** show a supercoiled DNA chain and an amyloid fibre adsorbed to a surface. However, it is much more difficult to model the viscoelastic response of materials close to surfaces with AFM due to the effects of lubrication hydrodynamics, and the cantilever geometry causes the sensitivity to be reduced compared to the optical/magnetic tweezer techniques (an important consideration for

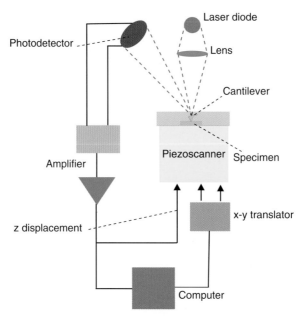

Figure 19.44 *Detailed schematic diagram of an atomic force microscope. A feedback scheme can be used to hold the cantilever at a constant force on the sample, i.e. noncontact imaging is possible. [Reproduced with permission from J. Yang, L.K. Tamm, A.P. Somlyo, Z. Shao, Journal of Microscopy, 1993, 171, 183–198. Copyright 1993, John Wiley & Sons.]*

single-molecule experiments). Soft surfaces can be perturbed (indented) during the process of AFM image collection, due to the contact with the cantilever and sensitive feedback systems have been implemented to reduce this damage (the so-called 'noncontact mode'). The magnitude of the cantilever displacement is given by

$$F_{\text{bond}} = K_{\text{cantilever}}\Delta x \tag{19.46}$$

where Δx is the displacement of the tip and $K_{\text{cantilever}}$ is the spring constant of the cantilever. In the simplest approximation, Hooke's law is obeyed by the cantilever. If the tip displacement and the spring constant are known the force used to extend a biomolecule can be calculated. A typical value for $K_{\text{cantilever}}$ is $10^{-3}\,\text{N m}^{-1}$ and a typical tip radius is 30 nm.

Atomic force microscopes can be used to probe the mechanical properties of single biomolecules (**Figure 19.46**). **Figure 19.47** shows some typical unwinding data from a multidomain protein. The unfolding of each separate domain is shown as an extra sawtooth in a force/extension curve.

Surface-force apparatus (SFA) measures the forces between surface areas that are of macroscopic dimensions. The technique involves the measurement of the distance of separation as a function of the applied force of crossed cylinders coated with molecularly cleaved mica sheets (**Figure 19.48**). The separation between the surfaces is measured interferometrically to a precision of 0.1 nm and the surfaces are driven together with piezoelectric transducers with a resolution of $10^{-8}\,\text{N}$. Much of the most accurate fundamental information that concerns mesoscopic forces has been established using SFA.

The *split photodiode detector* is a critical piece of technology for a series of force-probe techniques that include AFM, glass fibres and optical tweezers. The detector allows fast accurate measurement of light intensities and can provide subnanometre resolution of probes' positions on the time scale of $100\,\mu\text{s}$–$100\,\text{s}$. The mean photocurrent ($<i>$) at time t measured by a section of the split detector is

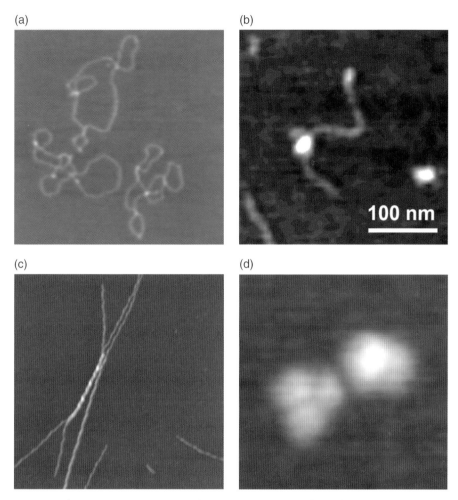

Figure 19.45   *Atomic force microscopy images of (a) supercoiled closed circular DNA, (b) a DNA gyrase protein bound to a linear DNA, (c) amyloid fibrils and (d) two individual antibody molecules. [Reproduced with permission from Neil Thomson, University of Leeds, 2005.]*

$$\langle i \rangle = n \int_0^\infty g(t)dt = nze \qquad (19.47)$$

where $g(t)$ is the photocurrent detected at a time (t) given by

$$g(t) = \frac{ze}{\tau_0} \exp\left(-\frac{t}{\tau_0}\right) \qquad (19.48)$$

where z is the total number of charges displaced upon absorption of a photon on the detector, e is the electronic charge, τ_0 is the time constant of the detector and n is the total number of photons collected. The position of a probe (e.g. the cantilever with an AFM or the colloidal probe with optical tweezers) measured using a split photodiode is found by comparison of the difference in current signals (Δi) between the two photodiodes.

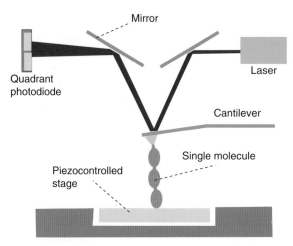

Figure 19.46 *An atomic force microscope (AFM) can be used to measure the force/extension properties of single molecules.*

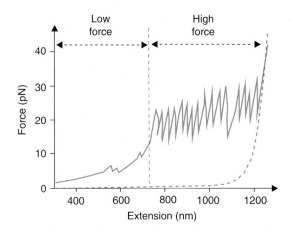

Figure 19.47 *Schematic diagram of unwinding curves for a multidomain single-molecule experiment, e.g. IG domains in a protein or DNA combined with histones. The applied force (applied for example by an AFM) is shown as a function of the chain extension. The saw teeth correspond to discrete transitions in the structure of the molecule, e.g. the unfolding of an IG domain or dissociation of a histone complex from a DNA chain.*

The displacement noise on the determination of the probe position as a function of frequency (standard deviation $\sigma_x(f)$) quantifies the accuracy of the split diode and can be calculated as

$$\sigma_x^2(f) = \frac{d^2}{2qn} \tag{19.49}$$

The resolution of the split diode experiment thus depends on the total number of photons collected by the detector (n), the efficiency of the detector for absorbing photons (q) and the spatial width of the detector (d). It does not depend on the electronic charge (e), the instrument amplification (z) or the magnification.

Modern optical tweezer designs have started to use fast digital cameras (>4 pixels), since they can be used in imaging modes to help with beam alignment, although their temporal response is slightly slower than with split

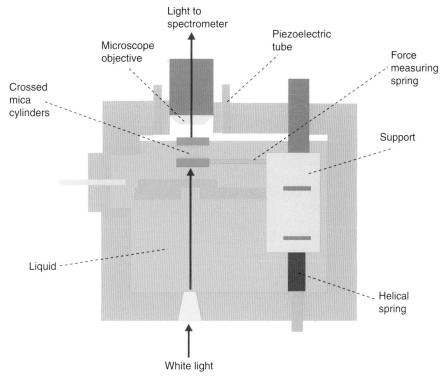

Figure 19.48 *A schematic diagram of a surface-force apparatus that can be used to measure mesoscopic forces. The distance between the two mica cylinders is calculated using interferometry and the force is gauged with a finely calibrated spring. [Reproduced with permission from Anal. Chem., 1989, 61 (7), pp 498A–498A. Copyright © 1989, American Chemical Society.]*

diodes. Furthermore, balanced detectors can also be used instead of split photodiodes to allow fast dynamic photon measurements that approach the quantum noise limit (squeezed light sources can even break the quantum noise limits that occur with conventional light sources).

With *micropipette* force measurements, the pressure in the pipettes is controlled and so too are their displacements using a piezonanomotor. They find applications in studies of membrane forces.

Examples of specific applications of force probes to the mechanical properties of single biological molecules include studies of myosin, dynein, kinesin, ATP synthase, bacterial flagellae, mechanics of single polymer chains, enzymatic relaxation of DNA supercoils, single enzyme catalysis, chromosomes (DNA complexed with histones), and IG domains in proteins (titin).

It should be stressed that Bell's equation is required to explain the forces measured in single molecule experiments. Forces experienced by single molecules depend on both the temperature and the time scale at which they are measured (**Section 4.7**).

19.8 Electron Microscopy

Electrons are easier to produce in the laboratory than X-rays or neutrons. Electron microscopy is thus a well-established technique and can be used in both diffraction and imaging modes. It offers exceptionally high

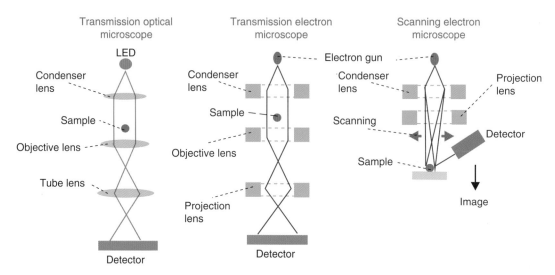

Figure 19.49   *Comparison of the arrangement of a transmission optical microscope (visible-wavelength photons) with that of a transmission electron microscope and a scanning electron microscope.*

Angstrom (0.1 nm) level resolution images of the structure of biological molecules. The main problem with electron microscopy is that electrons often interact too strongly with samples. The samples therefore need to be ultrathin, and need to be held under high vacuum conditions, which constrains the applications in biology. Furthermore, samples experience intense beam damage and sample stability is important to create high-resolution images. Lenses can be created for electron microscopes, so direct imaging is possible, but they are intermediate in quality between those of optical microscopes (very good) and X-ray microscopes (very poor), i.e. they have significant aberrations.

Three-dimensional reconstruction from two-dimensional images is also possible in transmission electron microscopy using a tilt series and the inverse Radon transformation (a standard method also found in MRI and X-ray CAT scans). Samples do not have to be crystalline, but some axes of symmetry are useful (they help the averaging process). Such techniques have been used to reconstruct three dimensional images of noncrystalline viruses, e.g. HIV.

The two main varieties of electron microscope are transmission electron microscopy (TEM) and scanning electron microscopy (SEM). **Figure 19.49** shows a comparison of TEM and SEM microscopes with a transmission optical microscope. In TEM the transmitted beam is detected, whereas with SEM it is the reflected beam that is measured.

19.9   Nuclear Magnetic Resonance Spectroscopy

The basic principle of nuclear magnetic resonance (NMR) involves two sequential steps. First, the degeneracy in the energy levels of magnetic nuclear spins in a population of atoms is lifted by the application of a constant applied magnetic field, B (a nuclear Zeeman effect, the nuclear energy levels become split, **Figure 19.50**). The nuclei in a range of biologically important atoms (^{1}H, ^{2}D, ^{13}C, F, etc.) have quantised nonzero spin states, which can be aligned in the magnetic field. As the second step this alignment of the nuclear spins is perturbed by the use of a pulse of radio-frequency photons. Radio-frequency photons are employed, since the energy differences induced by the static magnetic field in the nucleus are relatively small and match the energy range

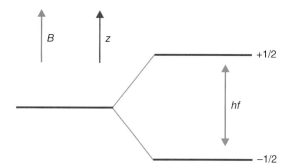

Figure 19.50   *The nuclear magnetic resonance (NMR) effect with a spin ½ nucleus can be thought of as a nuclear Zeeman effect. The magnetic field (B) is aligned along the z-axis. Application of the magnetic field lifts the degeneracy of the nuclear energy levels of the spin ½ nucleus. The difference in energies between the two spin states is equal to the energy (E) of a radio frequency photon of frequency f (E = hf). Thus, irradiation of the samples with radio-frequency photons with a range of energies allows absorption spectroscopy experiments to be performed.*

of these photons. The required perturbing frequency and thus the photon absorption efficiency, is dependent upon the static magnetic field (B) and the specific nuclei observed (and their molecular environments).

Fourier transform (FT) NMR is a standard NMR technique. It is used as a common tool in organic chemistry and biochemistry to elucidate patterns of molecular bonding in unknown molecules. Chemically discriminating spectroscopy is possible due to the dependence of nuclear absorption frequencies on the nuclear shielding of the surrounding atoms. A wide range of frequencies are explored simultaneously in the FT method, which provides a large improvement in signal-to-noise compared to the more naïve method in which the absorption radio frequency is gradually scanned. Furthermore, a wide range of radio-frequency pulse sequences can be applied to a sample in pulsed FT NMR to enable more sophisticated manipulation of the nuclear spin states. Each type of pulse sequence can provide useful information and probe different aspects of the molecule's nuclear environment.

The nuclear Overhauser effect (NOE) is particularly useful in NMR, since the distance between nuclei can be measured in absolute terms. This can provide *ab initio* structural elucidation of biological molecules.

Magnetic resonance imaging (MRI) has also become a standard technique. It uses a gradient in the magnetic field to align the nuclear spins, which allows spatial information in the sample to be detected as a shift in the absorption frequency of the radio waves. MRI allows noninvasive imaging of live organisms with up to a ~10 μm level of resolution. Variants of MRI, such as functional magnetic resonance imaging (fMRI), allow the metabolic activity in live human brains to be explored, e.g. blood-oxygen-level dependent (BOLD) contrast.

19.10   Static Scattering Techniques

The field of scattering encompasses a vast range of fundamental physical processes and techniques. The basic geometry of a scattering experiment is shown in **Figure 19.51**. Incident radiation or particles interact with the molecules in a sample and are deflected through an angle (θ). The *momentum transfer* (q) of the scattering process is simply related to the reciprocal of the length scale (d) probed,

$$q = \frac{2\pi}{d} \tag{19.50}$$

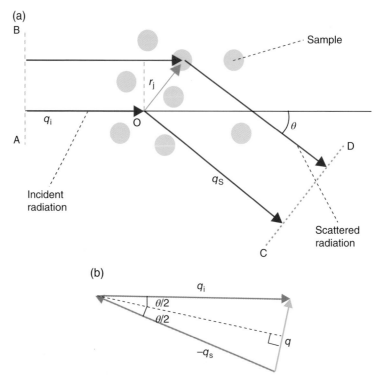

Figure 19.51 *(a) The elastic scattering processes involved with a wide range of radiation (e.g. X-rays, light, neutrons and electrons) can all be understood using the same schematic diagram. The radiation strikes the sample and is deflected through an angle θ. (b) The momentum transfer can be calculated from the wavelength and scattering angle using a simple geometrical construction.* $\left|\underline{q}_i\right| = \left|\underline{q}_s\right| = \dfrac{2\pi}{\lambda}$ *and* $q = \dfrac{4\pi}{\lambda} \sin\left(\dfrac{\theta}{2}\right)$, *where λ is the wavelength.*

This leads to the notion of reciprocal space, i.e. diffraction patterns where reciprocal length scales are probed. For a particular form of radiation the momentum transfer for an elastic (energy is conserved) scattering process can be calculated from

$$q = \frac{4\pi}{\lambda} \sin\left(\frac{\theta}{2}\right) \tag{19.51}$$

where λ is the wavelength. The use of momentum transfers rather than scattering angles allows the results of experiments with a range of different forms of radiation (X-rays, light, neutrons, etc.) to be more easily compared. The varieties of radiation that are typically used in biological scattering experiments are shown in **Table 19.1**, which includes the wavelength of the radiation and the length scales in the sample that can be typically probed. Specialised detectors and optics are required for each different form of radiation.

The small wavelength of X-rays and neutrons implies high-resolution measurements are possible, relatively unconstrained by diffraction limits. However, there are some caveats to the improvements in resolution that are in fact possible. Soft X-rays are absorbed strongly by water and thus are less useful in biology. It is very hard to make lenses for all forms of penetrating radiation (X-rays, γ-rays and neutrons). Historically, most X-ray

Table 19.1 *Comparison of scattering techniques that can be used to study biological samples.*

Technique	Typical Wavelength (nm)	q values (nm^{-1})	Real space distances (nm)	Sample contrast	Comment
Small-angle X-ray scattering	0.15	0.009–6.3	1–700	Electron density	Beam damage can be a problem
Small-angle neutron scattering	0.4	0.003–6.3	1–200	Scattering length density	Samples can be labelled with deuterium
Light Scattering	450	0.0003–0.13	50–2000	Refractive index	Multiple scattering is a problem at high concentrations
Wide-angle X-ray scattering	0.15	6.3–63	0.1–1	Electron density	Beam damage can be a problem
Wide-angle neutron Scattering	0.4	6.3–63	0.1–1	Scattering length density	Useful to measure hydrogen bonds
Electron scattering	0.0037	0.0063–6.3	0.1–1000	Electron density	Very thin sections are required due to the strong interaction with samples

imaging experiments have been restricted to length scales larger than a micrometre using absorption tomography technique, due to the poor quality of the available X-ray lenses.

X-rays are predominantly scattered by electrons, and neutrons are scattered by nuclei. Typically X-rays provide higher-resolution structures in crystallography experiments, since higher-intensity sources are available, but neutrons allow hydrogen atoms to be observed. The *contrast* that is measured during a scattering process can be varied in both neutron and X-ray scattering experiments. With neutrons isotopic substitution can be used to label biomolecules with deuterated atoms. This labelling scheme is particularly attractive when electronically light atoms need to be located in a crystalline structure, e.g. the elucidation of the structure of hydrogen bonds. With X-rays the wavelength of the radiation can be matched to the absorption edge of a heavy atom that exists in a biological structure and the contrast varied to elucidate both crystalline- and solution-state structures with much improved resolution.

Short-wavelength photons can damage biomolecules (it is an ionising form of radiation). UV light can cause sunburn, since energetic photons rip up the DNA molecules (in addition to many others) in skin cells, and need to be constantly repaired by the organism. More energetic photons (X-rays and γ-rays) tend to do even worse damage, but relatively low-energy photons can also be biologically destructive if they are tuned to a strong resonance in a biomolecule (think of microwave ovens).

Small-angle scattering is the scattering of X-rays, light or neutrons through small angles. Small angles imply big distances are probed in the sample, which is useful to observe structures on the nanoscale (as opposed to the measurement of atoms on Angstrom length scales with wide-angle scattering).

Historically, *X-ray crystallography* has been vitally important to the development of molecular biology since it can provide structures of biological molecules with Angstrom resolution. However, many molecules and complexes are not crystalline, and the method can then not be used. This problem can be circumvented using coherent X-ray techniques and is a motivating factor for the development of free-electron lasers and new synchrotron light sources, i.e. structures can be created of noncrystalline aperiodic specimens. However, it is

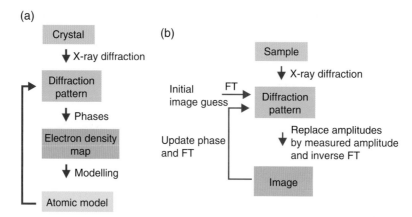

Figure 19.52 *Comparison of modelling strategies for (a) low-coherence X-ray crystallography (crystals are required), and (b) high-coherence X-ray imaging (samples can be aperiodic).*

first important to appreciate the basis of standard low-coherence crystallography techniques, since they form the backbone of more sophisticated high-coherence methods.

X-ray crystallography determines the structural arrangement of atoms within a crystal using diffraction (**Figure 19.52**). In a crystallography experiment, a beam of X-rays strikes a crystal and scatters into many different directions. From the angles and intensities of these scattered beams, a crystallographer can produce a three-dimensional picture of the density distribution of electrons within the crystal. From this electron-density distribution, the mean positions of the atoms in the crystal can be determined. Other information on chemical bonds, disorders, defects and impurities can also be obtained.

In general, seven different symmetry systems can divide three-dimensional space into regular units. With *cubic* symmetry there are three axes at right angles, all equal in length. *Hexagonal* symmetry has two equal axes that subtend each other at 120°, and each is at right angles to a third axis of different length. *Tetragonal* symmetry has three axes at right angles, two of equal length. *Trigonal* (rhombohedral) symmetry has three equally inclined axes, which are not at right angles, e.g. calcite crystals. *Orthorhombic* symmetry has three axes at right angles, all of different lengths. *Monoclinic* symmetry has three axes, one pair that is not at right angles, and the sides are all of different lengths. *Triclinic* symmetry has three axes, all at different angles, none of which is a right angle, and all the sides are of different lengths.

An intuitive understanding of Bragg's law can be motivated. A given X-ray reflection is associated with a set of evenly spaced planes that run through the crystal, and usually pass through the centres of the atoms of the crystal lattice (**Figure 19.53**). The orientation of a particular set of planes is identified by its three Miller indices (h, k, l), and the spacing distance is d. William Bragg proposed a model in which the incoming X-rays are scattered specularly (mirror-like) from each plane; X-rays scattered from adjacent planes will combine constructively (constructive interference) when the angle θ between the plane and the X-ray results in a path-length difference that is an integer multiple n of the X-ray wavelength λ,

$$2d\sin\theta = n\lambda \tag{19.52}$$

Bragg's law describes the condition for constructive interference from successive crystallographic planes (h,k,l) of the crystalline lattice. The equation holds for elastic scattering of coherent X-rays with the same wavelength λ (most standard low-coherence X-ray sources have sufficient coherence to allow constructive interference between a reasonable number of neighbouring lattice planes to observe a diffraction pattern). A reflection is *indexed* by identification of its Miller indices. Such indexing gives the unit-cell parameters, the lengths and angles of the unit cell, as well as its space group. However, because Bragg's law does not

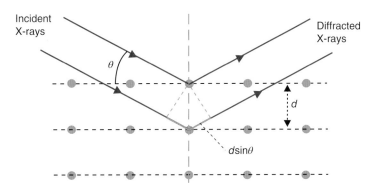

Figure 19.53 *X-ray diffraction from lattice planes in a crystal give rise to constructive interference at positions predicted by the Bragg equation (19.52), 2d sinθ = nλ, λ is the wavelength.*

interpret the relative intensities of the reflections, it is generally inadequate to solve for the arrangement of atoms within the unit cell; instead a Fourier transform method is needed.

The main goal of X-ray crystallography is to determine the density of electrons $\rho(\underline{r})$ throughout a crystal, where $\underline{r}$ represents the three-dimensional position vector within the crystal (**Figure 19.52**). To do so, X-ray scattering experiments collect data about the electron density's Fourier transform $F(\underline{q})$, which is inverted mathematically to obtain the density defined in real space, using the formula

$$\rho(\underline{r}) = \frac{1}{(2\pi)^3} \int F(\underline{q}) e^{i\underline{q}.\underline{r}} d\underline{q} \tag{19.53}$$

where the integral is taken over all values of $\underline{q}$. The three-dimensional real vector $\underline{q}$ represents a point in reciprocal space. The corresponding formula for an inverse Fourier transform is

$$F(\underline{q}) = \int \rho(\underline{r}) e^{-i\underline{q}.\underline{r}} d\underline{r} \tag{19.54}$$

where the integral is summed over all possible values of the position vector $\underline{r}$ within the crystal. The Fourier transform $F(\underline{q})$ is generally a complex number, and therefore has a magnitude $|F(\underline{q})|$ and a phase $\phi(\underline{q})$ related by the equation

$$F(\underline{q}) = |F(\underline{q})| e^{i\phi(\underline{q})} \tag{19.55}$$

The intensities of the reflections observed in X-ray diffraction can provide the magnitudes $|F(\underline{q})|$, but not the phases $\phi(\underline{q})$. To obtain the phases, full sets of reflections can be collected with known variations of the scattering, either by modulation of the wavelength past a certain absorption edge (MAD, multiwavelength anomalous dispersion) or by the addition of strongly scattering (electron dense) metal atoms such as mercury to the crystal. Combination of the magnitudes and phases yields the full Fourier transform $F(\underline{q})$, which may be inverted to obtain the electron density $\rho(\underline{r})$.

Crystals are often idealised as being perfectly periodic. In the ideal case, the atoms are positioned on a perfect lattice. The electron density is perfectly periodic, and the Fourier transform $F(\underline{q})$ is zero, except when $\underline{q}$ belongs to the reciprocal lattice (the so-called Bragg peaks). In reality, however, crystals are not perfectly periodic. Atoms vibrate about their mean position, and there may be disorder of various types, such as mosaicity, dislocations, point defects, and heterogeneity in the conformation of crystallised molecules

(Section 6.3). Therefore, the Bragg peaks have a finite width and there may be significant diffuse scattering, a continuum of scattered X-rays that falls between the Bragg peaks.

As seen previously in the section on microscopy, plane waves can be represented by a wave vector $\underline{k}_{in}$, and so the strength of the incoming wave is given by $Ae^{i\underline{k}_{in}\cdot\underline{r}}$. At position $\underline{r}$ within the sample, the density of scatterers is given by $\rho(\underline{r})$. These scatterers produce a scattered spherical wave of amplitude proportional to the local amplitude of the incoming wave times the number of scatterers in a small volume dV about $\underline{r}$,

$$A^* = Ae^{i\underline{k}\cdot\underline{r}}S\rho(\underline{r})dV \tag{19.56}$$

where A^* is the scattered wave (a complex number) and S is a proportionality constant. Consider the fraction of scattered waves that leave with an outgoing wave-vector of $\underline{k}_{out}$ and strike the detector at $\underline{r}_{det}$. Since no energy is lost (elastic scattering), the magnitudes of the wave-vectors $|\underline{k}_{in}| = |\underline{k}_{out}|$ before and after scattering are equal. From the time that the photon is scattered at $\underline{r}$ until it is absorbed at $\underline{r}_{det}$, the photon undergoes a change in phase given by

$$e^{i\underline{k}_{out}\cdot(\underline{r}_{det}-\underline{r})} \tag{19.57}$$

The net radiation that arrives at $\underline{r}_{det}$ is the sum of all the scattered waves throughout the crystal that may be written as a Fourier transform,

$$AS\int\rho(\underline{r})e^{i\underline{k}_{in}\cdot\underline{r}}e^{i\underline{k}_{out}\cdot(\underline{r}_{det}-\underline{r})}d\underline{r} = ASe^{i\underline{k}_{out}\cdot\underline{r}_{det}}\int\rho(\underline{r})e^{i(\underline{k}_{in}-\underline{k}_{out})\cdot\underline{r}}d\underline{r} \tag{19.58}$$

$$ASe^{i\underline{k}_{out}\cdot\underline{r}_{det}}\int\rho(\underline{r})e^{-i\underline{q}\cdot\underline{r}}d\underline{r} = ASe^{i\underline{k}_{out}\cdot\underline{r}_{det}}F(\underline{q}) \tag{19.59}$$

where $\underline{q} = \underline{k}_{out} - \underline{k}_{in}$. The measured intensity, I, of the reflection will be square of this amplitude,

$$I(\underline{q}) = A^2S^2\left|F(\underline{q})\right|^2 \tag{19.60}$$

To ensure that $\rho(\underline{r})$ is given real values, the Fourier transform $F(\underline{q})$ must be such that the Friedel mates $F(-\underline{q})$ and $F(\underline{q})$ are complex conjugates of one another. Thus, $F(-\underline{q})$ has the same magnitude as $F(\underline{q})$. $|F(\underline{q})| = |F(-\underline{q})|$, but they have the opposite phase, i.e. $\varphi(\underline{q}) = -\varphi(\underline{q})$,

$$F(-\underline{q}) = \left|F(-\underline{q})\right|e^{i\phi(-\underline{q})} = F^*(\underline{q}) = \left|F(\underline{q})\right|e^{-i\phi(\underline{q})} \tag{19.61}$$

The equality of their magnitudes ensures that the Friedel mates have the same intensity $|F|^2$. This symmetry allows the full Fourier transform to be measured from only half the reciprocal space, e.g. by rotation of the crystal by slightly more than a 180° instead of a full turn. In crystals with additional symmetries, even more reflections can have the same intensity (Bijvoet mates); in such cases, less of the reciprocal space needs to be measured, e.g. slightly more than 90°. The Friedel-mate constraint can be derived from the definition of the inverse Fourier transform,

$$\rho(\underline{r}) = \frac{1}{(2\pi)^3}\int F(\underline{q})e^{i\underline{q}\cdot\underline{r}}d\underline{q} = \frac{1}{(2\pi)^3}\int\left|F(\underline{q})\right|e^{i\phi(\underline{q})}e^{i\underline{q}\cdot\underline{r}}d\underline{q} \tag{19.62}$$

Euler's formula states that $e^{ix} = \cos(x) + i\sin(x)$, so the inverse Fourier transform can be separated into a sum of a purely real part and a purely imaginary part,

$$\rho\left(\underline{r}\right)=\frac{1}{\left(2\pi\right)^{3}}\int\left|F\left(\underline{q}\right)\right|e^{i\left(\phi+\underline{q}.\underline{r}\right)}d\underline{q}=\frac{1}{\left(2\pi\right)^{3}}\int\left|F\left(\underline{q}\right)\right|\cos\left(\phi+\underline{q}.\underline{r}\right)d\underline{q}+i\frac{1}{\left(2\pi\right)^{3}}\int\left|F\left(\underline{q}\right)\right|\sin\left(\phi+\underline{q}.\underline{r}\right)d\underline{q}=I_{\cos}+iI_{\sin}$$

$$(19.63)$$

where $I_{\cos}$ and $I_{\sin}$ are the cosine and sine components of the fourier transform respectively. The function $\rho(\underline{r})$ is real, if and only if the second integral $I_{\sin}$ is zero for all values of $\underline{r}$. In turn, this is true if and only if the above constraint is satisfied,

$$I_{\sin}=\frac{1}{\left(2\pi\right)^{3}}\int\left|F\left(\underline{q}\right)\right|\sin\left(\phi+\underline{q}.\underline{r}\right)dq=\frac{1}{\left(2\pi\right)^{3}}\int\left|F\left(-\underline{q}\right)\right|\sin\left(-\phi-\underline{q}.\underline{r}\right)dq=-I_{\sin} \qquad (19.64)$$

since $I_{\sin}=-I_{\sin}$ implies that $I_{\sin}=0$.

The data collected from a diffraction experiment is a reciprocal-space representation of the crystal lattice. The position of each diffraction spot is governed by the size and shape of the unit cell, and the inherent symmetry within the crystal. The intensity of each diffraction spot is recorded, and this intensity is proportional to the square of the *structure factor* amplitude.

The structure factor is a complex number that contains information that relates to both the amplitude and phase of a wave. In order to obtain an interpretable *electron density map*, both the amplitude and the phase must be known (an electron-density map allows a crystallographer to start to build a model of the molecule). The phase cannot be directly recorded during a diffraction experiment and this is called the *phase problem*.

Initial phase estimates can be obtained in a variety of ways that include *ab initio* phasing, molecular replacement, anomalous X-ray scattering and heavy atom-methods. While all four of the above methods are used to solve the phase problem for protein crystallography, small-molecule crystallography can yield data suitable for structural analysis using direct methods (*ab initio* phasing). Once the initial phases are obtained, an initial model can be built. This model can be used to refine the phases, which leads to an improved model, and the solution can be iterated (**Figure 19.52**). A model of the atomic positions and their respective Debye–Waller factors (which account for the thermal motion of the atom) can be refined to fit the observed diffraction data, which can yield a better set of phases. A new model can then be fit to the new electron-density map and a further round of refinement is carried out. This continues until the correlation between the diffraction data and the model is maximised. The progress of the iterative minimisation procedure can be measured by an R-factor defined as

$$R=\frac{\sum_{\text{all\_reflections}}\left|F_{\text{o}}-F_{\text{c}}\right|}{\sum_{\text{all\_reflections}}\left|F_{\text{o}}\right|} \qquad (19.65)$$

where F_{o} is the observed structure factor and F_{c} is the structure factor calculated from the model.

Many biological molecules are semicrystalline (**Chapter 6**), which makes crystallography more challenging, since their diffraction patterns are averaged over a range of orientations and contain the effects of many defect structures. With polymeric molecules such as DNA fibre diffraction patterns have been important in structure determination. Instead of sharp delta functions in reciprocal space, fibre diffraction patterns are averaged over an orientational angle perpendicular to the chain axis due to the many orientations that occur in a bundle of fibres. For helical macromolecules (a very common scenario) fibre diffraction patterns have characteristic cross patterns and the periodicity along the chain (they constitute a one-dimensional crystal) results in layer lines. Structure factors can be constructed using models with helical symmetry. It is thus possible to measure the helical width (r_{h}) and the helical pitch (P) relatively simply from a fibre diffraction pattern (**Figure 19.54**). For the fibre diffraction pattern from a continuous helical line,

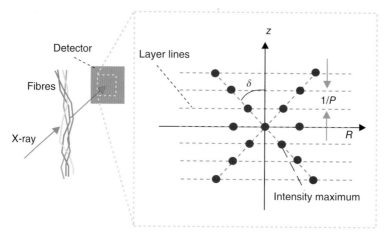

Figure 19.54 *In X-ray fibre diffraction experiments from helical macromolecules, a crossed diffraction pattern is observed. The separation of the layer lines allows the helical pitch (P) to be calculated. The angle δ allows the helix diameter to be measured.*

$$\tan\delta = \frac{P}{2\pi r_{\mathrm{h}}} \qquad (19.66)$$

where δ is the angle the helical cross pattern makes with the vertical axis. **Figure 19.54** shows the typical cross pattern for an alpha helix. The cross pattern in reciprocal space is due to the cylindrical symmetry of Bessel functions (the Fourier transform of a helix can be expressed in terms of Bessel functions). Accurate modelling of helical macromolecules is possible based on a full Fourier theory of diffraction. However, such models need lots of additional information from other sources to constrain them and they can be ambiguous.

Synchrotron X-ray sources produce a broad range of X-ray wavelengths. Thus, scanning a monochromator to select different wavelengths allows X-ray spectroscopy experiments to be performed. There are a wide range of different possibilities. In anomalous X-ray scattering the scattering contrast can be varied through a choice of wavelengths close to the absorption edge of an element (e.g. a counterion such as rubidium or bromine), which allows contrast variation experiments to be performed. Anomalous X-ray scattering (AXS) is particularly useful for the study of the structure of counterion clouds around biological molecules. X-ray fluorescence provides an unwanted background for AXS experiments, but it is characteristic of the elements illuminated and can be used to fingerprint unknown materials. Inelastic X-ray scattering experiments are also possible with synchrotron sources. These can provide dynamic information on molecular motion at very small length scales. For example, phonons (quantised lattice vibrations) have been measured in DNA fibres as they travel along the backbone of the polymeric chains.

X-rays and neutrons can experience mirror like reflection from surfaces and this can be used to probe the structure of noncrystalline biological materials attached to surfaces. *Reflectivity* curves require careful modelling to probe both the out-of-plane structure (the standard arrangement, which considers structures perpendicular to the surface) and the inplane structure (structures parallel to the surface), since the phase problem still occurs.

There are a number of novel scattering techniques that use the coherence of electromagnetic radiation and in particular that of X-rays. Two promising new coherent methods are *X-ray diffraction imaging* and *X-ray photon correlation spectroscopy*. Images of completely aperiodic viruses and cells have now been reconstructed using coherent X-ray diffraction to solve the phase problem with 50-nm resolution and X-ray photon correlation spectroscopy offers dynamic measurements from soft-matter systems with high sensitivity to length scale.

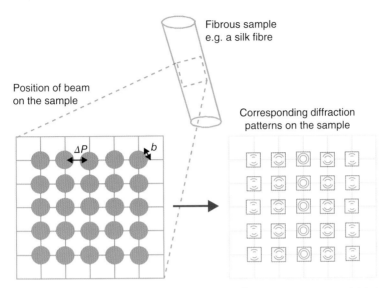

Figure 19.55 *Scanning X-ray microdiffraction across a semicrystalline anisotropic material. b is the diameter of the X-ray beam. Two-dimensional diffraction images are recorded at micrometre-spaced steps (ΔP) across the sample and provide detailed information on the molecular structure of the fibre.*

A further modern advance with X-ray and neutron scattering is the introduction of effective *focusing techniques*. Focusing of X-rays is routinely made to submicrometre levels at third-generation synchrotron sources (microfocus X-ray scattering) and micrometre-sized beams can even be made with laboratory-based microfocus sources. Such beams can be rastered across heterogeneous biological materials and the molecular structure probed as a function of position on the sample (**Figure 19.55**). New focusing devices have now been made that include capillaries, fresnel lenses, mirrors, and even simple compound lenses (e.g. lenticular holes in a block of aluminium). Such scanning X-ray microdiffraction techniques are helping to revolutionise the field of fibrous carbohydrates and proteins.

A series of other technical advances are helping to revolutionise the possibilities for X-ray imaging in biology (**Figure 19.56, Figure 19.57**). These include the development of new imaging modalities with high-coherence X-rays such as *diffractive imaging, ptychography*, and *femtosecond holography*; and *Talbot phase contrast tomography* with conventional laboratory-based low-coherence X-ray sources, e.g. standard rotating anodes. A crucial insight is that the available phase contrast with synthetic organic and biological colloids can be two orders of magnitude stronger than the absorption contrast with X-rays, which provides large improvements in the signal-to-noise ratio in the resultant images. Furthermore, new developments with the sources of X-rays increase the possibilities for this research as the available coherence, flux and collimation are improved, e.g. third-generation high-brilliance *synchrotrons*, and *free-electron lasers*.

The improved availability of both soft (0.3–50 nm) and hard (0.01–0.3 nm) wavelength high-coherence X-ray sources has been an important advance. The scattering of highly coherent light from materials that are rough on the length scale of the light's wavelength is observed as speckle. Grainy blobs are found in the scattering patterns and this was one of the first experimental observations made when the visible-wavelength laser sources were originally created in the 1960/1970s. It is this speckle that contains additional information about the object morphology and can allow direct imaging of materials. As well as the advantage of working with images in real space, X-ray imaging allows structures to be found for noncrystalline (nonperiodic) specimens, albeit at lower resolutions than possible with X-ray crystallography of their

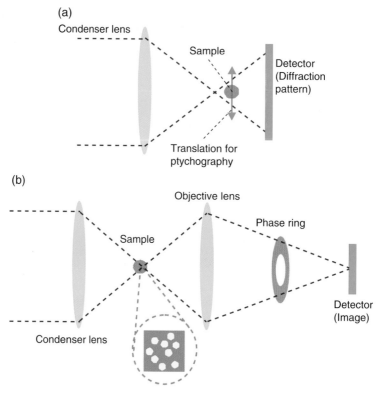

Figure 19.56 *Examples of the configurations of hard X-ray microscopes. (a) Fresnel coherent X-ray diffraction imaging setup. Perpendicular translation of the sample allows the ptychography technique to be performed. (b) Phase-contrast microscope with a phase ring and two Fresnel lenses, similar to the Zernicke microscope in* **Figure 19.24**.

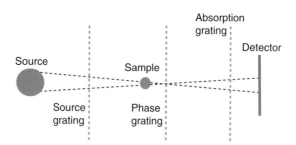

Figure 19.57 *The Talbot effect can be used to provide phase-contrast imaging with X-rays. The source grating creates a series of separate line light sources. The phase and absorption gratings allow the Talbot effect to create a phase-contrast image on the detector.*

crystalline equivalents; currently resolutions are ~10–50 nm for X-ray high-coherence diffractive imaging versus ~0.1 nm for X-ray crystallography. Conventional X-ray crystallography does depend on coherent scattering between adjacent atoms in a lattice, but not across the grain of a whole crystal and thus functions well with low-coherence sources.

X-ray imaging is sensitive to differences in the refractive index of a material, which contains both real and imaginary components. The refractive index is primarily determined by photons scattering from electrons in the sample and thus the electron density of a material. It is also possible to define a complex refractive index with neutrons. Historically, it has been the absorption contrast (β) that has been the most important factor in providing the contrast in X-ray/neutron imaging techniques. Mathematically, the complex refractive of a material ($n(\omega)$) is the sum of the two components,

$$n(\omega) = \alpha(\omega) + i\beta(\omega) \tag{19.67}$$

where α is the real component, ω is the frequency of the incident light, and β is the imaginary (absorptive) component. It is the high β values of bone that make human skeletons stand out in medical X-ray images. However, new phase-contrast techniques use the real part ($\alpha(\omega)$) in equation (19.67) to augment the absorption contrast ($\alpha(\omega)$ can be a factor of 100 bigger than $\beta(\omega)$ for thin noncrystalline samples). Furthermore, many of the emerging imaging technologies allow pixel intensities on images to be replaced by absolute refractive indices, which provide invaluable information for model building. The resultant images thus have their intensities placed on an absolute scale and allow quantitative comparison between different specimens and different experimental set ups. Refractive indices can subsequently be directly linked to electron densities or neutron scattering length densities with X-ray or neutron imaging respectively.

Coherent X-ray imaging was first suggested in the 1950s by oversampling a diffraction pattern to allow both the phase and amplitude of the object to be calculated. However, in practice the method needed to wait for high-coherence synchrotron sources and the development of robust iterative phase retrieval methods (the Feinup method, **Figure 19.52b**) to be of wide practical use. Phase retrieval has been found to profit from focusing of the X-rays onto the sample (otherwise reconstruction methods suffer from image artefacts), and images of a wide range of noncrystalline particles have now been constructed, e.g. viruses.

Ptychography is an inline high-coherence X-ray diffraction technique that allows lensless imaging. High-coherence X-ray diffraction patterns are collected at each of a number of equally spaced positions in a sample as it is transversely translated through the X-ray beam. Overlapping information in the neighbouring diffraction patterns can be quickly and robustly inverted to create extended high-resolution images using modern ptychography algorithms (an adapted Feinup method).

Holography was first introduced by the Nobel Laureate Dennis Gabor as a method to improve the resolution of electron microscopes (**Section 19.5.6**). With the advent of cheap highly coherent laser sources at visible wavelengths, holography at optical wavelengths has become common place. Both the phase and the amplitude of scattered waves can be calculated using holographic techniques, which provides much more detailed information on imaged materials. With large coherence length X-ray sources, inversion of images is relatively straightforward and can be performed with a single two-dimensional Fourier transform, e.g. with out-of-line Fourier transform holography. Practically however, iterative algorithms are often still important to provide improved resolution.

Holography is possible in a variety of different forms that include *inline*, *out-of-line* and *Fourier* holography. All varieties can be performed with either continuous (e.g. synchrotron) or pulsed (e.g. FELS or table top) high-coherence X-ray sources. Inline holography is technically the simplest to implement with X-rays. A coherent reference wave is transmitted to the detector that interferes with the sample scattering. This interference process encodes additional information on the phase of the waves and thus on the sample morphology. Out-of-line holography solves the practical problem of how to separate the scattered and unscattered beams on the detector which are often many orders of magnitude different in intensity and removes an additional conjugate image. It can be simply experimentally realised using a mask with two holes; a larger sample hole and a reference pinhole.

19.11 Dynamic Scattering Techniques

Once the structure of a biological sample is well understood, quantification of the dynamics of the components of the material poses some important questions. A challenge is to examine the dynamics of the material without perturbation of the sample morphology. With soft biological materials the dynamics can be studied with scattering methods by observation of the time decay of stimulated emission (*fluorescence techniques*) or measurement of the change in energy of scattered particles (*quasi-* or *inelastically scattered*).

Fluorescence intensity correlation spectroscopy (FCS) with visible-light wavelengths is a modern example of a scattering technique that has been adapted to single-molecule experiments. Fluorescent probes are added to the biological molecules whose dynamics are of interest and they are made to fluoresce using a tightly focused laser beam under an optical microscope (**Figure 19.58**), similar to that used with fluorescence microscopy.

The intensity of the fluorescently emitted radiation ($I_f(t)$) in the FCS experiment is proportional to the concentration of fluorescent molecules in the scattering volume ($c(\underline{r},t)$), the fluorescent yield (Q) and the intensity of the incident laser beam ($I(r)$),

$$I_f(t) = Q\varepsilon \int I(\underline{r}) c(\underline{r},t) d^3\underline{r} \tag{19.68}$$

where ε is the extinction coefficient of the molecular species for the light, r is the fluorophore position, t is the time and the integral over d^3r is over the complete scattering volume. The dynamic information is contained in the fluctuations of the emitted fluorescent intensity with time, which are directly related to the fluctuations of the concentration of the fluorescent molecules ($<\delta c(\underline{r},t)\delta c(r',0)>$),

$$\left\langle \delta I_f(t)\delta I_f(0) \right\rangle = (\varepsilon Q)^2 \int \int I(\underline{r}) I(\underline{r}') \left\langle \delta c(\underline{r},t)\delta c(\underline{r}',0) \right\rangle d^3\underline{r} d^3\underline{r}' \tag{19.69}$$

where <> denotes an average over time and δ is a small fluctuation in a quantity. $\delta c(\underline{r},t)$ is a concentration fluctuation at position $\underline{r}$ and time t. For *translational diffusion* the intensity autocorrelation function has a

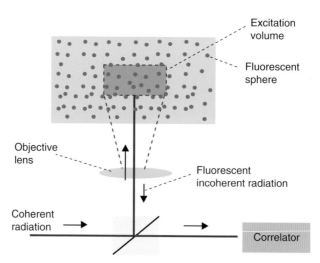

Figure 19.58 *Schematic diagram of a fluorescence intensity correlation spectroscopy experiment (FCS). Incident coherent light excites fluorescence in a small volume of the sample. Fluctuations in the emitted fluorescence are measured with a correlator and can be related to the dynamics of biomolecules to which the fluorescent tags are attached.*

simple exponential form. Thus, a fit of an exponential to the autocorrelated fluorescent signal from a fluorescently tagged biomolecule gives the characteristic time constant (τ) to diffuse out of the scattering volume (e.g. a cube of side b) and hence the diffusion coefficient (D) of the molecules can be calculated ($D = b^2/6\tau$). Fluctuations in the intensity of the fluorescent light emitted as particles move across the irradiated volume can thus be related to the diffusion coefficients of the fluorescent species (fluorescence correlation spectroscopy). Unfortunately, information on the momentum transfer (equation (19.51)) is lost in this stimulated emission method and, for larger scattering volumes of biological material, data from quasielastic scattering contains much more information.

Fluorescence depolarisation experiments are a further powerful tool of dynamics. A pulsed laser excites fluorescent probes attached to biological molecules, whose motion can be detected by the change in polarisation of the re-emitted photon. If the molecule reorients itself a considerable amount over the picosecond time scale of fluorescent emission, the polarisation state of the emitted photon is changed. The utility of fluorescence depolarisation stems from the fact that the technique can probe dynamics in the ultrafast picosecond time regime due to the availability of intense ultrafast pulsed laser sources (**Figure 19.59**). The high yield of fluorescent re-emission makes the experimental measurement of correlation functions over ultra-fast time scales feasible.

Quasielastic scattering experiments cover a wide range of techniques that monitor small energy changes in scattered radiation due to the motion of a sample, i.e. a Doppler shift of the energies of scattered photons. Typically it is the normalised intermediate scattering function ($F(\underline{q},t)$) that is measured in a quasielastic scattering experiment (**Figure 19.60**). The intermediate scattering function is a useful general tool, which is amenable to accurate quantitative theoretical analysis and its origins can be found in studies of quantum optics. For the quasielastic scattering of coherent light (also called photon correlation spectroscopy or dynamic light scattering) the quantity examined experimentally ($g_1(t)$, the field correlation function) relates to the correlation function of the scattered electric field. In terms of the scattered electric field (E) at time (t) the field correlation function measured at a certain angle is defined as

$$g_1(t) = \frac{\langle E^*(0)E(t) \rangle}{\langle I \rangle} \tag{19.70}$$

where <> is the ensemble-averaged quantity over the array of scatterers, and I is the intensity measured on the detector. The electric-field strength ($E(t)$) scattered by the collection of moving particles in the sample is given by

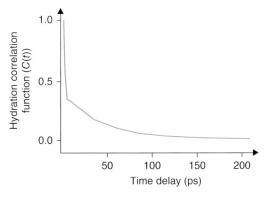

Figure 19.59 *Ultrafast dynamics can be probed using fluorescence depolarisation pulsed laser experiments with the protein subtilisin Carlsberg. A correlation function is plotted as a function of the delay time. [Reproduced with permission from S.K. Pal, J. Peon, B. Bagchi, A.H. Zewail, J. Phys. Chem. B 2002, 106, 12376–12395. Copyright © 2002, American Chemical Society.]*

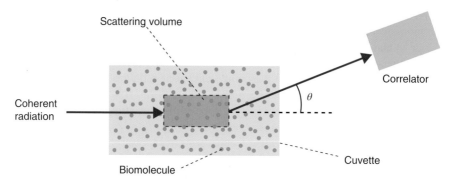

Figure 19.60 *Schematic arrangement of a dynamic light scattering experiment. Coherent light is Doppler shifted by the motion of the biomolecules in the illuminated scattering volume and this shift is subsequently detected using the correlator placed at an angle θ.*

$$E(t) = \sum_{j=1}^{N} A_j e^{i\underline{q} \cdot \underline{r}_j} E_0 e^{-i\omega_0 t} \tag{19.71}$$

where E_0 is a constant, A_j is the scattered amplitude, q is the momentum transfer, t is the time, and ω_0 is the frequency of the incident radiation. The sum from $j = 1,...,N$ is over all the scatterers in the sample. The electric field can be substituted in equation (19.70) and the correlation function can therefore be written

$$g_1(t) = \left\langle e^{i\underline{q} \cdot [\underline{r}_j(t) - \underline{r}_j(0)]} \right\rangle e^{-i\omega_0 t} \tag{19.72}$$

For particles that experience a simple process of diffusion, an analytic equation for the field correlation function can be calculated. Fick's second law of diffusion (equation (7.18)) in three dimensions is

$$\frac{\partial c(\underline{r},t)}{\partial t} = D\nabla^2 c(\underline{r},t) \tag{19.73}$$

where $c(\underline{r},t)$ is the concentration of molecules in the scattering volume and D is the translational diffusion coefficient. Let $P(O/\underline{r},t)$ be the conditional probability that a particle can be found in volume element $d^3\underline{r}$ at time t. For low particle concentrations the conditional probability $P(O/\underline{r},t)$ also obeys the diffusion equation,

$$\frac{\partial P(O|\underline{r},t)}{\partial t} = D\nabla^2 P(O|\underline{r},t) \tag{19.74}$$

The Fourier transform of either side of (19.74) can be taken, which allows the equation to be solved,

$$\int_0^\infty e^{i\underline{q} \cdot \underline{r}} \frac{\partial P(O|\underline{r},t)}{\partial t} d^3 r = D \int_0^\infty e^{i\underline{q} \cdot \underline{r}} \nabla^2 P(O|\underline{r},t) d^3 r \tag{19.75}$$

A general property of Fourier transforms is that the Fourier transform of any nth-order differential ($\partial^n/\partial y^n$) is equal to $(-iq)^n$ times the Fourier transform of the argument of the differential,

$$\int_{-\infty}^\infty e^{iqy} \frac{\partial^n}{\partial y^n} z(y) dy = (-iq)^n \int_{-\infty}^\infty e^{iqy} z(y) dy \tag{19.76}$$

where $z(y)$ is an arbitrary (but well-behaved) function. The *intermediate scattering function* $(F(q,t))$ is normally the quantity most amenable to theoretical calculation in a quasielastic scattering experiment. It is equal to the Fourier transform of the probability distribution $(P(O|\underline{r},t))$,

$$F_s\left(\underline{q},t\right) = \int_0^\infty P\left(O|\underline{r},t\right)e^{i\underline{q}\cdot\underline{r}}d^3r \tag{19.77}$$

This definition and the Fourier transform identity (19.76) allows equation (19.75) to be simplified in terms of the intermediate scattering function,

$$\frac{\partial F_s\left(\underline{q},t\right)}{\partial t} = -Dq^2 F_s\left(\underline{q},t\right) \tag{19.78}$$

This is a simple variables-separable differential equation that has the solution

$$F_s\left(\underline{q},t\right) = F_s\left(\underline{q},0\right)e^{-Dq^2t} \tag{19.79}$$

where the initial condition is given by $F_s(\underline{q},0) = 1$. From equation (19.72) the field correlation function for a collection of diffusing particles can therefore be written as

$$g_1(t) = F_s\left(\underline{q},t\right)e^{-i\omega_0 q} = e^{-Dq^2t}e^{-i\omega_0 t} \tag{19.80}$$

The intensity correlation function $(g_2(t))$ is the quantity typically measured by time correlation of the Doppler shifted signal on a detector and is related to the electric field correlation function $(g_1(t))$ by the Siegert relationship,

$$g_2(t) = 1 + |g_1(t)|^2 \tag{19.81}$$

which is obeyed by most samples if they are irradiated by Poisson distribution light sources, i.e. most LEDs, lamps and lasers, but not more exotic squeezed-light sources. By substitution of equation (19.80) in equation (19.81) it is seen that diffusion introduces a e^{-Dq^2t} term in the intensity correlation function, $g_2(\tau)$. **Figure 19.61** shows a typical correlation function for a fibrous protein that experiences translational diffusion and can be described by such an exponential decay. A qualitative understanding of the form of a correlation function can be achieved. At short times the particles have not moved anywhere, they do not dephase the scattered light, and the correlation is nearly perfect $g_1(\tau) \sim 1$. At longer times the motion of the sample decorrelates the phase of the scattered radiation and eventually the correlation function reduces to zero, which indicates a completely random decorrelation of the scattered photons.

A wide range of other radiation can be used in quasielastic scattering experiments in addition to visible light including X-rays (time scales $\sim 10^{-7}$–1000 s)), infrared (time scales $\sim 10^{-9}$–1000 s) and neutrons (time scales $\sim$0.1–100 ns) to probe a large number of dynamic processes in biological materials at the nanoscale. X-ray and infrared quasi-elastic techniques are similar in conception to dynamic light scattering (which also probes time scales $\sim 10^{-9}$–1000 s), whereas neutron spin-echo measurements use the spin of scattered neutrons to clock the dynamics of the scattering process.

Many more recent developments have occurred in photon correlation spectroscopy. Phase-locked optoelectronics allow sensitive measurement of zeta potentials (the electric potential around charged particles at their hydrodynamic diameter, i.e. the slipping plane) in photon correlation spectroscopy electrophoresis zeta sizers. Ultrafast optoelectronics provides methods to measure the phase and intensity of scattered light all the way down to femtosecond time scales, e.g. frequency-resolved optical grating (FROG), which uses a nonlinear

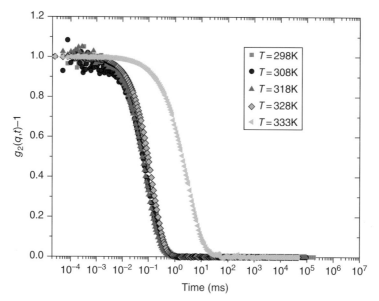

Figure 19.61 *The intermediate scattering function (intensity correlation, $g_2(q,t)$) as a function of delay time from a quasielastic light scattering experiment on titin, a giant protein from skeletal muscle. The dynamics slow down as the protein unfolds with increasing temperature (T) due to the increase in protein length. [Reproduced from E. Di Cola, T.A. Waigh, J. Trinick, L. Tskhovrebova, A. Houmeida, W. Pyckhout-Hintzen, C. Dewhurst, Biophysical Journal, 2005, 88, 4095–4106. With permission from Elsevier.]*

optical medium to create a spectrally resolved autocorrelation function. Spatially resolved imaging combined with photon correlation spectroscopy is also possible, using the correlation of speckle patterns on optical microscopy images created with laser illumination.

19.12 Osmotic Pressure

It is important to understand the phenomena associated with osmotic pressure in biological processes, since it determines how cell metabolisms are regulated (animal cells walls are ruptured if the external osmotic pressure is too high or too low), how intermolecular forces are mediated by solvent molecules, and the molecular crowding of the intracellular environment. Consider an idealised experiment with a semipermeable membrane that separates two polymer solutions shown in **Figure 19.62**. This apparatus is an example of a membrane osmometer, a device for the measurement of osmotic pressure. From standard thermodynamic theory the partial differential of the Gibbs free energy ($G = F–TS + PV$) with respect to the pressure (P) is equal to the volume (V) at constant temperature (T),

$$\left(\frac{\partial G}{\partial P}\right)_T = V \tag{19.82}$$

The chemical potential (μ, the Gibbs free energy per particle) with respect to one of the components (μ_1) is

$$\left(\frac{\partial \mu_1}{\partial P}\right)_T = \bar{V}_1 \tag{19.83}$$

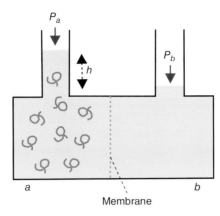

Figure 19.62   *A membrane osmometer can be used to measure the difference in osmotic pressure between two solution filled compartments, a and b, which have external pressures P_a and P_b applied, respectively. The difference in fluid heights (capillary rise) provides the osmotic pressure ($\pi = h\rho g$), where ρ is the fluid density and g is the acceleration due to gravity.*

where $\bar{V}_1$ is the partial molar volume of these particles. Consider the chemical potentials on either side of the membrane in sections a and b (**Figure 19.62**). *Section b* only contains solvent molecules and the chemical potential is that of the solvent (μ_1^0),

$$\mu_1^b = \mu_1^0 \tag{19.84}$$

In *section a* the chemical potential has two additional effects, that due to the solute and that due to the reservoir used to measure the pressure,

$$\mu_1^a = \mu_1^0 - (\text{solute\_effect}) + (\text{pressure\_effect}) \tag{19.85}$$

The effect of the solute on the chemical potential is given by a van der Waals type expansion in the component concentration and the pressure effect is given by an integral,

$$\mu_1^a = \mu_1^0 - RTV_1^0\left(\frac{c}{M} + Bc^3 + ..\right) + \int_{P_0}^{P_0+\pi} \bar{V}_1\,dP \tag{19.86}$$

where c is the concentration of the species too large to permeate the membrane (the solute), $V_1^0 = \bar{V}_1$ is the molecular volume of solvent at one atmosphere pressure, M is the molecular weight of the solute and B is the second virial coefficient of the solute. In thermal equilibrium the chemical potentials on each side of the membrane are equal and $\int_{P_0}^{P_0+\pi} \bar{V}_1\,dP = V_1^0\pi$, ($\pi$ is the osmotic pressure) therefore

$$\mu_1^0 = \mu_1^0 - RTV_1^0\left(\frac{c}{M} + Bc + ...\right) + V_1^0\pi \tag{19.87}$$

This expression can be solved for the osmotic pressure (π) of the solution,

$$\pi = RT\left(\frac{c}{M} + Bc^2 + ...\right) \tag{19.88}$$

For a dilute solution the second virial coefficient is very small ($B = 0$), so to a good approximation

$$\pi = \frac{RTc}{M} \tag{19.89}$$

The osmotic pressure of a solution is directly proportional to the number of solute molecules (c/M) and knowledge of the concentration allows an accurate determination of the molecular weight. The osmotic pressure of a protein solution is shown in **Figure 19.63**. With the simple osmometer shown in **Figure 19.62**, the osmotic pressure (π) can be calculated using

$$\pi = hg\rho \tag{19.90}$$

where h is the difference in fluid height, g is the acceleration due to gravity and ρ is the fluid density. Thus, a simple measurement of the height of the fluid in the capillary leads to a direct calculation of the solution's osmotic pressure, which can subsequently be connected with the molecular weight.

The effects of osmotic pressure are extremely important for the determination of the physical state and morphology of a biological material. For example, when a charged polyelectrolyte gel is placed in a dilute solution it can expand many times in volume due to the contribution of the counterions to the osmotic pressure. Simple macroscopic measurements of gel sizes thus provide a useful tool to understand their molecular structure (**Figure 19.64**).

There are a number of other methods typically used to measure osmotic pressure. These include the *vapour pressure osmometer* (depression of the boiling point by the osmotic effect), and *optical tweezers* (the piconewton pressures on colloidal particles can be measured directly).

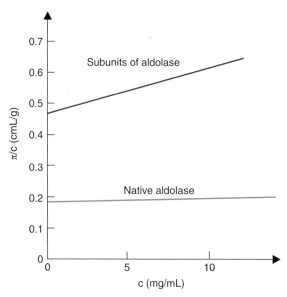

Figure 19.63 *The osmotic pressure of a solution of globular proteins. A number of aldolase subunits assemble into the native structure that causes a decrease in the osmotic pressure of the solution. The osmotic pressure per gram of protein (π/c) is plotted as a function of protein concentration (c). [Reproduced with permission from F.J. Castellino, O.R. Baker, Biochemisty, 1968, 7, 2207–2217. Copyright © 1968, American Chemical Society.]*

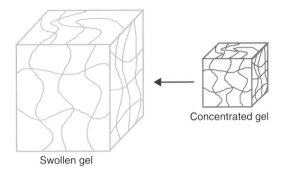

Concentrated gel

Swollen gel

Figure 19.64   *The swelling of polymer gels is driven by the osmotic pressure of the network and the swelling is resisted by the elasticity of the polymer chains. The osmotic pressure of counterions is the dominant force that swells polyelectrolyte gels.*

19.13   Chromatography

Chromatography is a standard method in organic chemistry and is commonly used in biochemistry. In liquid chromatography a mixture of solutes is flowed through a packed arrangement of beads in a column. The slightly different interaction strength between the different solutes and the beads leads to different residence times in the column, which in turn lead to different arrival times on a detector placed downstream. Thus, the mixture is separated up into fractions, which can then be studied in more detail with another analytical technique, e.g. light scattering, electrophoresis, or mass spectroscopy.

19.14   Electrophoresis

Electrophoresis is a cheap, powerful tool for the analysis and separation of charged biological molecules such as proteins and nucleic acids (**Section 2.12**). Electrophoresis can be used to measure the size of biopolymer molecules and deduce the chemical sequence of the chains. The force experienced by a particle (F) in an electric field (E) is given by Coulomb's law,

$$F = ZeE \tag{19.91}$$

where Z is the number of charges on the particle, and e is the electronic charge. The mobility of a charged particle in an electric field is proportional to the ratio of the net charge on the particle (which provides a Coulombic force) to its frictional coefficient. Electrophoresis can be used to obtain information about either the relative charge or the relative size of charged molecules. For steady-state electrophoretic motion the frictional force (the frictional coefficient (μ) multiplied by the velocity (v), μv) is balanced by the force due to the electric field. The electrophoretic mobility (U) with colloids is defined as

$$U = \frac{v}{E} = \frac{Ze}{\mu} \tag{19.92}$$

The Stokes law for the frictional force (equation (7.11)) can be inserted into this equation to give

$$U = \frac{Ze}{6\pi\eta R} \tag{19.93}$$

Thus, the mobility of the colloids measured in an electrophoresis experiment can be related to the charge fraction (Z) and the radius of the particles (R).

Conceptually, the simplest method to measure the mobility of a colloidal particle is by using *moving-boundary electrophoresis* (**Figure 19.65**). Particle velocities are measured directly with an optical microscope as they move in an electric field in a dilute solution. However, this technique suffers from artifacts such as convection and multicomponent interactions. It is possible to circumvent these problems using gels and ion-exchange papers, and these are the electrophoresis methods that are predominently used.

The charge on a protein depends on the pH of the buffer. Electrophoretic motion can be studied as a function of pH to calculate the isoelectric point. The isoelectric point is the pH at which the average net charge on a macromolecule is zero (**Figure 11.2**). Isoelectric focusing can provide quantitative molecular information on charged macromolecules using simple table top apparatus.

A standard method to measure the size of DNA chains is with *gel electrophoresis* (**Figure 19.66**, as introduced briefly in **Chapter 2**). The use of the gel removes the problems with convection of free boundary electrophoresis. The gel is placed across a constant applied voltage in a salty solution and the DNA chains are loaded onto the gel near the negative electrode. The gel is 'run' for a fixed period of time and the mobility of the chains (the time taken to travel a certain distance) can be simply related to their size. Surprisingly, detailed

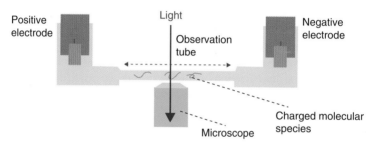

Figure 19.65 *A moving-boundary apparatus for the examination of free-solution electrophoresis. The motion of the colloids that experience electrophoresis is measured with an optical microscope.*

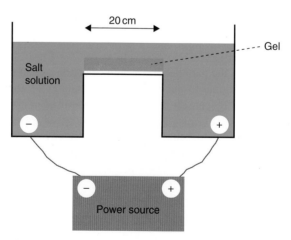

Figure 19.66 *Arrangement of a gel used in a simple electrophoresis experiment to measure the mobility of polyions loaded on the gel, e.g. DNA. The potential difference across the gel is maintained using a battery.*

information on the complex topological nature of the gel is not required for quantitative predictions to be made on the molecular weight of DNA chains as they move through the gel. This is due to the statistics of the reptation dynamics of the DNA chains in the gel and is discussed in more detail later in the section.

There are two common ways to locate DNA chains on a gel to measure the distance they have travelled. Ethidium bromide can be used to label the chains and these labels fluoresce strongly under ultraviolet light. Alternatively, it is possible to incorporate radioactive phosphorus atoms into DNA, at one of its ends, that will darken a photographic film.

An illustrative example is provided by electrophoresis with supercoiled bacterial DNA molecules. Gel electrophoresis is a relatively easy method to separate closed supercoiled DNA from the relaxed (cut) molecules. There is a large increase in the mobility of the supercoiled DNA due to its compact form and it therefore experiences a reduced frictional coefficient (μ, equation (19.92)) compared with the extended relaxed form.

For detailed sequencing of DNA chains restriction enzymes are used. These enzymes can cut the DNA chains whenever they find the GAATTC sequence. If there are n such sequences there are $n + 1$ bands. Other specific enzyme/DNA reactions allow individual DNA molecules to be cut in different places and the resultant information can be combined to sequence chains of up to lengths of around 400 base pairs.

Isoelectric focusing is also possible with gel electrophoresis and it can be used as a useful separation technique if the bands that contain the required charged molecules are cut out of the gel.

The theory of *reptation* is used to explain the ability of gel electrophoresis to separate DNA chains of different length (**Figure 19.67**, **Section 10.5**). The components of the electric force perpendicular to the axis of the tube are cancelled by the tube reaction force and the longitudinal components induce an electrophoretic motion of the chain along the tube (forced electrophoretic reptation).

In the limit of *very strong* electric fields the front end of the DNA chain moves forward and creates new parts of the tube (**Figure 19.68a**). The stretching force (f) is proportional to the number of monomers (N), since the total electric charge (Z) on the chain is proportional to N. The coefficient of friction (μ) for the whole chain is also proportional to N (as for reptation, $\mu \propto N$). Thus, the speed of motion (v) in a strong electric field is

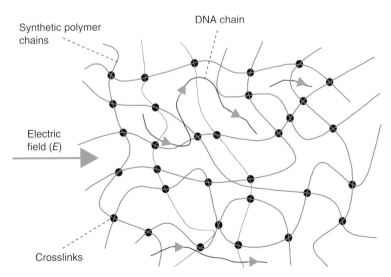

Figure 19.67   *The electrophoresis of DNA fragments across a crosslinked gel is driven by an electric field. Smaller chain fragments migrate more quickly than longer chain fragments. A process of driven reptation occurs.*

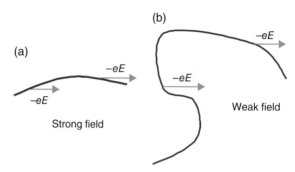

Figure 19.68 *Schematic diagram that indicates the difference in conformation of charged macromolecules during electrophoresis in a gel. (a) In a strong electric field the chains are completely stretched and electrophoresis is not a sensitive measure of the mobility. (b) In a weak field the chains adopt a Gaussian conformation and mobility measurements are much more successful. eE is the electrostatic force on a section of a chain, e.g. a single dissociated phosphate group on a DNA chain.*

independent of N ($v \sim Z/\mu \sim N^0$, equation (19.92)). The method of strong-field electrophoresis is therefore not useful for the separation of DNA fragments, since it is independent of chain length.

However, in a *weak* electric field, gel electrophoresis (**Figure 19.68b**) is much more successful, since the DNA molecules remain Gaussian coils. The force the electric field exerts on the DNA molecules (f) is proportional to the displacement of the chain parallel to the electric field ($N^{1/2}$), since only the motion of segments of the chain parallel to the electric field are not restricted by the crosslinks of the gel. The speed of reptation (v_r) is

$$v_r = \frac{f}{\mu} \sim \frac{N^{1/2}}{N} \sim N^{-1/2} \qquad (19.94)$$

where N is the number of monomers in the chain. The speed of the centre-of-mass motion (v) is a factor of $N^{1/2}$ slower than the speed of reptation, and the speed of centre-of-mass reptation if therefore inversely proportional to the chain length ($v \sim 1/N$) in a weak field. Weak-field electrophoresis thus provides a practical method for the separation of DNA chains. A more accurate calculation of the velocity of the DNA fragments in a weak field gives

$$\underline{v} = \frac{q}{3\eta}\left(\frac{1}{N} + \text{const}\left(\frac{EqL}{kT}\right)^2\right)\underline{E} \qquad (19.95)$$

where $\underline{E}$ is the electric field vector, q is the charge per unit length of the DNA, N is the length of the DNA chain, L is the length of the Kuhn segment, η is the viscosity of the medium, and the constant is of the order of unity. Gel electrophoresis is therefore not very sensitive for the separation of long DNA chains (**Figure 19.69**), which is a big problem if micrometre-long pieces of genomic DNA require sequencing.

A trick to increase the sensitivity of electrophoresis for the separation of long DNA chains is to periodically switch off or rotate by $90°$ (pulsed electrophoresis), the applied external field at the time scale for the renewal of the reptation tube ($\tau^* \sim N^3$, **Section 10.5**). In this case electrophoretic motion only occurs for chains of the length (N) defined by the periodicity of rotation of the electric field and this effect can be used to select the longer chains.

The *polymerase chain reaction* (PCR) is a biochemical technique for the amplification of short (10 k base pairs) stretches of nucleic acid (**Section 2.13**). It is often used in conjunction with electrophoresis methods in

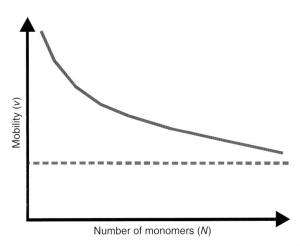

Figure 19.69   *Dependence of the mobility of DNA fragments (v) on the number of monomers in a chain (N) during a gel electrophoresis experiment. The mobility becomes an insensitive function of N as the chains increase in length.*

the process of genetic fingerprinting. Thus, a single DNA molecule can be amplified to provide sufficient quantities of DNA to be sequenced using electrophoresis by means of a PCR technique.

SDS electrophoresis can be used to obtain the molecular weights of proteins. SDS is a surfactant that is an effective protein denaturant. It binds to all proteins approximately to the same degree and causes them to adopt extended conformations. The apparent electrophoretic mobility ($u(c)$) of a denatured protein at a particular gel concentration (c) is phenomenologically given by

$$\ln u(c) = -k_x c + \ln u(0) \tag{19.96}$$

where k_x depends on the extent of crosslinking of the gel and $u(0)$ is a constant for a particular protein. The mobility ($u(0)$) is related to the molecular weight (M) through a simple relationship

$$u(0) = b - a\log M \tag{19.97}$$

where b and a are standard constants. The molecular weight of a denatured protein can therefore be calculated from the measurement of its mobility on a gel at a series of different gel concentrations.

There have been a number of modern developments in electrophoretic techniques. Problems with convection in free-boundary electrophoresis can be reduced by using a very fine bored capillary in *capillary electrophoresis* (**Figure 19.70**). This is a useful microanalytic separation technique. The importance of convection in a fluidic system is described by a dimensionless group, the Rayleigh number (Ra), which is defined as

$$\mathrm{Ra} = \frac{R^4 g}{\eta \alpha}\left(\frac{\Delta\rho}{\Delta r}\right) \tag{19.98}$$

where R is the radius of the channel, g is gravity, η is the viscosity, α is the thermal diffusivity of the medium, and ($\Delta\rho/\Delta r$) is the density change per unit radial distance caused by heating. For small Rayleigh number ($\mathrm{Ra} < 1$) convection is suppressed in an electrophoresis tube and this corresponds to a small capillary bore (R). Typically capillary diameters for electrophoresis experiments are of the order of a few micrometres to provide low Rayleigh number dynamics.

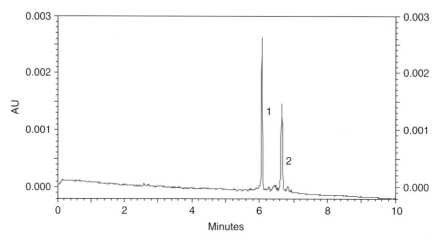

Figure 19.70 *Typical data from a capillary electrophoresis experiment that shows the separation of two double-stranded DNAs (1) is AAATTATATTAT/ATAATATAATTT and (2) is GGGCCGCGCCGC/GCGGCGCGGCCC. Sample (1) travels down the column faster than sample (2) due to a higher electrophoretic mobility. [Reproduced from I.I. Hamden, G.G. Skellern, R.D. Waigh, Journal of Chromatography, 1998, 806, 1, 165–168. With permission from Elsevier.]*

Etched obstacle arrays on silicon chips can also be used for electrophoresis (**Figure 19.71**). Silicon microarrays offer a number of advantages over standard gel techniques; smaller samples can be explored, and the microstructure of the etched silicon can be better defined than with gels and consequently so too can the microfluidics.

Other high-throughput technologies have been developed around microfluidic devices. Electrophoresis through membrane nanopores, allows single chains to be sequenced by the measurement of the change in current that accompanies DNA chain motion through a single nanopore. As electrophoresis technologies improve, so too does the cost of whole-genome sequencing, which has wide implications for human health.

19.15 Sedimentation

Sedimentation is a key separation technique used to extract a particular biomolecule of interest from the complex soup of species found in the cell. Separation by sedimentation is a standard first step in a molecular biophysics experiment. An external force acts on a mixture of suspended particles to separate them by means of their varying buoyancies with respect to the solvent in which they are suspended. In an analytical ultracentrifuge the radial acceleration provides the external force and causes the molecules to be separated as a function of both their density and shape (**Figure 19.72**).

From simple Newtonian mechanics the radial force (F) on a suspended particle rotated in a centrifuge is given by

$$F = m^* \omega^2 r \qquad (19.99)$$

where m^* is the particle's effective mass, ω is the angular velocity at which it is rotated, and r is the radial distance of the particle from its centre of motion. From Archimedes' principle, the mass of a particle (m) suspended in a

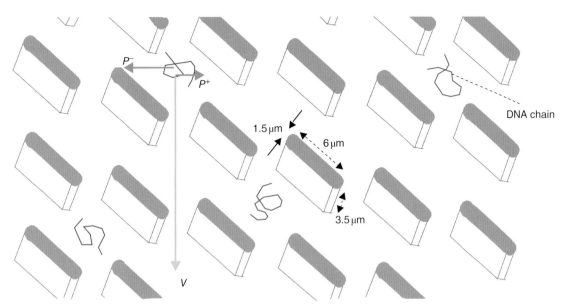

Figure 19.71 *Etched microarrays used for electrophoresis experiments. An electric field moves DNA molecules vertically downwards with a velocity v. Due to the anisotropic nature of the obstacle orientation there is a larger probability for the molecules to change channels to the right (P⁺) than to the left (P⁻), and this difference in probabilities is a function of the chain size. [Reproduced with permission from C.F. Chou, R.H. Austin, O. Bakajin et al, Electrophoresis, 2000, 21, 81–90. Copyright © 2000 Wiley-VCH Verlag GmbH, Weinheim, Germany.]*

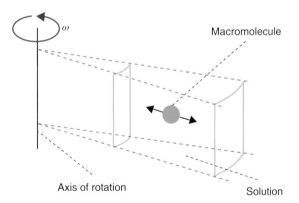

Figure 19.72 *Schematic diagram of a sedimentation experiment. The sample is rotated about an axis at an angular velocity ω and macromolecules adjust their radial position in the solution due to their relative densities.*

solvent needs to be corrected by the density of the surrounding solvent and the effective mass of the particle (m^*) is given by

$$m^* = m(1 - \nu\rho_0) \tag{19.100}$$

where ν is the partial specific volume of the molecule, and ρ_0 is the solvent density. The velocity at which the particles move in the centripetal force is given by equation (19.99) divided by the frictional resistance,

$$v = \frac{dr}{dt} = m^* \omega^2 \frac{r}{f\eta_0} \qquad (19.101)$$

where f is a frictional coefficient and η_0 is the solution viscosity. The variation of the sedimentation velocity (v) with particle size and density forms the basis of a method to separate particles using sedimentation. When a centrifugal field is applied to a solution of molecules, a moving boundary is formed between the solvent and the solute. This boundary travels down the cell with a velocity determined by the sedimentation velocity of the macromolecules. Concentration gradients can be accurately measured using ultraviolet absorption (**Figure 19.73**) and therefore the sedimentation velocities can be calculated. The velocity of sedimentation (dr_b/dt) is equal to the rate of motion of the boundary,

$$\frac{dr_b}{dt} = r_b \omega^2 s \qquad (19.102)$$

where s is the sedimentation coefficient, equal to the velocity of sedimentation divided by the centrifugal strength ($\omega^2 r_b$). Integration of equation (19.102) provides an expression for the position of the boundary as a function of time,

$$\ln\left[\frac{r_b(t)}{r_b(t_0)}\right] = \omega^2 s(t - t_0) \qquad (19.103)$$

Diffusion broadens the boundary as it progresses down the column (**Figure 19.74**) and the rate of motion provides the sedimentation coefficient from equation (19.103). The sedimentation coefficient depends on the size, shape and degree of hydration of a macromolecule. For globular proteins there is a well defined relationship between the sedimentation coefficient and the molecular weight.

It is also possible to make focusing measurements with sedimentation experiments if the particles are suspended in a gradient of dense salt, e.g. a solution of CsCl or $CsSO_4$. Particles collect together in a narrow

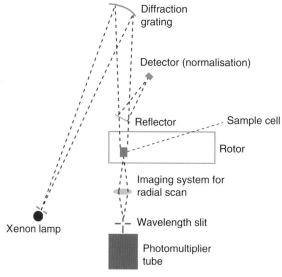

Figure 19.73 *Experimental arrangement of a modern analytic centrifuge with a Xenon lamp. UV absorption allows particle concentrations to be measured. The centrifuge can separate particles by relative size and relative density.*

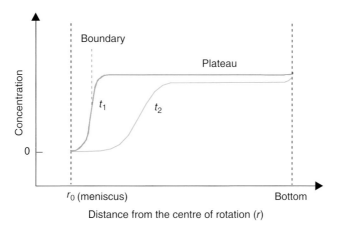

Figure 19.74 *Schematic diagram that shows the progress of a moving-boundary sedimentation experiment at different time steps (t_1 and t_2, $t_2 > t_1$). Particle concentration is shown as a function of the distance from the centre of rotation (r). The boundary moves towards the centre of rotation (the cell bottom) as a function of time.*

band at the point of matching buoyancy. The calculation of the sedimentation profile during a centrifugation experiment is an elegant illustration of the predictive power of equilibrium statistical mechanics (**Chapter 3**). The work required ($E(r)-E(r_0)$) to lift a particle from a radius r_0 to a radius r in a centrifugal force field is equal to the work done against the centrifugal force,

$$E(r) - E(r_0) = -\int_{r_0}^{r} m^* r \omega^2 dr = m^* \omega^2 \frac{(r_0^2 - r^2)}{2} \tag{19.104}$$

where m^* is the effective mass of the particles adjusted for the solvent density. In thermal equilibrium, the range of concentrations ($c(r)$) as a function of the radius is given by a simple Boltzmann distribution ($e^{-E'/kT}$) and therefore

$$\frac{c(r)}{c(r_0)} = \exp\left(-m^* \omega^2 \frac{[r_0^2 - r^2]}{2kT}\right) \tag{19.105}$$

The density near the radius of a particular band (r_b) can be expressed as a Taylor expansion,

$$\rho(r) = \rho(r_b) + \frac{\partial \rho}{\partial r}\big|_{r=r_b}(r-r_b) + \dots \tag{19.106}$$

Substitution of the density expansion in equation (19.100) allows equation (19.105) to be expressed as

$$\frac{c(r)}{c(r_b)} = \exp\left[-m r_b \omega^2 \bar{v} \rho'(r_b) \frac{(r-r_b)^2}{2kT}\right] \tag{19.107}$$

where $\rho'(r_b)$ is the density gradient defined by

$$\rho'(r_b) = \frac{\partial \rho}{\partial r}\big|_{r=r_b} \tag{19.108}$$

and m is the true mass of the particles. The concentration profile at radius (r_b) at which the band of particles occurs is therefore a Gaussian distribution with standard deviation (σ_r) and is given by

$$\sigma_r = \frac{kT}{\left[mr_b\omega^2\bar{v}\rho'(r_b)\right]^{1/2}} \qquad (19.109)$$

The band of particles is narrow and well focused for particles of large mass (m), in high centrifugal fields (large $r_b\omega^2$) and in steep density gradients (large $\rho'(r_b)$). Sedimentation focusing is thus another extremely useful technique for particle separation.

19.16 Rheology

All materials demonstrate flow behaviour intermediate between solids and liquids (**Chapter 15**). Rheology is the study of this phenomenon of *viscoelastic* flow and rheometers are instruments for the measurement of the rheology of materials.

There are two broad categories of techniques to measure the viscoelasticity of a material. First, there are *bulk methods* where the response of a macroscopic amount of a material to an externally applied stress or strain is recorded. These bulk methods have traditionally been used to examine the viscoelasticity of biological samples. Secondly, there is the measurement of the viscoelasticity of a sample as a function of length scale using *microrheology techniques*. Typically probes are injected into the system of interest that are passive (e.g. marker colloids) or active (e.g. magnetic colloids). The motion of the probes is recorded with a CCD camera or measured with light scattering, and the resultant fluctuation spectrum of the particle displacements is related to the viscoelasticity of the material in which they are embedded.

In *bulk rheology* experiments a series of standard geometries are often used and each tends to have different advantages in terms of the mechanism of sample loading, the time window that can be explored and the sensitivity of the measurements (**Figure 19.75**). Different geometries for rheometers require different corrections to analyse the dependence of the stress on the strain. How the geometry grips the sample is also important and, combined with the type of force or displacement transducers, this determines the sensitivity of the measurements. Bulk rheometers measure the large-scale viscoelastic properties of assemblies of biological molecules. Rheometers can function in linear (**Chapter 15**) or nonlinear modes. Nonlinear rheology corresponds to large deformations and deformation rates, in which both Deborah and Peclet numbers can be appreciable. Turbulent phenomena can occur at high shear rates.

In *drag-flow bulk rheometry* the velocity or displacement of a moving surface is measured simultaneously with the force on another surface that moves in response to its motion. Couette's original concentric-cylinder drag-flow rheometer was a controlled-strain drag-flow rheometer. The angular velocity of the outer cup was fixed and the torque on the inner cylinder was measured from the deflection of a suspended wire. The measured variable in a controlled-strain rheometer is the torque. Couette measured the twist in a torsion bar, whereas modern electronic rheometers use a linear variable differential transducer to do the same job (**Figure 19.76**).

More sophisticated modern rotary drag-flow rheometers measure normal stresses (the stresses normal to the direction of shear). It is an experimental challenge to find steady-state behaviour with normal stresses and they are easily disturbed by fluctuations in the temperature and axis of rotation. Rheometers thus need to be machined to high precision. Often, commercial rheometers are mechanically accurate to within 2 μm over the 25 mm cup diameter. Control of the torque combined with measurement of the angular motion in a controlled-stress rheometers, is also a standard technique in rotational rheometry. Furthermore, it is important to control the temperature, pressure and humidity to make accurate rheological measurements with biological specimens.

The most commonly measured linear viscoelastic material function is the complex shear modulus, $G^*(\omega)$ (**Section 15.2**). There are three standard techniques to measure $G*$; in the *shear wave propagation* method the time for a pulsed deformation to travel through a sample is measured, alternatively the sample can be made to

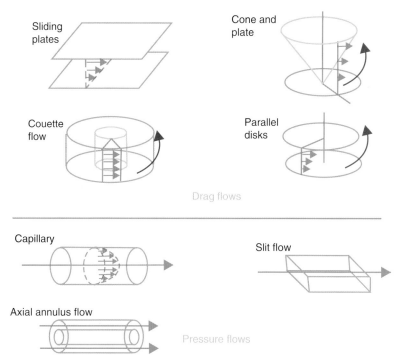

Figure 19.75 *Schematic diagram of the flow geometries commonly encountered in rheology experiments. The geometries are separated between* drag flows, *where the surfaces move relative to one another, and* pressure flows, *in which the flow rate is determined by the pressure drop across the pipe. [Adapted with permission from C.W. Macosoko, Rheology: Principles, Measurements and Application, 1994, Wiley, VCH. Copyright © 1994, John Wiley & Sons.]*

oscillate at its *resonant frequency* and the response at this single frequency observed, or the *forced response* to a sinusoidal oscillation in stress/strain can be measured in terms of the resultant strain/stress (**Figure 19.77**). Forced-response devices are better suited to low elasticity materials, such as polymer solutions and soft biomaterials.

There are two basic design types of *pressure-driven rheometers* (**Figure 19.75**). One features the control of the pressure and the measurement of the flow rate (e.g. capillary rheometers) and the other uses controlled flow rate and measures the pressure drop. Such capillary-type geometries have direct analogues in biological circulatory systems (e.g. blood flow), which motivates their use in biology.

Microrheology has experienced a number of important recent developments. Often, biological samples whose mechanical response is homogeneous on the macroscale, are inhomogeneous on the micrometre scale (e.g. living cells) and a range of microrheology techniques have been developed to measure this behaviour. The range of frequencies and moduli that can typically be accessed using the different microrheology techniques are shown in **Figure 19.78**.

Particle-tracking microrheology is practically the simplest microrheology technique to implement. It requires a CCD camera, an optical microscope, a high-magnification objective lens and some digital recording apparatus. The fluctuation–dissipation theory is used to relate the fluctuations in the displacements of tracer particles embedded in a material to the material's viscoelastic response. The fluctuation spectrum of the mean square displacements of colloidal particles embedded in a viscoelastic material is calculated as a function of

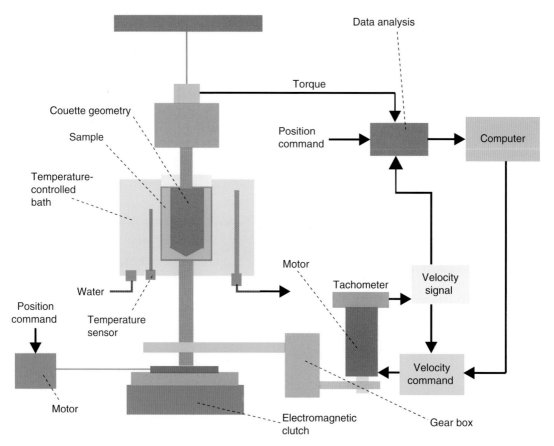

Figure 19.76 *Schematic diagram of a modern shear rheometer for the measurement of viscoelasticity. The sample is sheared in a Couette geometry. [Reproduced from L. Bohlin, Progress and trends in rheology II, ed H. Grieseku, M.F. Hibberd, Steinkopf, 1988, 161, Supplement to Rheologica Acta 1988. With kind permission from Springer Science + Business Media B.V..]*

time (**Figure 19.79**). For simple viscous liquids one would expect (**Section 7.1**) a linear dependence of the mean square displacement of the embedded probes ($<r^2>$) on the lag time (t),

$$\langle r^2 \rangle = 4Dt \tag{19.110}$$

In viscoelastic materials a sublinear diffusive process is observed at short times ($<r^2> \sim t^\alpha$, $\alpha < 1$, **Section 7.6**) and this includes information on the linear viscoelasticity of the material. The linear complex viscoelastic shear moduli (G^*) of the material can be subsequently calculated from the mean square displacement using the generalised Stokes–Einstein equation (compare with equation (7.10)),

$$G(s) = \frac{kT}{\pi a r^2(s)} \tag{19.111}$$

where $r^2(s)$ is the Laplace transform of the mean square displacement ($<r^2(t)>$) and $G(s)$ is the Laplace transform of $G(t)$. The Laplace frequency (s) has been introduced to provide a compact solution of the

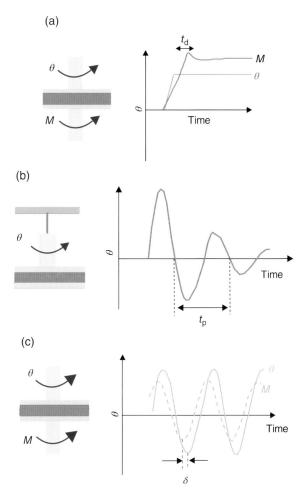

Figure 19.77 *Methods for the measurement of the complex shear modulus of a viscoelastic material include (a) wave speed, (b) resonance and (c) forced oscillations. [Adapted with permission from C.W. Macosoko, Rheology: Principles, Measurements and Application, 1994, Wiley-VCH. Copyright © 1994, John Wiley & Sons.]*

fluctuation–dissipation theorem. The complex shear moduli $G'(\omega)$ and $G''(\omega)$ can be determined mathematically by Fourier transform of the relaxation modulus as a function of time ($G(t)$). It is also possible to determine the linear viscoelastic spectrum by analysis of the mean square fluctuations in angular displacement of a probe as a function of time, and again a generalised Stokes–Einstein equation, this time for rotational motion, is used to calculate the complex shear modulus.

Laser deflection techniques allow the high-frequency viscoelastic behaviour of materials to be probed, again by measurement of the mean square displacements. Back focal-plane interferometry is a particularly sensitive laser deflection method for the measurement of small fluctuations (nm) of probe spheres that occur at high frequencies (**Figure 19.80**).

Multiply scattered laser light from colloidal spheres can also be used to yield the fluctuation spectrum of the displacement of colloidal spheres embedded in biological specimens using the technique of *diffusing wave*

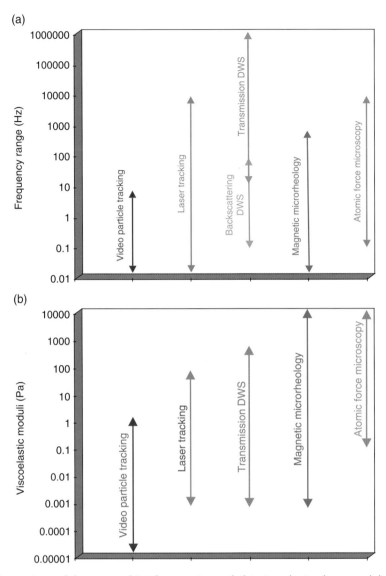

Figure 19.78 *Comparison of the range of (a) frequencies and (b) viscoelastic shear moduli that can typically be measured using different microrheology techniques. These are with reference to current standard setups, e.g. state-of-the-art laser tracking with balanced detectors and squeezed light offers orders of magnitude improvements in the highest frequency that can be measured. [Reproduced with permission from T.A. Waigh, Reports on Progress in Physics, 2005, 68, 685–742 © IOP Publishing. Reproduced with permission. All rights reserved.]*

spectroscopy (DWS, **Figure 19.81**). The intensity correlation function is measured as the autocorrelation of the scattered intensity (**Section 19.2**) and can be used to construct the mean square displacement ($<r^2(t)>$) of the probe spheres. The viscoelastic moduli can then be calculated in a similar manner to the particle tracking technique. DWS microrheology is useful for high-frequency viscoelastic measurements, since the multiple-scattering amplifies the sensitivity of the measurements to small particle displacements (Å) and thus to particle

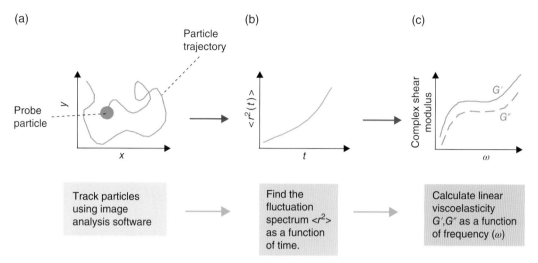

(a) (b) (c)

Particle
trajectory

Probe
particle

Track particles
using image
analysis software

Find the
fluctuation
spectrum $<r^2>$
as a function
of time.

Calculate linear
viscoelasticity
G', G'' as a function
of frequency (ω)

Figure 19.79 *The strategy used in particle-tracking micorheology experiments. (a) The trajectory of a fluctuating colloidal sphere is recorded, (b) the mean square displacement fluctuations ($<r^2>$) are calculated as a function of the lag time (t) and the (c) the complex shear moduli (G', G'') are found as a function of frequency (ω) using the generalised Stokes–Einstein equation (19.11).*

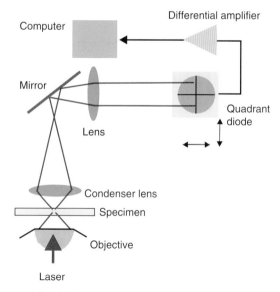

Figure 19.80 *The fluctuations of the displacement of a single bead embedded in a biological specimen can be followed with scattered laser light projected on to a quadrant diode. This provides a sensitive laser scattering microrheology technique.*

motions at high frequencies (MHz). Single scattering photon correlation spectroscopy techniques are also useful at lower colloidal concentrations and provide information on particle motions at slightly lower frequencies. DWS microrheology methods enable a very wide range of frequencies for the linear rheology of solution state biological materials to be accessed (**Figure 19.82**). Both transmission and backscattering

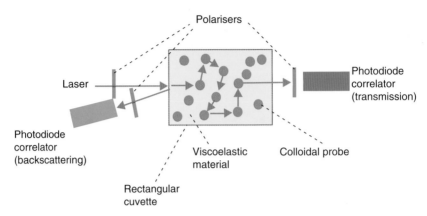

Figure 19.81   *Schematic diagram of a diffusing wave spectroscopy experiment. Coherent laser light is multiply scattered from a dense suspension of colloidal particles. Analysis of the resultant correlation functions can provide the high-frequency viscoelasticity of a biological specimen. Experiments can either be performed in transmission or backscattering geometries.*

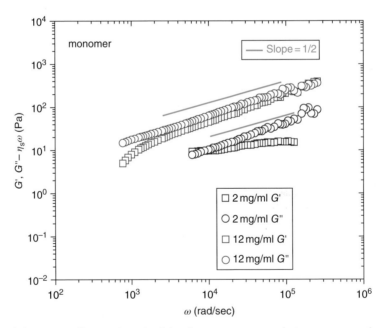

Figure 19.82   *High-frequency linear viscoelasticity from aggrecan solutions measured using transmission diffusing wave spectroscopy microrheology. The slope of the complex shear moduli $G' \sim G'' \sim \omega^{1/2}$ indicates that Rouse modes are present in these flexible polyelectrolyte solutions at high concentrations. Two separate concentrations are shown, 2 mg/mL and 12 mg/mL. [Reproduced with permission from A.P. Papagiannopoulos, T.A. Waigh, T. Hardingham, M. Heinrich, Biomacromolecules, 2006, 7, 2162–2172. Copyright © 2006, American Chemical Society.]*

geometries are possible for DWS experiments (**Figure 19.81**). Optical coherence tomography can also be used for microrheology experiments and allows large reductions in sample volumes with similar time sensitivity to DWS. Optical tweezers also find many applications in microrheology studies and are particularly well suited to the measurement of the viscoelasticity of membranes.

Further reduction in sample volumes for *nano-* and *picorheology* are possible, but data analysis often becomes more difficult due to the reduced sensitivity of the methods. Examples of submicrolitre rheometers that are currently being investigated are optical coherence tomography picrorheology with optical fibres (sensitive to picolitre volumes), fluorescent correlation spectroscopy (sensitive to picolitre volumes) and oscillatory atomic force microscopy (sensitive to nanolitre volumes).

Other micromechanical techniques solely specialise in the measurement of the elasticity of biological systems and neglect the behaviour of the viscosity. These include micropipette aspiration, steady-state deformation using AFM, and the use of internal markers to drive or record the deformation of the cytoplasm, e.g. magnetic beads or fluorescent markers.

19.17   Tribology

A range of tribometers have been developed to measure frictional behaviour at surfaces. In a typical modern device adapted for the measurement of solid–solid friction of a thin viscoelastic film, forces are obtained in a direct manner through the measurement of the deflection of a spring with nm resolution (**Figure 19.83**). The stiffness of the bending beam is exactly known in the traction machine. The instrument is calibrated both in the normal and tangential directions. The force measurement has a resolution of nN in the range nN to mN in both directions. The springs for force measurement are often made of photo-structurable glass and the spherical ball probes are made of silicon or steel with a well-defined diameter. Interferometers can be used to measure the deflection of the spring. With biological specimens challenges are presented by their nonplanarity and the requirement for hydrated environments, e.g. the cartilage in articulated joints. AFMs are sometimes used to measure frictional forces, since they are not confined to planar specimens, but quantitative measurements of frictional coefficients are challenging, since it is difficult to infer both the normal and frictional forces simultaneously using light scattered from the back of a cantilever. Drag-flow rheometers can also be adapted to provide precision frictional measurements (e.g. a plate–plate rheometer with a section of material attached to either plate), but the utility of the

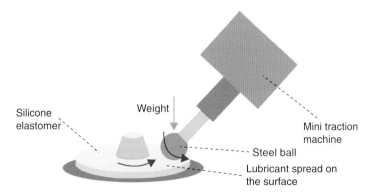

Figure 19.83   *Modern ball-on-plate tribometer used to measure the friction coefficient of thin viscoelastic films as a function of shear rate (Stribeck curves, Figure 9.20). [Reproduced from J. de Vicente, J.R. Stokes, H.A. Spikes, Tribology International, 2005, 38, 515–526. With permission from Elsevier.]*

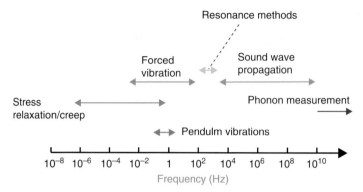

Figure 19.84 *Methods for the measurement of the linear viscoelasticity of solid materials over different frequency ranges.*

technique requires that the geometry of the specimens matches that of the rheometer cell which is not always practical, e.g. cartilage in ball and socket joints.

19.18   Solid Mechanical Properties

Solid materials with high elasticity and minimal flow behaviour require a separate set of techniques for their measurement, since extremely large forces must be applied to provide significant sample displacements (**Figure 19.84**). Dynamic mechanical testing apparatus (DMTA) are compressional analogues of oscillatory shear rheology that are often used with solid biomaterials. DMTA can provide the complex Young's modulus (E^*) of a material in compression as a function of frequency.

Highly anisotropic biomaterials provide a challenge for the experimentalist, since a large number of parameters need to be measured to fully characterise the stress response of the oriented material (**Section 13.1**). The relative orientation of the applied stress and resultant strain needs to be carefully calculated. Other material properties are also important for high-elasticity biomaterials, such as how samples buckle under compressive stress (measured using three point Euler buckling apparatus), and indentation tests for fracture mechanics.

Suggested Reading

If you can only read one book, then try:

Top recommended book: Serdynk, I.N. & Zaccai, N.R. (2007) *Methods in Molecular Biophysics*, Cambridge University Press. It contains a very comprehensive coverage of experimental methods.

Als-Nielsen, J. & McMorrow, D. (2011) *Elements of Modern X-ray Physics*, 2nd edition, Wiley. The field of coherent X-ray scattering/imaging is rapidly developing and this book covers much of the relevant physics.
Berne, B.J. & Percora, R. (2003) *Dynamic Light Scattering*, Dover. Classic text on photon correlation spectroscopy techniques.
Cantor, C.R. & Schimmel, P.R. (1980) *Biophysical Chemistry Part II, Techniques for the Study of Biological Structure and Function*, W.H. Freeman, Old fashioned, but well-explained account of a range of biophysical experiments. X-ray crystallography and fibre diffraction are particularly well covered.

Coey, P. (2010) *Magnetism and Magnetic Materials*, Cambridge University Press. Introduction to the physics of magnetic phenomena, e.g. for magnetic tweezers and nuclear magnetic resonance.

Drexler, W. & Fujimoto, J.G. (2008) *Optical Coherence Tomography: Technology and Applications*, Springer. OCT is a rapidly developing technique that should have a big impact on soft-matter and biological physics.

Gell, C., Brockwell, D. & Smith, A. (2006) *Handbook of Single Molecule Fluorescence Spectroscopy*, Oxford University Press. Good practical introduction to single-molecule techniques.

Gu, M., Bird, D., Day, D., Fu, L. & Morrish, D. (2010) *Femtosecond Biophotonics*, Cambridge University Press. Interesting selection of nonlinear microscopy techniques.

Hobbs, P.C.D. (2009) *Building Electrooptical Systems, Making it all Work*, Wiley. Developments in optoelectronics can allow world-class apparatus to be built from scratch within a reasonable budget. This book introduces some practical considerations.

Kane, S.A. (2003) *Introduction to Physics in Modern Medicine*, Taylor and Francis. Covers some of the modern medical physics techniques not found in the current chapter.

Lakowicz, J.R. (2006) *Principles of Fluorescence Spectroscopy*, Springer. Classic text on fluorescence techniques in biology.

Lipson, A., Lipson, S.G. & Lipson, H. (2009) *Optical Physics*, Cambridge University Press. This has a good introduction to super-resolution microscopy techniques as well as a wide range of other optical physics phenomena useful for biological physicists.

Mertz, J. (2009) *Introduction to Optical Microscopy*, Roberts and Co. Well-balanced physical over view of different optical microscope technologies including holography, optical coherence tomography, fluorescence and nonlinear varieties.

Popescu, G. (2011) *Quantitative Phase Imaging of Cells and Tissues*, McGraw Hill. Excellent introduction to holographic microscopy and related techniques for biological applications at visible wavelengths.

Saleh, B.E.A. & Teich, M.C. (2007) *Fundamentals of Photonics*, Wiley. A solid background in optical physics is a good place to begin to build new apparatus. This book covers the fundamental physics of photonics technologies.

Sheehan, D. (2009) *Physical Biochemistry Principles and Applications*, 2nd edition, Wiley. Good coverage of more biochemical characterisation techniques.

Viovy, J.L., Duke, T. & Caron, F. (1992) *Contemporary Physics*, **33** (1), 25–40. Useful introduction to the physics of gel electrophoresis.

Waigh, T.A. & Rau, C. (2012) X-ray and neutron imaging with colloids, *Current Opinion in Colloid and Interface Science*, **17**, 13–22. A short and simple introduction to X-ray and neutron imaging with nanoparticles.

Waigh, T.A. (2005) Microrheology of complex fluids, *Reports of Progress in Physics*, **68**, 685–742. Detailed introduction to the field of microrheology.

Tutorial Questions 19

19.1 Explain some of the practical challenges that prohibit the creation of a real gamma ray microscope.

19.2 Calculate the Abbe resolution limit for an optical microscope that functions with blue light ($\lambda = 400$ nm) with an objective lens with numerical aperture of 1.4. Make a list of the current resolution limits achieved with SIM, confocal, STORM, PALM and STED microscopes.

19.3 The discovery of green fluorescent protein (GFP) has had a huge impact in molecular biology (Nobel Prize 2008). Explain why this is true.

19.4 Describe the biological applications of terahertz radiation (**Figure 19.9**).

19.5 Consider the time scales at which biological processes take place. Make a list starting from the attosecond motion of electrons around nuclei to that of molecular dynamics on millisecond time scales. Explain which experimental techniques correspond to the faster time scales.

19.6 Explain how the force of a single optical laser trap could be calculated. The power spectral density of an optical trap is shown in **Figure 19.85**. Calculate the trap stiffness for a spherical particle of radius 1 μm in water (η is 0.001 Pa s). Explain how the force can be calibrated in an analogous magnetic tweezer experiment.

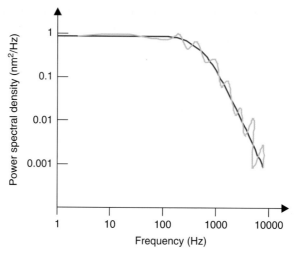

Figure 19.85   *Power spectral density as a function of frequency for a colloidal sphere in water trapped in a laser.*

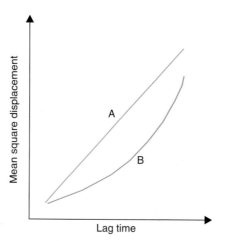

Figure 19.86   *Schematic diagram of the mean square displacements from particle tracking experiments of probe spheres embedded in materials A and B.*

19.7 Describe the advantages and disadvantages of working with single molecules, rather than an ensemble of Avogadro's number copies.

19.8 Image reconstruction in coherent X-ray diffractions with plane waves suffers from an ambiguity that causes the resultant images to be subject to random reorientations and translational offsets. Suggest methods to circumvent this problem.

19.9 The mean square displacements (MSD) of the probe particles in a video particle-tracking experiment embedded in two different fluids are shown in **Figure 19.86**. Explain which of the fluids (A or B) is viscoelastic over the time scale probed. Describe the effect of a static error in the measurement of the particle positions on the resultant mean square displacements.

19.10 Estimate the velocity of a DNA chain that contains one million base pairs in a polyacrylamide gel if the electric field is 2 V cm^{-1}. Calculate the factor by which this velocity would increase if DNA fragments of one tenth the size were chosen for the electrophoresis experiment. Assume the DNA chain is in the B form with 1.7 Å spacing between phosphate groups, there are no charge condensation effects, the electric field is in the weak-field limit, the solution viscosity is 0.002 Pa s and there is 300 monomers in a Kuhn segment.

19.11 Describe the power spectral density of a colloidal sphere diffusing in a purely viscous material. Explain how the power spectral density would change if a colloidal sphere was placed in a viscoelastic fluid that has a power-law mean square displacement on the lag time, i.e. $\langle r^2(t) \rangle \sim t^{\alpha}$, where $\alpha < 1$.

Part IV

Systems Biology

The human genome project was completed in 2003. However, although every gene is now known in the genome of a number of individuals (all 20 700 genes, combined with a large amount of regulatory material), there are huge gaps in our understanding of how they interact. Genes interact with one another in chemical circuits. Systems biologists work to uncover the meaning of these interactions and make quantitative sense of how genetics relates to an organism's metabolism and physiology.

Another perspective to the subject of systems biology is that of reverse engineering. The current state-of-the-art understanding of a cell is analogous to an extremely complicated piece of electronic equipment that lacks a circuit diagram and for which the owners have only a vague understanding of the equipment's capabilities. Thus an intensive programme of reverse engineering is required to construct the chemical circuits of the cell that determine its observed behaviour.

Cynics might say that the field of systems biology is no more than a rebranded form of biochemical kinetics. However, powerful new tools have been introduced into the field, borrowed from *network theory mathematics*, *synthetic biology*, *control engineering analysis* and *molecular biology*, which gives a unique flavour to the field beyond that found in traditional studies of enzyme kinetics. However since physics undergraduates typically lack any foundations in the areas of chemical and enzyme kinetics, these topics must first be covered, before the subjects of quantitative cellular physiology and systems biology can hope to be approached.

Control engineering analysis was developed by electrical engineers in earnest (simple mechanical regulators have existed for many hundreds of years) to force nonlinear electrical elements, first valves and then transistors, to be well-behaved linear elements in electronic equipment during the first half of the twentieth century. This was achieved by the introduction of negative feedback circuits that were combined with detailed mathematical tools for their computational analysis. Important advances in the mathematical foundations of the field were made by Norbert Weiner in the 1940s. Many ideas from control analysis that were originally applied to electrical circuits and robotics are equally applicable to chemical circuits inside cells.

Network theory is a branch of mathematics that traces its roots back to Euler's solution of a simple graphical problem (the 'Konigsberg bridge problem'), which led to the creation of 'graph theory'. Network theory is a subfield of graph theory. Metabolic pathways and gene circuits are full of networks, and network theory is thus very useful in systematically analyzing their behaviour (**Section 3.7**).

The Physics of Living Processes: A Mesoscopic Approach, First Edition. Thomas Andrew Waigh.
© 2014 John Wiley & Sons, Ltd. Published 2014 by John Wiley & Sons, Ltd.

Synthetic biology is a field of biology in which completely new biological systems are created. Often, this involves the insertion of recombinant DNA chains into existing cell lines that then express new functionalities, e.g. express a new molecule or a whole new metabolic circuit. Synthetic cells can be very useful for the development of models for quantitative physiology in systems biology, e.g. the repressilator.

Molecular biology, of course, describes the analysis of the molecules involved in cellular biology, and includes the subfield of systems biology. Modern molecular biology methods allow for high-throughput characterisation of chemical circuits and thus make their quantitative analysis practical for the first time. Huge information-rich data sets can be rapidly created, that require computational models for their analysis. It is an important goal for systems biology to create such models.

Suggested Reading

Noble, D. (2008) *The Music of Life*, Oxford University Press. Popular modern account of systems biology and physiology.

Church, G. & Regis, E. (2012) *Regenesis*, Perseus. Enthusiastic popular description of recent advances in synthetic biology.

20

Chemical Kinetics

Biological systems are subject to the same physical forces as the rest of Nature, although biological complexity can make the application of these concepts subtle to put into practice. Historically, this took a long time to establish, with people slow to discard the notion of vital forces unique to living processes. Countless experiments now demonstrate that all living phenomena (including consciousness) are due to the interaction of inanimate molecules that follow the standard physical laws of gravity, electromagnetism, quantum field theory, thermodynamics, etc., as with all other matter.

Complexity can rapidly evolve from even simple inorganic chemical reactions. Oscillatory chaotic chemical reactions came as a big shock to the physical chemistry establishment when they were discovered in the 1960s, since these reactions did not converge to a simple equilibrium state. For example the Briggs–Rauscher oscillating reaction acts as a chemical clock and the reaction mixture oscillates between clear and orange coloured states. Also, in the Belousov–Zhabotinsky reaction (**Figure 20.1**) oscillations occur in both time and space with expanding ring structures randomly nucleating and interacting across a reactant mixture. The first challenge in the analysis of these complex chemical systems was to deduce what was the source of energy driving the reactions, so that the law of conservation of energy was not broken, and indeed energy sources were quickly found. More intriguing, however, was how the oscillations were possible, since the theoretical paradigms at the time required a steady transformation to the equilibrium state. The answer lies in an understanding of how the conditions of equilibrium are broken. The rich varieties of dynamic phenomena that occur are due to the nonlinearities in the systems, which require some sophisticated new mathematical tools for their full analysis.

Once the theoretical concepts to understand simple synthetic reactions are established, more sophisticated analysis is possible of biochemical reactions. Enzyme kinetics can thus be studied and nonlinear phenomena based on their dynamics can also be developed (**Chapter 21**).

The Physics of Living Processes: A Mesoscopic Approach, First Edition. Thomas Andrew Waigh.
© 2014 John Wiley & Sons, Ltd. Published 2014 by John Wiley & Sons, Ltd.

Figure 20.1 *Optical image of a Belousov–Zhabotinsky oscillatory reaction using potassium, cerium sulfate, malonic acid, citric acid and sulfuric acid, i.e. a synthetic nonlinear oscillatory chemical reaction.*

20.1 Conservation Laws

A biological system must obey a series of fundamental physical laws that determine their dynamical behaviours, i.e. how their states change with time. These physical laws include those that govern chemical reactions, but also all of physics in general. Indeed, similar principles of reaction kinetics hold in more specialised areas of physics, such as nuclear physics and particle physics.

Both matter and energy obey conservation laws in biology; if negligible relativistic effects are assumed and no radioactive isotopes occur. Similarly, angular momentum and linear momentum are also conserved in biology, with readily observable effects, e.g. rotation during the swimming of micro-organisms or jet propulsion by squid. More exotic quantities such as parity are also conserved in biology, but the effects will not be considered in detail in the current text.

Thermodynamics can be applied to chemical systems (indeed it was a major motivation for its development during the industrial revolution), but the ideas apply to equilibrium systems and cells are rarely static so they require nonequilibrium extensions. Thus, care must be taken when applying thermodynamics directly to living processes, although the conceptual framework is of wide use for the construction of reference models to understand the action of purified biomolecules.

The *first law of thermodynamics* states that

$$\Delta U = \Delta Q + \Delta W \tag{20.1}$$

where ΔW is the work energy change (in principle it can result in the movement of a weight), ΔQ is the heat energy change associated with a change in the temperature of a substance and ΔU is the internal energy (the sum of kinetic energy and potential energy of individual internal components of the molecules).

The *second law of thermodynamics* states that the entropy always increases as a closed system evolves. It is integral to the principal of minimisation of free energy discussed in the next section. The increased ordering in molecular arrangements during the creation of life comes at the expense of a larger compensating entropy increase in the surrounding environments.

20.2 Free Energy

It is useful to recap some results from **Chapter 3**. Free energies are helpful quantities to determine the equilibrium state of biomolecules and biological systems. The Helmholtz free energy (equation (3.19)) was defined as $F = U–TS$, in systems where T, V, N are fixed, while E fluctuates. The entropy of a system, as always, is the Boltzmann constant multiplied by the natural logarithm of the number of microstates (Ω), $S = k\ln\Omega$, equation (3.15). The Gibbs free energy is $F = U–TS + pV$ (equation (3.18)), where p is the pressure and V is the volume. Here, T, V, μ are kept fixed in the system, while E and N fluctuate. The third and final potential constructed is the enthalpy, defined as $H = U + pV$, equation (3.21). Changes in enthalpy (ΔH) are useful for the determination of heats of reaction.

The principle of minimal free energy states that a system can alter its state if the change will result in a lower free energy. A system in its equilibrium state has a minimal free energy.

20.3   Reaction Rates

The reaction rate for a chemical transformation (say of a red molecule into a blue molecule) is of central importance in physical chemistry (**Figure 20.2**). An analysis of simple reaction rates will first be made. This can subsequently be extended to more sophisticated phenomena, such as biological clocks and oscillatory phenomena, which are in turn key for an understanding of whole-cell phenomena such as the timing of mitosis. This analysis can also lead to answers to larger-scale whole organism questions; such as why humans wake up in the morning as the sun rises or why they can experience jet lag.

The control of reaction kinetics has a vast range of applications. Some reactions need to be speeded up such as drug delivery, cooking (e.g. pressure cooking), paint drying, the destruction of air pollutants, and the breakdown of synthetic plastic waste. Conversely, some reactions need to be slowed down such as food decay/spoiling (e.g. staling due to recrystallisation of starches), discoloration of plastics, human aging (antioxidants), fading of clothes (detergent additives), fires (e.g. a Zeppelin going up in flames or flame retardants added to present-day consumer products), ozone layer destruction, and the rusting of metallic objects (the hulls of boats).

In a chemical reaction the concentration of the reacting system changes with time. How fast a reaction proceeds is called the reaction rate. Very simply, a reaction scheme can be formulated as

$$\text{Reactants} \rightarrow \text{Products} \tag{20.2}$$

Chemical reactions involve the formation and breakage of chemical bonds. Consider reactant molecules (say A and B) that approach one another, collide and then interact with the appropriate energy and orientation for a reaction to occur. Bonds are stretched, broken, formed and finally the product molecules (C) move away from the location of the reaction (**Figure 20.3**),

$$A + B \longrightarrow C \tag{20.3}$$

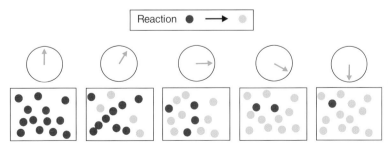

Figure 20.2   *A reaction occurs in which a red molecule is converted into a blue molecule. Over time all the red molecules in the reaction vessel become blue.*

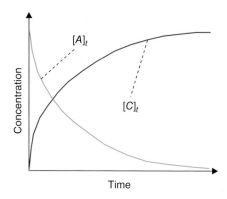

Figure 20.3   *An irreversible reaction of A molecules into C molecules. The concentration of reactant A decreases over time while that of reactant C increases.*

This is said to be a monomolecular reaction, since there is one product, while the reverse reaction,

$$C \rightarrow A + B \tag{20.4}$$

is said to be bimolecular, because there are two products. All chemical reactions are two way due to the principle of detailed balance (microscopic reversibility), but under some circumstances they can be considered to have a single direction, e.g. if the products are rapidly removed from the mixture. The rate of consumption (R_A) of species A in equation (20.3) is

$$R_A = -\frac{d[A]}{dt} \tag{20.5}$$

The concentration of A decreases as the reaction moves from the left to the right, so there is a negative sign in equation (20.5) and, from the stochiometry of the reaction, the reaction rate of A must be equal to that of B; $R_A = R_B$. The rate of formation of the product of equation (20.2), species C, is

$$R_C = \frac{d[C]}{dt} \tag{20.6}$$

The reaction rate constants, k_1 and k_2 are for the forward and reverse reactions respectively,

$$A + B \xrightarrow{k_1} C \tag{20.7}$$

$$C \xrightarrow{k_2} A + B \tag{20.8}$$

The reaction fluxes (J) are given by

$$J_C = k_2[C] \tag{20.9}$$

$$J_A = J_B = k_1[A][B] \tag{20.10}$$

The order of a reaction (n) is defined as the value of the exponent of the dependence of the reaction flux on the reactant concentration,

$$J = k[C]^n \tag{20.11}$$

Therefore, $J_C = k_2[C]$ describes a first-order reaction, whereas $J_A = J_B = k_1[A][B]$ defines a second-order reaction, since the two first-order exponents are added together. Reaction orders are motivated by experimental results. In general, reaction mechanisms do not require fluxes to follow power-law kinetics with integer exponents of the reactant concentrations, although they are a common occurrence.

The *chemical potential* is the free energy per particle (**Chapter 3**), and for a component i the chemical potential is defined as

$$\mu_i = \mu_i^0 + RT \ln a_i \tag{20.12}$$

where R is the ideal gas constant, T is the temperature, μ_i^0 is the chemical potential of component i in the standard state and a_i is the chemical activity of component i, an effective mole fraction (it provides the contribution from the mixing entropy), i.e. $a_i = \dfrac{n_i}{\sum n_i}$, where n_i are the number of molecules of component i.

Consider a more general reaction,

$$\upsilon_A A + \upsilon_B B \underset{\leftarrow}{\overrightarrow{}} \upsilon_C C + \upsilon_D D \tag{20.13}$$

where υ_i are the stoichiometric coefficients, i.e. the number of molecules of species i involved in the elementary reaction. The changes take place under constant pressure and temperature, and are limited to variations in the quantities of the reactants. The change in the Gibbs free energy upon reaction (ΔG) is given by the difference between the chemical potentials of the products and reactants, weighted by the number of molecules that participate in the reaction,

$$\Delta G = \upsilon_C \mu_C + \upsilon_D \mu_D - \upsilon_A \mu_A - \upsilon_B \mu_B \tag{20.14}$$

Using equation (20.12) for the chemical potentials gives

$$\Delta G = \Delta G^0 + RT(\upsilon_C \ln a_C + \upsilon_D \ln a_D - \upsilon_A \ln a_A - \upsilon_B \ln a_B) \tag{20.15}$$

where the changes have been referred to the molar standard reaction free energy (ΔG^0) given by

$$\Delta G^0 = \upsilon_C \mu_C^0 + \upsilon_D \mu_D^0 - \upsilon_A \mu_A^0 - \upsilon_B \mu_B^0 \tag{20.16}$$

Equation (20.15) can be rearranged to give van't Hoff's equation,

$$\Delta G = \Delta G^0 + RT \ln \left(\frac{a_C^{r_C} a_D^{r_D}}{a_A^{r_A} a_B^{r_B}} \right) \tag{20.17}$$

In the equilibrium state equation (20.17) predicts that the change in the Gibbs free energy will be zero. In a biochemical reaction that has reached equilibrium, the equilibrium constant (K_{eq}) is defined as

$$K_{eq} = \frac{[C]_{eq}^{r_C} [D]_{eq}^{r_D}}{[A]_{eq}^{r_A} [B]_{eq}^{r_B}} = e^{-\frac{\Delta G^0}{RT}} \tag{20.18}$$

This establishes a relationship between the equilibrium constant and the activation energy (ΔG^0) for the reaction. Thus, rate constants tend to vary exponentially with temperature, which in general makes reaction kinetics very sensitive to temperature. An alternative physical interpretation for K_{eq} is that it is equal to the rate constant for the forward reaction (k_+) divided by the rate constant for the backward reaction (k_-),

$$K_{eq} = \frac{k_+}{k_-} \tag{20.19}$$

For the reaction shown in equation (20.3) a net *reaction rate* (R) can be calculated

$$R = -\frac{d[A]}{dt} = \frac{d[C]}{dt} \tag{20.20}$$

where $[A]$ is the reactant concentration, and $[C]$ is the product concentration. More generally, the reaction rate is quantified in terms of changes in the concentration $[A]$ of a reactant or product species A with respect to time,

$$R_j = \frac{1}{\nu_j}\frac{d[A]}{dt} \tag{20.21}$$

where ν_j denotes the stoichiometric coefficient of R_j.

The rate of reaction is often found to be proportional to the molar concentration of the reactants raised to a simple power (which need not be integral). This empirical relationship is called the *rate equation*. The manner in which the reaction rate changes in magnitude with changes in the concentration of each participating reactant species is called the *reaction order*,

$$R = -\frac{1}{x}\frac{d[A]}{dt} = -\frac{1}{y}\frac{d[B]}{dt} = k[A]^{\alpha}[B]^{\beta} \tag{20.22}$$

where α and β are the reaction orders with respect to $[A]$ and $[B]$, respectively. An example is the oxidation of hydrogen to form water (e.g. the explosion of a zeppelin),

$$2H_2 + O_2 \rightarrow 2H_2O \tag{20.23}$$

The reaction rate is therefore

$$R = -\frac{1}{2}\frac{d[H_2]}{dt} = -\frac{d[O_2]}{dt} = \frac{1}{2}\frac{d[H_2O]}{dt} \tag{20.24}$$

Consider the biochemical/physical reaction

$$xA + yB \xrightarrow{k} \text{Products} \tag{20.25}$$

where k is the rate constant, and x and y are stoichiometric coefficients. The empirical equation for the rate of reaction is called the *law of mass action*,

$$R = k[A]^{\alpha}[B]^{\beta} \tag{20.26}$$

where α and β are the reaction orders for the reactants, which can be determined experimentally. To determine the reaction rate experimentally, the concentration $[A]$ is varied while $[B]$ is kept constant and the reaction rate is measured, which allows α to be calculated from a log–log plot. Then, the concentration $[B]$ needs to be varied while $[A]$ is kept a constant and β can be determined from another log–log plot (**Figure 20.4**).

A classic illustrative example of similar chemical reaction equations that give rise to widely varying reaction kinetics is that of molecular hydrogen combined with a range of different halides. The chemical reactions look very similar (as implied by the similar positions of the reactants in the periodic table), but the rate equations are completely different, which implies different quantum mechanical processes are involved. The general chemical reaction is

$$H_2 + X_2 \longrightarrow 2HX \tag{20.27}$$

where X can be iodine (I), bromine (Br), or chlorine (Cl). The rate law provides an important guide to reaction mechanism, since any proposed mechanism must be consistent with the observed rate law. A complex rate

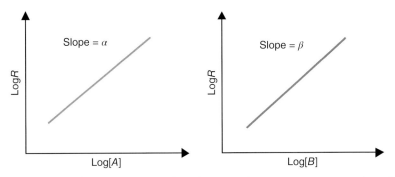

Figure 20.4 *The logarithm of the reaction rate plotted against the reactant concentration allows the reaction orders α and β to be calculated with respect to species A and species B, respectively.*

equation will imply a complex multistep reaction mechanism guided by the underlying quantum mechanics. The three experimentally determined rate laws are

$$R = \frac{d[\text{HI}]}{dt} = k[\text{H}_2][\text{I}_2] \tag{20.28}$$

for iodine,

$$R = \frac{d[\text{HBr}]}{dt} = \frac{k[\text{H}_2][\text{Br}_2]^{1/2}}{1 + \dfrac{k'[\text{HBr}]}{[\text{Br}]}} \tag{20.29}$$

for bromine,

$$R = \frac{d[\text{HCl}]}{dt} = k[\text{H}_2][\text{Cl}_2]^{1/2} \tag{20.30}$$

for chlorine. This emphatically illustrates the point that in general it is impossible to deduce the rate law from a chemical reaction equation such as equation (20.27) (without a detailed quantum mechanical model of the reaction). Practically, the experiments need to be performed if you want to be confident of a rate law, although sometimes educated guesses can be made. All three halide reactions have the same reaction scheme (equation (20.27)), but completely different kinetics. However, once the rate law is determined the rate of reaction can be predicted for any composition mixture.

Analytic solutions to rate laws will be considered next. With *zeroth-order kinetics* the reaction proceeds at the same rate (R) regardless of the concentration of the reactants (A, **Figure 20.5**),

$$A \xrightarrow{k} \text{products} \tag{20.31}$$

The rate equation is

$$R = -\frac{da}{dt} = k \tag{20.32}$$

where the units of k are mol dm^{-3} s^{-1} and a is shorthand for [A]. The boundary conditions are assumed to be $a = a_0$ when $t = 0$. Integration of equation (20.32), using these initial conditions, gives

$$a(t) = -kt + a_0 \tag{20.33}$$

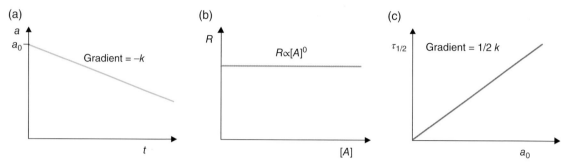

Figure 20.5 *Zeroth-order reaction kinetics. (a) The concentration of A (a) decreases linearly with time. (b) The reaction rate (R) is independent of the concentration of A ([A]). (c) The half-life ($\tau_{1/2}$) is proportional to the initial concentration of a (a_0).*

The half-life is the time ($\tau_{1/2}$) for the concentration to decay to half its initial value ($\frac{a_0}{2}$). Substitution in equation (20.33) gives

$$\tau_{1/2} = \frac{a_0}{2k} \tag{20.34}$$

Therefore, it is observed that the half-life is proportional to the initial concentration ($\tau_{1/2} \propto a_0$). An example of zeroth-order kinetics was seen in Oosawa's model for the disassembly of the end groups of self-assembling fibres (**Section 16.1**), i.e. the rate equation did not depend on monomer concentration.

For *first-order kinetics* for the reaction in equation (20.31), a different first-order differential rate equation is needed for the concentration dynamics,

$$-\frac{da}{dt} = ka \tag{20.35}$$

where k is the first-order rate constant with units of s^{-1}. Again the initial condition is taken as $a = a_0$ for $t = 0$. The solution to this simple variables separable differential equation is

$$a(t) = a_0 e^{-kt} \tag{20.36}$$

This gives the reactant concentration as a function of time. First-order kinetics are observed in lots of places in physics such as radioactive decay, Einstein absorption of photons by atoms or the population dynamics of micro-organisms. They are a sign of the underlying Poisson statistics in the processes (**Section 3.1**).

The *mean lifetime* of reactant molecule (τ) can be calculated from the concentration distribution for first-order kinetics,

$$\tau = \frac{1}{a_0} \int_0^\infty a(t)dt = \frac{1}{a_0} \int_0^\infty a_0 e^{-kt} dt = \frac{1}{k} \tag{20.37}$$

The mean lifetime of a reactant molecule is thus the inverse of the rate constant (k) in a first-order reaction.

By definition, in each successive period of duration equal to the half-life ($\tau_{1/2}$) the concentration of a reactant decays to half its value at the start of that period. After n such periods, the concentration is $(1/2)^n$ of its initial value (**Figure 20.6**). The half-life for first-order reactions is independent of the initial reactant

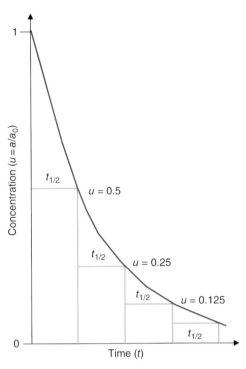

Figure 20.6 *In first-order reaction kinetics the concentration of A decays exponentially as a function of time* (u=a/a$_0$). *The half-life is defined as the time for half the molecules to decay.*

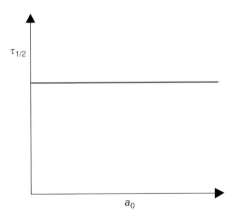

Figure 20.7 *The half-life as a function of the initial concentration of A (a$_0$) for a first-order reaction. The half-life is independent of the initial concentration of A.*

concentration (**Figure 20.7**). From equation (20.36), substitution of $a = \dfrac{a_0}{2}$ gives the half-life of a first-order reaction to be

$$\tau_{1/2} = \frac{\ln 2}{k} \tag{20.38}$$

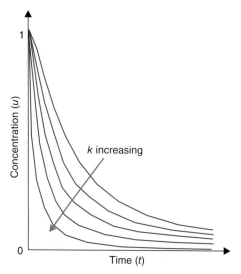

Figure 20.8   *The concentration of A as a function of time for a first-order reaction ($u = a/a_0$). An increase in the rate constant (k) causes a faster decay of concentration.*

To continue this analysis of kinetic equations it is more algebraically elegant to normalise the concentration of a to form a new variable u ($u = a/a_0$), where u is bounded by 0 and 1. It will facilitate comparison of the concentration in reactions with multiple species. Therefore, equation (20.36) becomes

$$u(\theta) = e^{-\theta} \tag{20.39}$$

where the time has also been rescaled as $\theta = kt$ and the equation is plotted in **Figure 20.8**.
    More generally for *nth-order kinetics* the reaction is

$$nA \xrightarrow{k} P \tag{20.40}$$

The rate equation is

$$-\frac{da}{dt} = ka^n \tag{20.41}$$

Again the initial conditions are taken as $a = a_0$ when $t = 0$. The solution requires another variables-separable integration and gives

$$\frac{1}{a^{n-1}} = (n-1)kt + \frac{1}{a_0^{n-1}} \quad n \neq 1 \tag{20.42}$$

where $n = 0, 2, 3, \dots$ The rate constant (k) can be experimentally determined from the slope of a plot of reciprocal concentration versus time (**Figure 20.9**) and the half-life is

$$\tau_{1/2} = \frac{2^{n-1} - 1}{(n-1)ka_0^{n-1}} \tag{20.43}$$

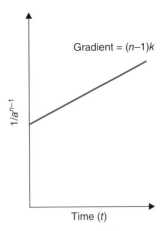

Figure 20.9 *For an nth-order chemical reaction a plot of $1/a^{n-1}$ against time (t) gives a linear plot. The gradient of the line is equal to n–1 multiplied by the rate constant (k).*

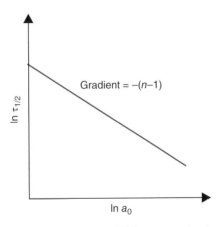

Figure 20.10 *A plot of the natural logarithm of the half-life against the logarithm of the initial concentration for an nth order chemical reaction. The reaction order can be calculated from the gradient (=–(n–1)).*

The half-life depends on the initial concentration since $\tau_{1/2} \propto a_0^{1-n}$, unlike that for a first-order reaction (**Figure 20.10**). The reaction order (n) can be determined from the slope of a graph of the half-life as a function of the reactant concentration.

20.4 Consecutive Reactions

The previous discussion of chemical reactions just considered a single step in a reaction scheme. However, many reactions in both synthetic chemistry and biology have multiple reaction steps, so it is important to extend the analysis to more sophisticated chain reactions, e.g. reactant A that transforms to a product X, which then transforms to a product P in an irreversible manner,

$$A \xrightarrow{k_1} X \xrightarrow{k_2} P \tag{20.44}$$

where k_1 and k_2 are two rate constants. Mathematically, the analysis reduces to a study of coupled ordinary differential equations, and they are analytically soluble in the simple case of two consecutive irreversible reactions. Such consecutive reactions also occur in nuclear and particle physics where there is an identical methodology for the analysis. For example, in a radioactive decay scheme for the decay of polonium into lead and then bismuth,

$$^{218}Po \rightarrow {}^{214}Pb \rightarrow {}^{214}Bi \tag{20.45}$$

where k_1 is $5 \times 10^{-3}\,s^{-1}$ and k_2 is $6 \times 10^{-4}\,s^{-1}$. Mass balance implies $p = a_0 - a - x$, where a_0, p, a and x are the initial concentration of a and the concentrations of p, a and x, respectively. Three coupled ordinary differential equations define the system,

$$\frac{da}{dt} = -k_1 a, \quad \frac{dx}{dt} = k_1 a - k_2 x, \quad \frac{dp}{dt} = k_2 x \tag{20.46}$$

These coupled equations can be solved analytically using an integrating factor (see tutorial question 20.10) to give

$$a(t) = a_0 \exp[-k_1 t] \tag{20.47}$$

$$x(t) = \frac{k_1 a_0}{k_2 - k_1} \{\exp[-k_1 t] - \exp[-k_2 t]\} \tag{20.48}$$

$$p(t) = a_0 - a_0 \exp[-k_1 t] - \frac{k_1 a_0}{k_2 - k_1} \{\exp[-k_1 t] - \exp[-k_2 t]\} \tag{20.49}$$

20.5 Case I and II Reactions

Different qualitative kinetic behaviours result from the two consecutive reactions given by equation (20.44), that depend on the ratio of the rate constants k_1 and k_2. For *Case I* consecutive reactions the intermediate formation is fast, whereas the intermediate decomposition is slow,

$$\kappa = \frac{k_2}{k_1} << 1, \quad k_2 << k_1, \quad A \xrightarrow{k_1} X \xrightarrow{k_2} P \tag{20.50}$$
$$\text{fast} \qquad \text{slow}$$

The Arrhenius equation (20.18) implies $\Delta G_I^* << \Delta G_{II}^*$, i.e. the free energy change of reaction I is much smaller than that of reaction II (**Figure 20.11**). Step II is the *rate-determining step*, since it has the highest activation energy barrier. The reactant species A is more reactive than the intermediate X. Introduction of normalised concentration parameters facilitates the analysis,

$$u = \frac{a}{a_0}, \quad v = \frac{x}{a_0}, \quad w = \frac{p}{a_0} \tag{20.51}$$

The progress of a Case I reaction is shown on **Figure 20.12**. The concentration of the intermediate is significant over the time course of the reaction.

For *Case II* consecutive reactions the intermediate formation is slow, whereas the intermediate decomposition is fast,

$$A \xrightarrow{k_1} X \xrightarrow{k_2} P \tag{20.52}$$
$$\text{slow} \qquad \text{fast}$$

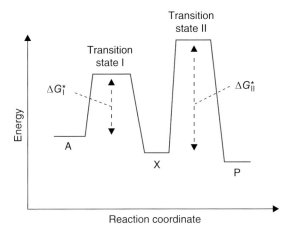

Figure 20.11 *Transition states for two consecutive reactions (the energy is plotted against the reaction coordinate) in which the energy barrier for the first reaction is smaller than that for the second reaction ($\Delta G_{II}^* \gg \Delta G_I^*$).*

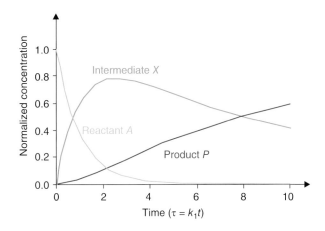

Figure 20.12 *Concentrations of the reactant A, intermediate X and product P as a function of time for a two-stage consecutive reaction. The energy barrier for the first reaction is smaller than that for the second reaction as shown in Figure 20.11.*

Furthermore, $\kappa = \dfrac{k_2}{k_1} \gg 1$, which implies from Arrhenius kinetics that $\Delta G_I^* \gg \Delta G_{II}^*$ (**Figure 20.13**). The intermediate X is fairly reactive so $[X]$ will be small at all times. The free energy of the first reaction I is much larger than the second reaction II. Step I is rate determining, since it has the highest activation energy barrier. The intermediate concentration (ν) is approximately constant after an initial induction period (**Figure 20.14**).

Thus, when reactant A decays rapidly, the concentration of intermediate species X is non-negligible for much of the reaction and product P concentration rises gradually, since the rate of transformation of X to P is slow. However, when reactant A decays slowly, the concentration of intermediate species X will be low for the duration of the reaction and to a good approximation the net rate of change of intermediate concentration with time is zero. Hence the material will be formed as quickly as it is removed. This motivates the quasisteady-state

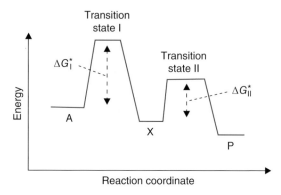

Figure 20.13 *Transition states for two consecutive reactions (energy plotted against reaction coordinate) in which the energy barrier for the first reaction is larger than that for the second reaction ($\Delta G_I^* \gg \Delta G_{II}^*$).*

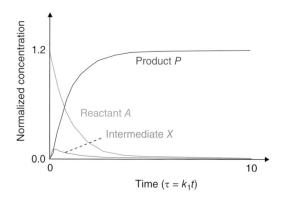

Figure 20.14 *Concentrations of the reactant A, intermediate X and product P as a function of time for a two stage consecutive reaction. The energy barrier for the first reaction is larger than that for the second reaction as shown in* **Figure 20.13**.

approximation (QSSA, $[X] \approx 0$ and $\dfrac{d[X]}{dt} \approx 0$), which is often used to simplify the analysis of biochemical reactions.

The analysis of consecutive reactions makes a strong argument for computer simulation in biology, since the analysis of even the simplest enzymatic reactions (which are based on consecutive reaction schemes) clearly profits from the use of a computer. Invariably models of realistic coupled biochemical reactions require numerical solution to sets of differential equations, since no simple analytic solutions are available.

20.6 Parallel Reactions

Concurrent parallel reactions are also possible where a single reactant species forms two distinct products or when two distinct mechanisms form a single reaction product. Considering the first of these two scenarios, it is assumed that each of the two parallel reactions exhibit irreversible first-order kinetics,

$$A \xrightarrow{k_1} X \tag{20.53}$$

$$A \xrightarrow{k_2} Y \qquad (20.54)$$

where k_1, k_2 are both first-order rate constants. The rate equation is

$$R = -\frac{da}{dt} = (k_1 + k_2)a = k_\Sigma a \qquad (20.55)$$

where a cumulative rate constant $(k\Sigma)$ is defined as $k_\Sigma = k_1 + k_2$. Equation (20.55) is the first-order reaction kinetics equation seen previously as equation (20.35), except with a rescaled reaction rate equation. By analogy, the solution is deduced to be

$$a(t) = a_0 \exp[-k_\Sigma t] = a_0 \exp[-(k_1 + k_2)t] \qquad (20.56)$$

Similarly, by analogy the half-life of the reaction is seen to be

$$\tau_{1/2} = \frac{\ln 2}{k_\Sigma} \qquad (20.57)$$

All of this is just an extension of simple first-order kinetics. The rate equation for the concentration of reactant x is

$$\frac{dx}{dt} = k_1 a \qquad (20.58)$$

Combination with equation (20.56) and integration gives

$$x(t) = \frac{k_1 a_0}{k_1 + k_2} \{1 - \exp[-(k_1 + k_2)t]\} \qquad (20.59)$$

Similarly for y the solution is

$$y(t) = \frac{k_2 a_0}{k_1 + k_2} \{1 - \exp[-(k_1 + k_2)t]\} \qquad (20.60)$$

In the limit of long times the ratio of the two product concentrations equals the ratio of the two rate constants,

$$\underset{t \to \infty}{\mathrm{Lim}} \frac{x(t)}{y(t)} = \frac{k_1}{k_2} \qquad (20.61)$$

The behaviour of the parallel mechanism when $k_1 = 10k_2$ is shown in **Figure 20.15**.

20.7 Approach to Chemical Equilibrium

Countless experiments with chemical systems have shown that in a state of equilibrium, the concentrations of reactants and products no longer change with time. The oscillating reactions described in the chapter introduction are exceptions to the rule and are not observed in the vast majority of synthetic chemical reactions. This apparent cessation of activity in equilibrium reactions occurs because the reactions are microscopically reversible and the reaction kinetics are well behaved (linear systems). Equilibrium does not require that the reactions have actually stopped on the molecular level.

As an example, consider the reversible decomposition of dinitrogen tetraoxide (a colourless gas) into nitrogen oxide (a brown gas, **Figure 20.16**),

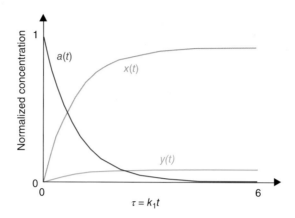

Figure 20.15   *Concentration of the reactant a(t) and the products x(t) and y(t) for a parallel reaction scheme as a function of the rescaled time ($\tau = k_1 t$). The rate constant for the creation of x (k_1) is ten times that of y (k_2).*

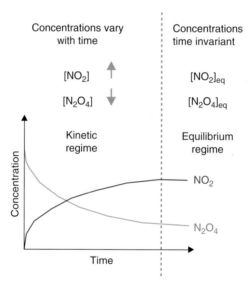

Figure 20.16   *The concentration of nitrogen dioxide (red) and dinitrogen tetroxide (blue) as a function of time in a reversible equilibrium reaction. There is no nitrogen dioxide at the start of the reaction in the vessel.*

$$N_2O_{4(g)} \underset{k'}{\overset{k}{\rightleftharpoons}} 2NO_{2(g)} \tag{20.62}$$

Nitric oxide (laughing gas) is an important signalling compound in blood and its surface chemistry has been explored to reduce stent rejection in heart operations. Kinetic analysis gives two rate equations for the reactants and products,

$$\overrightarrow{R} = k[N_2O_4], \quad \overleftarrow{R} = k'[NO_2]^2 \tag{20.63}$$

which are valid for all time (t). In the limit of long times ($t \rightarrow \infty$) the reaction will reach equilibrium ($\overrightarrow{R} = \overleftarrow{R}$) so

$$k[\text{N}_2\text{O}_4]_{\text{eq}} = k'[\text{NO}_2]_{\text{eq}}^2 \tag{20.64}$$

And therefore

$$\frac{[\text{NO}_2]_{\text{eq}}^2}{[\text{N}_2\text{O}_4]} = \frac{k}{k'} = K \tag{20.65}$$

where K is the equilibrium constant.

More generally, consider the approach to chemical equilibrium for a first-order reversible reaction,

$$A \underset{k'}{\overset{k}{\rightleftarrows}} B \tag{20.66}$$

The rate equation can be simply constructed as

$$\frac{da}{dt} = -ka + k'b \tag{20.67}$$

and the initial conditions are taken as $a = a_0$ and $b = 0$ when $t = 0$. The condition of mass balance gives $a + b = a_0$. To provide a more elegant mathematical formulation it is useful to recast the equations in terms of normalised units. These allow easier comparison between experiment and theory, and the comparison of consecutive experiments with different absolute concentrations. Therefore, the variables u, v, τ and θ are defined as

$$u = \frac{a}{a_0}, \ v = \frac{b}{a_0}, \ \tau = (k + k')t, \ \theta = \frac{k}{k'} \tag{20.68}$$

where $u + v = 1$ and the boundary conditions are $u = 1$ and $v = 0$ when $\tau = 0$. The rate equation (20.67) in normalised form is

$$\frac{du}{d\tau} + u = \frac{1}{1 + \theta} \tag{20.69}$$

This ordinary differential equation can be solved to give the two required concentration of A and B (in rescaled units using $u + v = 1$),

$$u(\tau) = \frac{1}{1 + \theta}\{1 + \theta e^{-\tau}\}, \ \ v(\tau) = \frac{\theta}{1 + \theta}\{1 - e^{-\tau}\} \tag{20.70}$$

The reaction quotient (Q) is defined as the ratio of the two reactants as a function of time and is a useful parameter to quantify the approach to equilibrium. For the reaction described by equation (20.67) it is

$$Q(\tau) = \frac{v(\tau)}{u(\tau)} = \theta\left\{\frac{1 - e^{-\tau}}{1 + \theta e^{-\tau}}\right\} \tag{20.71}$$

The *reaction quotient* (Q) is only equal to the *equilibrium constant* (K) at very long time scales (**Figure 20.17**),

$$Q(\tau \rightarrow \infty) = K = \theta = \frac{k}{k'} = \frac{v(\infty)}{u(\infty)} \tag{20.72}$$

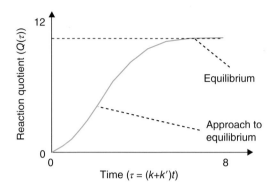

Figure 20.17   *The reaction quotient (Q) as a function of rescaled time (τ = (k + k′)t) for a reaction that approaches equilibrium. At short times the concentration evolves in the kinetic regime. At long times the mixture reaches an equilibrium state equal to K.*

20.8   Quasi-Steady-State Approximation

Detailed mathematical analysis of complex reaction mechanisms is difficult. Some useful methods for the solution of sets of coupled linear differential rate equations include matrix methods and Laplace transforms.

In many cases use of the quasisteady-state approximation (QSSA) leads to a considerable simplification in the kinetic analysis of consecutive reactions. The QSSA assumes that after an initial induction period (during which the concentration of the intermediates rises from zero), and for most of the duration of the reaction, the rate of change of concentration of all reaction intermediates are negligibly small. Mathematically, the QSSA implies

$$\frac{dx}{dt} = R_{\text{Xformation}} - R_{\text{Xremoval}} \cong 0, \quad R_{\text{Xformation}} = R_{\text{Xremoval}} \tag{20.73}$$

where $R_{\text{Xformation}}$ and R_{Xremoval} are the rate of formation of X and the rate of removal of X, respectively, for the reaction given in equation (20.44).

Consecutive reaction mechanisms where the first step is reversible and the second step is irreversible are found to be a reasonable model for enzyme kinetics (*Michaelis–Menten* kinetics),

$$A \underset{k_{-1}}{\overset{k_1}{\rightleftarrows}} X \overset{k_2}{\longrightarrow} P \tag{20.74}$$

Michaelis–Menten models are important for describing a wide range of enzymatic reactions. The rate equations in rescaled units are

$$\frac{du}{d\tau} = -u + \kappa v \tag{20.75}$$

$$\frac{dv}{d\tau} = u - (\kappa + \phi)v \tag{20.76}$$

$$\frac{dw}{d\tau} = \phi v \tag{20.77}$$

where $u = \dfrac{a}{a_0}$, $v = \dfrac{x}{a_0}$, $w = \dfrac{p}{a_0}$, mass balance gives $u + v + w = 1$ and the initial conditions are $u = 1$, $v = w = 0$ when $\tau = 0$. Furthermore, $\kappa = \dfrac{k_{-1}}{k_1}$, $\phi = \dfrac{k_2}{k_1}$, and $\tau = k_1 t$ are defined. The coupled ordinary differential equations (20.75)–(20.77) can be solved analytically using Laplace transforms. The solutions are

$$u(\tau) = \frac{1}{\beta - \alpha}\left\{(\kappa + \phi - \alpha)e^{-\alpha\tau} - (\kappa + \phi - \beta)e^{-\beta\tau}\right\} \tag{20.78}$$

$$v(\tau) = \frac{1}{\beta - \alpha}\left\{e^{-\alpha\tau} - e^{-\beta\tau}\right\} \tag{20.79}$$

$$w(\tau) = 1 - \frac{1}{\beta - \alpha}\left\{\beta e^{-\alpha\tau} - \alpha e^{-\beta\tau}\right\} \tag{20.80}$$

where α and β are composite quantities that contain the individual rate constants; $\alpha\beta = \phi$ and $\alpha + \beta = 1 + \kappa + \phi$. Application of the QSSA implies

$$\frac{dv}{d\tau} \cong 0 \tag{20.81}$$

And therefore from equation (20.76),

$$u_{SS} - (\kappa + \phi)v_{SS} = 0 \tag{20.82}$$

This equation can be rearranged to give

$$v_{SS} \cong \frac{u_{SS}}{\kappa + \phi} \tag{20.83}$$

Substitution in equation (20.75) gives

$$\frac{du_{SS}}{d\tau} = -\frac{\phi}{\kappa + \phi}u_{SS} \tag{20.84}$$

This variables separable differential equation has the solution

$$u_{SS} \cong e^{-\frac{\phi}{\kappa + \phi}\tau} \tag{20.85}$$

From equation (20.83) the equation for v can also be calculated,

$$v_{SS} \cong \frac{e^{-\left(\frac{\phi}{\kappa + \phi}\right)\tau}}{\kappa + \phi} \tag{20.86}$$

Using the QSSA simple rate equations can be developed which may be integrated to produce approximate expressions for the concentration profiles as a function of time. The QSSA will only hold if the concentration of the intermediate is small and effectively constant. From equation (20.77),

$$\frac{dw_{SS}}{d\tau} \approx \frac{\phi}{\kappa + \phi}e^{-\left(\frac{\phi}{\kappa + \phi}\right)\tau} \tag{20.87}$$

This can be integrated to give

$$w_{SS} \approx 1 - e^{-\left(\frac{\phi}{\kappa + \phi}\right)\tau} \tag{20.88}$$

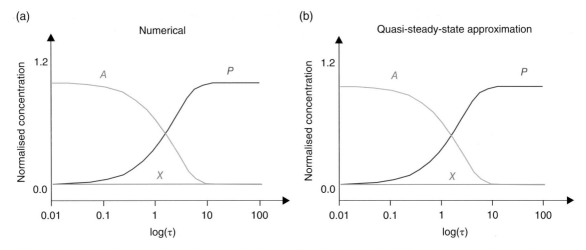

Figure 20.18 *(a) Concentrations of reactant A, intermediate X and product P for two consecutive reactions as a function of time calculated numerically. The intermediate concentration is low and relatively constant. (b) Identical two consecutive reaction concentrations as a function of time calculated using the quasisteady-state approximation. There is good agreement between the numerical calculation and the quasi-steady-state approximation.*

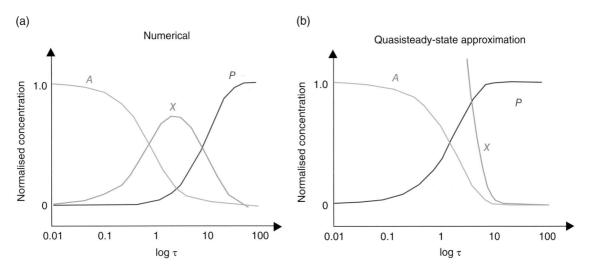

Figure 20.19 *(a) Concentrations of reactant A, intermediate X and product P for two consecutive reactions as a function of time calculated numerically. The intermediate concentration is large and varies with time. (b) Identical two consecutive reactions calculated using the quasisteady-state approximation. There is poor agreement between (a) and (b) at intermediate concentrations due to the failure of the quasisteady-state approximation.*

Concentration versus log time plots are shown in **Figure 20.18a** for reactant A, intermediate X and product P, when full set of coupled rate equations are solved without any approximation, i.e. equations (20.75)–(20.77) where $k_{-1} \gg k_1$, $k_2 \gg k_1$ and $k_{-1} = k_2 = 50$. The concentration of intermediate X is very small and approximately constant throughout the time course of the experiment. Concentration versus log time curves for reactant A, intermediate X and product P when the rate equations are solved using the QSSA are shown in **Figure 20.18b**. Values for the rate constants are the same as those used above. QSSA reproduces the concentration profiles very well and it is an accurate approximation.

QSSA will hold when the concentration of the intermediate reactants is small and constant. Hence, the rate constants for reduction of the intermediate reactants (k_{-1} and k_2) must be much larger than that for intermediate generation (k_1). Concentration versus log time curves for reactant A, intermediate X and product P can be calculated when a full set of coupled rate equations are solved without any approximation using $k_{-1} \ll k_1, k_2$ and $k_{-1} = k_2 = 0.1$. The concentration of intermediate is high and it is present throughout much of the duration of the experiment (**Figure 20.19a**). The QSSA is not good for the prediction of how the intermediate concentration varies with time and so it does not apply under the condition where the concentration of intermediate is high and the intermediate is long lived. The QSSA is thus a poor model of the reaction kinetics in **Figure 20.19b**.

20.9   General Kinetic Equation Analysis

More sophisticated reaction schemes in general can be modelled using matrices,

$$\frac{dx}{dt} = \underline{S}.k\left(\underline{x}\right) \tag{20.89}$$

where S is a stoichometric matrix, k is the reaction rate (flux) vector and x is the vector of the reactant concentrations. For example, for the four-step chemical reaction

$$x_1 \xrightleftharpoons{k_1} x_2 \xrightleftharpoons{k_2} x_3 \xrightarrow{k_3} x_4 \tag{20.90}$$

the reaction kinetics can be calculated using

$$\underline{x} = \begin{pmatrix} x_1 \\ x_2 \\ x_3 \\ x_4 \end{pmatrix}, \quad S = \begin{pmatrix} 1 & 0 & 0 \\ 1 & -1 & 0 \\ 0 & 1 & -1 \\ 0 & 0 & 1 \end{pmatrix}, \quad \text{and} \ \ \underline{k}(\underline{x}) = \begin{pmatrix} k_1 \\ k_2 \\ k_3 \end{pmatrix} \tag{20.91}$$

Reference should be made to more specialist systems biology texts for details, e.g. 'Systems Biology' by O. Palsson. There are lots of efficient numerical packages for the evaluation of solutions to such matrix equations, e.g. in MatLab, that can be found as freeware on the World Wide Web.

Suggested Reading

If you can only read one book, then try:

Atkins, P.W. & de Paulo, J. (2011) *Physical Chemistry for Life Sciences*, 2nd edition, Oxford University Press. Useful introduction to physical chemistry.

Keener, J. & Sneyd, J. (2008) *Mathematical Physiology*, Springer. Mathematically advanced treatment of chemical reactions that are important in physiology.

Palsson, B.Ø. (2011) *Systems Biology: Simulation of Dynamic Network States*, Cambridge University Press. Good robust chemical engineering approach to systems biology.

Price, N.C., Dwek, R.A., Ratcliffe, R.G. & Wormald, M.R. (2009) *Physical Chemistry for Biochemists*, Oxford University Press. Simple introduction to the principles of physical chemistry.

Tutorial Questions 20

20.1 State the law of mass action.

20.2 For a first-order reaction, derive the reaction equation and the half-lifetime.

20.3 For a first-order reaction, describe the relationship between the reaction rate and the reactant concentration.

20.4 For a chemical reaction

$$ZnS_{(s)} + 2HCl_{(aq)} \rightarrow H_2S_{(g)} + ZnCl_{2\,(aq)}$$

write down the reaction rate for ZnS, HCl, H_2S and $ZnCl_2$. Write down the stoichiometric coefficient for each element.

20.5 Outline an experimental method to determine the reaction order and sketch the relationship between the reaction rate and reaction order.

20.6 State the fundamental physics laws that apply to biological systems.

20.7 State the definitions of Gibbs and Helmholtz free energy. Derive the Gibbs free energy from the Helmholtz free energy.

20.8 For a biophysical/biochemical reaction: $xA + yB \xrightarrow{k} C$

   a. Write down the rate of reaction and state the law of mass action.
   b. Design an experimental method to determine the reaction order for reactant A and B.

20.9 State the differences between the 0-, 1st- and 2nd-order biophysical/biochemical reactions in terms of reaction equation.

20.10 Consider a reaction $A \xrightarrow{k_1} X \xrightarrow{k_2} P$
where k_1 is the reaction constant for A to X and k_2 is the reaction constant for X to P. Write down the reaction equations and derive solutions for $a(t)$, $x(t)$ and $p(t)$.

20.11 Solve equation (20.69) for a first-order reversible reaction to give equation (20.70). Use the boundary conditions $u = 1$ and $v = 0$ when $\tau = 0$.

21

Enzyme Kinetics

Many reactions inside live cells are carefully controlled by enzymes. A key defining aspect of an enzyme is that it modifies a reaction rate (or rates) without being used up in the reaction. A single enzyme can be used consecutively in millions of reactions and its activity can be modulated, subject to the requirements of a cell, at a particular instant in time. Enzymes can either speed up or slow down reactions dependent on the particular regulatory mechanism that is in action. The discussion of enzyme kinetics will begin with an analysis of classical models for their dynamics that are around one hundred years old and finishes with a modern perspective based on experiments with single enzyme molecules, e.g. single motor proteins.

21.1 Michaelis–Menten Kinetics

Reversible reaction kinetics are the most generally observed reaction mechanism between two reactants (due to the condition of detailed balance),

$$A + B \underset{k_-}{\overset{k_+}{\rightleftarrows}} C \tag{21.1}$$

Since the quantity A is removed by the forward reaction and produced by the reverse reaction, the rate of change the concentration of A (a) for this bidirectional reaction is

$$\frac{da}{dt} = k_- c - k_+ ab \tag{21.2}$$

where c and b are the concentration of C and B, respectively. Furthermore, k_+ and k_- are the forward and backward rate constants, respectively. At equilibrium, the concentrations are constant,

$$\frac{da}{dt} = \frac{db}{dt} = \frac{dc}{dt} = 0 \tag{21.3}$$

The Physics of Living Processes: A Mesoscopic Approach, First Edition. Thomas Andrew Waigh.
© 2014 John Wiley & Sons, Ltd. Published 2014 by John Wiley & Sons, Ltd.

Substitution in equation (21.2) leads to a simple relationship for the equilibrium concentration of C,

$$c = \frac{k_+}{k_-} ab \tag{21.4}$$

There are no other reactions that involve A and C, and therefore conservation of mass implies

$$a + c = a_0 \tag{21.5}$$

where a_0 is a constant, equal to the initial concentration of A, i.e. $a = a_0$ when $t = 0$. Combination of equations (21.4) and (21.5) gives

$$c = a_0 \frac{b}{K_{eq} + b} \tag{21.6}$$

where the *equilibrium constant* (K_{eq}) is given by $K_{eq} = k_- / k_+$. By inspection, equation (21.6) implies when the concentration of B is equal to the equilibrium constant ($b = K_{eq}$), half of A is in the bound state.

The Michaelis–Menten mechanism for enzyme kinetics requires a slightly more sophisticated reaction scheme than equation (21.1) (**Figure 21.1**),

$$E + S \underset{k_{-1}}{\overset{k_1}{\rightleftharpoons}} ES \xrightarrow{k_2} E + P \tag{21.7}$$

where k_1, k_{-1} and k_2 are reaction constants. E is the enzyme, S is the substrate and P is the product. Using the law of mass action equation (20.26), it is assumed that the rate of reaction is proportional to the reactant concentrations. Therefore, the reaction equations are

$$\frac{ds}{dt} = k_{-1}x - k_1 se \tag{21.8}$$

$$\frac{de}{dt} = (k_{-1} + k_2)x - k_1 se \tag{21.9}$$

$$\frac{dx}{dt} = k_1 se - (k_2 + k_{-1})x \tag{21.10}$$

$$\frac{dp}{dt} = k_2 x \tag{21.11}$$

where $s = [S]$ is the substrate (reactant) concentration, $e = [E]$ is the enzyme concentration, $x = [ES]$ is the substrate/enzyme complex concentration and $p = [P]$ is the product concentration. The *quasisteady-state approximation* implies that the intermediate concentration is a constant ($x = x_{ss}$ and $dx_{ss}/dt = 0$), so that equation (21.10) becomes

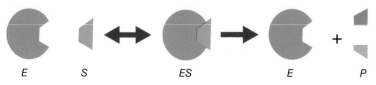

$$E \qquad S \qquad\qquad ES \qquad\qquad E \qquad P$$

Figure 21.1 *Reaction of an enzyme (E) with a substrate (S) to form an intermediate complex (ES) and finally the products (P). In this example the substrate (S) is split into two products (P).*

$$\frac{dx_{ss}}{dt} = k_1 es - k_{-1} x_{ss} - k_2 x_{ss} \approx 0 \tag{21.12}$$

Conservation of the enzyme concentration implies

$$e + x = e_\Sigma = e_0 \tag{21.13}$$

where e_Σ is the total enzyme concentration and e_0 is the initial value of the unbound enzyme concentration. Combination with equation (21.12) gives

$$k_1(e_\Sigma - x_{ss})s - k_{-1} x_{ss} - k_2 x_{ss} \approx 0 \tag{21.14}$$

which can be rearranged as

$$x_{ss} \approx \frac{k_1 s e_\Sigma}{k_{-1} + k_2 + k_1 s} \tag{21.15}$$

From equation (21.11) the rate of production of P is

$$R_\Sigma \approx k_2 x_{ss} = \frac{k_1 k_2 s e_\Sigma}{k_{-1} + k_2 + k_1 s} = \frac{k_2 e_\Sigma s}{\dfrac{k_{-1} + k_2}{k_1} + s} = \frac{k_C s e_\Sigma}{K_M + s} = k_E e_\Sigma \tag{21.16}$$

The Michaelis constant (mol dm^{-3}) is therefore defined as $K_M = \dfrac{k_{-1} + k_2}{k_1}$, the catalytic rate constant (s^{-1}) is $k_C = k_2$ and the enzyme rate constant (s^{-1}) is $k_E = \dfrac{k_C s}{K_M + s}$. Combination of these definitions gives

$$\frac{1}{k_E} = \frac{1}{k_C / K_M} s^{-1} + \frac{1}{k_C} = \frac{1}{k_U} s^{-1} + \frac{1}{k_C} \tag{21.17}$$

where

$$k_U = \frac{k_C}{K_M} = \frac{k_1 k_2}{k_{-1} + k_2} \tag{21.18}$$

The time taken for E and S to combine to form the enzyme complex ES is defined as τ_U and the time taken for complex ES to generate and release a product is defined as τ_C. The total time for the enzymatic reaction to proceed (τ_E) is therefore

$$\tau_E = \tau_U s^{-1} + \tau_C \tag{21.19}$$

Experimentally, the measured rate of reaction is

$$R_\Sigma = -\frac{ds}{dt} = \frac{\alpha s}{\beta + s} \tag{21.20}$$

where α and β are constants, and s is the reactant concentration. This is in agreement with the model equation (21.16).

A *modified mechanism* for the Michaelis–Menten reaction includes an intermediate step that is equivalent to a reversible final step,

$$E + S \; \underset{\longleftarrow}{\longrightarrow} \; ES \; \underset{\longleftarrow}{\longrightarrow} \; EP \longrightarrow E + P \tag{21.21}$$

A mathematically equivalent equation is

$$E + S \underset{k_{-1}}{\overset{k_1}{\rightleftharpoons}} ES \underset{k_{-2}}{\overset{k_2}{\rightleftharpoons}} E + P \tag{21.22}$$

Therefore, the only modification required in the rate equations (21.8)–(21.11) is in the final (now reversible) step, i.e. equation (21.11) is replaced by

$$\frac{dp}{dt} = k_2 x - k_{-2} e p \tag{21.23}$$

As before $e_\Sigma = x + e$, where e_Σ is the total initial enzyme concentration, x is the bound enzyme concentration ES, and e is the free enzyme concentration. Usually, $e_\Sigma << s$ and after mixing there is an initial period during which $x = [ES]$ builds up. Then, the equilibrium concentration of ES is assumed to be rapidly attained and reaches a constant low value during the course of the reaction. This requirement satisfies the QSSA,

$$\frac{dx}{dt} = \{k_1 s + k_{-1} + k_2 + k_{-2} p\} x_{SS} - k_1 s e_\Sigma - k_{-2} p e_\Sigma \approx 0 \tag{21.24}$$

The equation can be rearranged and the QSSA analysis thus implies

$$x_{ss} = \frac{(k_1 s + k_{-2} p) e_\Sigma}{k_1 s + k_{-2} p + k_{-1} + k_2} \tag{21.25}$$

And the rate of production of P is

$$R_\Sigma = -\frac{ds}{dt} = \frac{\{k_1 k_2 s - k_{-1} k_{-2} p\} e_\Sigma}{k_1 s + k_{-2} p + k_{-1} + k_2} \tag{21.26}$$

The measurement of the reaction rate is assumed to occur during a time period when only a small percentage (1–3%) of the substrate is transformed in to the product. This leads to $p \approx 0$ and $s \approx s_0$ for the initial substrate concentration. Therefore

$$R_\Sigma \approx R_{\Sigma,0} \approx \frac{k_1 k_2 e_\Sigma s_0}{k_{-1} + k_2 + k_1 s_0} = \frac{k_2 e_\Sigma s_0}{\dfrac{k_{-1} + k_2}{k_1} + s_0} \tag{21.27}$$

This has the same form as the empirical rate equation observed experimentally, equation (21.20). $K_M = \dfrac{k_{-1} + k_2}{k_1}$ is defined for the Michaelis constant (mol dm^{-3}) and the catalytic rate constant (s^{-1}) is $k_C = k_2$, which leads to the *Michaelis–Menten* (MM) equation for steady-state enzyme kinetics,

$$R_{\Sigma,0} = \frac{k_C e_\Sigma s_0}{K_M + s_0} \tag{21.28}$$

where K_M describes the enzyme/substrate binding and k_C describes decomposition of enzyme/substrate complex. Thus a number of different reaction mechanisms give rise to a similar type of Michaelis-Menten equation.

21.2 Lineweaver–Burke Plot

The *Michaelis constant* (K_M) and the rate constant (k_C) can be evaluated experimentally using a least squares fit to Michaelis–Menten equation (21.28). Linearisation of the MM equation facilitates this process. Equation (21.28) can be rearranged to give

$$\frac{1}{R_{\Sigma,0}} = \frac{K_M}{k_C e_\Sigma} \frac{1}{s_0} + \frac{1}{k_C e_\Sigma} \tag{21.29}$$

Thus, a plot of $1/R_{\Sigma,0}$ (reciprocal rate of production of P at low concentrations) versus $1/s_0$ (reciprocal of the initial substrate concentration) is often used and named after *Lineweaver and Burke*. The gradient of the plot is defined to be S_{LB} and the y-intercept is I_{LB}, which are identified as

$$S_{LB} = \frac{K_M}{k_C e_\Sigma}, \quad I_{LB} = \frac{1}{k_C e_\Sigma} \tag{21.30}$$

A composite rate constant k_Σ can be defined to give

$$k_\Sigma = \frac{k_C s_0}{K_M + s_0} \tag{21.31}$$

This in turn simplifies equation (21.29) for the Lineweaver–Burke plot (**Figure 21.2**) and gives

$$\frac{1}{k_\Sigma} = \frac{K_M}{k_C s_0} + \frac{1}{k_C} \tag{21.32}$$

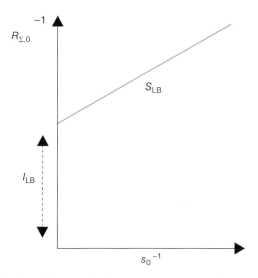

Figure 21.2 *Lineweaver–Burke plot of the Michaelis–Menten reaction of an enzyme with a substrate. s_0 is the initial substrate concentration and $R_{\Sigma,0}$ is the reaction rate. S_{LB} is calculated from the gradient and I_{LB} from the y intercept (see equations (21.30) for the definitions of the symbols).*

21.3   Enzyme Inhibition

An enzyme inhibitor is a substance that inhibits the catalytic action of the enzyme. Enzyme inhibition is a common feature of enzyme reactions and is an important mechanism to control enzyme activity. For example, cyanide and nerve gases are irreversible inhibitors that can decrease enzyme activity to zero (affecting cytochrome C and ion pumps respectively), with catastrophic effects for the organism.

An enzyme is usually a large protein molecule, considerably larger than the substrate molecule whose reaction is catalysed, although there are occasional exceptions to the rule, e.g. autocatalysis of RNA chains. Embedded in the large enzyme protein are one or more *active sites*, to which the substrate can bind to form a complex. An enzyme catalyses a single reaction, or similar reactions on a family of substrates with similar structures. The specificity of the enzyme for its substrate is a steric property of the enzyme that results from the three dimensional shape of the enzyme, and allows it to fit in a *lock-and-key* fashion with a corresponding substrate molecule.

The rate of enzyme activity can be reduced using *competitive* inhibitors or *allosteric* inhibitors. *Competitive* inhibitors occur when another molecule has a similar shape to that of the substrate molecule. Thus, the inhibitor may also bind to the active site, which prevents the binding of a substrate molecule, and thus inhibits the reaction. The inhibitor competes with the substrate molecule for the active site, which gives it its name (a competitive inhibitor). An example is cyanide compounds (found naturally in many plants to reduce predation) which inhibit cytochrome C in mitochondria and thus shut down energy metabolism. Similarly, nerve gases and snake poisons can block enzymes that control neurotransmitters. Competitive inhibition has also been important for fundamental biochemical studies, e.g. specific ion channels in membranes can be switched off, so a detailed model of their individual contributions to whole-cell electrical behavior can be created.

Allosteric inhibition occurs when the enzyme molecule has other binding sites, distinct from the active site, and the binding to these sites affects the activity of the enzyme at the active site. These binding sites are called *regulatory sites,* since binding to them can regulate the catalytic activity of the protein. The ligand that binds at the allosteric site is called an *effector* or *modifier.* If it increases the activity of the enzyme it is called an allosteric activator; conversely if it decreases the activity of the substrate it is called an allosteric inhibitor. The allosteric effect is thought to arise because of a conformational change inside the enzyme, e.g. a change in the folding of its polypeptide chain.

21.4   Competitive Inhibition

The simplest example of a competitive inhibitor, is where the reaction is stopped when the inhibitor is bound to the active site of the enzyme. There is thus an additional reaction that occurs in parallel with the Michaelis–Menten scheme,

$$S + E \underset{k_{-1}}{\overset{k_1}{\rightleftarrows}} C_1 \xrightarrow{k_2} E + P \tag{21.33}$$

$$E + I \underset{k_{-3}}{\overset{k_3}{\rightleftarrows}} C_2 \tag{21.34}$$

The concentrations are defined as $s = [S]$ for the substrate, $i = [I]$ for the inhibitor, $c_1 = [C_1]$ for the active enzyme/substrate complex, $c_2 = [C_2]$ for the inactive enzyme inhibitor complex and from the law of mass action

$$\frac{ds}{dt} = -k_1 se + k_{-1} c_1 \tag{21.35}$$

$$\frac{di}{dt} = -k_3 ie + k_{-3} c_2 \tag{21.36}$$

$$\frac{dc_1}{dt} = k_1 se - (k_{-1} + k_2) c_1 \tag{21.37}$$

$$\frac{dc_2}{dt} = k_3 ie - k_{-3} c_2 \tag{21.38}$$

From the law of conservation of mass, $e + c_1 + c_2 = e_0$. To be systematic, the next step is to introduce dimensionless variables, and identify those reactions that are rapid and equilibrate rapidly to their quasisteady states. The enzyme to substrate ratios are assumed to be small, the fast equations are those for c_1 and c_2. Hence, the quasisteady states are found by setting $\frac{dc_1}{dt} = 0$, $\frac{dc_2}{dt} = 0$ and solutions are found for c_1 and c_2. This does not mean that c_1 and c_2 are invariant, rather that they change in a quasisteady-state fashion, where the right-hand sides of these rate equations is kept close to zero. This leads to

$$c_1 = \frac{K_i e_0 s}{K_m i + K_i s + K_m K_i}, \; c_2 = \frac{K_m e_0 i}{K_m i + K_i s + K_m K_i} \tag{21.39}$$

where

$$K_m = \frac{k_{-1} + k_2}{k_1}, \; K_i = \frac{k_{-3}}{k_3} \tag{21.40}$$

Thus, the velocity of the reaction is

$$V = k_2 c_1 = \frac{k_2 e_0 s K_i}{K_m i + K_i s + K_m K_i} = \frac{V_{max} s}{s + K_m(1 + i/K_i)} \tag{21.41}$$

The solution resembles a rescaled Michaelis–Menten equation (21.28). The effect of the inhibitor is to increase the effective equilibrium constant of the enzyme by a constant factor, i.e. from K_m to $K_m(1 + i/K_i)$. Thus, the inhibitor decreases the velocity of reaction, while the maximum velocity is left unchanged.

21.5 Allosteric Inhibition

When the inhibitor can bind to an allosteric site, it is possible that the enzyme can bind both the inhibitor and the substrate simultaneously. In this case, there are four possible binding states for the enzyme, and transitions occur between them,

$$
\begin{array}{ccccc}
E & \underset{k_{-1}}{\overset{k_1 s}{\rightleftharpoons}} & ES & \overset{k_2}{\longrightarrow} & E + P \\
k_3 i \uparrow\downarrow k_{-3} & & k_3 i \uparrow\downarrow k_{-3} & & \\
EI & \underset{k_{-1}}{\overset{k_1 s}{\rightleftharpoons}} & EIS & &
\end{array}
\tag{21.42}
$$

The simplest analysis of this reaction scheme is for the equilibrium state. The constants $K_s = k_{-1}/k_1$, $K_i = k_{-3}/k_3$ are defined. x, y and z denote, respectively, the concentrations of ES, EI and EIS. From the law of mass action in the steady state the following equations hold

$$(e_0 - x - y - z)s - K_s x = 0, \quad (e_0 - x - y - z)i - K_i y = 0, \quad ys - K_s z = 0, \quad xi - K_i z = 0 \qquad (21.43)$$

This is a linear system of equations for x, y, and z, where $e_0 = e + x + y + z$ is the total amount of enzyme. Although there are four equations, one is a linear combination of the other three (the system is of rank three), so x, y, and z can be determined as functions of i and s,

$$x = \frac{e_0 K_i}{K_i + i} \frac{s}{K_s + s} \qquad (21.44)$$

It follows that the reaction rate, $V = k_2 x$, is given by

$$V = \frac{V_{max}}{1 + i/K_i} \frac{s}{K_s + s} \qquad (21.45)$$

where $V_{max} = k_2 e_0$. In contrast to the competitive inhibitor, the allosteric inhibitor decreases the maximum velocity of the reaction, while K_s, the equilibrium constant, is left unchanged.

21.6 Cooperativity

For many enzymes, the reaction velocity is not a simple hyperbolic curve, as predicted by the Michaelis–Menten model, but often has a sigmoidal character. This can result from cooperative effects, in which the enzyme is able to bind to more than one substrate molecule at a single time, and the binding of one substrate molecule affects the binding of subsequent ones. Suppose that an enzyme can bind two substrate molecules, so that it can exist in one of three states, namely as a free molecule E, as a complex with one occupied center C_1, and as a complex with two occupied centers C_2. The parallel reaction mechanism is therefore represented by two equations,

$$S + E \underset{k_{-1}}{\overset{k_1}{\rightleftarrows}} C_1 \overset{k_2}{\to} E + P, \qquad S + C_1 \underset{k_{-3}}{\overset{k_3}{\rightleftarrows}} C_2 \overset{k_4}{\to} C_1 + P \qquad (21.46)$$

Using the law of mass action, the rate equations for the concentrations $[S]$, $[E]$, $[C_1]$, $[C_2]$, and $[P]$ can be determined. The amount of product $[P]$ can be calculated by integration, and because the total amount of enzyme molecule is conserved, only three equations are needed for the three quantities $[S]$, $[C_1]$, and $[C_2]$. The concentrations $s = [S]$, $c_1 = [C_1]$, $c_2 = [C_2]$ are defined, and from the law of mass conservation of the enzyme, $e + c_1 + c_2 = e_0$. Therefore, the three rate equations are

$$\frac{ds}{dt} = -k_1 se + k_{-1} c_1 - k_3 s c_1 + k_{-3} c_2 \qquad (21.47)$$

$$\frac{dc_1}{dt} = k_1 se - (k_{-1} + k_2) c_1 - k_3 s c_1 + (k_4 + k_{-3}) c_2 \qquad (21.48)$$

$$\frac{dc_2}{dt} = k_3 s c_1 - (k_4 + k_{-3}) c_2 \qquad (21.49)$$

As before, the quasisteady-state assumption is used, so $dc_1/dt = dc_2/dt = 0$. The equations can be solved for c_1 and c_2,

$$c_1 = \frac{K_2 e_0 s}{K_1 K_2 + K_2 s + s^2}, \quad c_2 = \frac{e_0 s^2}{K_1 K_2 + K_2 s + s^2} \tag{21.50}$$

where $K_1 = \dfrac{k_{-1} + k_2}{k_1}$ and $K_2 = \dfrac{k_4 + k_{-3}}{k_3}$. The reaction velocity is thus given by

$$V = k_2 c_1 + k_4 c_2 = \frac{(k_2 K_2 + k_4 s) e_0 s}{K_1 K_2 + K_2 s + s^2} \tag{21.51}$$

Consider two extreme cases of these reaction kinetics. In the *first case* the active sites act independently and identically,

$$k_1 = 2k_3 = 2k_+, \quad S + E \underset{k_{-1}}{\overset{k_1}{\rightleftharpoons}} C_1 \xrightarrow{k_2} E + P \tag{21.52}$$

$$2k_{-1} = k_{-3} = 2k_-, \quad S + C_1 \underset{k_{-3}}{\overset{k_3}{\rightleftharpoons}} C_2 \xrightarrow{k_4} C_1 + P \tag{21.53}$$

where $2k_2 = k_4$. k_+ and k_- are the forward and backward reaction rates for the individual binding sites, respectively. The factors of two occur because two identical binding sites are involved in the reaction, which doubles the amount of the reactant. Therefore, equation (21.51) becomes

$$V = \frac{k_2 e_0 (K + s) s}{K^2 + 2Ks + s^2} = 2\frac{k_2 e_0 s}{K + s} \tag{21.54}$$

where $K = k_-/k_+$ is the equilibrium constant for the individual binding site. As expected, the rate of reaction is exactly twice that for the individual binding site.

For the *second case*, suppose that the binding of the first substrate molecule is slow, but with one site bound, binding of the second is fast (this is large positive cooperativity). This can be modeled by letting $k_3 \to \infty$ and $k_1 \to 0$, while keeping $k_1 k_3$ constant, in which case $K_2 \to 0$ and $K_1 \to \infty$, while $K_1 K_2$ is constant. In this limit, the velocity of the reaction is

$$V = \frac{k_4 e_0 s^2}{K_m^2 + s^2} = \frac{V_{\max} s^2}{K_m^2 + s^2} \tag{21.55}$$

where $K_m^2 = K_1 K_2$, $V_{\max} = k_4 e_0$, $K_1 = \dfrac{k_{-1} + k_2}{k_1}$ and $K_2 = \dfrac{k_4 + k_{-3}}{k_3}$. In general, if n substrate molecules can bind to the enzyme, there are n equilibrium constants, K_i, where $i = 1, \ldots, N$. In the limit as $K_n \to 0$ and $K_1 \to \infty$, while $K_1 K_n$ is kept fixed, the rate of reaction is

$$V = \frac{V_{\max} s^n}{K_m^n + s^n} \tag{21.56}$$

where $K_m^n = \prod_{i=1}^{n} K_i$, i.e. K_m^n is the product of the K_is. This rate equation (21.56) is known as the *Hill equation* and can be used to describe reactions with cooperative enzymes.

21.7 Hill Plot

Typically, the Hill equation is used for reactions whose detailed intermediate steps are not known, but for which cooperative behavior is suspected. The exponent n and the parameters V_{max} and K_m for the Hill plot are usually determined from experimental data. The logarithm of the Hill equation (21.56) can be taken

$$n \ln s = n \ln K_m + \ln\left(\frac{V}{V_{max} - V}\right) \tag{21.57}$$

A plot of $\ln(V/(V_{max} - V))$ against $\ln s$ (called a *Hill plot*) will result in a straight line of slope n. Although the exponent n suggests an n-step process (with n binding sites), in practice it is not unusual for the best fit for n to be noninteger. Thus, more detailed modeling is necessary to develop more realistic models in these cases.

An enzyme can also exhibit negative cooperativity, in which the binding of the first substrate molecule *decreases* the rate of subsequent binding. This can be modeled by decreasing k_3. In **Figure 21.3** the reaction velocity against the substrate concentration is plotted for the cases of independent binding sites (no cooperativity), extreme positive cooperativity (the Hill equation) and negative cooperativity. For positive cooperativity, the reaction velocity is a sigmoidal function of the substrate concentration, while negative cooperativity primarily decreases the maximum velocity.

21.8 Single Enzyme Molecules

Advances in instrumentation , principally fluorescence microscopy, enable kinetics of enzymes to be observed on a single molecule basis, e.g. the reaction of motor proteins enzyme with single ATP molecules. **Figure 21.4** shows the fluorescence yield from a single cholesterol oxidase enzyme. The data can be segmented into *on* and *off* phases, where each on phase corresponds to a completed enzymatic cycle. For single enzyme kinetics the

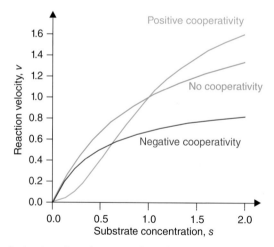

Figure 21.3 The reaction velocity (v) plotted against the substrate concentration (s) for an enzyme reaction. Negative cooperativity (orange), positive cooperativity (blue) and no cooperativity (green) can be observed. The behaviour can be modelled using the Hill equation.

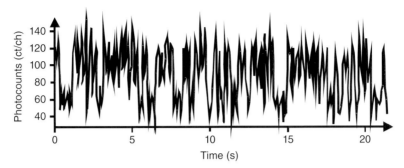

Figure 21.4 *Fluorescent emission photocounts from a single cholesterol oxidase molecule as a function of time. On–off cycles can be observed that correspond to a completed enzymatic reaction. [From H.P. Lu, L. Xun, S. Xie Science, 1998, 282, 5395. Reprinted with permission from AAAS.]*

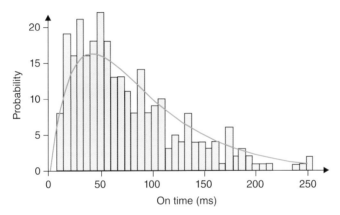

Figure 21.5 *Histogram for the on times of a single cholesterol oxidase molecule shown in Figure 21.4 with a fit of equation (21.58) for a Poisson process. [From H.P. Lu, L. Xun, S. Xie Science, 1998, 282, 5395. Reprinted with permission from AAAS.]*

concentration of the species in the Michaelis–Menten equation needs to be replaced by the probability of a cycle being completed. The dwell time distribution for the *on* phases gives information about the enzyme kinetics (**Figure 21.5**). It is found that the dwell time probability distribution ($p(\tau)$) with n rate-limiting internal states (each a Poisson process) is given by

$$p(\tau) = \frac{k^n \tau^{n-1} e^{-k\tau}}{(n-1)!} \tag{21.58}$$

where k is the reaction rate for each step. In the limit of large n, $p(\tau)$ becomes a Gaussian distribution. The number of rate-limiting states in a single enzyme reaction can thus be calculated from equation (21.58). An alternative strategy uses the randomness parameter (r) to calculate the number of internal states, defined as

$$r = \frac{\langle \tau^2 \rangle - \langle \tau \rangle^2}{\langle \tau \rangle^2} \tag{21.59}$$

For n independent Poisson internal processes it is found that $r \sim 1/n$. For example, the randomness parameter can be used to determine how many ATP molecules are needed for a kinesin nanomotor (an ATPase, i.e. an enzyme that hydrolyses ATP molecules) to move 8 nm. Equation (21.58) is, however, preferable if the data is sufficiently well resolved to allow its use, since fewer assumptions are required for its derivation. Furthermore, more sophisticated models of single enzyme wait times enable many of the assumptions used in deriving equation (21.58) to be relaxed, e.g. that the lifetimes of each internal state are equal.

Suggested Reading

If you can only read one book, then try:

Top recommended book: Cornish-Bowden, A. (2004) *Fundamental of Enzyme Kinetics*, 3rd edition, Portland Press, Simple, but authoritative account of enzyme kinetics with a relatively low mathematics content.

Cook, P.F. & Cleland, W.W. (2007) *Enzyme Kinetics and Mechanisms*, Garland Science. Tour of the physical chemistry of enzymes.
Keener, J. & Sneyd, J. (2008) *Mathematical Physiology*, Springer. Advanced mathematical account of models for enzyme kinetics.
Xie, S. (2001) Single molecule approach to enzymology, *Single Molecule*, **2** (4), 229–236. Compact account of some recent developments with single-molecule enzyme kinetics.

Tutorial Questions 21

21.1 Given a reversible reaction: $A + B \underset{k_-}{\overset{k_+}{\rightleftharpoons}} C$

    i. Based on the law of mass action, write down the reaction equation for reactant A.

    ii. Write down the constraints of the reaction based on the mass conservation law (assume that $[B]$ does not change before and after reaction).

    iii. Derive the equilibrium state of $[C]$ based on the answers to (i) and (ii).

    iv. State the physical meaning of K_{eq}.

22

Introduction to Systems Biology

The living cell cannot be viewed as just a bag filled with a random assortment of chemical reactants and their products. Intracellular chemistry is under strict control by genetic machinery. Complex gene networks regulate the cellular environment.

A challenge is to reverse engineer the chemical circuits that exist in live cells, i.e. discover what Nature has developed and then find out how. Much of the control engineering inside cells occurs at the genetic level (DNA expression) and thus uncovering the action of genetic transcription networks is an important task.

Systems biology has spawned a range of confusing specialist vocabulary. One of these is the 'ohmics' family of words, e.g. *proteo-ohmics* is the study of large-scale structures (that include networks) of proteins. Similarly, there is *cytomics* (the interaction of cells), *genomics* (the interaction of genomes), *lipidomics* (the interaction of lipids), *metabolomics* (the interaction of metabolites), *phosphoproteomics* (the interaction of phosphorylated proteins, which are key for some signalling tasks), and *secretohmics* (the interaction of a subset of secreted extracellular proteins). Thus, a modern task for a systems biologist might be to map the entire lipidome of a cell, i.e. all the lipids in a particular cell type combined with their interactions.

Systems biology seeks to understand how the global behaviour of a whole organism emerges from the local interactions between its multiple molecular components. Cellular processes involve interactions between proteins, genes, metabolites, and other molecules in cell structure, metabolism, gene regulation, and signal transduction.

22.1 Integrative Model of the Cell

An integrative model for the cell aims to describe the mapping between environmental signals, transcription factors inside the cell, and gene regulation. The environmental signals activate specific transcription factor proteins. The transcription factors, when active, bind to DNA, change the transcription rate of specific target genes, and thus alter the rate at which mRNA is produced. The mRNA is then translated into proteins. Hence, transcription factors regulate the rate at which proteins encoded by genes are produced. These proteins affect

The Physics of Living Processes: A Mesoscopic Approach, First Edition. Thomas Andrew Waigh.
© 2014 John Wiley & Sons, Ltd. Published 2014 by John Wiley & Sons, Ltd.

the environment both internally and externally. Some proteins are themselves transcription factors that activate or repress other genes.

The study of large complex gene regulatory networks requires powerful experimental tools that must be high throughput, low cost, reliable, and precise. Detailed information can be obtained from experiments in genomics and bioinformatics such as DNA/gene sequences (various chemical sequencing methods), interactions between proteins and DNA (microarray techniques), and temporal variation of gene products (microarrays and mass spectrometry).

The huge number of parameters involved and their complex interactions, mean that current measurements of gene regulation are largely qualitative. A big challenge for the future is to put gene regulation measurements on a more quantitative footing. Computational modelling is indispensable for the dynamical analysis of gene regulatory networks. Simulations allow more precise and unambiguous descriptions of a network to be made that provide systematic predictions of its behaviour.

A variety of mathematical formalisms have been developed to describe integrative models of the cell such as logical algorithms (Boolean functions and directed graphs) and stochastic nonlinear differential equations. For example, intracellular concentrations of proteins, mRNAs, and other molecules at a time-point (t) can be represented by a variable $x_i(t)$. Regulatory interactions can then be modelled by differential equations such as

$$\frac{d\underline{x}}{dt} = f\left(\underline{x}\right) \tag{22.1}$$

where $\underline{x} = [x_1, \ldots, x_n]$ are the concentrations of the n interacting components ($x_i(t)$), and $f(\underline{x})$ is a rate law that can be nonlinear. No analytical solutions exist for most nonlinear differential equations. Approximate solutions can be obtained by numerical simulation, given reasonable parameter value estimates and the initial concentrations of the interacting components. Steady-state solutions can be found and their stability can be estimated using methods of dynamical systems analysis.

22.2   Transcription Networks

A transcription network is a description of the interactions between genes that are transcribed into proteins or regulatory RNA. In **Figure 22.1** a network of metabolic genes (blue) involved in energy metabolism is shown. Genes that control biosynthesis are connected by protein transcription factors (green), which act as network hubs. These transcription factors can also regulate other genes associated with nonmetabolic processes. Nodes on the figure represent genes (or groups of genes called operons). An edge directed from node X to node Y indicates that the transcription factor encoded in the operon X regulates the operon Y. To establish the complete gene network for an organism is very labour intensive and the creation of such networks depends on the results of many experiments from many different labs.

22.3   Gene Regulation

Genetic codes for proteins are essential for the development and functioning of an organism. Gene expression involves a number of stages. Transcription of DNA from RNA is followed by translation of RNA into protein and finally post-translational modification of the proteins is possible, e.g. phosphorylation of proteins is important in signalling.

The expression of a gene during development can be measured in a number of ways. Enzyme activity can be quantified when the promoter of interest drives the expression of an enzyme that can cleave a molecule that in the cleaved state is coloured. Also the promoter of interest can drive the expression of a fluorescent protein such as GFP. The amount of fluorescence in the sample then reports on the extent of expression of a gene.

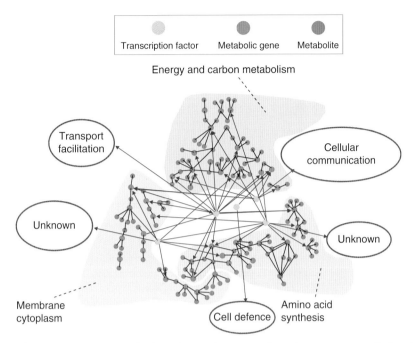

Figure 22.1   *E.coli transcription network that shows the interactions between transcription factors (green), genes (blue) and metabolites (orange).*

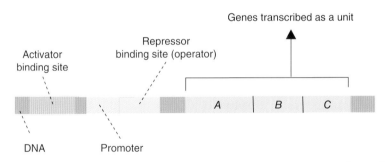

Figure 22.2   *Control regions before a gene (or a sequence of genes A, B, C) on a chain of DNA can include activators, promoters and repressors. The genes are transcribed as a unit (A, B, C) called an operon.*

In vivo gene transcription regulation is usually controlled by a regulatory DNA region called the *promoter* that is found before the protein/regulatory RNA expressing gene (**Figure 22.2**). The promoter contains a specific site (a DNA sequence) that can bind RNA polymerase (RNAP), a complex of several proteins that forms an enzyme that can synthesise an mRNA chain that corresponds to the gene coding sequence. *Activators* are transcription factor proteins that increase the rate of mRNA transcription when they bind the promoter (**Figure 22.3**). The activator typically transfers rapidly between active and inactive forms. In its active form, it has a high affinity to a specific site (or sites) on the promoter DNA. Repressors are transcription factor proteins that decrease the rate of mRNA transcription when they are bound to the promoter. There are a wide variety of different repressor molecules that interact with DNA.

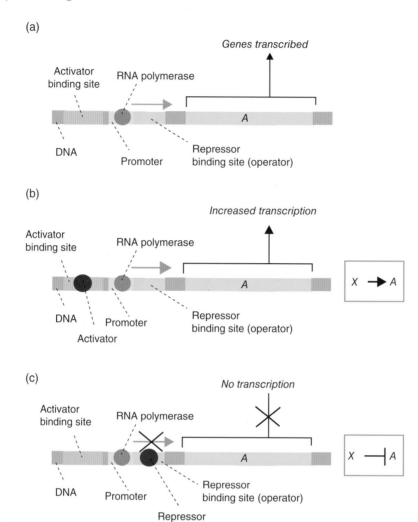

Figure 22.3 *(a) Simple regulation of a gene* A *on a chain of DNA. (b) Binding of an activator increases the expression of gene* A. *(c) Binding of a repressor reduces the expression of gene* A.

The transcription rate of genes can thus be speeded up or slowed down with activators or repressors respectively. As a reference calculation it is initially useful to calculate the response time for a simple gene regulation process, which involves no activators or regulators. Consider a part of a gene network in which gene X regulates gene Y,

$$X \to Y \tag{22.2}$$

Production of Y is balanced by protein degradation and dilution,

$$\alpha = \alpha_a + \alpha_b \tag{22.3}$$

where α is the total degradation/dilution rate (unit s^{-1}), α_a is the dilution rate, and α_b is the degradation rate. The cell is assumed to produce a protein (concentration Y) at a constant rate β (units M s^{-1}), i.e. zeroth-order reaction kinetics occur with respect to protein creation and first order kinetics occur with respect to protein degradation. Therefore, the total reaction rate is

$$\frac{dY}{dt} = \beta - \alpha Y \tag{22.4}$$

For steady state conditions $dY/dt = 0$, and therefore the steady-state concentration of Y (Y_s) is

$$Y_s = \frac{\beta}{\alpha} \tag{22.5}$$

The steady-state concentration is thus the ratio of the production and degradation/dilution rates. When no additional Y is made ($\beta = 0$) there is no input signal to the regulation process and the solution is just simple first-order kinetics for the decay of Y (equation (20.36)),

$$Y(t) = Y_s e^{-\alpha t} \tag{22.6}$$

The response time for regulation is given by the standard half life for the first order reaction (equation 20.38),

$$T_{1/2} = \ln(2)/\alpha \tag{22.7}$$

Thus, the production rate (β) affects the steady state level of Y, but not the response time. A large increase in Y production, which corresponds to a strong input to the network, implies the solution of equation (22.4) is

$$Y(t) = Y_s(1 - e^{-\alpha t}) \tag{22.8}$$

The concentration of Y rises from zero signal and gradually converges to a steady state again given by

$$Y_s = \frac{\beta}{\alpha} \tag{22.9}$$

and the response time is still given by the same expression for the half-life, equation (22.7). The larger the degradation/dilution rate (α) the faster the changes in concentration. Typical timescales for the reactions in the transcription network of the bacterium E.coli are shown in **Table 22.1**.

22.4 Lac Operon

One of the landmark studies in the understanding of transcription networks was Jacques Monod's work on the lac operon in the 1950s, the gene network that handles lactose metabolism in E.coli. In principle, it describes phenomena that occur to E.coli in a person's stomach when they drink a glass of milk. E.coli ignore the lactose

Table 22.1 *Time scales for gene regulation in E.coli cells.*

Process	Time scale
A small molecule (a signal) binds to a transcription factor.	~1 ms
An active transcription factor binds to its DNA site.	~1 s
A gene is transcribed and translated.	~5 min
A 50% change in concentration of the translated protein occurs,	~20–60 min
A DNA chain replicates,	~20–60 min
A whole cell replicates,	~20–60 min

from the milk when glucose is available in the environment, since lactose is an inferior energy source for them. However, upon the loss of glucose in the environment, gene circuits are switched on for lactose metabolism, and the regulatory network is sufficiently simple and well characterised that a quantitative analysis of the components can be made.

The lac operon is an example of *negative regulation*. The operon is defined as the cluster of genes that are involved in the same general process (**Figure 22.4**). The genes share the same promoter and are transcribed together. The *lac* operon consists of three genes that are required for the metabolism of lactose (**Z, Y, A**). In addition a gene **I**, which codes for the repressor, lies nearby the *lac* operon and is always expressed. When

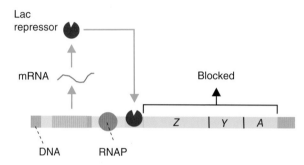

Figure 22.4 *Negative autoregulation is used in the lac operon of E.coli on its DNA chain to control the expression of genes Z, Y and A needed to metabolise lactose. Z, Y and A are called an operon.*

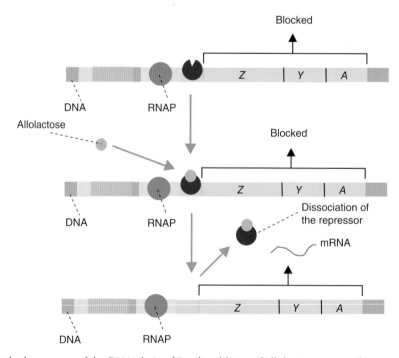

Figure 22.5 *In the lac operon of the DNA chain of E.coli, addition of allolactose causes the repressor to dissociate and allows expression of the genes Z, Y and A by RNA polymerase (RNAP) required for lactose metabolism. Z, Y and A are called an operon.*

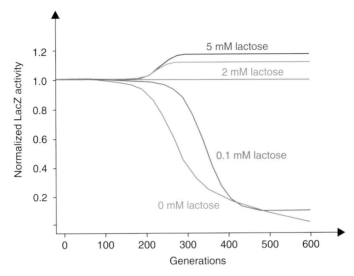

Figure 22.6 *Adaptation of the activity of the LAC operon in E.coli to different concentrations of lactose in their environment as a function of the number of cell generations. The LAC operon activity is monitored through the action of gene X [Reprinted by permission from Macmillan Publishers Ltd: Nature, E. Dekel, U. Alon, Nature, 2005, 436, 588–592, copyright 2005.]*

lactose is missing from the growth medium, the repressor binds very tightly to a short DNA sequence downstream of the promoter near the start of the *lac* gene **Z,** called the operator (**O**). The promoter **P** is bound by RNA polymerase, whereas operator **O** is bound by the *lac* repressor. The operator **O** blocks the movement of the polymerase along the DNA chain and prevents the transcription of genes **Z**, **Y** and **A** (**Figure 22.5**).

Gene-regulation networks converge again and again to the same transcription network activity. Laboratory evolution experiments show that bacterial cells reach optimal *lac* **Z** levels in a few hundred generations (**Figure 22.6**). In the experiments, E.coli cells are grown in tubes filled with specific levels of lactose. The *lac* **Z** activity of the cells is measured as a function of the Lactose concentration in the environment, relative to wild-type cells. The *lac* **Z** activity quickly approaches a constant value over a wide range of lactose concentrations; it is a robust carefully controlled parameter.

22.5 Repressilator

Real gene networks can be extremely complicated, which makes the isolation of the specific activity of individual components hard to explore. The repressilator represents a simple experimental model for oscillatory processes in a gene network, e.g. for circadian rhythms in sleeping/waking cycles. The repressilator can provide relatively robust timekeeping and is a classic example of synthetic biology, i.e. the creation of a new biochemical circuit in a living organism. A simple gene network is constructed from three genes and inserted into the genome of an E.coli cell (**Figure 22.7**). The resultant genetically modified bacterium has a GFP reporter that causes the bacterium to fluoresce once a specific gene in the network is switched on. The set of three genes inhibit one another in a loop (a continuous rock, paper, scissors game, where the GFP highlights one of the states). Thus, a movie of a repressilator bacterium in a fluorescence microscope shows the bacterium flash on and off in a regular manner.

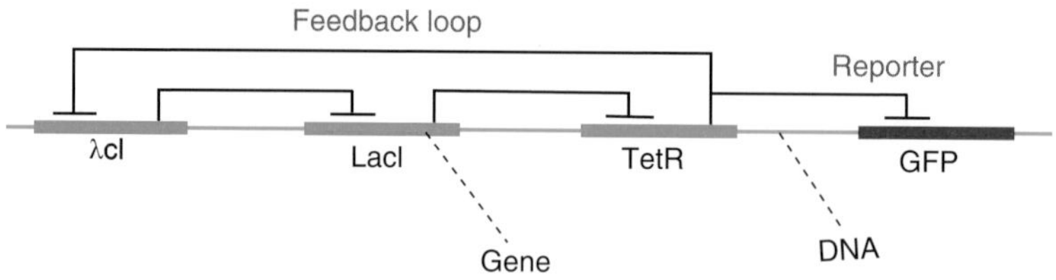

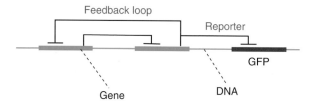

Figure 22.7 *The gene network for a repressilator with a GFP reporter gene along a single strand of DNA. λcl represses Lacl, which represses TetR, which in turn represses λcl and the GFP in a feedback loop.*

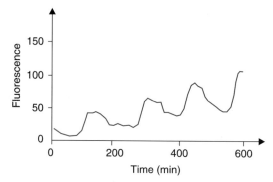

Figure 22.8 *The gene network for a toggle switch with a GFP reporter gene along a single section of DNA.*

Figure 22.9 *Oscillations in the fluorescence of an E.coli cell as a function of time in which a repressilator circuit has been incorporated. [Reprinted by permission from Macmillan Publishers Ltd: Nature, M.B. Elowitz, S. Leibler, Nature, 2000, 403, 335–338, copyright 2000.]*

Creation of a model of the repressilator represents an interesting theoretical problem. It is an example of a cross-inhibition network and can be described by two coupled differential equations,

$$\dot{u} = \frac{\alpha_1}{1+v^\beta} - u, \qquad \dot{v} = \frac{\alpha_2}{1+u^\gamma} - v \qquad (22.10)$$

where u and v are the concentrations of repressors 1 and 2, respectively. α_1 and α_2 are the rate constants for the synthesis of the two repressors. β and γ are cooperativity parameters that describe the enzyme kinetics of promoters 2 and 1, respectively. Equation (22.10) was first developed to describe the activity of the gene network of a toggle switch (**Figure 22.8**), but can also be used to calculate the dynamics of the repressilator. Toggle switches can cause long-term activation or depression of a gene, whereas the repressilator continually oscillates. Bistability and hysteresis are observed in the model described by equation (22.10), as well as the repressilator oscillations, all of which are experimentally observed. **Figure 22.9** shows oscillations in the fluorescence of an E.coli cell in which a repressilator circuit has been incorporated into its gene control network.

22.6 Autoregulation

An example of negative autoregulation is shown in **Figure 22.10**. A gene X is negatively autoregulated when it is repressed by its own gene product, the repressor X. The gene is also simply regulated by A. The repressor X binds to a site on its own promoter and thus acts to repress its own transcription. The symbol $\perp$ stands for repression on the figure.

Regulation of a gene by its own gene product is called *autoregulation* and it is a network motif. *Negative autoregulation* is found to reduce the response time of gene circuits. Consider the negative autoregulation of a gene Y, which creates a protein of concentration Y. Similar to equation (22.4) the reaction rate is given by the sum of two terms, but the enzymatic activity of the promoter causes Y to be created with a rate described by a function $f(Y)$,

$$\frac{dY}{dt} = f(Y) - \alpha Y \tag{22.11}$$

where $f(Y)$ to a good approximation for many promotors is described by the Hill equation (21.56), which can be rewritten in the form

$$f(Y) = \frac{\beta}{1 + (Y/K)^n} \tag{22.12}$$

The repression threshold is K, which is defined as the concentration of Y needed to reduce the promoter activity to 50%. To simplify the maths assume $f(Y) \approx \beta\theta(Y < K)$ and therefore

$$\frac{dY}{dt} = \beta - \alpha Y \tag{22.13}$$

where θ is a function that is equal to unity when $X < K$ and zero otherwise. At early times $\alpha Y << \beta$ so

$$Y(t) \sim \beta t \tag{22.14}$$

Thus, when $Y < K$, $Y << \beta/\alpha$. This model for the dynamics of a negatively autoregulated gene product is shown in **Figure 22.11**. The results are shown for when production starts at $t = 0$, the maximal production rate (β) *is* 5, the autorepression threshold (K) is 1 and the degradation/dilution rate (α) is 1. The dashed blue line shows a process of simple regulation which approaches a higher, unrepressed steady state, $Y_{st} = \beta/\alpha = 5$. At long times with negative autoregulation Y effectively locks into a steady state level (Y_{st}) equal to the repression coefficient of its own promoter,

$$Y_{st} = K \tag{22.15}$$

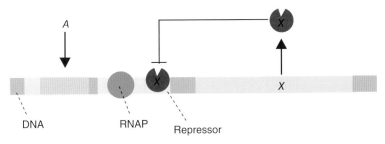

Figure 22.10 *Simplified schematic diagram of negative autoregulation motif along a DNA chain (also shown in more detail in **Figure 22.4**). The gene X represses its own expression.*

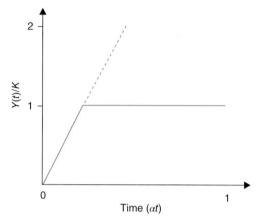

Figure 22.11   *Dynamics of the negative autoregulation of the rescaled concentration of protein Y (green curve) as a function of the rescaled time (αt). The concentration of the protein Y increases linearly with rescaled time (αt) and then saturates at long times, so that Y = K. The blue dashed curve shows the results of simple regulation.*

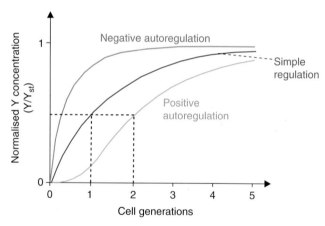

Figure 22.12   *Comparison of the concentration of a protein Y as a function of the number of cell generations that is negatively autoregulated, simply regulated or positively autoregulated. [Reprinted from U.Alon, An Introduction to Systems Biology, Chapman, 2007. With kind permission from Springer Science + Business Media B.V.]*

The resultant dynamics show a rapid rise and a sudden saturation. The response time $(T_{1/2})$ for negative autoregulation $(Y/Y_{st} = 0.5)$ is found to be

$$T_{1/2} = \frac{K}{2\beta} \tag{22.16}$$

The dynamics of a negatively autoregulated gene, a simply regulated gene and a positively autoregulated gene are compared in **Figure 22.12**. The negatively and positively autoregulated genes are assumed to have a Hill input function with a Hill coefficient (n) of 1 (equation (22.12)). The protein concentration is normalised by its steady-state value Y/Y_{st}, after an increase in the production rate. The response time is found by the intersection of the dynamics with a horizontal line at $Y/Y_{st} = 0.5$. Models for negative autoregulation are in reasonable agreement with experiment.

Thus, in conclusion, *negative autoregulation* is seen to speed up the response of gene circuits. A further advantage is that negative autoregulation promotes robustness to fluctuations in the production rate of a protein. In contrast, *positive autoregulation* slows the responses of gene networks and can lead to bistability.

22.7 Network Motifs

To form a self-edge from a node in an autoregulation motif, the edge needs to choose its own node of origin as a destination out of the N possible target nodes in the network. Therefore, the probability of a self-edge (p_{self}) is

$$p_{\text{self}} = \frac{1}{N} \tag{22.17}$$

When E edges are placed at random in the network, the probability of k self-edges ($p(k)$) chosen at random is given by a binomial distribution (equation (3.11)),

$$P(k) = \frac{E!}{k!(E-k)!} p_{\text{self}}^k (1 - p_{\text{self}})^{E-k} \tag{22.18}$$

The average number of self-edges ($<N_{\text{self}}>$) is equal to the number of edges (E) placed on a graph multiplied by the probability of a self-edge (equation (22.17)),

$$\langle N_{\text{self}} \rangle \sim E p_{\text{self}} \sim \frac{E}{N} \tag{22.19}$$

And the standard deviation (σ_{rand}) for such a probability distribution is the square root of the mean,

$$\sigma_{\text{rand}} \sim \sqrt{\frac{E}{N}} \tag{22.20}$$

For a transcription network in E.coli the number of network nodes (N) is 424 and the number of edges (E) is 519, so equations ((22.19)) and ((22.20)) predict $\langle N_{\text{self}} \rangle \sim 1.2$ and $\sigma_{\text{rand}} \sim 1.1$. In contrast the real network has 30 self-edges, which exceeds the expected random network value by many standard deviations. Autoregulation is thus seen to be a *strong network motif*.

The autoregulation motif has one edge and one node, but motifs with larger patterns of nodes and edges also exist. In general, network motifs are called subgraphs. Two examples of three node subgraphs are the three-node feedback loop and the three-node feedforward loop, which are shown in **Figure 22.13**. There are 13 distinct ways to connect three nodes with directed edges, 199 four-node subgraphs and 9364 five-node subgraphs.

For E.coli transcription networks there are 42 feedforward loops and no feedback loops. For a random network there would be expected to be about 1.7 loops with a standard deviation $\sim \sqrt{1.7}$. The 3-node *feedforward loop* is thus also seen to be a strong network motif in E.coli. In sensory transcription networks, it is found that the feedforward loop is the only significant network motif of the 13 possible three-node patterns.

The eight possible sign combinations of 3-node feedforward loops (FFLs) are shown in **Figure 22.14**. In coherent FFLs, the indirect path has the same overall sign as the direct path. The reverse is true for incoherent FFLs, where the sign of the indirect path is opposite to that of the direct path.

Figure 22.15 shows the relative abundance of the eight FFL types in the transcription networks of yeast and E.coli. The most abundant FFL is type 1 coherent followed by type 1 incoherent.

Feedforward loops have a number of uses in transcription networks. They can provide a sign-sensitive delay to protect against brief input fluctuations. Also, they can act as a pulse generator. Furthermore, they can

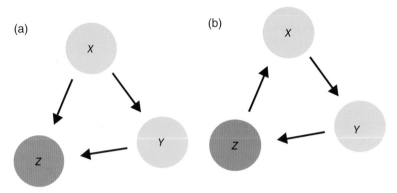

Figure 22.13   *Two examples of three-node subgraphs, (a) three-node feed-forward loop, and (b) three-node feedback loop between the genes X, Y and Z.*

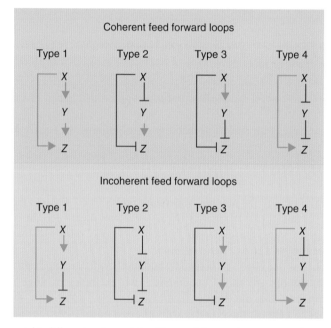

Figure 22.14   *There are eight different coherent feedforward loops that commonly occur in gene transcription networks. Four of the loops are coherent (direct and indirect paths have the same sign), whereas four of the loops are incoherent (direct and indirect paths have different signs). [Reprinted from U.Alon, An Introduction to Systems Biology, Chapman, 2007. With kind permission from Springer Science + Business Media B.V.]*

increase response rates (using an incoherent FFL). As an example, the construction of the flagellar motor of E. coli is regulated by a multioutput FFL.

A *genetic switch* can be constructed from network motifs, e.g. coherent feedforward loops. Such a switch turns a gene 'on' or 'off', and they are particularly useful in the determination of the long-term outcomes of developmental programmes in organisms. In **Figure 22.16** there are two promoters that are under the transcriptional control of the gene product of the partner promoter and the gene architecture acts as a *toggle switch*. It is similar in action to the gene network shown in **Figure 22.8**.

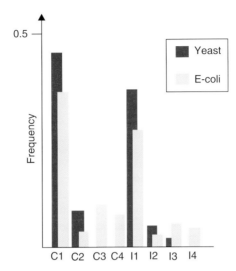

Figure 22.15   *The relative abundance (frequency) of the eight feedforward loops shown in* **Figure 22.14** *in yeast and E.coli cells. [Reproduced from S.S. Mangen et al, J.Mol.Biol, 2006, 356, 10731081. With permission from Elsevier.]*

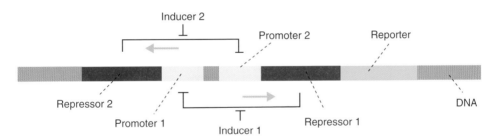

Figure 22.16   *A toggle switch in a gene transcription network (compare with* **Figure 22.8**).

There are a number of strong network motifs in cells (**Figure 22.17**), but only coherent feedforward and autoregulation loops will be examined in detail here.

Logic functions can be performed by network motifs. The coherent type-1 FFL can provide an AND input function (**Figure 22.18**). The transcription factor X activates the gene encoding transcription factor Y, and both X and Y are required to jointly activate gene Z (AND logic). There are 8 types of FFL that are possible, and each can appear with at least 2 types of input function (AND, OR).

In type-1 coherent FFL, the protein X is activated by an input signal X, which causes it to assume the active conformation X^* (**Figure 22.19**). It then binds to the promoters of genes Y and Z. As a result, protein Y accumulates, and, in the presence of an input signal Y, adopts an active state Y^*. When the Y^* concentration crosses the activation threshold K_{yz}, Y^* binds to the promoter of gene Z. The protein Z is produced when both X^* and Y^* bind the promoter of gene Z (an AND input function). It is assumed that initially ($t = 0$) a strong signal to X triggers the activation of X, which causes a step-like stimulation of X. The transcription factor X transits to its active form X^*. X^* then binds to the promoter of gene Y, which initiates production of protein Y (the second transcription factor in the FFL). In parallel, X^* also binds

Motif	Diagram	Functions
Negative autoregulation	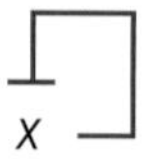	Can speed the response time and reduce cell–cell variability of the gene concentration.
Positive autoregulation	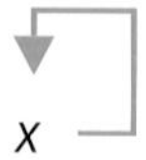	Slows the response time of gene expression and can exhibit bistability.
Coherent feedforward loop (C1-FFL)		Provides a sign-sensitive delay. Filters out brief ON input pulses when the Z-input function is AND logic and OFF pulses when the input function is OR logic.
Incoherent feedforward loop (I1-FFL)		Can generate pulses of gene activity and provides sign-sensitive response acceleration.
Single-input module (SIM)	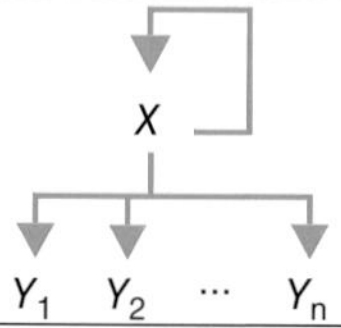	Allows coordinated control with temporal (LIFO) order of promoter activity.
Multioutput feedforward loop (multioutput FFL)	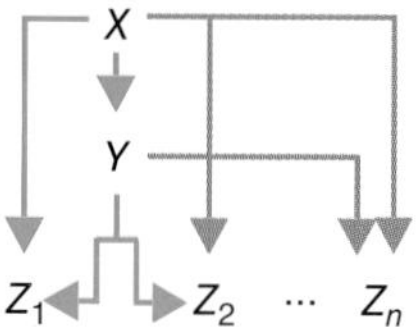	Acts as FFL for each input (with a sign-sensitive delay) with FIFO temporal order of promoter activity.
Bifan	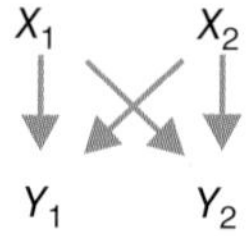	Provides combinatorial logic based on multiple inputs that depends on the input function of each gene.
Dense overlapping regulons (DOR)	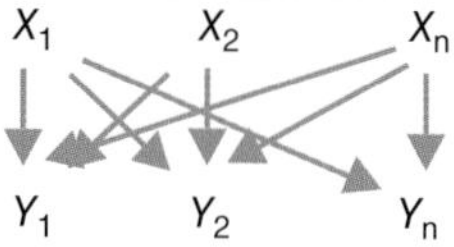	Also can provide combinatorial logic.
Two-node positive feedback	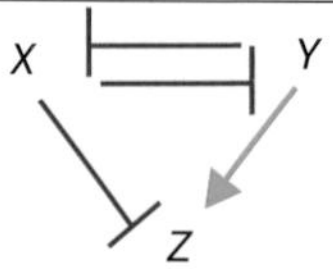	Can create a toggle switch used in developmental processes.

Figure 22.17 A range of examples of strong network motifs that are found in regulatory gene networks.

to the promoter of gene Z. Since the input of Z is AND logic, X^* alone cannot activate Z production. Production of Z requires both X^* and Y^*. Y must build up to the *activation threshold* for gene Z. This results in a delay for Z production. Mathematically, the equations needed to analyse the dynamics of the type-1 coherent FFL are

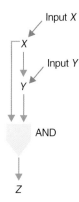

Figure 22.18   *Feedforward loops in gene networks can be constructed that provide logic functions such as AND (and OR, not shown). Both a* X *input and a* Y *input are required for there to be a* Z *output.*

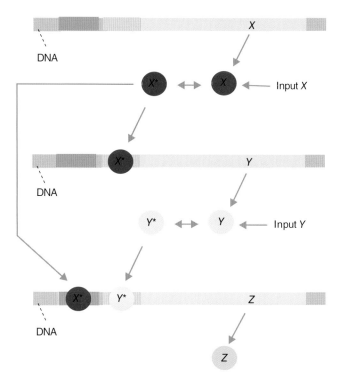

Figure 22.19   *A coherent feedforward loop formed from the interaction of genes* X, Y *and* Z, *also shown in **Figure 22.18**.*

$$\frac{dY}{dt} = \beta_y \theta\left(X^* > K_{xy}\right) - \alpha_y Y \qquad (22.21)$$

$$\frac{dZ}{dt} = \beta_z \theta(X^* > K_{xz}) \theta\left(Y^* > K_{yz}\right) - \alpha_z Z \qquad (22.22)$$

where Y and Z are the concentrations of protein Y and Z, respectively. θ is a step function. The C1-FFL is a *sign-sensitive delay element*. There is a delay time in switching on Z, but not in switching it off. It is possible to study the dynamics of the coherent type-1 FFL with AND logic which follow an ON step of X input at time $t = 0$. The activation threshold of Z by Y is K_{yz}. The production and degradation rates are $\alpha_y = \alpha_z = 1$, $\beta_y = \beta_z = 1$. The delay in Z production is T_{ON}. For simplicity assume throughout that a Y input is present. The transcription factor Y is thus in its active form Y^*. After the delay which follows an ON step of X input, Y^* begins to be produced at a rate β_y. Therefore, the concentration of Y begins to exponentially converge to a steady-state level (**Figure 22.20**), the solution to equation (22.21) is identical to equation (22.7),

$$Y^*(t) = Y_{st}(1 - e^{-\alpha_y t}) \tag{22.23}$$

$$Y_{st} = \frac{\beta_y}{\alpha_y} \tag{22.24}$$

Due to its AND logic functionality, Z begins to be expressed only after a delay time. The delay time (T_{ON}) is the time needed for Y^* to reach its threshold. From equation (22.23)

$$Y^*(T_{ON}) = Y_{st}\left(1 - e^{-\alpha_y T_{ON}}\right) = K_{yz} \tag{22.25}$$

This equation can be rearranged for the delay time,

$$T_{ON} = \frac{1}{\alpha_y} \ln\left[\frac{1}{1 - K_{yz}/Y_{st}}\right] \tag{22.26}$$

where α_y is the decay rate of protein Y and K_{yz} is the activation threshold. In bacteria K_{yz} is typically 3–10 times lower than Y_{st}, and T_{ON} is a few minutes to a few hours.

A system of genes in E.coli allows cells to grow on the sugar arabinose. The arabinose gene system creates proteins that transport the sugar arabinose into the cell and break it down as an energy and carbon source. Arabinose is only used when glucose is not present, since it is an inferior food source for E.coli. The arabinose system makes a decision based on two inputs: the concentration of arabinose and the concentration of glucose in

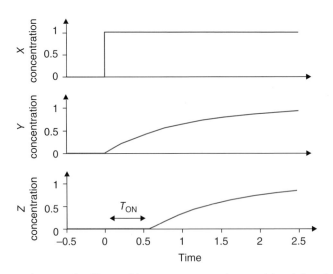

Figure 22.20 *The type 1 coherence feedforward loop can act as a sign sensitive delay element for the expression of a gene Z. The concentration of the proteins X, Y and Z is shown as a function of time.* T_{ON} *is the delay time.*

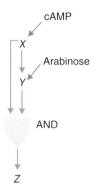

Figure 22.21 *The type 1 coherent feedforward loop found in the arabinose system of E.coli. X is activated by a cAMP signal and Y by an arabinose signal. The FFL provides AND logic for the expression of gene Z. Both cAMP and arabinose are needed for there to be an output at Z.*

its environment. Proteins are made when arabinose AND NOT glucose are present. A C1-FFL occurs as a sign-sensitive delay in the arabinose system of E. coli with the AND NOT logical function (**Figure 22.21**). This FFL is activated upon glucose starvation by the activator cAMP (a molecule produced upon glucose starvation), and in the presence of arabinose. The input function is an AND gate. The C1-FFL allows the system to wait a little to see if the production of the arabinose enzymes is really necessary and thus buffers it against temporary fluctuations in its food source. A delay is found after ON steps of cAMP, but not after OFF steps.

Sensory transcription networks predominantly are constructed from four families of network motifs (**Figure 22.17**). Two FFLs occur with multiple outputs, and two larger motifs, single-input module (SIM) and dense overlapping regions (DOR). Together they can create sophisticated temporal programs for gene expression. Developmental networks often include other motifs such as the two-node positive-feedback toggle switch.

Network motifs are a useful principle to guide the analysis of network structures. However, overemphasis on motifs can give a false picture of the design principles of regulatory networks. Network designs develop through a stochastic process of evolution and many functional *in vivo* designs diverge from the idealisations of perfect modular network motifs.

22.8 Robustness

Biological circuits are designed in a robust manner so they can function with a high level of independence with respect to both external and internal biochemical parameters. A large variation in such parameters is observed from cell to cell, although the cells continue to function well as integrated systems (they continue to live) and they are thus deduced to be robustly designed.

Noise in electrical circuits is often very bad news for electrical devices. Nature has evolved strategies to overcome chemical noise and in some cases use it to the cell's advantage. Furthermore, noise can determine how organisms adapt to their environment. Signalling molecules that appear in the environment of a cell could occur over a vast range of concentrations. Thus, a detector with a single set point is rarely useful, since it is not robust. Nature requires the behaviour should be adaptive. It is possible to follow this adaptation in real time in a wide range of biological systems.

Formally, noise is often separated in terms of *intrinsic* and *extrinsic* varieties, dependent on whether its origin is internal or external to the cell, respectively. Some stochastic events are thus due to cell to cell variability (extrinsic noise) and others are due to stochastic events inside individual cells (intrinsic noise).

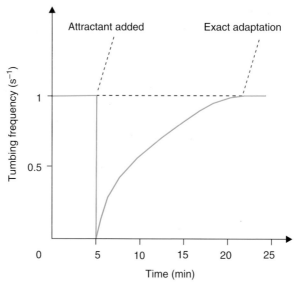

Figure 22.22 *The tumbling frequency of the E.coli as a function time. The tumbling frequency varies as the attractant is added at a time of 5 min. Adaptation occurs after the attractant is added and the tumbling frequency returns to its original level [Reprinted from U.Alon, An introduction to systems biology, Chapman, 2007. With kind permission from Springer Science + Business Media B.V.]*

Bacterial adaptation to variations in their environments can be observed in real time in the motion of bacteria in response to attractants. Runs and tumbles of the bacteria are related to the direction of rotation of the flagellar motors (**Section 7.3**), there can be up to 8 flagellae per bacterium. When all the motors spin counterclockwise (CCW), the flagella turn in a bundle and the cell is propelled forward. When one or more motors turn clockwise (CW), the cell tumbles and randomises its orientation. The switching dynamics of a single motor from CCW to CW and back again can be seen by tethering a cell to a surface by one flagellum hook, so that the motor spins the entire cell body (at frequencies of only a few hertz due to the large viscous drag of the body).

The average tumbling frequency of a population of E.coli cells exposed at a time of 5 min to a step increase in the concentration of saturating attractant (such as aspartate) is shown in **Figure 22.22**. After 5 min the attractant is uniformly present at a constant concentration. Adaptation means that the effect of the stimulus is gradually forgotten despite its continued presence. Exact adaptation is a perfect return to prestimulus levels, a steady-state tumbling frequency that does not depend on the level of attractant. The protein circuits quickly adapt to high concentrations of attractant (even higher concentrations would then be required to further decrease the tumbling rate). The process of adaptation is a robust one.

22.9 Morphogenesis

Patterning in development is an extremely important set of events in an organism's life cycle, i.e. it determines the shape in which the organism is formed. A lot is known quantitatively about the genetics involved in morphogenesis from studies of development in simple organisms such as fruit flies. Morphogenesis is found to be very robust against variations of the morphogen concentration (signalling molecules), which seems reasonable, since unstable pattern formation would lead to improperly constructed organisms and will eventually result in an organism's death.

Spatial patterns require positional information. This information is carried by gradients of the morphogen signalling molecules. Patterns created by the motion of morphogens across an organism can be modelled by a *reaction–diffusion* (diffusion–degradation) equation,

$$\frac{\partial M}{\partial t} = D\frac{\partial^2 M}{\partial x^2} - \alpha M \tag{22.27}$$

where M is the concentration of the morphogen, D is the diffusion coefficient, x is the position in the organism, t is the time and α describes the reaction rate at which the morphogen is used up. Equation ((22.27)) is just Fick's second law (equation (7.18)) combined with first-order kinetics for morphogen degradation (equation (20.35)). Consider a morphogen produced at the origin that diffuses into a field of cells. Under steady state conditions the morphogen concentration does not change with time, so

$$\frac{\partial M}{\partial t} = 0 \tag{22.28}$$

The solution of equation (22.27) is then an exponential decay

$$M(x) = M_0 e^{-x/\lambda} \tag{22.29}$$

where $\lambda = \sqrt{D/\alpha}$. The morphogen gradient can lead to a French flag pattern if there are two separate morphogen thresholds present and such a pattern is shown in **Figure 22.23**. Morphogen M is produced at the origin and diffuses into a field of cells. The morphogen is degraded as it diffuses, which results in a steady-state concentration profile that decays exponentially with distance from the source at the origin. Cells exposed to the morphogens assume fate A if the concentration of M is greater than a first threshold, fate B if M is between thresholds one and two, and fate C if M is lower than the second threshold. The reaction-diffusion mechanism can thus create a simple hypothetical organism with three stripes. Robustness is not so critical for stripe formation on the coating of an organism (e.g. stripes on a tiger), but for the creation of body parts extreme

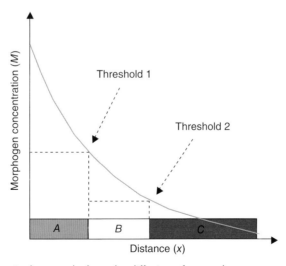

Figure 22.23   *A French flag tricolour results from the diffusion of a morphogen across a material in one dimension with two threshold levels. The morphogen concentration is shown as a function of perpendicular distance (x) across a two-dimensional sheet of the material.*

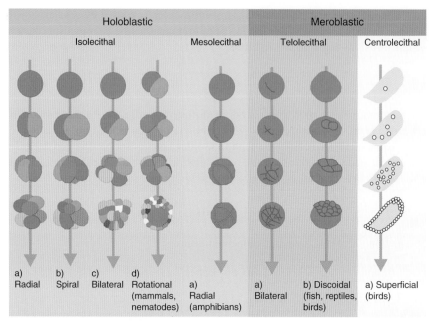

Figure 22.24 *Physical forces determine the development of animal embryos. A single fertilised cell divides into a blastula (ball of cells) and symmetry breaking then occurs that creates a template for the organism's final morphology. Different types of blastula are observed in different families of organism.*

sensitivity to the boundary conditions would be a disaster. Unfortunately, equation (22.27) is fairly sensitive to the choice of the parameters and boundary conditions, but nonlinear modifications can improve its robustness and thus the model's biological credibility.

Multicellular organisms all develop from a single fertilised cell. Although the final organism can be very complicated, it is known that the complexity observed in development is in some sense superficial, since the genome has a limited amount of stored information with which to program the different stages of growth. It is simply impossible for every stage in development to be programmed in molecular detail by information contained in the genome, instead Nature makes use of simple physical forces (surface tension, elasticity, gravity, etc.) to guide the developmental processes. The developmental process is found to be surprisingly robust. Removal of a single cell at an early stage in the developmental process does not necessarily affect the morphology of the final organism (it can for example result in identical twins).

Organisms are often surprisingly symmetrical (**Figure 22.24**). It is fascinating to observe how the developmental process both breaks and enforces the symmetries of the initial ball of cells (the blastula), e.g. the predominantly bilateral symmetry in humans is provided with a slight chiral perturbation by cilial motors during early stages of development, which leads to the position of the heart on the left side of the body and other slight left/right asymmetries. The initial stages of morphogenesis are evolutionary very well conserved over a wide range of vertebrate organisms.

22.10 Kinetic Proofreading

Biochemical recognition systems can often pick out a specific molecule in a sea of millions of other molecules that bind to it with only a slightly weaker affinity. Kinetic proofreading gives a possible physical mechanism to explain this process. For example kinetic proofreading of the genetic code in mRNA translation is thought to reduce error rates in molecular recognition of amino acids, and similarly it is a possible mechanism to explain the sensitivity of the immune response of organisms to pathogens.

During the translation process of an mRNA chain into a protein an equilibrium process might be expected to occur for the association of the mRNA/ribosome (C) complex with a new candidate amino acid (a) for the addition reaction,

$$a + C \underset{k'_c}{\overset{k_c}{\rightleftharpoons}} aC \rightarrow \text{Correct amino acid added} \tag{22.30}$$

where k_c is the association constant and k'_c is the dissociation constant of the amino acid. However, such a simple scheme can generate very high error rates and would create too many proteins with the wrong sequence. The error rate in an equilibrium process of recognition is determined by the ratio of dissociation constants (k'_c) for the correct and incorrect tRNAs which carry the candidate amino acids. It is the off rate that distinguishes the correct codes from the incorrect one. In fact an irreversible intermediate step with an additional off rate occurs *in vivo* (l'_c) and causes a large improvement in fidelity,

$$a + C \underset{k'_c}{\overset{k_c}{\rightleftharpoons}} aC \longrightarrow a^*C \longrightarrow \text{Correct amino acid added}$$

$$\downarrow l'_c$$

$$a + C \tag{22.31}$$

Similar ideas are thought to occur in the recognition of self and nonself by the immune system.

22.11 Temporal Programs

The single-input module (SIM) network motif is shown in **Figure 22.25**. The transcription factor X regulates a group of genes $Z_1,... Z_n$, with no additional transcription factor inputs. The gene for X usually regulates itself. An example of a SIM is the argininine biosynthesis pathway.

The SIM network motif is a simple pattern in which one regulator controls a group of genes. It can generate temporal programs of expression in which genes are turned on one by one in a well-defined order. This provides a 'just when needed' production strategy, where a protein is not made before it is required.

SIMs often regulate genes that participate in a specific metabolic pathway. These genes work sequentially to assemble a desired molecule in a kind of molecular assembly line. As the activity of X gradually rises, it crosses

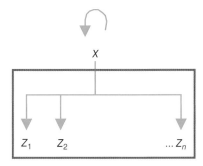

Figure 22.25 *The single-input module (SIM) network motif is used to create temporal programs in cells, e.g. the gene X is used to switch on the genes Z_1, Z_2, ..., Z_n in the correct order.*

the different thresholds for each target promoter in a defined order. When the activity of X declines, it crosses the thresholds in reverse order (last-in-first out or LIFO order). Genes are thus activated and deactivated in order as the activity of X increases and then decreases again. The correct temporal order is created in arginine biosynthesis system with minute waits between the expression of each consecutive gene. The earlier the protein functions in the pathway, the earlier its gene is activated. SIMs are found in damage repair systems, the timing of genes in the cell cycle, circadian clocks, and developmental processes. Other motifs such as multioutput FFLs can generate first-in-first-out (FIFO) temporal order in the expression of a sequence of genes and represent a useful alternative for cellular programming.

22.12 Nonlinear Models

Up to now the majority of the equations and theories presented in this book have depended on linearity for their analysis (with a few exceptions, e.g. the Poisson–Boltzmann equation). Linear equations have many advantages and the principle one is that they are relatively simple to solve analytically. However, they lead to a rather restrictive intuition about how the world acts. Robust stable nonlinear phenomena do occur in Nature, and more specifically in biology, such as the pulsatile flow behaviour of blood in capillaries or oscillations in the divisional clock of yeast cells. However, the nonlinear equations required to describe the phenomena are more mathematically challenging and require new techniques for their analysis. Other everyday physical examples of nonlinear phenomena include turbulence when water flows through a tap (a faucet), oscillations of a ball bearing on an oscillating table, Rayleigh–Bernard convection in weather patterns, and the Lorentz butterfly effect.

Lots of biologically important nonlinear phenomena occur in disease. Ventricular fibrillation kills 200 000 people in the USA every year (with similar rates occurring in Europe) and is thought to be due to chaotic nonlinear electrical activity of the heart. Epilepsy is another example of a nonlinear oscillatory diseased state. However, there are also many examples of nonlinear dynamics in healthy organisms, where the nonlinear mechanism is critical for their healthy functioning, such as cell signalling, e.g. spiral waves in dictostelium and the blood flow/divisional clock examples refereed to earlier.

Differential equations are often used for the creation of a model for genetic regulatory networks and they can be nonlinear. For example, the cellular concentration of proteins, mRNAs, and other molecules at a time point (t) can be represented by continuous variables $x_i(t)$ and the regulatory interactions are modelled by differential equations (equation (22.1)),

$$\frac{dx}{dt} = f(x) \tag{22.32}$$

where $x = [x_1, \ldots, x_n]$ and $f(x)$ is a nonlinear rate law. No analytical solution exists for most nonlinear differential equations. However, approximate solutions can often be obtained by numerical simulation, if reasonable estimates of parameter ranges and initial conditions are available.

Consider again the crossinhibition network for the interaction of two genes G_1 and G_2,

$$G_1 \underset{\longleftarrow}{\overset{\longrightarrow}{}} G_2 \tag{22.33}$$

A mathematical model is given by equations of the form of equations (22.10),

$$\dot{x}_1 = \kappa_1 f(x_2) - \gamma_1 x_1 \tag{22.34}$$

$$\dot{x}_2 = \kappa_2 f(x_1) - \gamma_2 x_2 \tag{22.35}$$

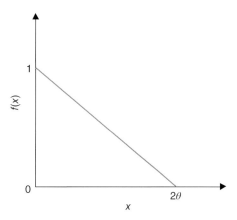

Figure 22.26 *A simple linear approximation to the function* f(x) *for the activity of an enzyme as a function of* x *the substrate concentration;* $f(x) = 1 - x/2\theta$ *for* $0 < x < 2\theta$.

where x_1 is the concentration of protein 1, x_2 is the concentration of protein 2. $\kappa_1 > 0$ and $\kappa_2 > 0$ are the production rate constants. $\gamma_1 > 0$ and $\gamma_2 > 0$ are the degradation rate constants. Initially, it is assumed that the rate-law function $f(x)$ (**Figure 22.26**) is given by a simple linear equation,

$$f(x) = 1 - x/(2\theta), \ \theta > 0, \ x < 2\theta \tag{22.36}$$

Therefore, equations (22.34) and (22.35) can be rewritten as

$$\dot{x}_1 = -\gamma_1 x_1 - \kappa_1 \theta_1 x_2 + \kappa_1 \tag{22.37}$$

$$\dot{x}_2 = -\kappa_2 \theta_2 x_1 - \gamma_2 x_2 + \kappa_2 \tag{22.38}$$

In general, such equations are hard to solve, but their behaviour can be understood by the analysis of steady states. If $\dot{x}_1 = 0$ then equation (22.34) becomes

$$x_1 = \frac{\kappa_1}{\gamma_1} f(x_2) \tag{22.39}$$

And similarly if $\dot{x}_2 = 0$ then equation (22.35) becomes

$$x_2 = \frac{\kappa_2}{\gamma_2} f(x_1) \tag{22.40}$$

The solution of these equations is shown in **Figure 22.27** and there is a single unstable steady state. Linear differential equations are too simple to capture dynamic phenomena of interest, since there is no bistability and no hysteresis. Nullclines, sometimes called zero-growth isoclines, are encountered in two-dimensional systems of differential equations

$$\dot{x} = F(x, y) \tag{22.41}$$

$$\dot{y} = G(x, y) \tag{22.42}$$

They are curves along which the vector field is either completely horizontal or vertical. A nullcline is a boundary between regions where $\dot{x}$ or $\dot{y}$ switch signs. Nullclines can be found by setting either $\dot{x} = 0$ or $\dot{y} = 0$. The intersections between x and y nullclines are equilibrium points, and thus the discovery of nullclines can be a useful way to identify such points, particularly when the system is not amenable to analytical solutions.

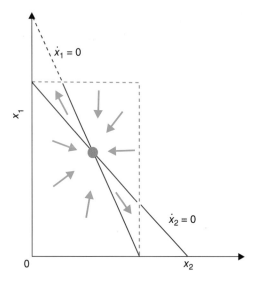

Figure 22.27 *Linear stability analysis indicates the existence of a single unstable steady state for the concentrations of proteins 1 and 2 (x_1 and x_2) in a crossinhibitory network.*

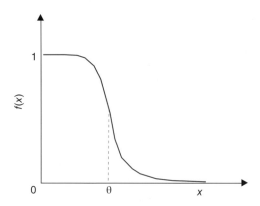

Figure 22.28 *Hill model for the functional form for f(x), the activity of an enzyme as a function of the substrate concentration (x).*

A more sophisticated model of a crossinhibition network is

$$
\begin{array}{c}
G_1 \xrightarrow{\;k_1\;} R_1 \longrightarrow P_1 \\[6pt]
G_1 \xrightarrow[\;k_2\;]{} R_2 \longrightarrow P_1
\end{array}
\tag{22.43}
$$

where x_1, x_2 are the concentration of mRNAs R_1, R_2, k_1, $k_2 > 0$ are mRNA production rates and γ_1, $\gamma_2 > 0$ are mRNA degradation rates. The proteins P_1, P_2 slow down the activity of genes G_1, G_2. The kinetics are again

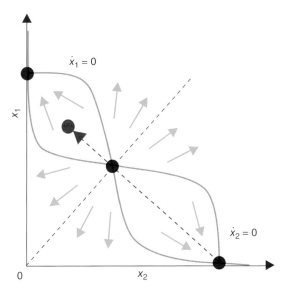

Figure 22.29 *The model of crossinhibition network with Hill enzyme kinetics has two stable and one unstable steady states (black circles) for the concentrations of proteins 1 and 2 (x_1 and x_2). The green lines indicate isoclines.*

described by equation (22.34) and (22.35). The function $f(x)$ is given by the nonlinear Hill form equation (21.56) (**Figure 22.28**),

$$f(x) = \frac{\theta^n}{\theta^n + x^n} \tag{22.44}$$

Analysis of the nullclines shows there are two stable and one unstable steady state (**Figure 22.29**). The system can switch between stable steady states and displays hysteresis, in agreement with experiments.

22.13 Population Dynamics

Population biology and ecology are fascinating areas of study. A population of organisms can interact in a vast number of ways and can demonstrate sophisticated emergent phenomena. This leads to such classic examples as the problem-solving behaviour of ants in ant colonies or the synchronisation of light emission by Southeast Asian fireflies. It is possible for the averaged stochastic interactions of millions of ants with one another (or the fireflies) to provide the colony with a single collective identity and characteristic behaviours.

Sir Ronald Ross received the 1902 Nobel Prize in medicine for the discovery of the life cycle of the malaria parasite. He also did some general mathematical modelling work to explain the patterns of outbreaks of infectious disease and universal features are observed over a wide variety of diseases as they travel through populations. Three principal temporal patterns are observed with infectious diseases. *Endemic diseases* have relatively small fluctuations in monthly counts and only a slight change in the number of cases over a number of years, e.g. leprosy or tuberculosis. *Outbreak diseases* are always present, but flare up in epidemic outbreaks at frequent intervals, e.g. measles, dysentery and malaria. *Epidemic diseases* have intense outbreaks followed by apparent disappearance, e.g. plague and cholera.

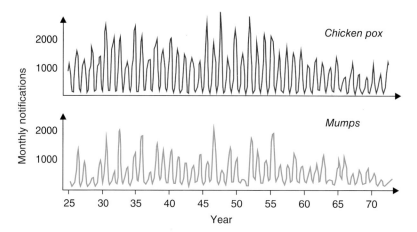

Figure 22.30   *Recurrent outbreaks of Chicken Pox and Mumps in New York City, USA follow an oscillatory epidemic model for their population dynamics. The number of disease cases per month is plotted as a function of the date. [Reproduced from W.P. London, J.A. Yorke, American Journal of Epidemiology, 1973, 98, 6, 453, by permission of Oxford University Press.]*

A simple model accounts for the periodic outbreaks of diseases

$$\frac{dS}{dt} = -\beta SI \tag{22.45}$$

$$\frac{dI}{dt} = \beta SI - \gamma I \tag{22.46}$$

$$\frac{dR}{dt} = \gamma I \tag{22.47}$$

where S is the number of susceptible individuals in a population, I is the number of infected individuals and R is the number of recovered or removed individuals (that are dead or have an immunity). β describes the rate of infection, whereas γ describes the rate at which infected individuals recover from the disease (or die). Such a model can describe the kinetics of outbreaks, and simple extensions can describe oscillatory phenomena such as that shown in **Figure 22.30** for the recurrent outbreak of Chicken Pox and Mumps in New York City.

Suggested Reading

If you can only read one book, then try:

Top recommended book: Alon, Uri (2007) *An Introduction to Systems Biology*, Chapman and Hall. A classic book on systems biology. Much of the contents of the chapter draw heavily on Alon's approach.

Aguda, B.D. & Friedman, A. (2008) *Models of Cellular Regulation*, Oxford University Press. Short compact account of systems biology.
Beard, D.A. & Qian, H. (2010) *Chemical Biophysics*, Cambridge University Press. Good introduction to computational systems biology.

Ellner, S.P. & Guckenheimer, J. (2006) *Dynamic Models in Biology*, Princeton University Press. Good pedagogic introduction to simple applied mathematics models of dynamical biological phenomena including disease epidemics.

Forgacs, G. & Newman, S.A. (2005) *Biological Physics of Developing Embryo*, Cambridge University Press. Fascinating discussion of the physical forces that determine the fates of embryos.

Keener, J. & Sneyd, J. (2008) *Mathematical Physiology*, Springer. Mathematically sophisticated account of molecular physiology.

Palsson, B.O. (2011) *Systems Biology: Simulation of Dynamic Network States*, Cambridge University Press. Modern chemical engineering approach to systems biology.

Strogatz, S. (1994) *Non-Linear Dynamics and Chaos*, Perseus. Good undergraduate introduction to the applied mathematics techniques required to analyse the nonlinear dynamic equations that are common in systems biology.

Winfree, A.T. (2010) *Geometry of Biological Time*, Springer. Classic iconoclastic study of nonlinear oscillatory phenomena in biology.

Tutorial Questions 22

22.1 Make sure you are familiar with DNA base pairing: A–T, C–G or T–A, G–C. In the case of RNA, T is replaced by U. A DNA template strand is $(3')$ GCGATATCGCAAA $(5')$. Write down its DNA coding strand and RNA transcript. Explain the structural difference between DNA and RNA, and hence the looser binding and more active pairing and translation.

22.2 Design a new genetic model to explore the phenomenon of gene regulation. Give an explanation for a biological collaborator so that they can construct the necessary plasmid to test your model. Assume that you wish to follow protein expression using fluorescence.

22.3 Describe some actual gene circuits found in biology, e.g. motility of bacteria.

22.4 Give some more examples from biology where reaction-diffusion phenomena may be important for pattern development.

22.5 Estimate the response time for a negative autoregulation motif if autorepression threshold K is 1 mM and maximal production rate β is 5 mM s^{-1}.

22.6 A gene Y with simple regulation is produced at a constant rate β_1. The production rate suddenly shifts to a different rate β_2. (a) Plot the concentration of $Y(t)$ as a function of time. (b) Calculate the response time (time to reach half-way between the steady states).

22.7 Show that the solution of equation (22.4) is indeed equation (22.7) subject to the boundary condition $Y = 0$ when $t = 0$.

22.8 Demonstrate that equation (22.29) is indeed a solution of equation (22.27) under steady state conditions.

Part V

Spikes, Brains and the Senses

The study of spikes in electrically excitable cells, the functioning of brains, and their interaction with the senses are all fascinating areas of study. Consciousness poses a huge unsolved problem in condensed-matter physics that underlies all of this research. However, within the space constraints imposed on this Part only a few of the pragmatic attempts currently being made to chip away at this 'hard problem' will be introduced. Thus areas where biological physicists have made a little progress and might still hope to extend our understanding during the twenty-first century will be considered.

The scale of the problem of brain exploration is enormous; there are around one hundred billion cells in a human brain and one hundred trillion connections occur between them. Currently, the wiring diagram of even a mouse brain has not been constructed[*], let alone one from a human. The idea that free will is associated with some remarkable phase transition of these colossally connected neuronal networks is therefore a fun one to consider philosophically, but will be avoided in what follows, since it does not yet directly lead to any simple experiments or any testable theories[†]. Thus pragmatic, practical lines of attack will be stressed in what follows.

The section also includes a general introduction to some of the methodologies needed to understand the control of an organism's physiology. This requires ideas from control theory and nonlinear dynamics to be applied on the macroscale. The microscopic equivalents of these techniques were introduced in the section on systems biology. Integration of ideas from control theory with the activity of nerve cells in networks is also important for understanding the physiology of brains.

[*] A first draft of a map of the mouse brain was published during the final stages of preparation of the book.
[†] MRI experiments have recently been performed that provide evidence for consciousness as a phase transition, but more research is required for there to be a broad consensus on the results in the community.

Suggested Reading

Koch, C. (2012) *Conciousness: Confessions of a Romantic Reductionist*, MIT Press. Short, but fascinating popular discussion of modern research into consciousness.

Crick, F. (1995) *The Astonishing Hypothesis: The Scientific Search for the Soul*, Simon and Schuster. Slightly older perspective on brain research from one of the world's most successful biological physicists.

Kandel, E. (2007) *In Search of Memory: The Emergence of a New Science of Mind*, W.W. Norton and Co. The cellular basis of memory is relatively well understood and Eric Kandel provides a readable account of the field's history.

23

Spikes

An introduction will be made to some of the main modelling tools used in neuroscience and cardiology. The creation of a good model for spiking potentials in electrically excitable cells is a reasonable place to start if the connections between ion channels and consciousness are to be explored. Similarly, to connect the genetics of ion channels to heart disease (**Figure 23.1**), the macroscopic electrical waves that travel around the heart need to be linked to the spiking potentials produced by individual ion channels. Other important cellular signalling events such as nonspiking intracellular electrical activity, and the encoding of sensory information will also be briefly considered.

23.1   Structure and Function of a Neuron

A neuron is an electrically excitable cell that processes and transmits information by a combination of electrical and chemical signalling (**Figure 23.2**). Neurons are found in a wide range of animals. An axon is a long, slender projection inside the neuron, that conducts electrical impulses from the dendrites of the neuron to its terminals. A spike describes the form of the voltage in a neuron as a function of time. There is a sharp increase in voltage that peaks and then a refractory period occurs, before the next increase in voltage can be initiated. Each spike contain approximately 1–3 bytes of information and this can be considered a constraint for data transmission from the senses (eyes, ears, noses, etc., **Chapter 24**) to the brain, or calculations that are performed by neural networks in the brain or central nervous system. Good quantitative models exist that allow spiking voltages to be calculated in terms of the activity of membrane proteins (the ion channels). Information is transferred between nerve cells chemically using neurotransmitters contained in vesicles that move across the synaptic cleft (**Figure 23.3**) and these initiate more spiking potentials in neighbouring cells.

23.2   Membrane Potential

The cell membrane separates the interior of all cells from the outside environment. Not only does it provide a mechanical barrier that shields intracellular organelles from damage, but it also serves as a battery, which

The Physics of Living Processes: A Mesoscopic Approach, First Edition. Thomas Andrew Waigh.
© 2014 John Wiley & Sons, Ltd. Published 2014 by John Wiley & Sons, Ltd.

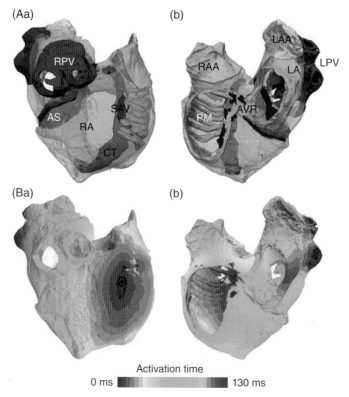

Figure 23.1 *(Aa) Anatomy of the surface of the heart and (Ab) the atrial cavities. (B) Simulation of the propagation of electrical activity across (Ba) the surface of a heart and (Bb) in the atrial cavities. The electrically active tissue is treated as a continuum. [Reproduced from M.A. Colman et al, J. Physiology, 2013, 591.17, 4249. © 2013 The Authors. The Journal of Physiology published by John Wiley & Sons Ltd on behalf of The Physiological Society. This is an open access article under the terms of the Creative Commons Attribution License, which permits use, distribution and reproduction in any medium, provided the original work is properly cited].*

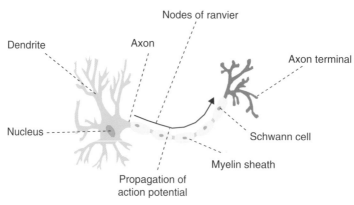

Figure 23.2 *Schematic diagram of a neuron. Dendrites transmit electrical activity along the axon to the axon terminals, where neurotransmitters are released to stimulate neighbouring dendrites.*

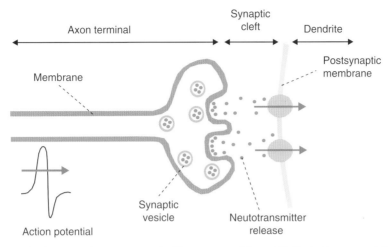

Figure 23.3 *Schematic diagram of an axon terminal in a neuron (**Figure 23.2**). Action potentials cause the release of neurotransmitters in vesicles that have an excitatory or inhibitory role on the spiking potentials formed by neighbouring dendrites.*

supplies power to operate various *molecular machines* embedded in the membrane. The energy for the battery is stored in the difference in ion concentrations between the inside and outside of the cell. The membrane is selectively permeable to ions and organic molecules, and controls the movement of many substances in to and out of cells. Predominantly, the membrane consists of a phospholipid bilayer that contains embedded proteins (such as ion channels) that are involved in cellular signalling processes, as well as many other specialised biomolecules (**Chapter 12**).

Consider the flux of ions through a membrane that are under the influence of both an ionic concentration gradient (∇c) and an electric field ($\underline{E}$). By definition, the electric field is equal to the spatial gradient of the potential (ϕ), $\underline{E} = -\nabla\phi$. For simplicity, the gradient is assumed one-dimensional, so $\nabla \equiv \partial/\partial x$, where the distance x is perpendicular to the membrane. According to Fick's first law of diffusion, the chemical flux (J^c) is related to the concentration gradient (equation (7.17)),

$$J^c = -D\frac{dc}{dx} \tag{23.1}$$

where c is the concentration of ions, the proportionality constant (D) is the diffusion coefficient and the total derivative has been taken rather than the partial derivative, since there is no ambiguity in one dimension. Electrically charged ions drift in the electric field, and quickly attain their terminal drift velocity (v_{drift}),

$$v_{\mathrm{drift}} = \mu E = -\mu\frac{d\phi}{dx} \tag{23.2}$$

where the proportionality coefficient (μ) is called the *mobility* of the ions. The drift of charged ions will result in an electric current (I),

$$I = v_{\mathrm{drift}}Q \tag{23.3}$$

where Q is the total charge per volume. Faraday's law relates the charge per unit volume to the concentration of ions (c), so

$$c = Q/zF \tag{23.4}$$

where F is the Faraday constant, and z is the valence of ions. The flux due to the electric field (J^e) is therefore given by

$$J^e = \frac{I}{zF} = v_{drift}\frac{Q}{zF} = v_{drift}c = -\mu c\frac{d\phi}{dx} \tag{23.5}$$

Lipid membranes have very low permeability to ions (Na^+, K^+), which limits the flow of ions into and out of the cell. Ions can cross the membrane through embedded proteins (called ion channels or transporters). Hence, the membrane can both separate electric charges and conduct electric currents. In physical terms, it acts as both a capacitor and resistor. Separation of different quantities of electrical charges inside and outside the cell results in a difference of electric potential across the membrane, called the *membrane potential*. The resting potential in most cells is about −80 mV. All cells have membrane potentials, but only a few specialist *electrically excitable* cells can actively modulate the voltage in a spiking potential (heart cells, brain cells, retinal cells, etc.).

23.3 Ion Channels

Ion channels exist in a wide variety of forms and many find specialist roles in specific cell types. New membrane channels continue to be discovered. Ion channels allow cells to actively modulate their membrane potentials as well as regulate the intracellular ionic environment. Patch clamp experiments (**Section 12.4**) enable the action of single ion channels to be followed, so it has been possible to create statistical models for their individual stochastic opening and closing events that integrate molecular information on their structure with their electrical activity. After studying each individual variety of ion channel in a patch of membrane it is then possible to connect the average membrane potential and current to the molecular statistics of all the underlying ion channels taken together.

The stochastic modelling of ion channels on the molecular level can be broken into two separate stages. First, the probability that an individual ion channel is actually open will be considered. Secondly, the average current from an ensemble of channels will be calculated.

To model the *first stage* consider the gating of a single ion channel as a Markov process, i.e. independent statistical events occur with constant probabilities,

$$C(\text{closed}) \underset{k^-}{\overset{k^+}{\rightleftarrows}} O(\text{open}) \tag{23.6}$$

There are two states for the channel, which are closed (C) or open (O). There is a rate constant (k^+) for a channel that was closed to become open, and vice versa, a rate constant (k^-) for an open channel to become closed. Since there are only two possibilities for the states of ion channels, the probability that a channel is open ($p_O(t)$) and the probability that a channel is closed ($p_C(t)$) at a time (t) must add up to one,

$$p_C(t) + p_O(t) = 1 \tag{23.7}$$

After a small time step (Δt) there are two ways for the channel to enter or leave the closed state. Therefore, the probability the channel is in a closed state at time, $t + \Delta t$ is

$$p_C(t + \Delta t) = p_C(t) - k^+ p_C \Delta t + k^- p_O(t)\Delta t \tag{23.8}$$

A similar expression holds for the opening probability, $p_O(t + \Delta t)$. In the limit that $\Delta t \to 0$ two differential equations are formed,

$$\frac{dp_C}{dt} = -k^+ p_C + k^- p_O \tag{23.9}$$

$$\frac{dp_O}{dt} = +k^+ p_C - k^- p_O \tag{23.10}$$

These coupled differential equations can be solved to give an exponentially decaying probability for switching between the open and closed states. It can be shown that the average times for the channels to be closed ($<\tau_C>$) or open ($<\tau_O>$), the decay times of the exponentials, are given by

$$\langle \tau_C \rangle = \frac{1}{k^+} \tag{23.11}$$

$$\langle \tau_O \rangle = \frac{1}{k^-} \tag{23.12}$$

Thus, measurement of the average time over which a channel is closed and that over which it is open allows the two rate constants for opening and closing to be calculated and these parameters can thus be deduced from single ion channel patch clamp experiments (**Figure 23.4**). This methodology can be extended relatively simply if the ion channel has additional well-defined internal transition states, through a calculation of the internal transition probabilities.

Now that a model for single ion channels has been created, a challenge is to extend the calculation to a population of channels (the *second stage*), since this is more relevant to the behaviour of the electrical activity of a whole cell, e.g. a neuron. It is possible to create a quantitative relation between the ionic current that flows through a population of channels, fluctuations in this current, the current through an individual channel, and the population size. The assumptions are that the voltage potential across the membrane is held constant (a voltage clamp is used), there are N channels in the membrane that randomly switch between open and closed states, the current through each channel is the same, and the state of each channel is independent of the others.

The probability that there are exactly k open channels ($P(k)$) chosen from the population of N channels in a membrane follows a binomial distribution (which assumes that the channels are independent and have a constant probability of being open, **Section 3.1**),

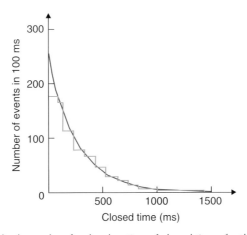

Figure 23.4   *Histogram of patch-clamp data for the duration of closed times for the action of a single ion channel.*

$$P(k) = \frac{N!}{k!(N-k)!} p^k (1-p)^{(N-k)} \tag{23.13}$$

where p is the probability that a channel is open. If the current is measured M times, an average current ($<I>$) can be calculated,

$$\langle I \rangle = \frac{1}{M} \sum_{j=1}^{M} I_j \tag{23.14}$$

where I_j are the currents for each individual ion channel j. Similarly, the variance (σ^2) of the current is also a useful experimentally measurable quantity, which is given by the standard expression (equation (3.3)),

$$\sigma^2 = \frac{1}{M} \sum_{j=1}^{M} I_j^2 - \langle I \rangle^2 \tag{23.15}$$

The probability that the jth channel is open is given by

$$p_j = \frac{1}{M} k_j \tag{23.16}$$

The mean current is $I = Npi$ (where i is the single ion channel current) and therefore an expression for the variance is

$$\sigma^2 = iI - piI \tag{23.17}$$

It is possible to rearrange this equation to give the single-channel current,

$$i = \frac{\sigma^2}{I(1-p)} \tag{23.18}$$

This expression relates the single-channel current (i) to the total population current (I), the single-channel variance (σ^2) and the probability that a single channel is open (p). Real cells contain a variety of different types of ion channels (sodium, potassium, chlorine, etc.), but such an analysis can be extended to accurate calculations of whole-cell current fluctuations.

23.4   Voltage Clamps and Patch Clamps

In order to vary a membrane voltage in a controlled manner, Hodgkin and Huxley used a feedback amplifier, which was connected to two electrodes, one inside and one outside the axon of a giant squid neuron. The amplifier automatically supplied the correct amount of current to maintain the membrane potential at a desired constant level, and prevented any runaway fluctuation of the potential (no action potentials can be generated for voltage-clamped cells). Such a voltage-clamp method can be used to determine ionic conductances of membrane channels, such as that for a chlorine (Cl) channel. Ohm's law for the channel implies ($g_{Cl} = 1/R_{Cl}$, where R_{Cl} is the electrical resistance of the chlorine ion channel at the voltage V_{clamp}),

$$I_{Cl} = g_{Cl} \left(V_{clamp} - V_{Cl} \right) \tag{23.19}$$

Here, V_{clamp} is the clamped constant voltage, g_{Cl} is the Cl^- channel conductivity at the voltage V_{clamp}, and the current measured is that which flows through the amplifier to counter-balance the respective ionic current of the same magnitude that flows through the membrane. Similar expressions hold for the sodium and potassium currents. Next, using Kirchoff's first law, values for g_{Na}, g_K, and g_{Cl} can be substituted to describe the rate of change of the membrane potential,

$$C_m \frac{dV_m}{dt} = -(I_{Na} + I_K + I_{Cl}) = -g_{Na}(V - V_{Na}) - g_K(V - V_K) - g_{Cl}(V - V_{Cl}) \tag{23.20}$$

where I_{Na} and I_K are sodium and potassium currents that are given by expressions similar to equation (23.19). The method was used to demonstrate that the conductances for Na^+ and K^+ currents vary with both the time and the membrane potential in squid nerve cells, whereas Cl^- conductance has a very small constant value. Thus, Hodgkin and Huxley were able to measure all the parameters included in their model for the action potential (**Section 23.8**) and similar methodologies continue to be useful for more advanced models of electrically excitable cells.

23.5 Nernst Equation

The *Nernst potential* is a purely thermodynamic effect. It is the potential set up across a membrane if there is a difference in the concentrations of ionic species (e.g. salt concentration, **Figure 23.5**) across the membrane. It is relatively easy to show that in thermodynamic equilibrium, when the ionic energies follow a Boltzmann distribution, the difference in voltages ($V_i - V_e$) between two compartments separated by a membrane is given by

$$V_i - V_e = \frac{RT}{F} \ln\left(\frac{c_o}{c_i}\right) \tag{23.21}$$

where RT is the ideal gas constant times the temperature, F is the Faraday constant ($F = N_a e$, Avogadro's number multiplied by the electronic charge), and c_o/c_i is the ratio of the concentrations of the ion X in the two compartments. The equation applies at equilibrium only. It gives no information about currents through the membrane that help establish the equilibrium or currents due to perturbations from the steady state. In fact, there is typically very little actual ion transfer from side to side in real cells, although their effects can be significant. The resultant Nernst potentials for multiple ions will be discussed later.

Nernst and Planck were the first to realise that chemical (J^c, equation (23.1)) and electrical (J^e, equation (23.2)) components of the ionic flux are additive and result in a total flux

$$J = J^c + J^e = -D\frac{dc}{dx} - \mu c\frac{d\phi}{dx} \tag{23.22}$$

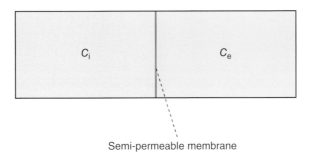

Figure 23.5 *A Nernst potential is setup across a semipermeable membrane when there is an inbalance in the concentration of a species of salt ions (c_i and c_e) on the two sides of the membrane.*

In thermodynamic equilibrium across the membrane, it is assumed that in the presence of an electric potential ($V(x)$) the ions have a Boltzmann distribution (equation (3.22)),

$$c(x) = Ae^{-qV(x)/kT} \tag{23.23}$$

where q is the charge, A is a constant, k is the Boltzmann constant and T is the absolute temperature. Equation (23.23) can be differentiated using the chain rule to give

$$\frac{dc}{dx} = c\left(-\frac{q}{kT}\right)\frac{dV}{dx} \tag{23.24}$$

Hence in homeostasis, when there is no net current through the membrane (steady-state scenario), equation (23.22) implies

$$0 = J^c + J^e = -D\frac{dc}{dx} - \mu c\frac{dV}{dx} = -D\left(-\frac{q}{kT}\right)c\frac{dV}{dx} - \mu c\frac{dV}{dx} \tag{23.25}$$

This equation must hold for any displacement (x), and therefore

$$D = \mu kT/q \tag{23.26}$$

This is another example of the fluctuation dissipation theorem introduced first in the discussion of diffusion (equation (7.9)), where it connected the diffusion coefficient to the frictional coefficient. Here, it connects the diffusion coefficient to the ion mobility (the ion's response to an electric field). The Einstein relation, equation (23.26), was obtained by the consideration of the ions as particles, and it used the Boltzmann distribution for their energies. Both approximations have their shortcomings, but they provide a useful estimate. Combination of equations (23.22) and (23.26) gives

$$J = -D\frac{dc}{dx} - \mu c\frac{dV}{dx} = -\mu\frac{kT}{q}\frac{dc}{dx} - \mu c\frac{dV}{dx} \tag{23.27}$$

For a steady flux, $J = 0$, across the cell membrane this equation becomes

$$0 = -\mu\left(\frac{kT}{q}\frac{dc}{dx} + c\frac{dV}{dx}\right) \tag{23.28}$$

Therefore,

$$\frac{dV}{dx} = -\frac{kT}{q}\frac{1}{c}\frac{dc}{dx} \tag{23.29}$$

This equation can be integrated across the displacement perpendicular to the membrane, from inside (i) to outside (o),

$$V_o - V_i = -\frac{kT}{q}\ln\frac{c_o}{c_i} \tag{23.30}$$

This is the Nernst equation for the equilibrium potential (V), equation (23.21) with $R/F = k/q$, as required.

For the well-studied case of a giant squid axon, the concentrations of ions inside and outside the membrane are shown in **Table 23.1**. At a room temperature of 23 °C, $kT/q = 26$ mV. From equation (23.30), Nernst potentials caused by the ions across the membrane can be calculated and are also shown in **Table 23.1**. The right-hand side of equation (23.30) becomes negative in the case of negative ions, because the charge (q) changes its sign. However, the total potential across a cell is not just the sum of the different

Table 23.1   *Typical ion concentrations measured in a giant squid axon at a room temperature of 23 °C.*

Ion	Internal concentration (mmol/L)	External concentration (mmol/L)	Nernst potential (mV)
$[Na^+]$	50	460	57
$[K^+]$	400	20	−77
$[Cl^-]$	40	550	−68

Nernst potentials, they need to be weighted by the different conductivities of the membrane for different ions and this is discussed in **Section 23.6**.

23.6   Electrical Circuit Model of a Cell Membrane

A very simple equivalent electrical circuit model of the cell membrane is shown in **Figure 23.6**, which consists of a capacitor and a resistor in parallel. Standard electrical circuit theory can be used for the circuit's analysis. Kirchoff's first law states that the currents that travel into a junction should sum to zero, so that the ionic current (I_{ionic}) balances that due to the capacitor ($Q = CV$, and if the capacitance is constant with time, $\frac{dQ}{dt} = C\frac{dV}{dt}$),

$$C\frac{dV}{dt} + I_{ionic} = 0 \qquad (23.31)$$

Kirchoff's second law states that the voltages in a loop should sum to zero,

$$V_i - V_e = V \qquad (23.32)$$

There are many different possible models for I_{ionic}, the amount of current that flows through a membrane. The assumption of a constant electric field across the membrane leads to the Goldman–Hodgkin–Katz (GHK) model. The Poisson and Nernst–Planck models are two other alternatives, but there are also barrier models, binding models, saturating models, etc. Hodgkin and Huxley used Ohm's law in their initial work. Of course (given no *a priori* structural evidence) the best model is provided by the best empirical fit of the

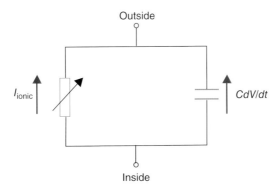

Figure 23.6   *An equivalent electrical circuit for a region of membrane that describes the current flow modulated by both its resistance (I_ionic) and capacitance (CdV/dt).*

current–voltage characteristics that has the minimum number of free parameters. The linear Ohm's law model for the current through a sodium ion channel (I_{Na}) is given by

$$I_{Na} = g_{Na}(V - V_{Na}) \tag{23.33}$$

where g_{Na} is the sodium ion channel conductivity (compare with equation (23.19)). In contrast the GHK model for the sodium channel is

$$I_{Na} = P_{Na}\frac{F^2}{RT}V\left[\frac{[Na^+]_i - [Na^+]_e e^{-VF/RT}}{1 - e^{-VF/RT}}\right] \tag{23.34}$$

where P_{Na} is the permeability of the membrane to sodium. A crucial property is that in electrically excitable cells the ion channel activities (given by g_{Na} or P_{Na}) are not constant, but are functions of voltage and time, which makes accurate modeling more challenging.

The membrane can separate electrically charged ions, and also conduct electric currents (carried by ions). Hence, it can be idealised as an equivalent electrical circuit comprised of a capacitor (C_m) and several resistors corresponding to the conductance to each variety of ion (Na^+, K^+, Cl^-, etc.). Equilibrium potentials, V_{Na}, V_K and V_{Cl} act as electrochemical batteries connected in series with the ionic conductances, g_{Na}, g_K, and g_{Cl}, respectively (**Figure 23.7**). The polarities and voltages of the imaginary batteries oppose the tendency of the ions to move in the direction of the concentration gradients. The membrane potential (V_m) across the membrane leads to a capacitive current $C_m dV_m/dt$, and the ionic currents are $I_{Na} = g_{Na}(V_m - V_{Na})$, $I_K = g_K(V_m - V_K)$, and $I_{Cl} = g_{Cl}(V_m - V_{Cl})$. Application of Kirchhoff's first law, requires that the sum of all currents in a closed circuit must be zero, therefore

$$C_m\frac{dV_m}{dt} + I_{Na} + I_K + I_{Cl} = 0 \tag{23.35}$$

Substitution of the Ohm's law expressions (e.g. equation (23.33)) for each of the ion channels gives

$$C_m\frac{dV_m}{dt} = -g_{Na}(V_m - V_{Na}) - g_K(V_m - V_K) - g_{Cl}(V_m - V_{Cl}) \tag{23.36}$$

In the resting steady state, the membrane potential does not change with time (t) and a constant resting potential ($V_m = V_r$) occurs. Therefore

$$C_m\frac{dV_m}{dt} = 0 \tag{23.37}$$

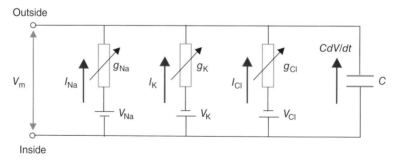

Figure 23.7 *Equivalent electrical circuit for the voltage across a membrane that contains sodium (Na), potassium (K), and chloride (Cl) ion channels, and has an effective capacitance C.*

Substitution in equation (23.35) gives an expression for the resting potential,

$$0 = -g_{Na}(V_r - V_{Na}) - g_K(V_r - V_K) - g_{Cl}(V_r - V_{Cl}) \tag{23.38}$$

This equation can be rearranged to give a relationship between the resting potential (V_r) of a cell membrane and the Nernst potentials for individual ions,

$$V_r = \frac{g_{Na}V_{Na} + g_K V_K + g_{Cl}V_{Cl}}{g_{Na} + g_K + g_{Cl}} = \frac{\sum_i g_i V_i}{\sum_i g_i} \tag{23.39}$$

This is a key analytic result. The resting potential of a cell is the sum of the Nernst potentials for all the ions in the cell's environment weighted by the conductivities of the cell membrane and normalised by the sum of the conductances. Molecular details of the ion channels and membrane lipids can enter into the calculations of potentials through these conductivities.

23.7 Cable Equation

A long axon can be idealised as an equivalent electrical circuit with an average membrane capacitance (C_m), variable membrane ionic currents (I_{ion}), and a constant longitudinal intracellular resistance (R, **Figure 23.8**). The membrane potential ($V_m(x)$) along an axon is spatially nonuniform. Kirchhoff's first law requires that the sum of all currents at any point must be zero. Consider the spatial dependency of the currents in a small piece of conductor of length Δx,

$$I(x) = I(x + \Delta x) + I_c \Delta x + I_{ion} \Delta x \tag{23.40}$$

The expression can be rearranged to give

$$I(x) - I(x + \Delta x) = (I_c + I_{ion})\Delta x \tag{23.41}$$

The charge accumulated by the capacitor (q) can be calculated from the definition of capacitance ($q = C_m V_m$) and the capacitive current is therefore

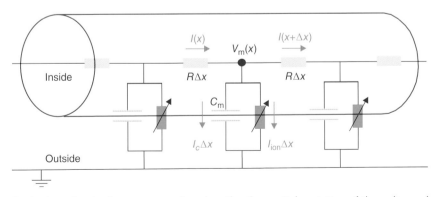

Figure 23.8 *Equivalent circuit of an axon used to describe the spatial variation of the voltage along its length. The time (t) and distance (x) dependence of the membrane potential ($V_m(x)$) is described by the cable equation, equation (23.47). R is the membrane resistance and C_m is the membrane capacitance.*

$$I_c = \partial q / \partial t = C_m \partial V_m / \partial t \tag{23.42}$$

$I(x + \Delta x)$ in equation (23.41) can be expanded in a Taylor's series, and substituted in equation (23.42). The current balance is found to be

$$I(x) - \left\{ I(x) + \frac{\partial I}{\partial x} \Delta x \right\} = \left(C_m \frac{\partial V_m}{\partial t} + I_{ion} \right) \Delta x \tag{23.43}$$

Therefore

$$\frac{\partial I}{\partial x} = - \left(C_m \frac{\partial V_m}{\partial t} + I_{ion} \right) \tag{23.44}$$

The longitudinal current in the circuit can be found from Ohm's law ($I = V/R$) as

$$I = \frac{V_m(x) - V_m(x + \Delta x)}{R \Delta x} = \frac{V_m(x) - \{V_m(x) + \partial V_m / \partial x \Delta x)\}}{R \Delta x} = -\frac{1}{R} \frac{\partial V_m}{\partial x} \tag{23.45}$$

Equation (23.44) and (23.45), can be combined to give the *cable equation* for the membrane potential,

$$\frac{\partial I}{\partial x} = -\frac{1}{R} \frac{\partial^2 V_m}{\partial x^2} = - \left(C_m \frac{\partial V_m}{\partial t} + I_{ion} \right) \tag{23.46}$$

Therefore,

$$C_m \frac{\partial V_m}{\partial t} = \frac{1}{R} \frac{\partial^2 V_m}{\partial x^2} - I_{ion} \tag{23.47}$$

The cable equation is a challenge to solve analytically. It was originally developed by Lord Kelvin in the 1850s to explain long-distance electrical telecommunication (telegrams), and analytic solutions do exist for some simple cases. Consider a steady-state solution for the passive linear cable, where $\partial V_m / \partial t = 0$ and $I_{ion} = (V_m - V_r)/R_m$. The cable equation becomes (where total derivatives are now used)

$$0 = \frac{1}{R} \frac{d^2 V_m}{dx^2} - \frac{(V_m - V_r)}{R_m} \tag{23.48}$$

Therefore

$$\frac{d^2 V_m}{dx^2} = \frac{R}{R_m} (V_m - V_r) \tag{23.49}$$

The characteristic length scale (λ) is defined as $\lambda = \sqrt{R_m / R}$ which leads to

$$\frac{d^2 V_m}{dx^2} = \frac{1}{\lambda^2} (V_m - V_r) \tag{23.50}$$

This equation has a well-known solution

$$V_m(x) = V_r + V_0 \exp(-x/\lambda) \tag{23.51}$$

This solution describes an exponential decay of the initial potential (V_0) with distance along the cable (x). The parameter λ (the *space constant*) characterises the spatial rate of decay of the passive voltage. The value of the space constant in neurons is normally ~1–5 mm.

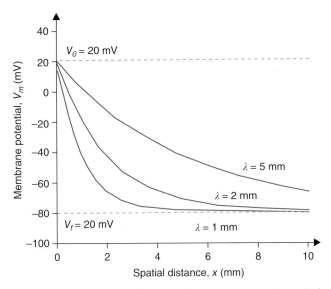

Figure 23.9   *Steady-state solution for the decay of the membrane potential (V_m) with distance along an axon (x), for different values of the space constant (λ).*

For a cylindrical conductor, the resistance (R) in the direction of the applied potential gradient is given by

$$R = \rho L / A = \rho L / \left(\pi r^2 \right) \tag{23.52}$$

where L is the length of the conductor, A is the cross-sectional area and ρ is the intracellular resistivity. Similarly for the resistance across the membrane (R_m),

$$R_m = \rho_m L / \left(2 \pi r L \right) \tag{23.53}$$

since $A = 2 \pi r L$ is the surface area of the cylindrical membrane. Here, ρ_m is the membrane resistivity, and r is the radius of the cylinder. Hence, $\lambda = (R_m/R_i)^{1/2} = (r\rho_m/2\rho)^{1/2}$. An axon with a 30-μm radius and resistivities of $\rho_m = 5000 \, \Omega \, \text{cm}^{-2}$ and $\rho = 50 \, \Omega \, \text{cm}^{-1}$ provides a length constant of ~2.7 mm (**Figure 23.9**).

For signalling in dendrites, the passive cable responses are sufficient. A typical dendritic ending in the brain extends over a distance of a few 100 μm. The human cerebral cortex is a sheet of neuronal tissue that is 3 mm thick, with a space constant of ~2.7 mm, so passive transport is adequate for the passage of the electrical signals along the dendrites in the cortex. However, in long axons (~1 m), the passive response is insufficient and active modulation of ion channels is required as described by the Hodgkin–Huxley model.

Furthermore, to model the behaviour of electrical activity in the brain in more detail the cable equation needs to be solved for cell branching, i.e. in dendritic cells. The equations are much more complicated in this case and require more sophisticated techniques for numerical solution.

23.8   Hodgkin–Huxley Model

Active modulation of the membrane potential of electrically excitable cells increases the length scales over which they propagate. Hodgkin and Huxley were able to create a successful model for the electrical activity of a giant squid neuron by the combination of the cable equation (23.47) with a more detailed model of the

Figure 23.10 *The grouping of the four internal states (S_{00}, S_{01}, S_{11}, S_{10}) for the kinetics of a potassium channel used to simplify their kinetics in terms of the three internal states (S_0, S_1 and S_2) with rate constants α and β.*

kinetics of sodium and potassium ion channels. The giant squid was crucial in the experiments, since it was sufficiently big to be manipulated in early voltage-clamp experiments, before micrometre-sized electrodes became practical. Giant squids should not be confused with colossal squids that are 10 metres in length, poorly understood zoologically and extremely challenging to catch. The first movies of live colossal squids were only obtained in 2012 off the coast of Japan. A classic quote from Hodgkin and Huxley was that the giant squid should have been a Nobel Prize cowinner, since without it the advances would have been impossible.

A two-state model was created previously for the stochastic switching of an ion channel (**Section 23.3**). However, it is known that the potassium ion channel can switch between four different conformations; S_{00}, S_{01}, S_{10} and S_{11} (**Figure 23.10**). This additional complication is introduced, since it is subsequently required to explain the experimentally determined form of the spiking potential of a squid neuron. The model can be simplified by pooling S_{01} and S_{10} as a single state (S_1) of the ion channel. First-order chemical kinetics are assumed, with α the opening rate constant and β the closing rate constant. This gives two differential equations for the rate of change of the closed (x_0) and open (x_2) states,

$$\frac{dx_0}{dt} = \beta x_1 - 2\alpha x_0 \tag{23.54}$$

$$\frac{dx_2}{dt} = \alpha x_1 - 2\beta x_2 \tag{23.55}$$

where $x_0 = [S_0]$, $x_1 = [S_1]$ and $x_2 = [S_2]$ and the condition of mass conservation requires $x_0 + x_1 + x_2 = 1$. These equations can be more easily solved by the definition of a parameter n such that $x_0 = (1-n)^2$, $x_1 = 2n(1-n)$, and $x_2 = n^2$. This gives a single differential equation to solve rather than both the equations (23.54) and (23.55),

$$\frac{dn}{dt} = \alpha(1-n) - \beta n \tag{23.56}$$

where n is called the potassium activation gate parameter whose value varies between 0 and 1.

For *sodium channel gating* the channel has one h subunit and two m subunits, where h describes inactivation of the channel and m describes channel activation. Thus, the S_{ij} channel has i open m subunits, and j open h subunits (**Figure 23.11**). Similar to the case of the K^+ channel, solution of the kinetic equations for the channel states requires a change of variables,

$$x_{12} = m^2 h = [S_{12}] \tag{23.57}$$

The kinetics for the activation of the ion channel are described by

$$\frac{dm}{dt} = \alpha(1-m) - \beta m \tag{23.58}$$

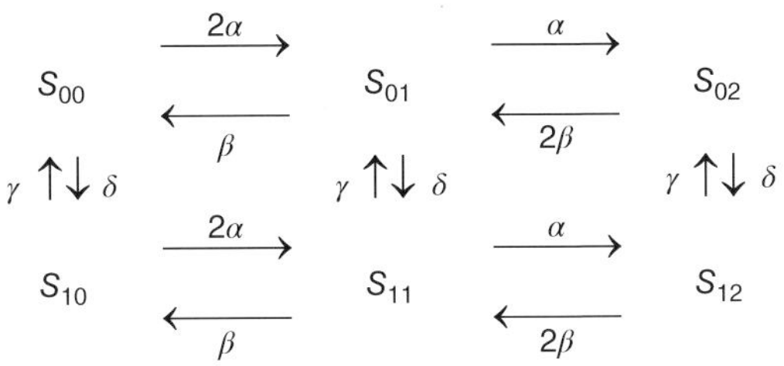

Figure 23.11   *Internal state kinetics of a sodium ion channel with six internal states (S_{00}, S_{01}, S_{02}, S_{10}, S_{11} and S_{12}) and four rate constants (α, β, γ and δ).*

Similarly, the kinetics for the inactivation of the ion channel are described by

$$\frac{dh}{dt} = \gamma(1-h) - \delta h \tag{23.59}$$

where α, β, γ, and δ are the rate constants for the chemical kinetics shown in **Figure 23.11**.

Returning to potassium (K^+) ion channels, experimental data can then be modeled for their conductance. If the voltage is stepped up and held fixed by a clamp, the conductance (g_K) increases to a new steady level. This model can be fit to the experimental data and the rise rate allows a time constant (τ_n) to be determined, whereas the steady state value gives n_∞. The conductance is found to follow the form

$$g_K = \bar{g}_K n^4 \tag{23.60}$$

where $\bar{g}_K$ is an average potassium ion conductance. This was justified in retrospect as originating from the action of four independent subunits on each potassium ion channel. Equation (23.56) can be rewritten with explicit voltage dependencies ($\alpha(V)$ and $\beta(V)$) as

$$\frac{dn}{dt} = \alpha(V)(1-n) - \beta(V)n \tag{23.61}$$

This equation can be rearranged to give

$$\tau_n(V)\frac{dn}{dt} = n_\infty(V) - n \tag{23.62}$$

where $\tau_n(V)$ is a voltage-dependent time constant ($\tau_n(V) = 1/\alpha(V)$). **Figure 23.12** shows a fit to this model. Next experimental data can be studied for the conductance of sodium (Na^+) ion channels. The voltage is stepped up and held fixed with a clamp, g_{Na} increases and then decreases according to

$$g_{Na} = \bar{g}_{Na} m^3 h \tag{23.63}$$

where $\bar{g}_{Na}$ is an average sodium ion channel conductance. In common with the potassium ion channel, there are also assumed to be four subunits on the sodium ion channel. However, three subunits switch on the ion channel and one sub unit switches it off. Two differential equations are required to describe the kinetics equations (23.58) and (23.59). One equation is for the inactivation parameter (h) and the other for the activation parameter (m), which can be rewritten in the form

$$\tau_h(V)\frac{dh}{dt} = h_\infty(V) - h \tag{23.64}$$

and

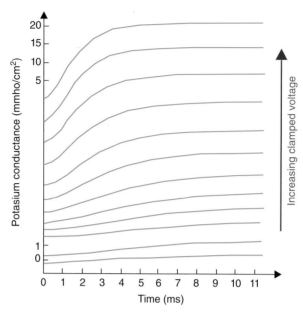

Figure 23.12 *The potassium ion channel conductance as a function of time for different values of voltage imposed by a clamp. [Reproduced with permission from A.L. Hodgkin, A.F. Huxley, Journal of Physiology, 1952, 117, 500–544. John Wiley & Sons, Ltd.]*

$$\tau_m(V)\frac{dm}{dt} = m_\infty(V) - m \tag{23.65}$$

The fit of the model to the data is a little more complicated, but it is still possible (**Figure 23.13**). Combination of the choices for the ion channel currents with the cable equation gives one of the *Hodgkin–Huxley equations*,

$$C\frac{dV}{dt} + \bar{g}_K n^4(V - V_K) + \bar{g}_{Na}m^3 h(V - V_{Na}) + g_L(V - V_L) + I_{app} = 0 \tag{23.66}$$

This is combined with equations (23.62), (23.64), and (23.65) to give the full model. **Figure 23.14** shows numerical solutions for the Hodgkin–Huxley equations. The Hodgkin–Huxley equations are phenomenological, i.e. they are not a unique solution to a well-posed physical problem and were mathematically motivated to describe giant squid experiments. The equations and simple extensions are now known to be a reasonably good description of less exotic cell types, e.g. human neurons and cardiac myocytes.

23.9 Action Potential

In biophysics, often the choice of the correct organism is crucial for the success of a research project. In early genetics peas were used (Gregor Mendel performed these experiments), more recently bacteria were an invaluable system for the study of molecular genetics, memory was first studied at the single-cell level in sea snails and with action potentials in nerve cells it was the giant squid axon that provided the first dramatic break through, due to its large size and thus convenience of use.

During an action potential in a giant squid axon the sodium conductivity (g_{Na}) increases quickly, but then inactivation kicks in and it decreases again. The potassium conductivity (g_K) increases more slowly, and only

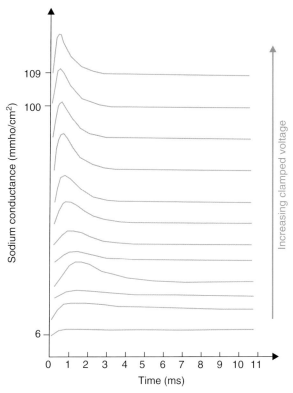

Figure 23.13   *The sodium ion channel conductance as a function of time with different voltages imposed by a clamp. [Reproduced with permission from A.L. Hodgkin, A.F. Huxley, Journal of Physiology, 1952, 117, 500–544. John Wiley & Sons, Ltd.]*

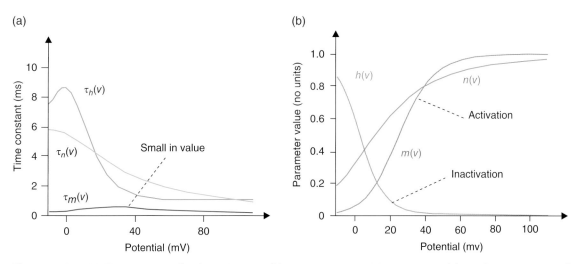

Figure 23.14   *(a) Time constants for the activation of the potassium gate (n), activation of the sodium gate (m) and inactivation of the potassium gate (h) as a function of the clamped membrane potential. (b) Magnitude of n, h and m as a function of the clamped membrane potential.*

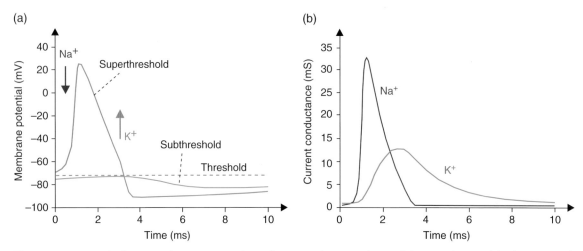

Figure 23.15   *(a) Spiking membrane potential as a function of time in the Hodgkin–Huxley model. The potential reaches a maximum due to the influx of sodium ions and then potassium ions flow out of the cell before the recovery phase. (b) Current conductance as a function of time for sodium and potassium ion channels.*

decreases once the voltage has decreased (**Figure 23.15**). The Na^+ current is autocatalytic and experiences positive feedback once it is above a threshold value. An increase in voltage (V) increases the activation parameter (m), which in turn increases the Na^+ current, which again increases the voltage (V) and so on, in a process of positive feedback. Hence, the threshold for action potential initiation is the point at which the inward Na^+ current exactly balances the outward K^+ current.

When the membrane potential is shifted from the resting value of about –80 mV towards positive potentials, the membrane becomes depolarised. If the degree of such depolarisation exceeds a threshold value, an action potential is generated. This is an all-or-nothing active response. The action potential has a standard amplitude of ~100 mV, determined by the conductive properties of the membrane, which is independent of the degree of initial depolarisation. Full depolarisation to about +20 mV is a self-sustained process, and it is based on the influx of Na^+ ions into the cell via voltage-sensitive ion channels. Additional positive charge shifts the potential to more positive values, which results in the opening of even more channels. A slower opening of K^+ channels leads to an outward flux of the excessive positive charge, which leads to repolarisation and restores the resting potential. The nerve cell then requires a refractory (recovery) time period (~2–3 ms) to return to the initial ion concentrations before another spiking potential can be created.

23.10   Spikes – Travelling Electrical Waves

It is useful to recap the picture for electrically excitable cells to consolidate our understanding. Electrical signals are passed between neurons via vesicles over very small distances in the synaptic cleft; the transfer is passive (diffusive) and involves molecules called neurontransmitters. Transmission of signals in neurons, along cable-like axons, is much faster than diffusion over the relatively large lengths involved, and is based on the propagation of nonlinear waves; the *action potentials* (**Figure 23.16**).

Propagation of an action potential through an axon can be described by the cable equation with the Hodgkin–Huxley descriptions of the ionic membrane currents. The system of four equations can be simplified by consideration of only the depolarising phase of the action potential, since propagation of the action potential wavefront along the axon is driven primarily by a polarisation gradient generated during this phase. In the

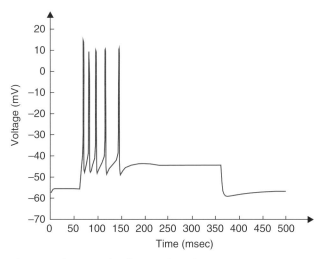

Figure 23.16   *Schematic diagram of a train of spikes produced in a neuron in response to a sensory stimulus. The membrane voltage is shown as a function of time.*

depolarisation phase, the slow inactivation process has not yet started, and the inactivation variable h maintains its initial value, $h(t) = h_0$. The Na$^+$ activation is a very fast process ($\tau_m \ll \tau_n, \tau_h$ and $\tau_m = RC_m$), and the activation variable rapidly reaches its steady-state level, $m(t) = m_\infty(V)$. Thus,

$$C_m \frac{\partial V_m}{\partial t} = \frac{1}{R}\frac{\partial^2 V_m}{\partial x^2} - \left\{ g_{Na}m^3(V_m)(V_m - V_{Na}) + \bar{g}_K n^4(V_m - V_K) \right\} = \frac{1}{R}\frac{\partial^2 V_m}{\partial x^2} + f(V_m, n) \tag{23.67}$$

Along with the equation for n (equation (23.62)), this gives a system of two equations, which can be studied using the standard steady-state kinetic analysis and nullcline methods for nonlinear equations.

The characteristics of the all-or-nothing action potential response are defined by the membrane properties only and the membrane structure along an axon is fairly uniform. The action potential is expected to propagate with a constant velocity (v), and maintains its shape as it propagates (the spatial and temporal profiles of V_m are similar). Therefore, a solution $V_m(x, t)$ to equation (23.67) is explored in a coordinate system $\xi(x, t) = x + vt$ that moves with the same velocity (v) as the action potential itself,

$$V_m(x, t) = u(x + vt) = u(\xi(x, t)) \tag{23.68}$$

where x is the actual distance along the axon and t is the time. The derivatives in equation (23.67) can be rewritten in the new coordinates using the chain rule,

$$\frac{\partial V_m}{\partial t} = v\frac{du}{d\xi} \tag{23.69}$$

$$\frac{\partial^2 V_m}{\partial x^2} = \frac{d^2 u}{d\xi^2} \tag{23.70}$$

Substitution of these equations into equation (23.67), gives an expression for the wavefront,

$$v\frac{\partial u}{\partial \xi} = D\frac{\partial^2 u}{\partial \xi^2} + f(u, n) \tag{23.71}$$

where $D = 1/RC_m$ is an effective diffusion coefficient for the membrane potential. Equation (23.71) is an ordinary differential equation, and can be analysed using standard methods. The membrane potential at the wavefront obeys the one-dimensional wave equation,

$$\frac{\partial^2 V_m}{\partial x^2} = \frac{1}{v^2}\frac{\partial^2 V_m}{\partial t^2} \qquad (23.72)$$

The travelling wave it describes is a *spike* that is both spatially and temporally localised.

It is possible to consider a piece of excitable nerve tissue as a continuum. For example, the normal contractions of the heart are triggered as the electrical action potential propagates through the cardiac tissue (**Section 24.4**). Correct signal transmission (the temporal ordering and the direction of the action potential propagation) allows efficient coordinated contraction of all four chambers of the heart, and maintains the blood-pumping function (**Figure 23.1**). Heart tissue can thus be modelled as a continuum of excitable material. Ventricular fibrillation (VF) is an arrhythmic condition in which there is uncoordinated contraction of the cardiac muscle in the ventricles of the heart, which makes them quiver randomly rather than contract in a well-defined manner. VF is a result of disorders in the sequence that the action potential propagates through the ventricular tissue. This condition, as well as many other cardiac arrhythmias, can be studied using mathematical models based on voltage-clamp data and the Hodgkin–Huxley equations.

It is useful to make an order of magnitude estimate of the number of Na^+ and K^+ ions that pass through each cm^2 of a neuron's membrane during a single action potential. The change in membrane voltage during a Na^+ influx is about 100 mV, the membrane capacitance per cm^2 is 1 μF, which gives the total electrical charge transferred per cm^2 as 10^{-7} C cm^{-2} ($q = C_m V_m$). The charge on each sodium ion is 1.6×10^{-19} C, so the number of sodium ions that pass through the membrane, during each action potential, is 0.6×10^{12} cm^{-2}. The number of potassium ions transported in the opposite direction, per cm^2 and per action potential, is the same because the two ion types bear the same charge, and the charge balance is maintained after the action potential. The density of sodium ion channels in the axon membrane is 1.2×10^{12} cm^{-2}. For potassium channels, it is 1.8×10^{12} cm^{-2}. Comparison of the number of ions that pass through the membrane (0.6×10^{12} cm^{-2}) to the number of ions shows that not every channel needs to be active during an action potential. There are approximately twice as many channels as are needed.

An additional question is whether the ions can reach the membrane channels over the total duration of the action potential. Under worst-case scenario conditions, the ion-channel distance is 0.8 nm, half of the ion–ion distance at the concentration at which they are maintained in the cell. The relevant ion mobility for Na^+ is 5.2 μm s^{-1} per V cm^{-1}. The voltage difference across a 5-nm thick membrane is about 100 mV, which results in a field of 2×10^5 V cm^{-1}. An ion will thus move at a velocity of 1 m s^{-1} and will travel a distance of ~1 cm over the duration of an entire action potential. Hence, there is a sufficient amount of time for the ionic movement and it is deduced that action potentials are not limited by ion mobilities.

Information is carried by nerve impulses. How much and how quickly this information is transferred limits the senses and ultimately all mental processes (**Chapter 25**). All spikes in nerve cells look very similar, they are stereotyped. Since the spike amplitudes carry no information it is deduced that the information must be contained in the spike timing, i.e. spike trains contain frequency-modulated signals. Typically the higher the frequency the higher the intensity of a stimulus (**Figure 23.17**), but the frequency can decrease over time with continued constant stimulation as the system becomes habituated. Organisms are found to react to stimuli after as few as 1–2 spikes, so something more than a naïve rate code (where rate = number of spikes/time) must be in effect, otherwise there would be huge errors in rapid responses to sensory information (which are not observed in many organisms).

Entropy and information are powerful tools for investigating spike trains. The Shannon expression (compare with the equivalent Boltzmann expression equation (3.17)) for the entropy is

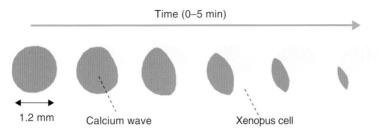

Figure 23.17 *Schematic diagram of spatial calcium waves in Xenopus cells induced by fertilisation at times of 0–5 min. [Reproduced with permission from R. Nuccitelli in C. Fall et al, Computational Cell Biology, Springer, 2002.]*

$$S(\text{spikes}) = -\sum_n p(n)\log_2 p(n) \tag{23.73}$$

where $p(n)$ is the probability of observing n spikes in a window of time width T. It is possible to calculate the information transmitted by a train of spikes. Consider in general some output signal (Y, the spikes in our case) that contain some information about the input signals (X), e.g. sensory information. The gain in information (I) due to the transmission of spikes can be calculated from differences in the entropy or disorder

$$I = \int \{P[Y](S(X) - S[X/Y])\}[dY] \tag{23.74}$$

where $S(X)$ is the logarithm (base 2) of the total number of possible events X, $P[Y]$ is the probability distribution of Y, and $S[X/Y]$ is the logarithm of the total number of Xs consistent with Y. From substitution of actual probability distributions in equation (23.74), it is thought that on average a single spike can transmit slightly more than a single bit of information (a '0' or a '1'). An estimate gives 1–3 bits of information per spike that depends on the context of the spike in a train of spikes, i.e. the signal-to-noise ratio, the exact probability distribution of the spikes, etc. On average additional information is contained within the time delays, beyond a simple binary '0' and '1' value associated with the existence of a spike.

Hyperacuity is the phenomenon by which sensors outperform the expectations of simple physical estimates. Typically, the high performance is due to sophisticated signal processing of spike trains by neural hardware (**Section 25.3**).

23.11   Cell Signalling

Electrical signalling with spikes and synaptic transmission of neurotransmitters are by no means the only methods cells and organisms use for internal signalling. There are a wide range of other mechanisms. For example, cells can be connected electrically using gap junctions, which allow ion transport directly between such electrically coupled cells, e.g. in cardiac muscle cells.

Notch signalling can occur when transmembrane receptors in adjacent cells make physical contact. Another possibility is when hormones are released in the blood and the circulatory system distributes the signals globally.

Neurotransmitters are important for signalling across the synaptic cleft in nerve cells. They can be excitatory or inhibitory or can activate more complex metabolic pathways. A wide range of molecules can modulate the activity of neurotransmitters such as nitric oxide, peptides, and amino acids. These molecules can thus be used to treat diseases associated with neuronal signalling.

Figure 23.18 *Schematic diagram of spiral calcium ion waves detected with a calcium-sensitive fluorophore observed inside Xenopus eggs. Such behaviour is often modelled with reaction diffusion equations (**Section 22.9**).*

It is a common motif in signalling systems for enzymes to carry information via regulation of their biochemical activities; these activities are controlled by covalent modifications or allosteric binding of effector molecules to the enzymes, e.g. phosphorylation of enzymes. It is a major success of modern mass spectroscopy equipment that minute mass changes associated with different signalling states of large globular proteins can be measured.

Intracellular calcium signalling can be followed using microfluorometry when fluorescent calcium indicator dyes are inserted into cells. Sperm-induced calcium waves in Xenopus eggs are observed in **Figure 23.18**. Spiral waves are also possible for calcium ion activity in Xenopus cells. These waves can be modelled using a reaction diffusion equation similar to that encountered in the discussion of morphogenesis (**Section 22.9**).

Suggested Reading

If you can only read one book, then try:

Top recommended book: Steratt, D., Graham, B., Gillies, A. & Willshaw, D. (2011) *Principles of Computational Modelling in Neuroscience*, Cambridge University Press. Good clear modern introduction to physical models of electrophysiology.

Cotterill, R. (2002) *Introduction to Biophysics*, Wiley. Good compact account of the physics of electrophysiology.

Greenspan, R.J. (2007) *An Introduction to Nervous Systems*, Cold Spring Harbour. Fascinating introduction to the minds of small organisms, e.g. flies, jelly fish, etc.

Hille, B. (2001) *Ion Channels of Excitable Membranes*, Sinauer. Classic text on ion channels (although could do with some updating).

Kandel, E., Schwartz, J.H. & Jessell, T.M. (2000) *Principles of Neural Science*, McGraw-Hill. Classic introduction to the biology of neural science.

Keener, J. & Sneyd, J. (2008) Mathematical Physiology I: Cellular Physiology and Mathematical Physiology II: Systems Physiology, Springer. Mathematically advanced account of electrophysiology.

Luckey, M. (2008) *Membrane Structural Biology, with Biochemical and Biophysical Foundations*, Cambridge University Press. More modern account of the structures of membranes.

Rieke, F., Warland, D., van Steveninck, R. & Bialek, W. (1999) *Spikes: Exploring the Neural Code*, MIT. Thought-provoking account of the inverse problem, i.e. how to reconstruct information transmitted through nerve cells by spike trains.

Squire, L.R., Berg, D., Bloom, F.E., du Lac, S., Ghosh, A. & Spitzer, N.C. (2008) *Fundamental Neuroscience*, Academic Press. Detailed introduction to the biology of mammalian brains.

Tutorial Questions 23

23.1 A nerve-cell membrane has a layer of positive charge on the outside and negative charge on the inside. These charged layers attract each other. The potential difference between them is 70 mV. Assume a dielectric constant $\kappa = 5.7$ for the membrane, an axon radius of 5 μm and a membrane thickness of 5 nm. Calculate the force per unit area that the charges on one side of the membrane exert on the other side. Express the answer in terms of b, v and κ.

23.2 The resistivity of the fluid within an axon is 0.5 Ω m. Calculate the resistance along an axon 5 mm in length with a radius of 5 μm. Repeat for a radius of 500 μm.

23.3 Calculate the Nernst potentials shown in **Table 23.1** from the ion concentrations and compare with the values in the table.

23.4 Concentrations of Na^+, K^+ and Cl^- inside a squid axon are determined to be 50, 400 and 40 mM, respectively. The corresponding concentrations outside the axon's membrane are 460 (Na^+), 20 (K^+) and 550 (Cl^-) mM. The membrane conductance for Na^+ ions is almost zero, while the conductances for K^+ and Cl^- ions are 2.7×10^{-6} and 1×10^{-6} mho cm^{-2}. Calculate the resting potential for the squid axon at 37 °C. (reminder: $kT/q = 26$ mV at 23 °C and 1 mho = 1 Siemens = Ω^{-1}).

23.5 Describe a voltage-clamp experiment and its relevance to understanding ion channels. Explain the exact meaning of the currents measured for the different ions.

23.6 Explain the difference between the passive subthreshold and the active action potential responses. Describe the origin of the threshold-dependent response.

24

Physiology of Cells and Organisms

Physiology is a central topic in the study of medicine. According to a classic medical textbook 'the goal of physiology is to explain the physical and chemical factors that are responsible for the origin, development and progression of life'. Subdivisions include viral physiology, bacterial physiology, cellular physiology, plant physiology and human physiology. The emphasis in the current book is on cellular physiology (under the guise of systems biology), but a brief foray will also be made into the area of human physiology in the current chapter.

Physiology is typically thought of as a subject for medics and vets. However, it is now increasingly coming under the remit of modern biological physics, as our understanding of the underlying molecular processes improves and more quantitative modelling becomes possible.

Physiology is a vast area of study, so only a few key illustrative areas will be discussed. First some generic ideas on how to analyse the physiology of organs will be introduced (control theory) and then the measurement of the electrical activity from hearts and brains (ECG and EEG respectively) will be considered. The heart is a well-controlled electromechanical pump and cardiovascular disease (when the heart malfunctions) is the biggest cause of death in the Western world. The pump can wear out in a number of different ways. For instance, the blood supply can be lost to the heart muscle (stents are inserted to solve this problem), blockage can occur to the main pipes (aneurisms due to accumulation of fats are treated with drugs such as statins), and the electrical waves that determine muscular contractions can be mistimed (ventricular fibrillation requires the insertion of pace makers). All these issues lead to interesting biological physics problems. EEGs allow large-scale electrical activity of the brain to be measured, are important medically for the detection of diseased states (e.g. epilepsy), and hint at a connection between electrical activity and the cohesive properties required for consciousness (**Chapter 26**).

Organisms need to control the complex fluctuations of their biological processes on length scales from the sizes of molecules to whole organs and on time scales from femtoseconds to many years. Biological control can be understood by reference to ideas developed by engineers and mathematicians to control machines. A good way to consider the role of control theory for physicists is by approaching the problem of sending a rocket to the moon. A naïve approach would be to use Newton's laws to calculate the trajectory and then just fire the rocket in the correct direction. However, there are a vast range of nonlinear effects that will perturb

The Physics of Living Processes: A Mesoscopic Approach, First Edition. Thomas Andrew Waigh.
© 2014 John Wiley & Sons, Ltd. Published 2014 by John Wiley & Sons, Ltd.

the path of the rocket (e.g. air resistance). The task of exact modelling the trajectory appears to be impossible. A practical solution is found in the form of a control loop that corrects the direction of the rocket if it begins to stray away from the required result (the Apollo rockets used Kalman filter control loops to achieve this goal). The position of the rocket relative to the moon needs to be measured and then the deviation of this signal from the required result is fed into a mechanical correction system, e.g. boosters rockets. Assuming the rocket flight is sufficiently stable (large-amplitude chaotic fluctuations in position will disrupt most control-loop mechanisms), highly accurate determination of trajectories is possible. Similar considerations are required to teach a robot to ride a bike or, more pertinently to biology, for the iris diaphragm in the eye to open by the correct amount to allow the optimal amount of light to illuminate the retina. In this chapter control theory of supracellular processes will be discussed, but similar mathematical ideas are used in systems biology on the molecular level inside cells (**Section D**). In systems biology the upregulation and downregulation of genetic information is finely controlled using feedback loops, but not everything is locally controlled at the molecular level in organisms and this is the subject of the current chapter, e.g. blood flow, temperature, etc. are globally controlled.

24.1   Feedback Loops

There are a wide range of variables such as temperature, oxygen concentration in the blood, cardiac output, number of red/white blood cells, blood concentrations of calcium, sodium, potassium and the blood concentration of glucose that the human body needs to regulate. They are regulated by feedback loops. *Negative feedback* loops are found in the generation of rhythmic activity of the heart, changes in the rate of breathing, daily variations in body temperature, hormone levels, hibernation, colouration, fur growth and reproductive activity. *Positive feedback* also can occur both usefully (e.g. depolarisation of axons) and destructively (e.g. blood-pressure regulation by diseased kidneys) in living organisms.

Feedback and control processes are a subject central to the quantitative study of physiology. Indeed, many properties of an organism are in a fine dynamic balance. For example, the weight of an individual often does not appreciably fluctuate over a period of years, it is subject to an accurate set point in energy metabolism (with a mild middle age paunch due to old age and lack of exercise!) Temperatures of warm-blooded mammals often need to be controlled to within ±1 °C or damage can occur that results in death.

The simplest control loop contains two processes; a process *Y* that depends on *X* and conversely a process *X* that depends on *Y* (**Figure 24.1**). Control loops occur in two main varieties. In *feedforward* loops two signals from *X* arrive at *Y* (normally there is a time delay between the signals or the dependence is trivial). *Feedforward* loops can occur in regulatory processes. They allow a guide signal to be sent forward in time and compared

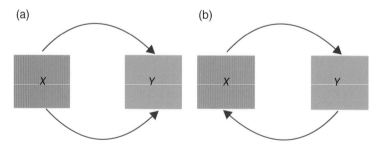

Figure 24.1   *Schematic diagram of simple (a) feedforward and (b) feedback loops that consist of two coupled processes* X *and* Y.

with another signal. They are a common feature of neural networks in the brain (**Section 26.2**). In *feedback loops* there is a two-way flow of information from each of the two interacting processes *X* and *Y*.

It is important to determine the equilibrium/steady-state behaviour of a feedback loop. A system is *stable* if it returns to the steady state when perturbed. Linear systems typically decay exponentially to steady state values. Nonlinear systems display much more complex behaviour, e.g. period doubling and chaotic behaviour can occur. Unfortunately nonlinearity is fairly common in biology and tractable linear approximations, although useful, should always be treated with a little scepticism during the creation of models for real biological processes.

In *negative feedback* the signal tends to reduce the changes that create it, whereas with *positive feedback* the opposite is true and the signal tends to increase the change that creates it. A more detailed example of linear negative feedback is given by the thyroid gland, which is used to control the rate of metabolism in humans (**Figure 24.2**). For a simple quantitative understanding of thyroid physiology the control loop can be considered as consisting of two processes in a negative feedback loop as shown in **Figure 24.1b**. In the thyroid gland an increase in thyroid hormone produces more thyroid stimulating hormone (*Y* depends on *X*, **Figure 24.2a**). In the pituitary gland the thyroid stimulating hormone level increases if the thyroid hormone falls (*X* depends on *Y*, **Figure 24.2b**). In healthy individuals the two processes balance and form a steady state.

A second example of negative feedback is given by the concentration of carbon dioxide in the lungs during breathing. Carbon dioxide needs to be removed from the lungs at a rate that depends on how fast the body consumes oxygen (*p*, **Figure 24.3**). A simple model relates the amount of CO_2 in the aveoli (air sacs in the lungs) to the rate of breathing (the ventilation rate). Again, there are two competing processes that control breathing and they form a negative feedback loop. For the first process, experiments show that the ventilation rate (*y*) and the alveoli CO_2 concentrations (*x*) are approximately related by

$$x = \frac{15p}{y-2} \tag{24.1}$$

where *p* is the body's oxygen consumption in mmol min^{-1}, the pressure *x* is measured in torr and *y* is measured in breaths per minute. In the second process that controls the rate of breathing, the brain senses the levels of CO_2 and completes the feedback loop by increasing the ventilation rate if necessary, which is shown

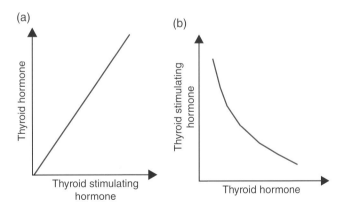

Figure 24.2   *The feedback loop that determines the activity of the thyroid gland in humans depends on the interaction between hormones created in the thyroid gland and the pituitary gland. (a) In the thyroid gland the thyroid hormone changes in response to the level of the thyroid stimulating hormone. (b) In the pituitary gland the thyroid stimulating hormone changes in response to the thyroid hormone.*

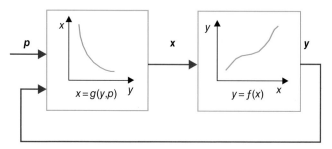

Figure 24.3 *Schematic diagram of a feedback loop based on the combination of two processes;* $x = g(y, p)$ *and* $y = f(x)$. p *is the input signal.*

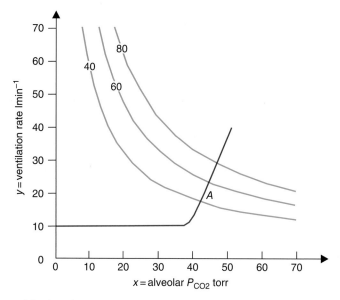

Figure 24.4 *Regulation of the breathing rate as a function of the carbon dioxide level in the alveoli. A change of metabolic rate (parameter* p*) causes a change in ventilation rate* y*, so that* $x = P_{CO_2}$ *does not change as much. Solution by inspection gives* $p = 60\,mmol\,min^{-1}$ *when operating point is at A. [Reprinted from R.K. Hobbie, B.J. Roth, Intermediate Physics for Medicine and Biology, Springer, 2007. With kind permission from Springer Science + Business Media B.V.]*

graphically in **Figure 24.4**. The ideal operating point (A) for the breathing process is simply where the plot of the two processes x as a function of y, and y as a function of x intersect.

Other examples of feedback loops include Cheyne–Stokes respiration (abnormal oscillatory breathing), hot tubs and heat stroke (positive feedback can occur for an over heated person in a hot tub, since perspiration and increased blood flow near the skin only makes matters worse), oscillatory white blood counts, and the regulation of the heart pumping rate (the Frank–Starling mechanism).

For feedback loops with several internal variables it is often possible to simplify them. For example, a three-function control loop with $x = f(w)$, $y = g(x)$, and $w = h(y)$ can be simplified to form $x = F(y)$ and $y = g(x)$. Thus, a macroscopic description of a physiological control loop can result from coarse graining of many internal variables due to specific molecular details at smaller length scales.

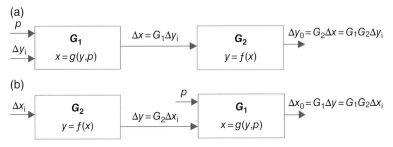

Figure 24.5 *The open-loop gain can be calculated by opening the loop at any point. (a) Loop opened in* y, *(b) Loop opened in* x. *In both cases the open-loop gain is the product,* G_1G_2.

By definition, the gain of a process in a control loop is the ratio of the change in the output variable to the change in the input variable. The gain of the first box in a two-process control loop (**Figure 24.5**) is given by

$$G_1 = \left(\frac{\Delta x}{\Delta y}\right)_{\text{boxg, pfixed}} = \left(\frac{\partial x}{\partial y}\right)_{\text{boxg, pfixed}} \tag{24.2}$$

where Δx is the change in x and Δy is the corresponding change in y. Similarly the gain of the second process is given by

$$G_2 = \left(\frac{\Delta y}{\Delta x}\right)_{\text{boxf}} = \left(\frac{\partial y}{\partial x}\right)_{\text{boxf}} \tag{24.3}$$

The product G_1G_2 is called the *open-loop gain*,

$$G_1G_2 = \frac{(\partial y/\partial x)_{\text{boxf}}}{(\partial y/\partial x)_{\text{boxg}}} \tag{24.4}$$

The open-loop gain can thus be calculated relatively simply if the functional form of the gain for the processes in each box is known. Alternatively, it is also possible to calculate the open-loop gain graphically using the two gradients for a particular set of parameters. It is important to calculate the gain in the direction that causality operates. Going around the loop the wrong way gives the reciprocal of the open-loop gain.

As an example of the open-loop gain consider the simple model of respiration that consists of equation (24.1) and the following expression for the second control process

$$y = 10 + 2.5(x - 40) \tag{24.5}$$

Therefore, the gain of each process can be calculated from equation (24.2) and (24.3),

$$G_1 = \frac{-15p}{(y-2)^2}, \quad G_2 = 2.5 \tag{24.6}$$

When x is 45 torr, p is 60 mmol min^{-1} and y is 23 min^{-1} the open-loop gain (G_1G_2) is –5.1. The effect of this negative feedback loop is to cause a change in y which reduces x by a factor of 1-OLG,

$$\Delta x' = \frac{\Delta x}{1 - \text{OLG}} \tag{24.7}$$

When OLG < 0 there is negative feedback and when OLG > 0 there is positive feedback. Thus, for the negative-feedback example given, x is reduced by a factor of 0.23. Furthermore, the response of any system

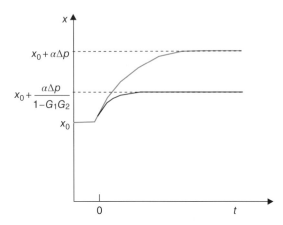

Figure 24.6   *The time dependence (t) of the physiological variable x with (orange) and without (green) feedback in response to a step change in an input parameter p. With feedback equilibrium is reached more quickly and a smaller change in the position of equilibrium occurs, i.e. it is reduced by a factor* $\dfrac{1}{1-G_1G_2}$. G_1G_2 *is the open loop gain.*

cannot be infinitely fast. Typically, it is found that equilibrium is established in an exponential manner (without feedback),

$$x - x_0 = a\Delta p \left(1 - e^{-t/\tau_1}\right) \tag{24.8}$$

where τ_1 is the time constant for the process, Δp is the change in the input parameter, a is a constant amplitude and x_0 is the initial value of x (**Figure 24.6**). The approach to equilibrium *without* a feedback loop is thus an exponential process with one time constant. *With* feedback another exponential dependence is found,

$$x - x_0 = \frac{a\Delta p}{1 - G_1 G_2}\left(1 - e^{-t/\tau}\right) \tag{24.9}$$

where τ is the new time constant for the process. This is similar in form to the system without feedback except the time constant is reduced,

$$\tau = \frac{\tau_1}{1 - G_1 G_2} \tag{24.10}$$

Negative feedback thus speeds up the rate at which the system achieves the steady state, as seen previously with gene regulation, **Section 22.6**. If the equations that reflect the input and output variables of each process of a negative feedback loop are known, then their simultaneous solution gives the steady-state values of the variables. When a single process in the negative feedback loop determines the time behaviour, the rate of return of a variable to equilibrium is proportional to the distance of that variable from equilibrium. The return of equilibrium is an exponential decay and the system can be characterised by a time constant. In a negative-feedback system one variable can change to stabilise another variable. The amount of stabilisation and the accompanying decrease in time constant depends on the open-loop gain. It is possible to have oscillatory behaviour with damped or constant amplitude if the two processes in the feedback loop have comparable time constants and sufficient open-loop gain, or if the processes depend on the value of its input at an earlier time or if the processes have three or more degrees of freedom. A nonlinear oscillatory system can have its phase reset by an external stimulus, e.g. contractions in the heart can be reset during defibrillation after a heart attack.

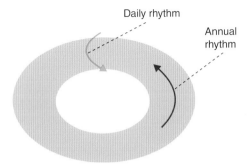

Figure 24.7   *All organisms on planet Earth experience physiological changes on a topological doughnut space. Daily rhythms (24 h) are superposed on annual rhythms (365.242 days).*

24.2   Nonlinear Behaviour

Temporal programmes are of primary importance in the organisation of the life cycle of cells, tissues and organisms. They are of particular importance during the development of organisms during morphogenesis. These temporal programmes are both externally and internally originated. The internal rhythms require chemical clocks inside cells and these cellular clocks are based around nonlinear oscillators.

Time takes an unusual form with biological systems. The physiology of an organism is typically entrained to the day and night cycle of the Earth with its twenty-four-hour periodicity. This is so engrained in the make up of organisms that the topology of their daily internal clocks can be thought of as circular. The superposition of the daily periodicity on an annual seasonal periodicity leads the topology of the annual clocks to be doughnut-shaped and this has unique consequences for their behaviour. Detailed analysis of the clock topology and dynamics provides insights into such phenomena as the phase resetting of biological clocks and jet lag. Thus, the physiology of all Earth-based organisms occurs on a topological doughnut space (**Figure 24.7**).

Detailed analysis of biological systems shows that much more sophisticated nonlinear phenomena are possible, which can be studied using bifurcation analysis. A point in parameter space at which the number of limit cycles or their stability changes, is called a bifurcation. A single pulse can reset the phase of a nonlinear oscillator, e.g. in cardiac muscle. Periodic pulses can entrain an oscillator to the driving frequency. This is how cardiac pacemakers function. Other phenomena include strange attractors for oscillatory motion, period doubling, and deterministic chaos (aperiodic oscillatory motion that never repeats).

24.3   Potential Outside an Axon

Consider the charge distribution in a resting nerve cell. The difference between the potentials inside and outside of a nerve cell is approximately 80 mV and the membrane thickness (b) is 6 nm. The electric field is assumed constant and from its definition as the gradient of the potential it can be calculated as

$$E = -\frac{dV}{dx} \approx \frac{V}{b} = \frac{-80 \times 10^{-3}\ V}{6 \times 10^{-9}\ m} = 1.33 \times 10^{7}\ Vm^{-1} \qquad (24.11)$$

The electric field in a capacitor with a dielectric of relative permittivity (κ) is

$$E = \frac{Q}{\kappa \varepsilon_0 S} \qquad (24.12)$$

where Q is the charge, ε_0 is the permittivity of free space, and S is the area. Therefore, from the definition of capacitance ($C = Q/V$),

$$C = \frac{\kappa \varepsilon_0 S}{b} \tag{24.13}$$

The relative permittivity (κ) of an axon membrane is approximately equal to 7. The capacitance per unit area (C/S) is therefore approximately 7×10^{-4} C m^{-2}. The membrane thickness for myelinated membranes is much bigger, typically 2000 nm instead of 6 nm and this reduces the capacitance by a factor of 333. This reduction in capacitance has important physiological implications. The cable model for an axon was introduced in **Section 23.6** and the solution for the voltage during passive transport was found to be an exponential decay with time,

$$V(t) = V_0 e^{-t/\tau} \tag{24.14}$$

where τ is a time constant that depends on the capacitance,

$$\tau = R_m C_m = \kappa \varepsilon_0 \rho_m \tag{24.15}$$

The time constant is independent of both area and thickness of the membrane. The natural nerve membrane resistivity is $\rho_m \approx 1.6 \times 10^7$ Ω m and the time constant is therefore around 1×10^{-3} s. Changes in capacitances therefore have important consequences for the rate of signaling. Unmyelinated fibres (which includes the grey matter in the brain) constitute about 2/3 of the fibres in the human body and have small radii in the range 0.05–0.06 µm. The conduction speed in m s^{-1} is $u \approx 1800\sqrt{a}$, where a is the axon radius in metres. Myelinated fibres (the white matter in nervous systems) in contrast are relatively large 0.5–10 µm (**Figure 24.8**). Their outer radius is approximately $1.67a$. The spacing between nodes (D) is proportional to the outer diameter ($D \approx 330a$). The conduction speed in the myelinated fibre is $u \approx 20 \times 10^6 a$. Therefore, there is saltatory ('jumping') conductance along a myelinated fibre; fast transport in the myelinated sections due to their high capacitance followed by slow transport at the unmyelinated regions, called the nodes of Ranvier. The nodes of Ranvier are needed to facilitate charge injection from the surrounding cytoplasm. Local anaesthetics such as procaine work by the prevention of permeability changes at the nodes of Ranvier, which reduces the number of available ions and damps the signal.

Nerve pulses occur on millisecond time scales or slower. Reaction times and the perception of the outside world cannot occur any faster than these time scales (although detection can happen slightly faster under some circumstances in sensory hyperacuity, **Chapter 25**).

Noninvasive measurements of axon electrical activity can detect the electrical fields that surround axons (magnetic field measurements are also possible using the MEG technique, but will not be considered here).

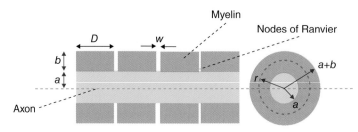

Figure 24.8 *The idealised structure of a myelinated neuronal fibre in longitudinal section and in cross section. The internodal spacing D is actually about 100 times the outer diameter of the axon. The nodes of Ranvier allow ions to enter the axon.*

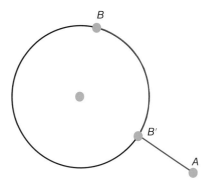

Figure 24.9 *A point current source is at the centre of a cylinder. The path of integration to calculate the potential difference between points A and B goes first from A to B′ and then from B′ to B.*

It is therefore useful to consider the calculation of electric potentials outside long cylindrical axons. The exterior potential changes outside an axon are small (<0.1% on interior), but non-negligible. Simple generic models apply to potentials observed outside all electrically excitable cells. For an axon, the potential change along *B′B* is zero (**Figure 24.9**), since there is no change in the radial distance. From the definition of the potential in terms of the electric field, the potential difference between points A and B is

$$V(B) - V(A) = \int_{r_A}^{r_B} E_r dr = \frac{i_0}{4\pi\sigma_0}\left(\frac{1}{r_B} - \frac{1}{r_A}\right) \tag{24.16}$$

where σ_0 is the conductivity, i_0 is the current, r_A is the radial distance at point A, and r_B is the radial distance at point B. In the limit that the point r_A is taken at very large distances ($r_A \to \infty$) the potential is

$$V(r) = \frac{i_0}{4\pi\sigma_0 r_0} \tag{24.17}$$

where r_0 is the distance from the fibre to the observation point. A nerve pulse corresponds to a current dipole and it can be shown that the potential far from the fibre is

$$V = \frac{\Delta V_i \sigma_i a^2}{4\sigma_0 r_0^2} \cos\theta \tag{24.18}$$

where a is the fibre radius, θ is the angle to the observation point and ΔV_i is the magnitude of the action potential. Such calculations form the basis of models of electrocardiograms and electroencephalograms where the exterior potential V depends on ΔV_i, but not on the length of the depolarisation region. The potential from a dipolar current falls off as $1/r^2$ instead of as $1/r$, as from a point source (equation (24.17)). The potential varies with angle, and is positive to the right of the transition region and negative to the left.

24.4 Electromechanical Properties of the Heart

To consider the action of the heart it is conceptually useful to separate up its activity under the headings of *pumping components* and *electrical components* (**Figure 24.10**). In humans, two pumps are attached side by side in series in a single integrated unit that constitutes the heart. Human babies only use one side of the heart

(a)

(b)

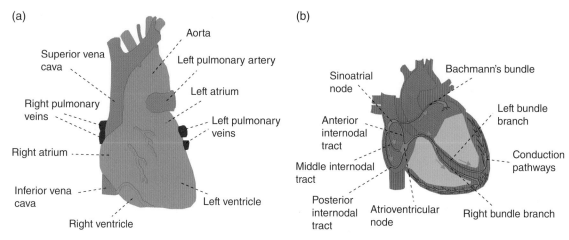

Figure 24.10 *The anatomy of a human heart can be decomposed into (a) the mechanical components and (b) the electrical components.*

before birth in the womb. They have a hole that bypasses the sections normally used for pumping blood around the lungs, which closes at birth. It helps to reduce the load on the mother's circulatory system. Some organisms have multiple pumps that are completely separated, e.g. squid have three hearts.

In mammals, the function of the right side of the heart is to collect deoxygenated blood. In the right atrium, blood from the body (via the superior and the inferior vena cava) is pumped via the right ventricle, into the lungs so that carbon dioxide can be dropped off and oxygen picked up. The left side collects oxygenated blood from the lungs and takes it into the left atrium. From the left atrium the blood moves to the left ventricle that pumps it out to the body (via the aorta). On both sides, the lower ventricles are thicker and stronger than the upper atria. The wall of muscle that surrounds the left ventricle is thicker than the wall that surrounds the right ventricle due to the higher force needed to pump the blood through the circulation system than through the lungs.

The heart's electrical components can next be considered. Some cardiac cells are self-excitable, they contract without any signal from the nervous system, even if removed from the heart and placed in culture. Indeed tissue made from cardiac stem cells can spontaneously begin electrical oscillations and mechanical contractions. Each of the cardiac cells in a heart has their own intrinsic contraction rhythm that is modified by interaction with their neighbours in a process of entrainment. A region of the human heart called the *sinoatrial node*, or pacemaker, sets the rate and timing at which all cardiac muscle cells contract. The SA node generates electrical impulses, much like those produced by nerve cells. Cardiac muscle cells are electrically coupled by intercalated disks between adjacent cells, so impulses from the SA node spread rapidly through the walls of the atria, and cause both the left and right atria to contract in unison. The impulses also pass to another region of specialised cardiac muscle tissue, a relay point called the *atrioventricular node*, located in the wall between the right atrium and the right ventricle. Here, the impulses are delayed for about 0.1 s before they spread to the walls of the ventricle. The delay ensures that the blood in the atria empties completely before the ventricles contract. Specialised muscle fibres called Purkinje fibres then conduct the signals to the apex of the heart through the ventricular walls.

24.5 Electrocardiogram

Noninvasive measurements of the electrical activity of hearts are possible by the attachment of leads to the surface of the human body and calculation of the potential of the skin at a number of different positions.

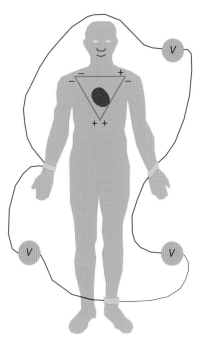

Figure 24.11 *Experimental arrangement for the measurement of electrocardiograms based on the calculation of three voltage differences.*

The simplest arrangement requires three leads (**Figure 24.11**). A schematic diagram of the electrocardiogram data collected is shown in **Figure 24.12**. The P, Q, R, S, and T phases of the electrical activity can be modelled in terms of travelling waves that traverse the three dimensional structure of the heart (**Figure 24.1**).

24.6   Electroencephalography

In an analogous manner to the ECG of hearts, electroencephalograms (EEGs) of brains can be constructed by the attachment of electrodes to the scalp of the head (**Figure 24.13**). The small electrical fields due to neural activity (much smaller than those of the heart) are then measured and mapped. Modelling of the origin of this activity is still at an early stage. Using standard analogue signal analysis techniques (Fourier transforms, wavelets, etc.) it is possible to extract standard periodic wave forms that can then be connected to a broad range of mental phenomena in the subject (**Figure 24.14**). **Table 24.1** gives a list of frequencies and mental phenomena. Eventually it is hoped to reverse engineer the collective neuronal patterns that give rise to periodic electrical wave activity to give a clear insight into brain disease. An enticing thought is that the electrical waves are also a signature of cohesive properties of the brain and perhaps consciousness.

The alpha wave is the best-understood EEG phenomenon and occurs with conscious healthy adult humans when their eyes are closed. It is thought to be due to the lowest-frequency resonance mode of activity of all the weakly coupled neural electrical oscillators in the brain. An analogy can be made to mechanical masses attached to springs encountered in introductory vibrations and waves courses. Normal mode analysis imply there are $3N-3$ (N is the number of coupled oscillators) different vibration modes in coupled mechanical mass–spring systems. The lowest frequency of the mass–spring system correspond to all the oscillators

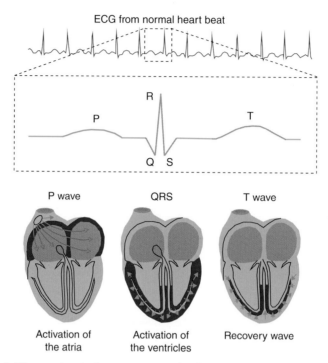

Figure 24.12 *Origin of different electrical events in a single electrocardiogram waveform. P corresponds to the activation of the atria, QRS is the activation of the ventricles and T is a recovery wave.*

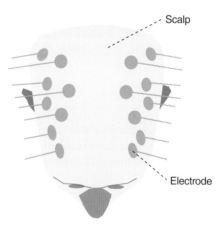

Figure 24.13 *Experimental arrangement of electrodes to measure brain activity using electroencephalography.*

vibrating together in unison, and this is analogous to the *alpha wave* (**Table 24.1**), due to coupling of nonlinear electrical oscillators in the brain.

The Nunez model for the resonant standing electrical wave frequencies of the brain (the dispersion relation) is

$$f_n = \frac{v}{L}\sqrt{n^2 - \left(\frac{\beta\lambda L}{2\pi}\right)^2} \tag{24.19}$$

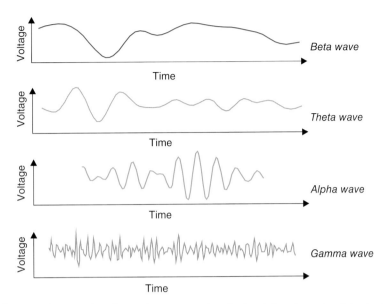

Figure 24.14 *Schematic diagram that shows the average electrical activity of a patient's brain decomposed (using Fourier filters) in terms of Alpha, Beta, Gamma and Theta waves from electroencephalography measurements.*

Table 24.1 *The frequency of different electrical waves in human brains measured with EEG and the corresponding mental attribute.*

Name	Frequency (Hz)	Mental attribute
Delta wave	4	Sleep
Theta wave	4–7	Seen in young children and meditating adults
Alpha wave	8–12	Seen when eyes are closed
Mu rhythm		Sensorimotor cortex
Beta wave	12–30	Motor behaviour
Gamma wave	30–100	Binding of different populations of neurons

where $n = 1,2,3\ldots$, v is the propagation speed of brain waves (600–900 cm s^{-1}), L is the front-to-back distance of a cortical hemisphere, λ is a parameter related to the fibre density, β is a neuromodulator parameter ≈ 1 and f_n is the temporal frequency of the normal mode. $v/L \sim 7$–15 Hz for n is 1, so equation (24.19) gives a reasonable value for the alpha wave frequency of the human brain (compare with **Table 24.1**). Such models of alpha waves have experienced some other qualitative successes, e.g. how the alpha wave frequency scales with the brain size.

Suggested Reading

If you can only read one book, then try:

Top recommended book: Hobbie, R.K. & Roth, B.J. (2010) *Intermediate Physics for Medicine and Biology*, 4th edition, Springer. Excellent introduction to aspects of control theory in biological physics.

Astrom, K.J. & Murray, R.M. (2008) *Feedback Systems an Introduction for Scientists and Engineers*, Princeton University Press. Good introduction to mathematical techniques in control theory.

Bechhoefer, J. (2005) Feedback for physicists: a tutorial essay on control, *Review of Modern Physics*, **77**, 783–836. Another perspective on physical phenomena involved in control theory.

Buzsaki, G. (2011) *Rhythms of the Brain*, Oxford University Press. Fascinating popular discussion of the possible origins of electroencephalography (EEG) wave forms.

Feller, J. (2012) *Quantitative Human Physiology: An Introduction*, Academic Press. Modern introduction to human physiology.

Guyton, A.C. & Hall, J.E. (2004) *Medical Physiology*, 11th edition, Elsevier. Classic medical textbook on physiology.

Herman, I.P. (2008) *Physics of the Human Body: A Physical View of Physiology*, Springer. Large selection of physical examples in human physiology.

Keener, J. & Sneyd, J. (2009) *Mathematical Physiology*, 2nd edition, Springer. Mathematically advanced tour of quantitative models of physiology.

Nunez, P. & Srinivasan, R. (2006) *Electric Fields of the Brain: The Neurophysics of EEG*, Oxford University Press. Comprehensive textbook account of EEG measurement which includes the development of models for the electrical phenomena involved.

Silbernaql, S. & Despopoulos, A. (2008) *Color Atlas of Physiology*, 6th edition, THIEME. Often assimilation of the huge amount of data involved in medical physiology textbooks is challenging. This short textbook has a refreshing compact colourful style.

Strogatz, S. (2000) *Non-linear Dynamics and Chaos*, Perseus. Classic introduction to applied maths techniques for the analysis of nonlinear problems.

Tutorial Questions 24

24.1 Describe some nonlinear physiological phenomena in biology.

24.2 Find the open-loop gain for the system described by

$$x = [(y-p)/2]^{1/2}, \quad y = 5 - x^2$$

24.3 Find the open-loop gain for the control system shown below:

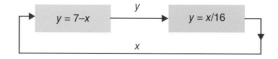

24.4 MRI is able to image the activity of live brains. Describe some classic experiments in this area.

24.5 Describe the state-of-the-art understanding of the function of glial cells in the brain.

24.6 Electrical waves in excitable tissue are involved in heart conditions and epilepsy. Describe some of the conditions observed in terms of the waves across the tissues involved.

24.7 Explain how magnetoencephalography (MEG) can be used to measure mental activity.

25

The Senses

Detectors, and their associated instrumentation, are a mainstay of the physical sciences. Evolution has created a panoply of sensory modules in living organisms. These sensory modules integrate detectors, amplifiers, feedback loops, and data compression, with advanced analysis algorithms calculated by neural hardware (wetware).

Indeed, the performance of biological detectors is often remarkable even when compared to modern synthetic silicon-based instrumentation. For example, cells in the mammalian retina can detect single photons, and cochlea inside the human ear are sensitive to Angstrom-level displacements produced by sound waves. Coincidence detection is used by bats in echo location to measure microsecond delays and insects have evolved hyperspectral imaging across ultraviolet, visible and infrared parts of the electromagnetic spectra.

In addition to exploring the physics of detection, a separate parallel approach towards an understanding of biological senses is to consider the transport of information. In animals the creation and processing of spike trains in nerve cells give rise to clear thresholds for a detection system's performance. The refractory period of a nerve cell limits the time period between spikes to ~2 ms (the wait time between spikes) and each spike contains ~1–3 bytes of information, so an information transfer limit can be readily estimated for a single nerve cell as ~1500 bits/s. All of the senses of higher organisms are constrained by such limits. In many cases sensory information is sacrificed, so that the most critical messages are transmitted at reasonably fast rates, e.g. visual information is lost as signals are compressed in the optical nerves connected to mammalian eyes.

25.1 Biological Senses

There are six main types of sensor that occur in biological systems at the cellular level; *mechanoreceptors*, *chemoreceptors*, *photoreceptors*, *thermoreceptors*, *electroreceptors*, and *magnetoreceptors*. Mechanoreceptors measure mechanical displacements, chemoreceptors monitor molecules in the receptors' environments, photoreceptors are sensitive to photons, electroreceptors can measure both static and time-varying electric fields, and magnetic receptors measure magnetic fields.

The Physics of Living Processes: A Mesoscopic Approach, First Edition. Thomas Andrew Waigh.
© 2014 John Wiley & Sons, Ltd. Published 2014 by John Wiley & Sons, Ltd.

(a)   (b)

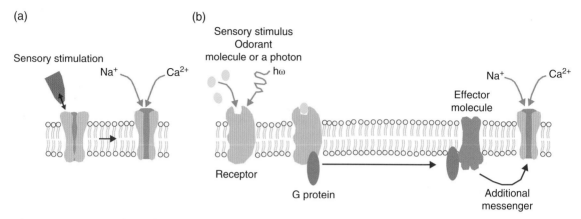

Figure 25.1   *General mechanism for sensory transduction at the molecular level in cells. (a)* Ionotropic transduction, *in which direct gating on an ion channel occurs in response to a stimulus on it. (b)* Metabotropic transduction *in which a receptor activates a G protein that leads to a transduction cascade that eventually gates an ion channel. [Reproduced with permission from G.L. Fain, Sensory Transduction, Sinauer, 2003. Copyright Sinauer Associates, 2014.]*

In humans, *chemosensitivity* is observed in the sense of smell and gustation. In smell, each individual odour gives a particular spatial map of excitation in the olfactory bulb. It is thought that through this spatial coding the brain can distinguish between the specific odours. *Photosensitivity* is observed in the phenomenon of human vision, where photon wavelengths in the range 390–700 nm are detected. A wider range of photon energies from infrared to ultraviolet are measured in other organisms using a wide variety of specialised biological instrumentation. *Thermal sensitivity* also occurs in some organisms. *Mechanosensitivity* is found in humans with the response to direct pressure changes due to contact (touch), but also in the detection of sound waves (hearing) and in balance. *Magnetism* is a minority sense that is known to be used for navigation by bacteria, insects, birds, and fish; a list that continues to be extended, although many of the exact mechanisms are still disputed. Sensitivity to *electrical fields* is another minority sense in some sharks, rays, dolphins, duck-billed platypuses, many salt and fresh water fish; a list that again continues to be extended. *Pain* is an additional medically important sense. It is a mechanism by which organisms monitor damage to the integrity of their cellular tissues.

At the molecular level the two main strategies for detection are *ionotropic* and *metabotropic* (**Figure 25.1**). With ionotropic transduction an ion channel is gated directly in response to a stimulus, whereas with metabotropic transduction there is a G protein that is released in response to a stimulus that directs a sequence of signalling reactions that eventually gate an ion channel.

25.2   Weber's Law

Biological detection often occurs over many orders of magnitude of signal strength for a particular sense. This wide ranging sensitivity is described by Weber's Law,

$$p = k \ln \frac{S}{S_0} \tag{25.1}$$

where p is the strength of perception in the organism's brain, S is the stimulus strength, S_0 is the threshold stimulus when the signal begins to be detected and k is a constant. A fundamental motivation for the logarithmic scaling is that it is due to the finite bandwidth for information transmission in the nerves and the need to measure sensory information over many orders of magnitude of stimulus strength, i.e. approximately logarithmic data compression occurs during spike transmission for efficient information transmission. Weber's law is approximately true for hearing and vision. This is why sound is measured in logarithmic units, i.e. decibels, to match the range of human hearing. E.coli are also thought to have a sensitivity described by Weber's law for the sensory module used in chemotaxis. Here, information transfer efficiency is also thought to be the guiding principle that gives rise to a sensory response equivalent to equation (25.1).

25.3 Information Processing and Hyperacuity

The senses are detection systems that are confronted with the effects of both noisy input signals and noisy transmission systems. However, in many cases the sensors perform superbly well and can surpass some of the apparent physical limits. For example, simple geometric optics arguments imply that 0.01° is the highest angular spacing accuracy that can be resolved by the human eye (**Section 25.6**) due to limits of diffraction effects, since the resolution is mathematically well defined by the geometry of the eyeball and the retina. However, through observation of a Vernier scale (parallel lines on white paper, **Figure 25.2**) volunteers can measure angular spacings of 0.002°, five times better than the diffraction limit and these spacings correspond to displacements smaller that the spacing between the photoreceptors in the eye. The explanation of this conundrum is based on the image analysis software in the eye. Integration of many photons along the parallel lines allows the diffraction limit to be beaten in the calculation of the lines' optical centres of mass to provide subdiffraction and subpixel (subcone size) resolution. The explanation is similar to that of the Heisenberg gedanken microscope used to explain superresolution microscopy (**Section 19.5.2**).

Human colour vision depends on the sensitivity of cones in the eye to red, green, and blue light (**Figure 25.3**). Since there are only three varieties of photoreceptor the spectral resolution might be expected to be fairly poor. However, a vast range of colours can be detected by the human eye through the synthesis of these three broad band measurements and ±5 nm spectral resolution is routinely achieved. Thus, hyperacuity occurs in colour vision and is thought to be due to sensitive integration of measurements from a large number of cones.

Temporal hyperacuity is observed in the echo location of bats (**Figure 25.4**). Bats can discriminate sub-30-µs timing in ultrasound echoes, although it is known that a single neuron cannot perform much quicker than its refractory period (~2 ms). The bats are thought to do this using a process of coincidence detection. Coincidence detection of discrete temporal events is a standard technique in high-resolution physical instrumentation, e.g. in particle/nuclear physics. Furthermore, electric fish are thought to use a similar coincidence detection design in their neural architectures to respond to 100 ns shifts of phase in the oscillation of electric fields in their environment.

25.4 Mechanoreceptors

Mechanoreceptors occur in a number of different forms in a huge range of organisms. In cell membranes, protein channels can measure stretches of the membrane structure, e.g. in a single E.coli cell or the detection of osmotic swelling in mammalian cells. *Kinaesthesia* involves the detection and control of the dynamic

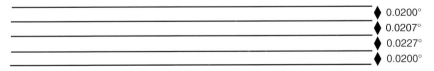

Figure 25.2   *The position of lines drawn on white paper can be resolved with superior angular resolution than expected from the diffraction limit. It is an example of visual hyperacuity.*

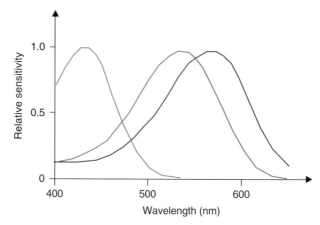

Figure 25.3 *The sensitivity of the red, green and blue cones from the macaque retina as a function of wavelength. Colour vision is created by the synthesis of the signals from the three different varieties of cones. [Reprinted by permission from Macmillan Publishers Ltd: Nature, B.J. Nunn et al, Nature, 1984, 309, 264–267, copyright 1984.]*

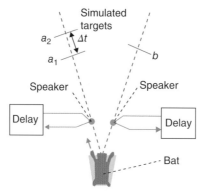

Figure 25.4 *Temporal hyperacuity can be demonstrated in the echolocation ability of bats. By simulating the motion of a target using two loudspeakers it has been shown that bats are sensitive to microsecond time delays (Δt). [Reproduced from J.A. Simmons, Cognition, 1989, 33, 155–19. With permission from Elsevier.]*

position and orientation of an organism, e.g. the stretch receptors in mammalian muscles or the sensilla in insects. *Touch* is observed in worms (C.Elegans), insects, and mammalian skin. An example of the voltage measured from a patch clamp attached to a paramecium (a single-celled micro-organism) when it is touched is shown in **Figure 25.5**.

A specialised form of mechanoreception is found in the human ear (**Figure 25.6**). The ear has three different sections. The *external ear* initially gathers the sound. The *middle ear* transfers energy from the air (low acoustic impedance) to the liquid of the inner ear (high acoustic impedance). Finally, the *inner ear* transforms the pressure signal into nerve impulses that travel to the brain. The *middle ear* transforms the acoustic impedance using two mechanisms. The first uses a simple area change. A sound wave (pressure amplitude, p_{air}), impinges on the ear drum (area, $S_{eardrum}$). The total excess force (F) on the ear drum due to the sound wave is

$$F = p_{air} S_{eardrum} \qquad (25.2)$$

(a)

Paramecium

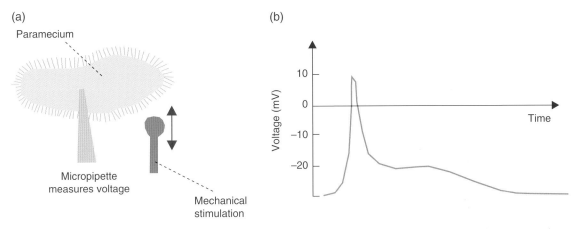

(b)

Micropipette
measures voltage

Mechanical
stimulation

Figure 25.5 *The touch response of a single-celled micro-organism (a paramecium) invokes a spiking potential in a simultaneous patch clamp measurement. (a) Experimental arrangement, (b) voltage as a function of time measured with the micropipette. [From R. Eckert, Science, 1972, 176, 473–481. Reprinted with permission from AAAS.]*

The bones transmit a force to the oval window membrane that has a much smaller area, $S_{ovalwindow}$. The pressure induced in the liquid in the inner ear ($P_{innerear}$) is therefore amplified,

$$p_{innerear} = p_{air} \frac{S_{eardrum}}{S_{ovalwindow}} \approx 20 p_{air} \tag{25.3}$$

There are also levers that connect the drum and oval window. These provide some additional amplification and adjust the acoustic impedance. The *inner ear* contains three semicircular canals used for balance, and the cochlea that converts sound waves into nerve impulses. The cochlea is a small fluid-filled spiral. Sound pressure wave travels through the liquid in the cochlea and produce a displacement of the basilar membrane. There are two types of hair cell in the cochlea, attached to the basilar membrane, that are arranged as one row of inner hairs and three rows of outer hairs. Each of these hairs has further very fine hairs attached to them that are called cilia. The motion of the fluid moves the cilia and they then provide an electrical response that is transmitted as a spike train by specialist nerve cells (cochlear nerves, **Figure 25.7**).

25.5   Chemoreceptors

Chemoreception occurs in a diverse range of forms in organisms. Examples include silk moths that can detect pheromones at extremely low concentrations in air (numbers as low as 7 molecules per second can be detected). Receptors are found for the respiration gases oxygen and carbon dioxide in mammals, and are important for the control of breathing (they form part of the control loop, **Chapter 24**). Gustation is found in the taste buds of mammals, and chemosensory sensilla of insects. Humans can detect five primary flavours in the mouth; sweetness, bitterness, sourness, saltiness and umami (**Figure 25.8**).

Olfaction is the sense of smell, which is demonstrated by insect antennas and mammalian noses. The sense of smell in humans allows odorants to be detected that are inhaled through the nose. The olfactory receptors are membrane proteins that display affinity for a range of odour molecules. Olfactory neurons then send signals to the brain along the olfactory nerve (**Figure 25.9**).

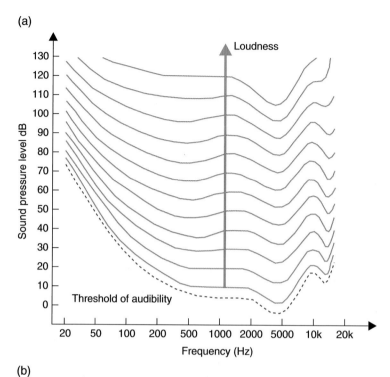

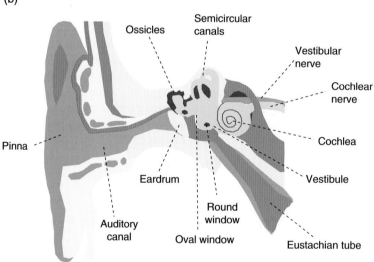

Figure 25.6 *(a) Frequency-dependent pressure response of the human ear as a function of the average loudness. (b) The anatomy of a human ear. The pinna collects the sound, which is mechanically amplified and then converted into nerve impulses by the cochlea, which are transmitted as spikes in the cochlear nerve.*

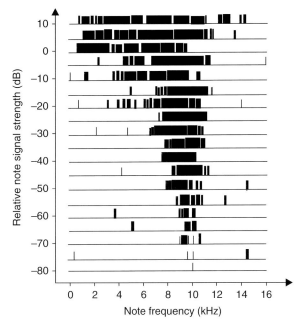

Figure 25.7 *Schematic diagram of the recording of electrical signals (spikes) from a guinea pig ear in response to different frequencies of sound waves. More spikes are observed over a wider range of frequencies for louder sound waves. The resultant response is similar to **Figure 25.6**. [Reproduced from E.F. Evans, Handbook of Sensory Physiology, 1975, Vol V/2 ed W.D. Keidel, W.D. Neffs, Springer, 1–108. With permission from Elsevier.]*

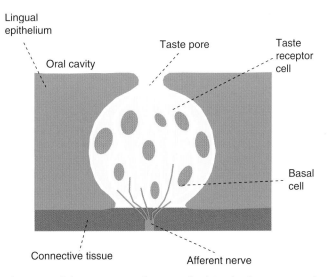

Figure 25.8 *Schematic diagram of the structure of a taste bud in the human mouth. Chemical changes are detected by the taste receptor cells and the information is transmitted by spike trains to the afferent nerve.*

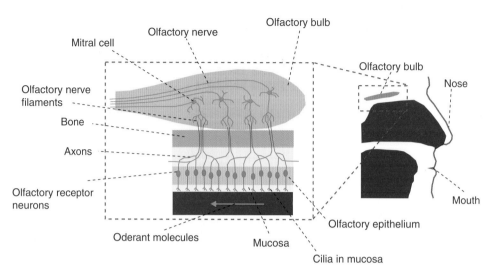

Figure 25.9   *Structure of the neural circuits in the human nose that are specialised for smell. Different odours are thought to give different spatial patterns of excitation in the olfactory bulbs. The spatial patterns are transmitted by spike trains in the olfactory nerve.*

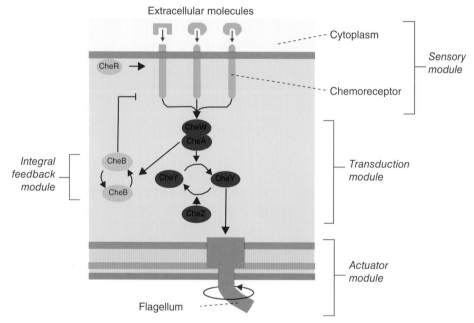

Figure 25.10   *Bacterial chemotaxis is directed by the interaction between* sensory, transduction *and* actuator *modules.*

Chemosensitivity is also observed in prokaryotes, e.g. E.coli. Bacterial chemotaxis is understood in molecular detail. In a Gram-negative bacteria such as E.coli, an attractant passes through the outer membrane (the sensory module) which initiates a transduction cascade (in the transduction module) that drives the motors (actuator module, **Figure 25.10**) attached to its flagellae. A twiddle, the incoherent motion of a number of flagellae (**Section 7.3**), changes the direction of motion of the bacterium. Forward motion of the bacterium is produced by counterclockwise rotation of the flagella, which is interrupted by a sudden rotation of the flagella in a clockwise direction that cause the tumbling twiddles. A twiddle lasts for about 0.1 s, and then the bacterium starts off again in a new direction with another straight line trajectory. Combination of this run and tumble motility strategy with chemoreception allow bacteria to run up or down chemical gradients and thus they can locate more favourable environmental niches.

25.6 Photoreceptors

Photosensitivity has evolved many times during evolution, but all animals detect visible photons using opsin molecules that lead to a G protein signalling cascade once the photons are detected (**Figure 25.1b**). At the simplest level bacteria contain single molecules of bacteriorhodopsin that make them light sensitive. More sophisticated single cellular eukaryotic organisms also use light-sensitive opsin proteins that allow them to make coarse measurements of light levels in their surrounding environments. Multicellular invertebrates have evolved vision with a wide range of optical strategies including eye spots, pinhole eyes, scanning eyes, mirror eyes, and compound eyes. This varied instrumentation, combined with arrays of opsin molecules, allows the eyes to form spatially resolved images of their surroundings and this has a clear evolutionary advantage for organisms.

The principal components of the human eye (**Figure 25.11**) are the cornea, aqueous humor, lens, vitreous humor, and the retina. The iris defines the area of the pupil that receives light. Light from an object must be refracted to form an image on the retina. Most of the refraction takes place at the surface between the air and the cornea. This can be seen from the relative differences in the refractive indices (*n*) of the material

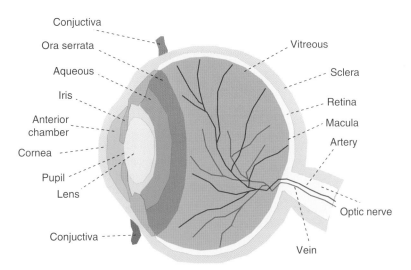

Figure 25.11 *Structure of the human eye, which is used to detect visible-wavelength photons. Light is primarily refracted by the cornea and small adjustments to the focus are made by the lens. Rods and cones in the retina allow spatial images to be constructed and the information is transmitted by spike trains in the optic nerve.*

involved; air ($n = 1$), cornea ($n = 1.38$), aqueous humor ($n = 1.34$), crystalline lens ($n = 1.4$) and vitreous lens ($n = 1.34$). In mammalian eyes muller cells are thought to act as waveguides through the eye to help transport light to the back of the retina to reduce losses due to absorption (the complex part of the refractive index).

The spatial resolution of the eyes of all organisms is determined by optical diffraction effects. The smallest spatial frequency detectable (x) is therefore

$$x = D/f\lambda \tag{25.4}$$

where D is the aperture size, f is the lens focal length and λ is the wavelength. Only objects of sizes greater than $1/x$ can be resolved by the eye. Thus, large lenses (and thus big apertures) are needed for high visual acuity, i.e. large D values in equation (25.4).

In human eyes the retina is divided into two regions. The fovea is the area of greatest visual discrimination and is entirely composed of cones. The peripheral region is where the concentration of rods is highest and is thus most sensitive to faint light. *Rods* are therefore used for night vision, since they provide high sensitivity, and no colour vision. *Cones* provide colour vision, and work best in brighter light.

A Poisson distribution (equation (3.10)) occurs for the number of photons emitted by the vast majority of naturally occurring light sources. The probability distribution for the number of photons (n) in a flash of light observed by an eye is thus

$$P(n) = e^{-\langle n \rangle} \frac{\langle n \rangle^n}{n!} \tag{25.5}$$

Let I be the mean intensity of the flash of the light source. The mean number of photons detected by an organism's eye from the light source is

$$\langle n \rangle = \alpha I \tag{25.6}$$

where α describes the efficiency of the detector. The organism will decide it has seen the flash if the number of photons is above a certian threshold, K. The probability that an organism observes a flash of intensity I ($P_{see}(I)$) is thus given by a combination of equations (25.5) and (25.6),

$$P_{see}(I) = \exp(-\alpha I) \sum_{n=K}^{\infty} \frac{(\alpha I)^n}{n!} \tag{25.7}$$

Fits of this model to data for the human eye imply that the threshold for detection of a flash of light by a rod is 5–7 photons. More detailed modelling of vision in a variety of organisms indicates that opsin molecules can reach the limit of one-photon sensitivity *in vitro*, but noise in the amplification and transmission of signals tends to reduce this performance *in vivo*.

A number of organisms are sensitive to the polarisation of light. These include shrimp, octopus, cuttlefish and insects (bees). In insects, polarisation sensitivity is due to a preferential orientation of the rhodopsin light-sensitive molecules.

The neural network responsible for information transfer from the mammalian retina has a complicated structure that comprises of six different types of neuron and is shown in **Figure 25.12**. Strangely, light is transmitted through a number of cell types before its arrival on the rods and cones. The types of transmission cell are mostly transparent, but haemoglobin inside them does absorb some of the light, and compromises the performance.

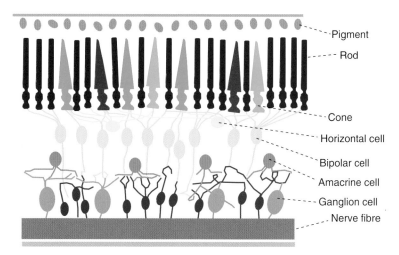

Figure 25.12 *Neural circuits in the human eye are connected to light-sensitive molecules in the retina (rods and cones). The retinal neural circuits perform some preprocessing functions before the image information is transmitted to the brain.*

Muller cells alleviate this problem by acting as waveguides, but there is still a blind spot due to the position of the optic nerve in mammals. Octopus eyes have the cellular wiring the sensible way round, with no blind spot, and thus their eyes demonstrate a separate evolutionary origin to mammals. Returning to **Figure 25.12** we see that the receptive fields of the ganglion cells in mammals are produced by the interplay of the receptors (rods and cones) and the ganglion cells with neurons of three other types; horizontal cells, bipolar cells, and amacrine cells. The rod cells are sensitive to brightness and the cone cells monitor contrast, motion, size and wavelength. The rod and cone cells feed signals to the *biopolar* cells, which then send signals to the *ganglion output* cells. Neurons of the two remaining classes are *horizontal* cells and *amacrine* cells that mediate lateral spread of signals, help eliminate noise and enhance contrast at discontinuities of illumination.

Some image compression occurs within the retina. A large number of rods often feed only one ganglion cell, so there is a pooling of signals. Furthermore, the absolute light levels are subtracted from the signal and replaced by a relative background contrast. Low spatial frequencies are also often subtracted from the signal. A small number of retina cells in humans are not used in vision, instead they are used for management of the body's internal clocks and entrain circadian rhythms to the ambient light levels.

25.7 Thermoreceptors

Snakes can sense infrared thermal radiation, at wavelengths of 5–30 μm. Pits in the head of the snakes can act as a simple low resolution pinhole camera, e.g. in rattlesnakes. The pits contain heat-sensitive ion channels (**Figure 25.13**). In rattle snakes there is a 60×60 array of detector elements with their sensitivity peaked at 37 °C (mammal body temperature). Rattlesnakes can strike in total darkness with an accuracy of about 5°.

Some beetles can detect infrared radiation as an indicator of forest fires, since they prefer to lay their eggs in burnt vegetation.

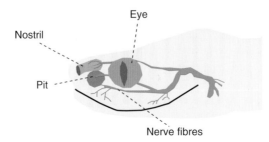

Figure 25.13   *Pits in the noses of rattlesnakes enable them to create relatively low-resolution infrared images of their environment. The wavelength response is optimised for small mammalian prey items, e.g. mice.*

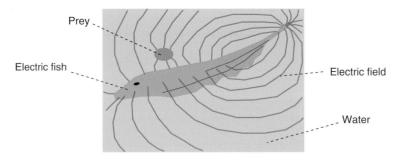

Figure 25.14   *Electric fish are able to both sense and create electric fields in their environments, and thus detect prey items.*

25.8   Electroreceptors

Electrolocation is found in a wide variety of fish that include sharks, rays, lung fish, sturgeons, and some mammals (dolphins and echidna). The sense is used to locate objects in low-light conditions. Electrolocation is possible using both active and passive mechanisms, i.e. the organism may create an electric field and measure the environmental response, or just measure the ambient electric fields that occur in its environment. Electroreceptors are also used in electrocommunication, although the detailed mechanisms of how information is encoded (the languages) are still not well understood. Some fish can use electricity to stun their prey and can create pulses of up to 500 V at 1 Amp (**Figure 25.14**), e.g. electric eels. Recent studies imply that insects (specifically bees) are also sensitive to electric fields, but the mechanism that allows them to do this needs to be studied in more detail.

25.9   Magnetoreceptors

Magnetism is used for orientation in the Earth's magnetic field by a large number of organisms, e.g. several species of bacteria contain chains of intracellular organelles that contain magnetite (a permanent magnet). Bacteria in the Northern hemisphere seek the earth's Northern magnetic pole and vice-versa. They are called magnetotactic bacteria (**Figure 25.15**).

Magnetoreception is also observed in birds, fungi, insects, turtles, lobsters and sharks. Many of the exact mechanisms are not well understood. Fruit flies have a light-sensitive molecule (cryptochrome) in

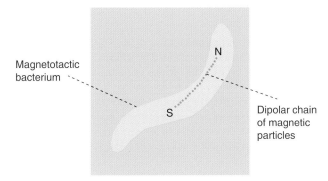

Figure 25.15 *Chains of internalised inorganic magnetite particles (the chains form due to dipolar interactions) allow bacteria to sense magnetic fields due to the resultant torque.*

the photoreceptor cells of their eyes that is affected by magnetic fields. In some birds magnetite is found in their beaks and cryptochrome also occurs in their eyes. It is not yet clear which of these mechanisms determines the birds' sense of magnetoreception. Better understood are sharks' electroreception organs that can sense variations in electric potential (V) and thus the magnetic fields that induce them. Faraday's law gives the relationship between the voltage and the rate of change of magnetic flux ($d\phi/dt$),

$$V = -\frac{d\phi}{dt} \tag{25.8}$$

where ϕ is the magnetic flux that passes through the organs. The rate of change of magnetic flux detected by a shark, can be either caused by the motion of the shark or the source.

Suggested Reading

If you can only read one book, then try:

Top recommended book: Bossomaier, T. (2012) *Introduction to the Senses: From Biology to Computer Science*, Cambridge University Press. Good readable modern account of the physics of human senses and how the information is coded.

Bialeck, W. (2012) *Biophysics*, Princeton University Press. Advanced tour of biological physics, which includes an excellent modern discussion of photon detection in living organisms.
Dowling, J.E. (2012) *The Retina: An Approachable Part of the Brain*, Harvard. Clear readable account of the action of neural networks in the brain.
Endres, R.G. (2013) *Physical Principles in Sensing and Signalling; With an Introduction to Modelling in Biology*, Oxford University Press. Compact introduction to the sensory modules of bacterial cells.
Fain, G.L. (2003) *Sensory Transduction*, Sinauer. Another useful biological guide to sensory systems.
Johnson, S. (2012) *The Optics of Life*, Princeton. Simple popular modern account of biological optics.
Land, M.F. & Nilson, D.E. (2012) *Animal Eyes*, 2nd edition, Oxford University Press, Classic text on the visual systems in a wide range of organisms.
Rieke, F., Warland, D., van Steveninck, R. & Bialek, W. (1999) *Spikes: Exploring the Neural Code*, MIT. Classic discussion of hyperacuity and how the electrical activity of neurons relates to the senses.
Smith, C.U.M. (2000) *Biology of Sensory Systems*, Wiley. Detailed biological introduction to sensory systems in a wide range of organisms.

Tutorial Questions 25

25.1 Describe three noninvasive methods to image brain activity.

25.2 Give some examples of hyperspectral imaging performed by animals.

25.3 Investigate the phenomenon of Mach bands in the human eye (you can perform these experiments on yourself). Explain how the phenomenon relates to the wiring of the neurons in the retina.

25.4 Investigate the work of Benjamin Libet. Describe his most famous experiment.

25.5 Investigate the work of Eric Kandel, a Nobel Prize winner in medicine. Describe some of his famous experiments with sea snails.

25.6 Nerve responses occur at ~1 ms time scale, but human hearing extends to ~20 kHz. Explain how this is possible.

26

Brains

This final short chapter makes an introductory discussion of some physical aspects of brain physiology. Following on from the consideration of spikes and the senses in the previous chapter, the *neural encoding inverse problem* will first be approached, i.e. how to reconstruct the sensory information transmitted by a series of nerve impulses to the brain. This is chosen as a starting point, since it is reasonably well understood and sets the stage for how the sensory information is subsequently processed in the brain (which is less well understood). The storage of memories in neuronal networks is then considered, because it provides insights into the learning behaviour of neural circuits, and theoretical models can be introduced that allow the fundamental underlying behaviours to be identified, such as *neural networks*. The discussion then moves on to motion control, which can be explored in terms of simple neuronal circuits and gives a good example of conscious volition. More modern topics will finally be presented that include the quest for the complete wiring diagram of the brain (the *connectome project*) and ideas on cohesive properties (*consciousness*). As was stressed in the Section introduction, tractable, physical problems will be approached, since the problem of a full understanding of consciousness is extremely difficult and predominantly remains unsolved.

How the information from nerve spikes is stored in memory is relatively well understood. The answer is found in the modulation of synaptic strengths between neurons in the brain, i.e. the connections between neurons in neural networks. Indeed, it is experimentally possible to observe the modulation of synaptic strengths on the single-cell level in simple organisms, and classic experiments have studied the learning behaviour and memory of sea slugs, e.g. their Pavlovian responses.

Brain cells involved in neural processing mainly occur in two different varieties; *neurons* (**Figure 26.1**) and *glial* cells (**Figure 26.2**). Historically, neurons have been at the forefront of brain research, whereas glial cells were thought to have just a structural role (their name is derived from this status as brain 'glue' cells). However, more modern research has found new roles for glial cells, such as their ability to trim neuronal connections during development (such connections are of primary importance in the determination of neuronal responses), and further important new results are expected in this area. Neurons are found in four primary varieties; *bipolar*, *unipolar*, *multipolar* and *pyramidal* cells (**Figure 26.3**). Bipolar neurons are important for the transmission of sensory information, unipolar neurons act as primary sensory neurons, multipolar neurons represent the majority of the cells in the brain, which includes motor neurons and interneurons, and pyramidal

The Physics of Living Processes: A Mesoscopic Approach, First Edition. Thomas Andrew Waigh.
© 2014 John Wiley & Sons, Ltd. Published 2014 by John Wiley & Sons, Ltd.

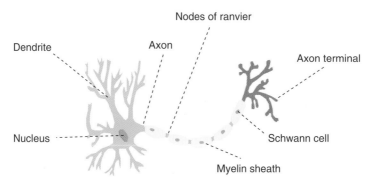

Figure 26.1 *The structure of a typical neuron in which electrical signals propagate from the dendrites along the axons and finally arrive at the axon terminals. The nodes of Ranvier allow ions to enter the axon and the myelin sheath reduces the axon capacitance. These cells appear white under an optical microscope with standard stains (silver/gold) due to the myelin sheaths ('white matter' as opposed to 'grey matter' for unmyelinated cells).*

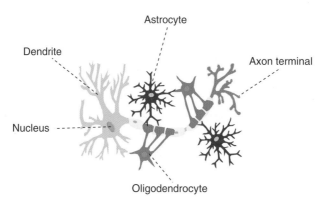

Figure 26.2 *A neuron associated with two types of glial cell (astrocytes and oligodendrocytes). The glial cells (red and dark blue) perform a number of roles that include pruning the dendritic connections.*

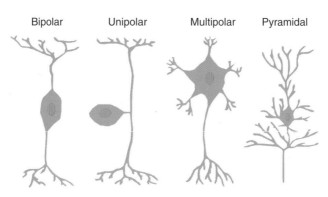

Figure 26.3 *Schematic diagram of the four basic neuron types: bipolar (interneuron), unipolar (sensory neuron), multipolar (motorneuron) and Pyramidal cell.*

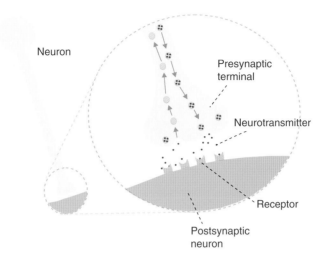

Figure 26.4 *Neurotransmitters transfer information between neurons and use a process of diffusion to bridge the very short distance across the synaptic cleft. The neurotransmitters can be* excitatory, inhibitory *or cause more subtle long-lasting changes.*

cells are found in specialised parts of the brain, e.g. the hippocampus that is important for the creation of new memories.

Electricity is not the whole story in nerve-cell communication. Historically, proponents of a full electrical theory were nicknamed the 'sparkers', as opposed to the 'soupers' who believed neuronal signalling was completely chemical. It is now known that in addition to electrical communication, many neurons affect one another through the release of a neurotransmitter that binds to the synapse membrane receptors. Receptors can be excitatory (increase the firing rate), inhibitory (decrease the firing rate) or modulatory (have long-lasting effects, **Figure 26.4**). Thus, a more accurate picture of nerve cell communication required a compromise between the sparkers' and soupers' perspectives.

Different areas of the brain have specific functions, i.e. specific regions of the brain perform specialised tasks (**Figure 26.5**). For example, part of the brain is specialised for the memory of music and another for the memory of visual images. Studies of localised damage of brains (e.g. localised loss of blood circulation in stroke victims, in which localised regions of the brain are damaged) have allowed people to map the functional anatomy of the brain.

As stated in the Section introduction, there are 100 trillion connections in the human brain and the creation of a complete wiring diagram is still not even partially complete. The *human connectome project* is currently underway and centres on the uses of high-resolution imaging experiments, such as electron microscopy and fluorescence microscopy, to map the complete neural network topology. These results are then combined with lower-resolution magnetic resonance imaging experiments, to relate the network activity to large-scale morphologies that are actually observed in live brains.

Much simpler neural networks than those in human brains can be explored in other organisms, which provide useful reference systems. Paramecium are single-celled organisms that consist of a single electrically excitable cell. The electrical activity allows paramecium to react quickly to external stimuli. Many examples of organisms with gradually more sophisticated neural architectures are available, and they represent evolutionary intermediates to more sophisticated mammalian brains. Round worms have around 300 neurons in their nervous system. Jelly fish have 1000 neurons, that allow them to perform rhythmic swimming manoeuvres. Sea snails have about 20 000 neurons, can memorise stimuli from their environment and demonstrate a variety

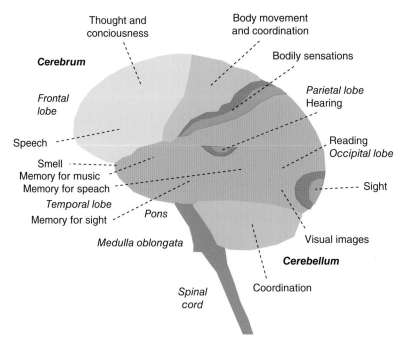

Figure 26.5 *Schematic diagram of the functional anatomy of a human brain. The orange regions predominantly receive messages, whereas the green regions send messages.*

of feeding responses. Fruit flies have 100 000 neurons that create internal physiological clocks, as well as allow them to perform sophisticated flying and behavioural responses. Mice have 75 million neurons, chimpanzees have 7 billion neurons, and humans have 85 billion neurons (more than dolphins and elephants).

The measurement of the activity of neural networks presents a big experimental challenge. Submicrometre-resolution images are required in three dimensions for the neural architectures and these measurements need to be noninvasive or they will irreversibly perturb the behaviour of the networks imaged. Thus, the main choices for brain imaging are magnetic resonance imaging, positron emission tomography, and optical coherence tomography. These can be combined with electroencephalography (EEG) and magnetoencephalography (MEG) for the measurement of spatially resolved electrical activity. The resolution of all these *in vivo* techniques falls short of that required to resolve individual synapses of individual cells. Higher-resolution methods such as electron microscopy or fluorescence microscopy need to be conducted *ex vivo*, and are typically performed with sliced brain tissue to remove confusing background signal artefacts. The networks that result from the combination of multiple tomographically sliced images are exceedingly complicated (data acquisition is normally fully automated) and the huge data sets require careful management.

26.1 Neural Encoding Inverse Problem

All the sensory information that arrives in a brain is encoded in the form of neuronal spiking potentials. The brain then creates a representation of the surrounding world by the reconstruction of sensory signals from these trains of spikes, i.e. it decodes the spikes trains (**Figure 26.6**). All individual spikes look more or less identical (they are stereotyped) and, since their amplitudes are the same, it is deduced that the information transfer must be due to frequency modulation, i.e. the timing of the spikes contains the information. However, based on

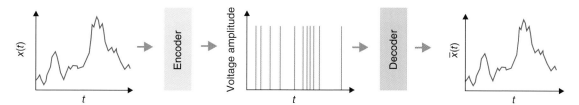

Figure 26.6 *Sensory information (x(t)) is encoded in spiking neurons that are then decoded to reconstruct a representation of the original information ($\bar{x}(t)$) in the brain.*

detailed experiments of spike-train data, a deeper understanding is possible. Single neurons can often produce only one spike during the characteristic time of variations in sensory stimuli. There is thus a sparse temporal representation of signals in the spike trains that are transmitted. Single neurons can transmit large amounts of information, with a number of bits encoded in a single spike. The factors that affect the robustness of neural computation include the reliability of single neuronal responses and fundamental physical limits imposed by noise in the sensory data.

Consider an input sensory signal ($x(t)$, e.g. the amplitude of a particular frequency of sound in an ear or the pressure on a touch sensor in the skin of a toe) that causes a single neuron to create a train of spikes. The neuron encodes the signal in the spike train using a function $a(x(t))$. This process of encoding can be separated into two separate consecutive functions; the creation of the neuronal current ($J(x(t))$) followed by the creation of spikes from the neuronal current $G(J(x(t)))$,

$$a(x(t)) = G[J(x(t))] = \sum_n \delta(t - t_n) \tag{26.1}$$

where G is the encoding function that produces delta function spikes, $\delta(t - t_n)$, at times t_n. Different models for the creation of the neuron current are possible. The simplest model is to assume that the current is linearly proportional to the input signal and there is a background signal of constant amplitude (spikes can occur when there is no input signal),

$$J(x(t)) = \alpha x(t) + J_{bias} \tag{26.2}$$

where $J(x(t))$ is the neuron current, J_{bias} gives the background firing rate and α is a constant. The inverse decoding problem can then be described by a convolution integral, and the decoded input signal ($\bar{x}(t)$) is constructed by an integral of the spikes with a linear decoder function $h(t-t')$,

$$\bar{x}(t) = \int_0^T h(t-t') \sum_n \delta(t' - t_n) dt' \tag{26.3}$$

where T is the duration of the spike train considered and n is the number of spikes in the train. The assumption that $h(t-t')$ is a linear decoder may at first sight seem to be unrealistic, since neurons are known to be highly nonlinear devices. However, neurons often act in pairs (a push–pull topology) that can improve the linearity of their response, so equation (26.3) is often a reasonable approximation. A challenge is to calculate $h(t-t')$ from some experimental data using the convolution integral in equation (26.3). In practice this is done most conveniently in frequency space; a Fourier transform is taken of equation (26.3) to convert the convolution into a product. The neurons' statistics are assumed to be stationary, i.e. an arbitrary time offset will not change the decoder function. The Fourier transform of equation (26.3) can then be minimised against the Fourier transform of the experimental data, to calculate an approximate functional form of the linear decoder ($h(t–t')$). This multidimensional minimisation problem can be satisfactorily solved using Monte Carlo estimation.

26.2 Memory

Up to now the discussion of nervous systems and action potentials gives no understanding of how information is stored in networks of neurons. The response of neurons to a given stimulus can depend upon whether the system has been exposed to that stimulus previously and this is the basis of memory.

Early on in the history of memory research, before the mechanisms of information transfer in nerve cells were understood, Donald Hebb suggested that the synaptic connections between two neighbouring nerve cells are strengthened if one of the cells persistently helps the other to emit nerve impulses. The opposite situation is also included in Hebb's hypothesis, if the firing of a cell happens rarely and is accompanied by frequent firing of another cell to which it is in synaptic contact, the contact becomes weakened. Strong experimental support has subsequently been found for Hebbian learning with snails. Large aquatic snails (Aplysia) were found to be an ideal experimental system for memory research, much as the giant squid was for the action potential. Sea snails have very simple neuronal circuits (**Figure 26.7**). Aplysia has a nervous system that comprises of only about 20 000 relatively large neurons, but it displays some of the types of memory conditioning seen in more sophisticated organisms and is thus a convenient subject for investigations of memory mechanisms. When the Aplysia's siphon is continuously irritated the gill becomes habituated and relaxes. Also, if the tail and the siphon are irritated at the same time, subsequent stimulation of the siphon alone produces a vigorous gill-withdrawal reflex. These Pavlovian responses can be demonstrated on a single-cellular level with the snails, and the responses and the stimuli are seen to be physiologically linked. Pavlov's original experiments were performed with dogs and he demonstrated that they could associate the sound of a bell with food, which caused them to salivate, i.e. a conditioning step of 'bell' and 'food'→'salivation' can cause 'bell'→'salivation' without any food. The corresponding conditioning step for the snail is 'tail irritation' and 'siphon irritation'→'gill withdrawal', which can cause 'siphon irritation'→'gill withdrawal' without any tail irritation.

The Pavlovian/Hebbian phenomena are now understood at the molecular level. *Habituation* develops because there is a gradual diminution in the membrane's ability to pass calcium ions in the synapse between the sensory and motor neurons upon prolonged stimulation. The opposite can also happen in the process of *sensitisation*, which involves an enhancement of the membrane's capacity for the passage of calcium due to under stimulation.

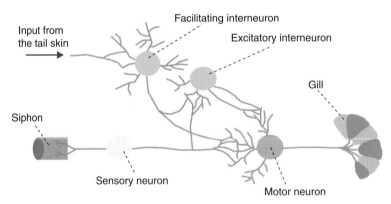

Figure 26.7   *Schematic diagram of some simple neural circuits inside a sea snail. A Pavlovian response can be observed for the correlation between tail and siphon irritation, and the gill-withdraw reflex.*

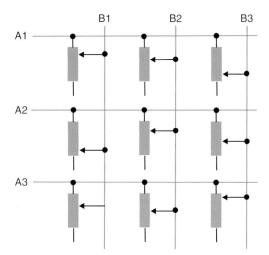

Figure 26.8 *A simple neural (Steinbuch) network can be constructed from parallel arrays of electrical conductors and variable resistors. Patterns of electrical activity (memories) can be stored in variable resistors and retrieved when voltages are applied to A1, A2, A3,... which results in voltages B1, B2, B3,... (the retrieved memories). Different learning rules can be used to update the strengths of the variable resistors.*

A next challenge is to theoretically describe the basis of memory in an artificial network. A small part of a Steinbuch neural network is shown in **Figure 26.8**. Many features of memory are well described by this model of a neural network. Initially, neural networks were electrical creations, built from batteries, conducting wires and resistors, but now it is possible to construct them in a computer or even inside a cell using synthetic biology. For the simplest electrical neural networks (**Figure 26.8**) the value of the resistance strength between the input and output wires can be varied in the network to represent synaptic strengths. The network allows correlations between the activity patterns in the horizontal conductors and the vertical conductors to be stored. Presentation of the same input pattern to the horizontal set of conductors would produce a required output pattern in the vertical set. The network can thus *learn* to associate input and output patterns. Furthermore, the network can store several sets of correlations simultaneously in a superimposed manner, which is an example of *parallel distributed processing*.

The simplest biologically relevant neural network (*pattern association networks*) consists of an array of neurons (i) that receive input from axons (j) through synapses of strength w_{ij}. r_i denotes the firing rate of the ith neuron and r'_j is the firing rate input into the jth neuron (r denotes a rate). The neuron calculates a linear summation of its input rates. Thus, the activation of neuron i (h_i) is given by

$$h_i = \sum_j r'_j w_{ij} \tag{26.4}$$

where the summation ($\sum_j$) occurs over all the input axons. This describes a process of memory recall. If h_i is above a threshold value neuron i will fire and the collection of all h_is constitutes a memory.

Hebbian learning can be described using a simple update rule for the matrix weights (w_{ij}). If both the input rate (r_i) and the output rate (r'_j) are well correlated there is a large increase in the corresponding matrix weight and *vice versa*. Therefore, a reasonable mathematical relationship is

$$\delta w_{ij} = k r_i r'_j \tag{26.5}$$

where δw_{ij} is the change of the synaptic weight due to simultaneous presynaptic firing r_j', postsynaptic firing r_i and k is the learning rate constant.

To quantitatively deduce the output firing rate recall response, a real network requires one additional step,

$$r_i = f(h_i) \tag{26.6}$$

where f is the activation function and h_i are the activation rates given by equation (26.4). Neurons tend to have threshold values above which they fire strongly and the threshold can thus be modelled with a binary response.

A simple choice for $f(h_i)$ was used by Longuet-Higgins, Buneman, and Willshaw for a neural-network model (**Figure 26.9a**). The algorithm for the calculation of the activity of each neuron is to sum the products

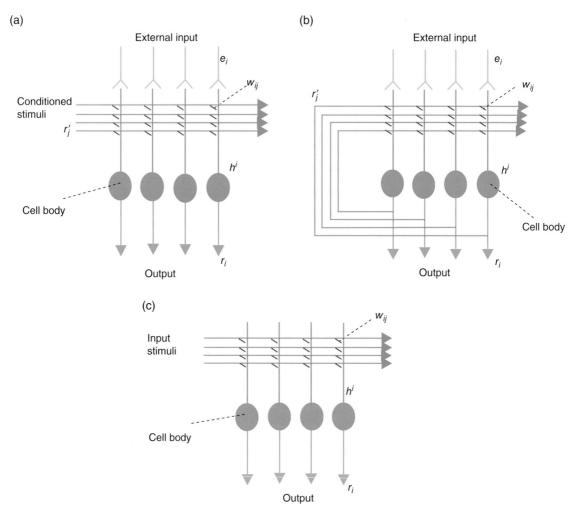

Figure 26.9   *Three neural network architectures that use local learning rules and are thus credible models for biological systems. (a) Pattern association network, (b) autoassociation network and (c) competitive network. [Reproduced from E.T. Rolls, A. Treves, Neural Networks and Brain Function, OUP, 2004. With permission from Oxford University Press.]*

of the activities of the other neurons using equation (26.4) and then compare them to a threshold value ($h_{threshold}$) of the neuronal activation rate,

$$r_i = 1 \quad \text{if} \quad h_i \geq h_{threshold} \tag{26.7}$$

and

$$r_i = 0 \quad \text{if} \quad h_i < h_{threshold}$$

A neuron is thus active if, and only if, the sum of the products in equation (26.4) is greater than or equal to the threshold. This situation is quite unlike that postulated by Hebb, since he thought only a small fraction of neurons were involved in any one memory. Another criticism is that it involves only binary neurons that are either active or inactive. Graded responses are not well described by this model, but they are a facet of the behaviour of real neurons.

Action potentials propagate along a neuron's axon only when the depolarisation of its bounding membrane exceeds a threshold value. The pulses are dispatched at a frequency that depends upon the amount by which the threshold has been exceeded. Sigmoidal curves are often measured in the input–output characteristics of neurons (**Figure 26.10**). Thus, the responses from real neurons are graded and sigmoidal models are more accurate approximations in equation (26.6) for the activation function than a binary 'on' or 'off' response.

Some neurons have an inhibitory effect (due to GABA signalling molecules), whereas other neurons have an excitatory effect (due to glutamate or acetylcholine signalling molecules). If it were not for the presence of the inhibitory neurons the activity level of a typical neural network would rapidly reach its saturation value, i.e. all the neurons would excite at the same time and memory retrieval would be compromised. Thus, more realistic neural networks include neurons that show both excitatory and inhibitory effects.

Hebb had a second less well known hypothesis, that was for a given set of sensory stimuli only a very small fraction of all possible neural pathways are in use and these active routes lie adjacent to one another, i.e. memories are represented in a sparse manner. The opposite extreme is called fully distributed memory, in which half of the neurons in a network are active during the process of memory retrieval and it is thought to be less biologically relevant due to its inefficiency and slow response.

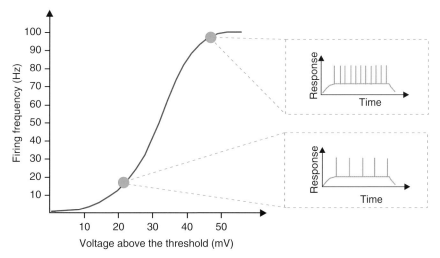

Figure 26.10 *A sigmoidal relationship is observed between the firing frequency of spikes in a neuron and the magnitude of the voltage above a threshold value.*

Feedforward neural networks are often found in real neuronal architectures, i.e. neural networks that use the feedforward control approach introduced in **Section 24.1**. They provide an early arrival inhibition process that sets up an oppositional field, through which only the strongest excitatory signals can penetrate. Feedforward neurons receive their activation from other neurons that lie at greater distances. Such architectures lead to a winners-take-all process for the most active excitatory neurons that can outcompete the inhibitory inter-neurons. The feedforward motif decides between competing outcomes in the future and thus it determines what the network will do at some later time. Feedforward inhibitory neurons are thus often found in the brain, e.g. in motor control regions.

The idea of localised learning is important for more realistic models of brain networks, since memories in real brains can be acquired and recalled quickly, without additional limits imposed by nonlocal data transfer. A strength of equation (26.5) is that it is a local learning rule. Other neural-network models, such as percep-trons, are popular in computer science, but lack the local update rule and are thus not a credible candidate to describe biological systems. There are three main types of model for learning networks that are thought to be biologically viable: *the pattern association networks* that have already been considered, *autoassociation networks* and *competitive networks*.

An *autoassociative* neural network has self-connected loops in its structure (recurrent collaterals, **Figure 26.9b**). Groups of neurons can thus autoassociate (a process that includes self-associations). Through this mechanism the network can identify input patterns to which it has been previously exposed. Individual memories can be superimposed and complete input patterns can be recovered even if the network is fed only partial versions of the input patterns. Self-association of nerve cells is observed with real nerve-cell network architectures. The autoassociative network is also robust to partial damage of its structure, which is another biologically appealing attribute. The output firing rate recall responses of the autoassociation networks depends on both the recurrent collateral effect (the outputs h_i are fed in as inputs) and the external input (e_i),

$$r_i = f(h_i + e_i) \tag{26.8}$$

This expression can be used to replace equation (26.6) as a mathematical model for self-association networks.

The third key example of a neural network is the *competitive network* shown in **Figure 26.9c**. The neuron makes modifiable excitatory synapses (w_{ij}) with the output neurons. The output firing rates of the cells compete with one another (e.g. by mutual inhibition) and only a subset succeed in the process of competition and are left firing strongly. The resultant network can categorise patterns of input firing rates and this can function as a preprocessor of sensory information.

26.3   Motor Processes

The central pattern generator circuit of the lamprey is shown in **Figure 26.11**. Only one of the roughly one hundred such circuits is shown. Excitatory neurons and their synapses are indicated by blue/yellow circles and green triangles, respectively. Inhibitory neurons are shown as gold/purple circles. The synapses that contact the boxes, exert the most widespread influence. Regular sequences of muscle movements in organisms (e.g. rhythmic motion of legs when an organism walks) are provided by such central pattern generators that are constructed as self-contained units. In the lamprey, the paired neuronal units are activated at mutually staggered times and this provokes forward or backward swimming that depends on the relative phase lag between the units. The phase lags are influenced by the excitability of each type of neuron and this is in turn determined by the interplay between different neurotransmitters. Such neuronal circuitry requires no conscious control for its continued activity.

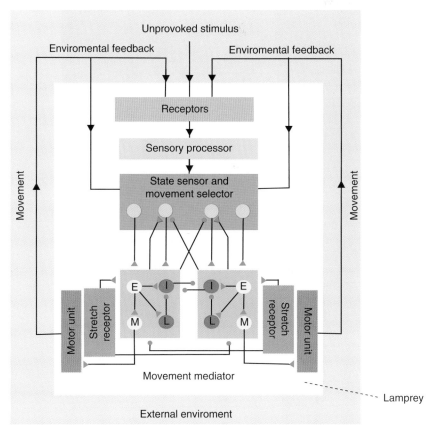

Figure 26.11 *A central pattern generator for the lamprey that creates the regular oscillatory muscular contractions needed for swimming. [Reproduced with permission from R. Cotterill, Biophysics: An Introduction, Wiley, 2002.]*

An alternative example of motion control is provided by the honeybee's nervous system (the proboscis extension response). The proboscis is extended when the honey bee smells a flower. This is a conditioned reflex that requires a learning process, i.e. feedback from the environment. This can be compared with the simpler feedforward units required for lamprey swimming, which are hard wired in the organism (directly genetically controlled).

26.4 Connectome

Attempts to measure the complete wiring diagram of the brain have a long illustrious history that stretches back to Raman y Cajal, who performed some of the first imaging experiments on neuronal cells using optical microscopy and silver stains. However, only recently, due to advances in computational processing and electron microscopy, has the problem experienced a reasonable chance of success with higher organisms. Completion of the human connectome project (i.e. mapping the complete connectome of a single individual) would be expected to have a huge impact on brain research, similar to the way the human genome project has revolutionised the field of genetics research.

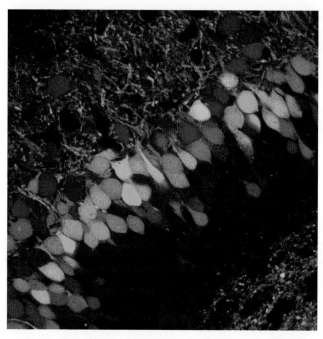

Figure 26.12 *Brainbow image of a mouse brain created using fluorescence microscopy. Individual neuronal cells express different combinations of genetically expressed protein fluorophores, which cause them to have separate clearly identifiable colours. [Reprinted by permission from Macmillan Publishers Ltd: Nature, J. Livet, et al, Nature, 2007, 450, 7166, 56–62.]*

A 'brain bow' from a mouse brain is shown in the **Figure 26.12**. This is a novel fluorescence micro-scopy approach to the challenge of finding neural wiring diagrams. Three different fluorescent proteins are introduced into a mouse's genome and are randomly expressed at different levels inside the brain cells, which produces multiple colours in combination. This allows easy identification of separate neuronal cells that can have tortuous branched morphologies and thus the measurement of network connectivity. Such an optical technique is relatively new and its use in large-scale brain mapping is still being investigated.

In addition to fluorescence techniques, brain mapping has been performed by serial electron microscopy imaging of sliced brains (high resolution, but very invasive) and magnetic resonance imaging (low resolution and almost completely noninvasive).

26.5   Cohesive Properties

The large-scale cohesive properties of neuronal architectures continue to provide the biggest challenges to brain biophysics, since they give rise to consciousness, the 'hard problem' discussed in the introduction. Many different aspects of the conscious collective behaviour of 100 trillion interacting synapses in the human brain have been explored experimentally and interested readers should pursue some of the vast range of relevant literature. Early experiments of Benjamin Libet and others during brain surgery implied that it takes about 40 ms for nerve signals to reach the somatosensory (touch) area that responds to stimulation of the hand, and similar delays apply to auditory and visual stimulation. However, the conscious experience of such unpro-voked stimuli by the patient requires 500 ms to occur. The experiments thus imply that unconscious neuronal processes can proceed and motivate volitional acts, which are retrospectively felt to have been consciously motivated by the subject. The demarcation between conscious and unconscious action is thus fuzzy and requires careful consideration.

Saccades are the jerky movements that human eyes perform when they track an object that has caught an individual's attention. Careful experiments on saccades also demonstrate the blurred division between conscious and unconscious control similar to those seen in the Libet experiments.

Several brain components are believed to be indispensable to consciousness. The anterior cingulate and the thalmic reticular nucleus together were termed the *searchlight of consciousness* by Francis Crick. There is also the amygdala that acts as an adjudicator that evaluates different high level choice decisions in the brain. Planned movements necessitate a clutch-control system that is found in the basal ganglia. Damage to this region causes movements to become erratic and is a symptom of Parkinson's disease.

Cohesive electrical properties of the brain can be measured in EEG and MEG experiments as explained in **Section 24.6**. Detailed predictions of global electrical patterns in brain tissue from single cellular units are still unavailable.

Suggested Reading

If you can only read one book, then try:

Top recommended book: Cotterill, R. (2002) *Biophysics, an Introduction*, Wiley. Fascinating iconoclastic account of brain physics aimed at a good level for undergraduate students.

Buzsaki, G. (2006) *Rhythms of the Brain*, Oxford University Press. Good popular discussion of cohesive phenomena and electrophysiology in brains.

Eliasmith, C. & Anderson, C.H. (2003) *Neural Engineering*, MIT. Useful compact account of neural engineering.

Eliasmith, C. (2013) *How to Build a Brain: A Neural Architecture for Biological Cognition*, Oxford University Press. Explores realistic computer simulations of brains including neural coding and neurotransmitters. Freeware code is available to simulate different brain functions.

Greenspan, R.J. (2007) *An Introduction to Nervous Systems*, Cold Spring Harbour. Biological introduction to neural networks in simple organisms.

Kandel, E.R., Schwartz, J.H. & Jessell, T.M. (2000) *Principles of Neural Science*, 4th edition, McGraw Hill. Comprehensive account of the biology of neurons.

Koch, C. (1999) *Biophysics of Computation*, Oxford University Press. Describes detailed mathematical models for computations performed by small number of neurons.

Koch, C. (2012) *Conciousness – Confessions of a Romantic Reductionist*, MIT. Up to date popular account of physical experiments on consciousness.

Rieke, F., Warland, D., van Steveninck, R. & Bialek, W. (1997) *Spikes: Exploring the Neural Code*, MIT. A classic text on the neural decoding problem.

Rolls, E.T. & Treves, A. (1998) *Neural Networks and Brain Function*, Oxford University Press. Detailed introduction to the theory of neural networks placed in a physiologically relevant context.

Sporns, O. (2011) *Networks of the Brain*, MIT Press. Popular account of connectome research, i.e. the application of network theory to brains.

Squire, L.R., Berg, D., Bloom, F.E., du Lac, S., Ghosh, A. & Spitzer, N.C. (2008) *Fundamental Neuroscience*, Academic Press. Another perspective on neuron biology similar to Kandel's in scope.

Sterratt, D., Graham, B., Gillies, A. & Willshaw, D. (2011) *Principles of Computational Modelling in Neuroscience*, Cambridge University Press. Well-written introductory account of computational models.

Swanson, L.W. (2012) *Brain Architecture; Understanding the Basic Plan*, Oxford University Press. Excellent popular account of brain anatomy.

Tutorial Questions 26

26.1 A simple Steinbuch type network has 6 input neurons and 4 output neurons. The network is described by a matrix

$$\begin{pmatrix} 1 & 1 & 0 & 0 \\ 0 & 0 & 0 & 0 \\ 1 & 1 & 0 & 0 \\ 0 & 0 & 0 & 0 \\ 1 & 1 & 0 & 0 \\ 0 & 0 & 0 & 0 \end{pmatrix}$$

An input vector $(1 \ 0 \ 1 \ 0 \ 1 \ 0)$ is applied to the network. The threshold value for a neuron in the network is 1. Calculate the output of the network, i.e. the recalled memory.

26.2 Estimate the number of bytes of information that can be stored in a human brain.

26.3 Google search 'Brain map' on the web and familiarise yourself with the structure of some mammalian brains, e.g. use the Allen brain atlas.

Appendix A

Physical Constants

Boltzmann's constant	$k_B = 1.38 \times 10^{-23} \, \text{J} \, \text{K}^{-1}$
Ideal gas constant	$R = N_A k_B = 8.314 \, \text{J} \, \text{mol}^{-1} \, \text{K}^{-1}$
Thermal energy at 295 K is	$4.1 \, \text{pN} \, \text{nm} = 4.1 \times 10^{-21} \, \text{J}$
Electronic charge	$e = 1.6 \times 10^{-19} \, \text{C}$
Permittivity of free space	$\varepsilon_0 = 8.9 \times 10^{-12} \, \text{C}^2 \, \text{N}^{-1} \, \text{m}^2$
Permittivity of water	$\varepsilon \approx 80 \varepsilon_0$
Avogadro's number	$N_A = 6 \times 10^{23}$
Bjerrum length (room temperature)	$7 \, \text{Å}$
Debye screening length	
1:1 electrolytes, e.g. NaCl	$\kappa^{-1} = \dfrac{0.304 \, \text{nm}}{\sqrt{[\text{NaCl}]}}$
2:1 electrolytes, e.g. $CaCl_2$	$\kappa^{-1} = \dfrac{0.176 \, \text{nm}}{\sqrt{[\text{CaCl}_2]}}$
Viscosity of water at 20 °C	$\eta = 1.002 \times 10^{-3} \, \text{Pa} \, \text{s}$
Viscosity of water at 37 °C	$\eta = 0.692 \times 10^{-3} \, \text{Pa} \, \text{s}$
Planck's constant	$h = 6.626 \times 10^{-34} \, \text{J} \, \text{s}$
Speed of light	$c = 2.998 \times 10^8 \, \text{m} \, \text{s}^{-1}$
Units of pressure	$1 \, \text{atmosphere} = 1.01 \times 10^5 \, \text{Pa}$
Units of mass	$1 \, \text{Dalton (Da)} = 1.66 \times 10^{-27} \, \text{kg}$

The Physics of Living Processes: A Mesoscopic Approach, First Edition. Thomas Andrew Waigh.
© 2014 John Wiley & Sons, Ltd. Published 2014 by John Wiley & Sons, Ltd.

Appendix B

Answers to Tutorial Questions

Chapter 1

1.1 There are millions of possibilities:

Blood clotting proteins (e.g. factor XIII) – haemophilia, strokes,
Hemoglobin – sickle cell anaemia,
Insulin – diabetes,
Vitamin D – rickets,
Vitamin C – scurvy,
Myosin II – heart disease,
Actin – heart disease,
Amyloid beta precursor protein – Alzheimer's disease,
Chlorine-ion channel – cystic fibrosis.

1.2 The reader is referred to a good biochemistry textbook. Metals occur in small quantities in a wide range of biological molecules, e.g.

Iron in haemoglobin (oxygen transport) and ferritin (storage),
Magnesium in hexokinases (ATP production) and chlorophyll (green plant photosynthesis),
Calcium in prothrombin (blood clots) and troponin (muscular contraction).

1.3 For this very restricted model there are $2^{199} = 8 \times 10^{59}$ permutations of the chain. This is clearly a lower limit (more than 2 distinct orientational angles are possible, although some permutations are impossible due to excluded-volume effects). The peptides must bias their search through the 8×10^{59} permutations or it will take an impossibly long time. 'Funnels' are invoked in theoretical models to improve the search time in the multidimensional conformational landscape for the correctly folded protein conformation.

1.4 There are many reasons. The majority of carbohydrates are noncrystalline, so atomic-level structures cannot be provided by X-ray crystallography. Carbohydrates are not directly coded for genetically, unlike

The Physics of Living Processes: A Mesoscopic Approach, First Edition. Thomas Andrew Waigh.
© 2014 John Wiley & Sons, Ltd. Published 2014 by John Wiley & Sons, Ltd.

most proteins, and their biochemical synthesis tends to result in much higher polydispersities than proteins. Many carbohydrates are not synthesised by animals and the creation of glycoconjugates (e.g. glycoproteins, glycolipids, etc.) depend on the recent diet of the organism. Carbohydrate monomers can often be joined together (polymerised) in many more ways than proteins, so they are structurally very heterogeneous.

1.5 The interaction of genes in transcription networks is only partially understood. This is a major motivation for the establishment of the field of systems biology (described in **Part IV**).

1.6 Carcinogenic materials either disrupt the genome of cells directly or other cellular metabolic processes that involve the genome. Many radioactive substances are carcinogenic, some viruses, tobacco, asbestos, dioxins and so too are a wide range of everyday chemicals. Cells thus constantly have to repair damage to their chromosomes. Cancers results from self-amplifying mutations (**Section 2.5**). Most mutations are not self-amplifying and are corrected by the organism's natural defences. See 'Biology of Cancer' by R.A. Weinberg for a detailed description.

1.7 The lengths of the different forms of helix are as follows

$$\text{length}_A = \frac{4 \times 10^8}{660} \times 2.6 \text{ Å} = 158 \, \mu\text{m},$$

$$\text{length}_B = \frac{4 \times 10^8}{660} \times 3.4 \text{ Å} = 206 \, \mu\text{m},$$

$$\text{length}_Z = \frac{4 \times 10^8}{660} \times 3.7 \text{ Å} = 224 \, \mu\text{m}.$$

1.8 The membrane has a fluid-like bilayer structure. The protein will arrange itself through the membrane, so that its hydrophobic and hydrophilic regions are correctly positioned. **See Figure 8.3**.

1.9 The pH can be calculated using the definition of the equilibrium constant (K_a),

$$K_a = \frac{[\text{H}^+][\text{Arg}^-]}{[\text{HArg}]},$$

$$[\text{H}^+] \approx [\text{Arg}^-],$$

$$K_a \approx \frac{[\text{H}^+]^2}{[\text{HArg}]},$$

$$[\text{H}^+] = K_a^{1/2}[\text{HArg}]^{1/2},$$

$$\text{pH} = \frac{1}{2}\text{pK}_a - \frac{1}{2}\log[\text{HArg}] = 6.60.$$

Chapter 2

2.1 This is a classic question posed to challenge Darwin's theory of evolution. See Richard Dawkins book 'The blind watchmaker', Penguin. Choosing the human eye as an example, many present-day organisms indicate the evolutionary steps were:

Light-sensitive proteins (opsins) developed in a cell's membrane, e.g. paramecium,
Light-sensitive proteins arranged themselves in a pit, which provides sensitivity to direction, e.g. limpets,

Light-sensitive proteins were used in a pin-hole camera, e.g. giant clams and Nautilus,
Light-sensitive proteins evolved with a transparent lens, e.g. human and octopus.
For more details on the evolution of eyes see M.F. Land, D.E. Nilsson, 'Animal eyes', OUP.

2.2 Forces that can contribute to cell adhesion include: covalent bonding, ionic bonding, hydrogen bonding, van der Waals forces, steric forces (attached flexible chains or membrane fluctuations), electrostatics and hydrodynamics.

2.3 Max Perutz invented X-ray techniques for studying large complicated globular proteins. He discovered that the phase problem in the analysis of X-ray diffraction patterns could be solved by *multiple isomorphous replacement*. This involves comparison of patterns from several crystals; one from the native protein and others that had been soaked in solutions of heavy metals.

2.4 The mixture phase separates with a morphology that depends on its composition, i.e. oil droplets suspended in water (if water is the majority component) or water droplets suspended in oil (if oil is the majority component). The structure can coarsen over time (small droplets stick together) as the surface free energy is minimised. This is a challenge for vinaigrette recipes. In **Chapter 3** the concept of free energy is introduced and it allows a more quantitative explanation. Demixing occurs because the strong energetic advantage (U) of partitioning the oil and water molecules into separate regions outweighs the entropic penalty ($-TS$).

2.5 Genetic engineering can be used to create:

Vaccines for hepatitis B,
Genetically modified foods,
Human growth hormone or human insulin,
Spider silk from bacteria,
New plant colourations.

2.6 Cells communicate with their neighbours in a variety of ways. They involve: direct contact (e.g. gap junctions connect two cellular cytoplasms, notch signalling – adhesive proteins), transmission over short distances (neurotransmitters, growth factor and clotting factor) and transmission over long distances (hormones in the blood). Electricity can transmit signals inside (and between) specialised electrically excitable cells (e.g. cardiac cells) and can be measured with a patch clamp (see **Part V**).

Chapter 3

3.1 Let the height of the vessel be h, the area is A and the total number of particles is N.

$$A \int_0^h k_1 e^{-m_{net}gz/kT} dz = N,$$

$$\frac{kT}{mg} \left(1 - e^{-mhg/kT} \right) = \frac{N}{Ak_1},$$

$$\frac{1}{e^{mgh/kT}} \approx 0.$$

Therefore, $k_1 = \dfrac{Nmg}{kTA}$.

Chapter 4

4.1 Need to differentiate the potential with respect to the displacement to calculate the force,

$$F(r) = -\frac{\partial V(r)}{\partial r} = \varepsilon \left(\frac{r_0}{r}\right)^{13} 12r_0 - 2\varepsilon r_0 \left(\frac{r_0}{r}\right)^7 6$$

$$F(r) = 12\varepsilon r_0 \left(\frac{r_0}{r}\right)^{13} - 12\varepsilon r_0 \left(\frac{r_0}{r}\right)^7$$

When $r = \dfrac{r_0}{2}$ we have

$$F(r_0/2) = 2.35 \times 10^{-4} N$$

A substantial force on a single atom.

4.2 Formulae for the quick calculation of the Debye screening length are included in the Appendix A.

For monovalent salts 0.001 M, $\kappa^{-1} = \dfrac{0.304}{\sqrt{0.001}} = 9.6 \, \text{nm}$.

Similarly for 0.01 M, $\kappa^{-1} = 3$ nm; for 0.1 M, $\kappa^{-1} = 1$ nm; for 1 M $\kappa^{-1} = 0.3$ nm.
The concentration of ions from spontaneous dissociation in pure water is

$$[H^+] = [OH^-] = 1 \times 10^{-7} M,$$

$$\kappa^{-1} = \frac{0.304}{[H^+]}.$$

Thus, the screening length of pure water is 3×10^6 nm, i.e. 3 mm, due to spontaneous dissociation! The length scale of the electrostatic interaction between molecules in water can thus at most be a million times bigger than their diameter.

For divalent salts 0.001 M, $\kappa^{-1} = \dfrac{0.176}{\sqrt{0.001}} = 5.6 \, \text{nm}$.

Similarly for 0.01 M, $\kappa^{-1} = 1.76$ nm; for 0.1 M, $\kappa^{-1} = 0.56$ nm; for 1 M $\kappa^{-1} = 0.176$ nm.

A rough estimate for the equivalent salt concentration for physiological conditions is 0.1 M monovalent salt and thus $\kappa^{-1} = 1$ nm.

4.3 For a sphere the potential follows the form $V(r) \sim 1/r$, whereas for a cylinder $V(r) \sim \ln(r)$. The electrostatic potential thus decreases more quickly with distance (r) from a sphere. Close to a plane surface, $V(r) \sim r$. The equations for the potential can all be derived from Gauss's law.

4.4 For the steric interaction the potential takes the form

$$V(r) \sim e^{-r/R_g}.$$

For the screened electrostatic interaction the potential takes the form

$$V(r) \sim e^{-kr}.$$

Assuming the prefactors are of a similar order of magnitude the steric force becomes significant when $\kappa^{-1} < R_g$ and $r_{sep} < R_g$, where r_{sep} is the separation distance.

4.5 The magnitude of the adhesion force measured depends on the lifetime of the bond and thus the duration of the experiment.

Chapter 5

5.1 The cooperativity of the phase transition increases with the length of the helix and thus tends to sharpen the DSC endotherm, i.e. the helix–coil transition occurs over a narrower temperature range.

5.2 Hysteresis behaviour has been observed for long polymeric chains as the quality of the solvent is reduced and the chain size is increased, e.g. the size of the globular chains depends on the route by which they were globularised.

5.3 The enthalpy change is

$$\Delta H_m = \frac{2\gamma_{sl}T_m}{r\Delta T} = \frac{2 \times 1.2 \times 10^{-3} \times 323}{50 \times 10^{-9} \times 1} = 15 \text{ MJ kg}.$$

Chapter 6

6.1 P_2 is the orientational order parameter, ψ is the lamellar order parameter and h is the helical order parameter. During heating it is possible that

(*Wet self-assembled*) $P_2 > 0$, $\psi > 0$, $h > 0$ becomes (*Gelatinised*) $P_2 = 0$, $\psi = 0$, $h = 0$.

(*Wet self-assembled*) $P_2 > 0$, $\psi > 0$, $h > 0$ becomes (*Intermediate*) $P_2 > 0$, $\psi = 0$, $h > 0$ becomes (*Gelatinised*) $P_2 = 0$, $\psi = 0$, $h = 0$.

Also. the state of self-assembly can be modified through the addition of water,

(*Dry unassembled*) $P_2 > 0$, $\psi = 0$, $h > 0$ becomes (*Wet self-assembled*) $P_2 > 0$, $\psi > 0$, $h > 0$.

Therefore, there is a third possibility for heat treatment,

(*Dry unassembled*) $P_2 > 0$, $\psi = 0$, $h > 0$ becomes (*Gelatinised*) $P_2 = 0$, $\psi = 0$, $h = 0$.

The process of staling (gelatinised starch forms small unpalatable crystallites) can also be parameterised in a similar manner,

(*Gelatinised*) $P_2 = 0$, $\psi = 0$, $h = 0$ becomes (*Stale*) $P_2 = 0$, $\psi = 0$, $h > 0$.

Steric constraints introduce a strong coupling between the orientation of the mesogens and the degree of mesogen helicity.

6.2 The orientational order parameter can be calculated as

$$\langle \cos^2\theta \rangle = \frac{\int_0^\pi \cos^2\theta p(\theta)d\theta}{\int_0^\pi p(\theta)d\theta} = \frac{2}{\pi}\int_{\pi/4}^{3\pi/4} \cos^2\theta d\theta = \frac{1}{\pi}\int_{\pi/4}^{3\pi/4}(1 + \cos 2\theta)d\theta = \frac{1}{2},$$

$$P_2(\cos\theta) = \frac{3}{2}\cos^2\theta - \frac{1}{2} = \frac{1}{4}.$$

You would expect there to be two brushes emanating from the point defect in a polarising microscope.

6.3 The Onsager calculation for the nematic/isotropic transition gives

$$\phi < 3.34\frac{D}{L} = 0.167 \text{ i.e. } 16.7\% \text{ volume fraction.}$$

6.4 The entropy of the side chains is antagonistic to the entropy of the backbone. This increases the rigidity of the backbone chain and can induce nematic ordering in the backbone.

Chapter 7

7.1 $Re = \dfrac{2vL\rho}{\eta} = \dfrac{2 \times 10^{-1} \times 10^{-3} \times 1.3}{1.8 \times 10^{-5}} = 14.4.$

The Reynold's number is not small, so the inertial forces could be quite considerable.

7.2 From the definition of one-dimensional diffusion, $\langle x^2 \rangle = 2Dt$. Rearranging for the characteristic time gives

$$t = \frac{\langle x^2 \rangle}{2D} = \frac{(2.7 \times 10^{-3})^2}{2 \times 1.35 \times 10^{-9}} = 2.7 \times 10^3 \text{ s}.$$

This mechanism is far too slow.

7.3 The rotational diffusion coefficient (D_θ) is

$$D_\theta = \frac{kT}{8\pi\eta a^3} = \frac{4.1 \times 10^{-21}}{8 \times 3.142 \times 0.001 \times (2 \times 10^{-6})^2} = 2.04 \times 10^{-2} \text{rad s}^{-1}.$$

$\langle \theta^2 \rangle \approx 2D_\theta t$ for small angles.

So the characteristic time for fluctuations of 90° $(\pi/2)$ is $t = 60$ s.

In three dimensions

$$\langle r^2 \rangle = 6Dt,$$

$$\langle \theta^2 \rangle = 6D_\theta t_\theta.$$

And therefore $(2\pi)^2 = 6D_\theta t_\theta$ and $(2\pi a)^2 = 6Dt$.

Substituting expressions for D_θ and D we have $t_\theta = \dfrac{4}{3}t$,

where t_θ is the characteristic time for rotation through 2π and t is the time for translation by $2\pi a$.

7.4 For the motile particle the effective diffusion coefficient (D) is

$$D = \frac{v^2 \tau}{3(1-\alpha)}$$

The average value of the cosine (α) is $\dfrac{11}{12}$, i.e. there is on average a 23.6° angle between successive runs.

Chapter 8

8.1 $\alpha = \dfrac{4\pi R^2 \gamma}{kT} = \dfrac{4 \times 3.142 \times (2 \times 10^{-9})^2 \times 20 \times 10^{-3}}{4.1 \times 10^{-21}} = 245.$

$\text{CMC} \approx e^{-\frac{\alpha}{N^{1/3}}} = e^{-\frac{245}{(10000)^{1/3}}} = 1.15 \times 10^{-5} \text{ M}.$

8.2 The critical concentration (c_c) for self-assembly is

$$c_c = K = e^{\frac{\Delta G_0}{kT}} = 0.01 \text{ M}.$$

The average degree of filament polymerisation (n_{av}) is

$$n_{av} = \sqrt{\frac{c_t}{K}} = 10.$$

The average filament length ($l_{av} = n_{av} \times$ monomer length) is

$$l_{av} = 10 \times 5 = 50 \text{ nm}.$$

Fairly short filaments are formed even at high monomer concentrations.

8.3 In 2D and 3D the surface free energy tends to encourage compact self-assembled aggregates. This is not true in 1D where aggregates can be long and polydisperse.

Chapter 9

9.1 The contact angle can be calculated from the Young–Laplace equation,

$$\gamma_{sg} = \gamma_{sl} + \gamma_{lg} \cos\theta,$$

$$18 = 73.2 + 72\cos\theta,$$

$$\theta = 140°.$$

The wetting coefficient is

$$k = \frac{18 - 73.2}{72} = -0.77.$$

The surface is unwetted, which provides a useful self-cleaning mechanism for the lotus leaf.

Chapter 10

10.1 The crosslinking density (v) is linearly related to the Young's modulus (E) for flexible rubbery networks,

$$E = 3kTv.$$

For elastin $v = \dfrac{1 \times 10^6}{3 \times 4.1 \times 10^{-21}} = 8.13 \times 10^{25} \text{m}^{-3}.$

For collagen $v = \dfrac{1 \times 10^9}{3 \times 4.1 \times 10^{-21}} = 8.13 \times 10^{28} \text{m}^{-3}.$

The collagen chains are semiflexible and are thus not well described by a purely flexible model for the elasticity.

10.2 $E_{thermal} = kT$ and $E_{bend} = E_{thermal}$.

Therefore

$$kT = \frac{kTl_p\langle\theta^2\rangle}{2s},$$

$$\langle \theta^2 \rangle = 2.3 \times 10^{-4} \text{rad s}^2,$$

$$\langle \theta^2 \rangle^{1/2} = 2.7°.$$

The mean square angular displacement of the filament is therefore 2.7°.

10.3 The length of the titin molecule trapped in the pore (see after equation (10.57)) is

$$R_{\parallel} \approx Na \left(\frac{D_b}{a} \right)^{-2/3} = 600 \, \text{nm}.$$

When stretched to 750 nm the tension blob size is smaller than the size of the pore and the conformation (and elasticity) of the chain is unaltered by the size of the pore.

10.4 The free energy of an ideal chain is

$$F(R) = -T \ln Z_N(R) = \text{const} + \frac{3TR^2}{2Nl^2}.$$

If the chain now experiences an extending force (*f*) on both ends then

$$f = \frac{\partial F}{\partial R} = \frac{3T}{Ll} R.$$

The polymer in a good solvent obeys Hooke's law.

In a bad solvent there is a strong increase in the force measured by the traps (DNA $\sim 0.3 kT/bp$). An unwinding globule–coil transition is now possible and will be observed by a sawtooth in the force/distance curves.

10.5 The size of the DNA chain according to the worm-like chain model is

$$\langle R^2 \rangle^{1/2} = \sqrt{2Ll_p} = \sqrt{(2 \times 60 \times 10^{-6} \times 450 \times 10^{-10})} = 2.32 \, \mu\text{m}.$$

Chapter 11

11.1 The ability of the amine groups to dissociate is reduced due to the interaction between neighbouring groups along the polylysine chain (there is an energetic penalty).

11.2 $\xi = \dfrac{l_b}{b} = \dfrac{7}{5}$

The effective charge fraction of the polylysine chain predicted by the Manning model is therefore $\dfrac{1}{\xi} = \dfrac{5}{7}$. 5/7 of the fully ionised charge.

11.3 The total persistence length (l_p) in the OSF model is equal to the intrinsic component added to the electrostatic component,

$$l_T = l_p + \frac{l_b}{4\kappa^2 A^2}$$

The second term is the electrostatic contribution, $A = \dfrac{a}{f} = \dfrac{1}{0.5} = 2 \, \text{nm}$ and $l_b = 0.7 \, \text{nm}$.

Thus $l_e = 0.7 \, \text{nm}$.

11.4 Critical properties of the material are:

Samples are extremely hydrophilic and swell many times their dry size (polyelectrolytes gels can be less than 1% polymer, with 99% water).

The materials are charged (reduce adhesion) and biocompatible.

Assuming all the charges on the polymer chain dissociate the osmotic pressure (π) is proportional to the number of charges per unit volume (n); $\pi = nkT$.

The number of charged units is $n = 0.001 \times 6 \times 10^{23} = 6 \times 10^{20} \text{dm}^{-3}$.

Thus the osmotic pressure is $0.404 \, \text{J dm}^3$.

The neutral polymer contribution is kT per blob. It is much smaller than the contribution of the counterions.

11.5 The charged blob size (D) is given as

$$D \sim a\sigma^{2/3} u^{-1/3}$$

where a is the monomer length, σ is the number of monomers between charged units ($=1$ fully charged) and u is the Bjerrum length.

$u = 7/3.6 \, \text{Å}$

$a = \text{peptide step length} = 3.6 \, \text{Å}$

$D = 2.9 \, \text{Å}$ for fully charged blobs (fully elongated)

$D = 13.4 \, \text{Å}$ for weakly charged blobs ($\sigma = 10$)

The charged chain forms a semiflexible rod of positively charged blobs (polylysine is a cationic polyelectrolyte),

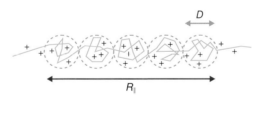

$$R_\| = D\left(\frac{N}{g}\right)$$

where N is the number of monomers in the chain, g is the number of monomers in a blob and D is the size of a blob.

Chapter 12

12.1 The line tension is given by

$$\lambda = \tau R^* = 2.6 \times 10^{-9} \times 0.03 = 8 \times 10^{-11} \text{J m}^{-1}.$$

12.2 The axial and hoop stresses are given by

$$\sigma_{axial} = \frac{rP}{2h} = \frac{1 \times 10^5 \times 1 \times 10^{-6}}{2 \times 10^{-9}} = 0.5 \times 10^8 \, \text{Pa},$$

$$\sigma_{hoop} = \frac{rP}{h} = 1 \times 10^8 \, \text{Pa}.$$

12.3 For the persistence lengths, $\xi_{p1} \sim be^{10\pi}$ and $\xi_{p2} \sim be^{20\pi}$.
      The ratio of the two persistence lengths is therefore

$$\frac{\xi_{p1}}{\xi_{p2}} = e^{10\pi}.$$

For the exponent measured with X-rays $\eta_{m1} \sim (BK)^{-1/2}$ and $\eta_{m1} \sim (B2K)^{-1/2}$.

The ratio of the two exponents is therefore $\dfrac{\eta_{m1}}{\eta_{m2}} = \sqrt{2}$.

Undulation forces may perturb the membrane structure.

12.4 $\dfrac{R_{g2}^2}{R_{g1}^2} = \dfrac{L_c^2}{L_c^{4/3}} = L_c^{2/3}$,

$\dfrac{R_{g2}}{R_{g1}} = L_c^{1/3}$,

$\dfrac{A_1}{A_2} = \dfrac{L_c^3}{L_c^2} = L_c$.

Chapter 13

13.1 In the parallel arrangement the Young's modulus of the mixture (E_m) is given by

$$E_m = E_c \phi_c + E_a(1 - \phi_c) = 50 \times 10^9 \times 0.9 \, \text{Pa} + 50 \times 10^6 \times 0.1 \, \text{Pa},$$

$$E_m \approx 45 \, \text{GPa}.$$

In the perpendicular arrangement, E_m is given by

$$E_m = \frac{E_c E_a}{E_a(1 - \varphi) + E_c \varphi} = 55 \, \text{MPa}.$$

The ratio of the Young's moduli is 818:1, parallel to perpendicular.

13.2 For the unfilled foam, $E_f \sim \left(\dfrac{t}{a}\right)^4 E$, where $E = 9 \, \text{GPa}$, $a = 20 \, \mu\text{m}$, $t = 1 \, \mu\text{m}$.
      Therefore

$$E_f \sim \left(\frac{1}{20}\right)^4 9 \, \text{GPa} = 0.56 \, \text{MPa}.$$

For the filled foam, $E_f \sim \left(\frac{t}{a}\right)^2 E$,

$$E_f \sim \left(\frac{1}{20}\right)^2 9\,\text{GPa} = 22.5\,\text{MPa}.$$

Chapter 14

14.1 Paramecium that swim at relatively high speeds have a vortex on either side when observed using microscopic particle imaging velocimetry.

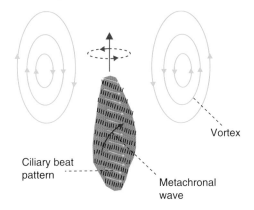

Other examples where vortices are known to be important include: blood flow in the left human ventricle, where they are needed for efficient transport, jet propulsion of squid (also jellyfish) and the lift on a bird's wing during flight.

The vorticity can be neglected in so-called potential (irrotational) flows, where the vorticity equals zero everywhere, $\underline{\omega} = \nabla \times \underline{u} = 0$. Bernouilli's equation for streamlines is an example of a potential flow.

14.2 The top speed of a duck is mainly limited by the drag force of surface gravity waves (bow waves, created by its motion). This determines its 'hull speed'.

Bow wave

For smaller organisms (e.g. whirligig beetles) their length is 1 cm and the hull speed (set in this case by capillary waves rather than gravity waves) is ~25 cm s^{-1}. This is much faster than the beetles typically need to travel.

14.3 From equation (14.15),

$$\Delta p = \frac{8\eta L \bar{v}}{R^2} = \frac{8 \times 0.001 \times 50 \times 0.001}{\left(20 \times 10^{-6}\right)^2} = 10^6\,\text{Pa}.$$

Chapter 15

15.1 Shear rate $= \dfrac{1 \times 10^{-2}}{10 \times 10^{-6}} = 10^{3} \mathrm{s}^{-1}$,

$$\mathrm{Pe} = \frac{6\pi \eta a^{3} \dot{\gamma}}{kT} = 6\pi \times 0.001 \times \left(10^{-6}\right)^{3} \frac{10^{3}}{4.1 \times 10^{-21}} = 4.6 \times 10^{3}.$$

The experiment is in the regime of high Peclet number dynamics. The shear rate could therefore be substantially affecting the microstructure.

15.2 For the Maxwell model, $\eta = \tau G$.

The characteristic relaxation times is therefore, $\tau = 10^{6}\,\mathrm{s} \sim 11.6\,\mathrm{days}$.

15.3 The viscosity is given by the Einstein relationship,

$$\eta = \eta_{b}\left(1 + \frac{5}{2}\phi\right) = 10^{-3}\left(1 + \frac{5}{2}0.02\right) = 1.05 \times 10^{-3}\mathrm{Pa\,s}.$$

15.4 Thixotropy is a time-dependent reduction in viscosity during shear, e.g. tomato ketchup and yoghurt. Not all shear-thinning fluids are time dependent (thixotropic).

Chapter 16

16.1 Molecular motors (many of which come in multiple varieties, e.g. Mysoin I-XVIII): kinesins, actins, microtubules, dyneins, myosins, RNA polymerase, DNA polymerase, F_{0} synthase rotary motor, flagellar motor, etc. Diseases include:

Kinesin – Charcot–Marie–Tooth disease and some kidney diseases,
Dynein – cilial malfunction, chronic infections of the respiratory tract, motor neuron disease,
Myosin – Usher syndrome, deafness, heart disease.

16.2 Factors: actin monomer concentration, at high forces the fibres buckle, monomers have a finite time to diffuse between the 2 fibre ends.

Chapter 17

17.1 Using equation (17.31) gives $E_{\mathrm{protein}} = 1\,\mathrm{MPa}$.

17.2 The characteristic time for stress relaxation is dependent on the motion of the interstitial water

$$\tau_{1} = \frac{\delta \delta_{\mathrm{eq}}}{\pi^{2} E k} = \frac{10^{-3} \times 0.95 \times 10^{-3}}{\left(3.142\right)^{2} \times 0.78 \times 10^{6} \times 6 \times 10^{-13}} = 2\,\mathrm{s}.$$

Chapter 19

19.1 The main challenge (as with X-rays) is to create efficient aberration-free γ-ray lenses. Synchrotrons can be used as relatively convenient γ-ray sources and CCD cameras can be adapted for high-energy photons.

19.2 The Abbe resolution limit is

$$d = \frac{\lambda}{2\mathrm{NA}} = \frac{400}{2 \times 1.4} = 143\,\mathrm{nm}.$$

The current resolution limits are SIM, $d \sim \lambda/4\mathrm{NA} \sim 80\,\mathrm{nm}$ (with pulsed laser excitation this can be improved on, i.e. nonlinear techniques).

For confocal microscopes, $d \sim \lambda/4\mathrm{NA} \sim 80\,\mathrm{nm}$,
STORM, $d \sim 10\text{--}20\,\mathrm{nm}$,
PALM, $d \sim 20\,\mathrm{nm}$,
STED, $d \sim 30\,\mathrm{nm}$.

19.3 Genetically expressed fluorescent molecules can be created using recombinant DNA technology and the GFP is attached to another molecule of interest. The GFP-tagged molecules can be simply observed in a fluorescence microscope to probe their mode of action. This is an extremely useful tool that has solved a huge number of problems in cellular structure and physiology.

19.4 Terahertz radiation is composed of electromagnetic waves of frequency $\sim 10^{12}\,\mathrm{Hz}$. It occurs between the microwave and infrared parts of the electromagnetic spectrum. Previously it was hard to create terahertz sources. There are now a wide range of possibilities, e.g. using femtosecond lasers to irradiate semiconductor crystals. There is a lot of interest in bioimaging applications (e.g. airport scanners) and spectroscopy (e.g. finger-printing biomolecules).

19.5 Attosecond, e.g. time scale for fluctuations in electron clouds.
Femtosecond, e.g. charge transfer,
Picosecond, e.g. time scale for the motion of cages of water molecules, water acts solid-like at time scales faster than a picosecond,
Nanosecond, e.g. rotational motion of globular proteins,
Microsecond, e.g. translational motion of globular proteins,
Millisecond, e.g. time scale for motor protein stepping or electrical impulses in nerve cells,
Atto-picosecond time scales can be studied with pulsed laser and electron sources.

19.6 Use the Stokes force or stochastic Langevin analysis to calibrate the apparatus. From the figure the cornering frequency (f_c) is given by $f_c \approx 500\,\mathrm{Hz}$.
  Also $\gamma = 6\pi\eta r = 1.88 \times 10^{-8}$.
  The trap stiffness (κ_x) is therefore

$$\kappa_x = 2\pi f_c \gamma = 5.91 \times 10^{-5}\,\mathrm{N\ m}^{-1}.$$

An analogous method can be used for magnetic tweezers, i.e. measure the power spectral density and model with the Langevin equation.

19.7 Advantage of single-molecule experiments: information content is much higher.
  Disadvantages: more sophisticated stochastic modelling is required, noninvasive measurements are difficult, experiments are more challenging (signal-to-noise ratio is often poor).

19.8 Two standard methods are currently used. Focusing the X-rays incident on the sample (using a Fresnel lens or KB mirrors) can be shown to lift the ambiguity in the image reconstructions. Ptychography experiments (which need lots of diffraction patterns at different sample positions) also are immune to the random reorientation artefacts.

19.9 *A* corresponds to a predominantly viscous material MSD $\sim t^1$.

B corresponds to a viscoelastic material, with a characteristic subdiffusive behaviour of the probe particles $\langle \Delta r^2(t) \rangle \sim t^\alpha$, $\alpha < 1$. Static errors tend to induce a plateau in the MSDs when the value of the MSD is comparable with the square of the static displacement error. Normally that corresponds to the short time limit. Thus, fluid B could be a purely viscous fluid if the static errors have not been properly accounted for.

19.10 Using equation (19.95) the velocity of electrophoresis is $v \approx \dfrac{q\bar{E}}{3\eta N}$.

There are 1.7 Å between the phosphate groups on the DNA, $q = \dfrac{1}{1.7} e\text{Å}^{-1}$ and $N = 10^6/300$,

$$v = \frac{(1/1.7) \times 1.6 \times 10^{-19} \times 10^{10} \times 2 \times 100}{3 \times 0.002 \times 10^6/300} = 9.4 \times 10^{-9} \text{m s}^{-1}.$$

Decreasing the size of the chains by a factor of 10, should increase the velocity by a factor of 10.

19.11 From equation (19.44) with $\omega_c = 0$, we have the required expression for the power spectral density with no trapping force,

$$\langle \Delta r^2(\omega) \rangle = \frac{kT}{\pi \gamma \omega^2}.$$

For a power law fluid, the frequency response is modified

$$\langle \Delta r^2(\omega) \rangle \sim \frac{1}{\omega^{2-\alpha}}$$

where α is a positive constant and the MSD $\sim t^\alpha$.

Chapter 20

20.1 The law of mass action states that the reaction rate is proportional to the concentration of the reactants raised to a power, i.e.

$$R = k[A]^\alpha [B]^\beta [C]^\chi [D]^\delta \ldots$$

It is only true for elementary reactions. A corollary in equilibrium, where the forward and backwards reaction rates are equal, is that the equilibrium constant can be expressed as

$$K = \frac{[C]^\chi [D]^\delta}{[A]^\alpha [B]^\beta}$$

where A and B are reactants, whereas C and D are products. In other words, when a reversible reaction has attained equilibrium at a given temperature the reaction quotient remains constant.

20.2 First-order reaction $\dfrac{d[A]}{dt} = -k[A]$.

Solve variables separable $\dfrac{d[A]}{[A]} = -k\,dt$,

$$\int_{[A]_0}^{[A]} \frac{d[A]}{[A]} = -k \int_0^t dt,$$

$$\ln\left(\frac{[A]}{[A]_0}\right) = -kt,$$

$$[A] = [A]_0 e^{-kt}.$$

The half-life by definition is

$$kt_{1/2} = -\ln\left(\frac{0.5[A]_0}{[A]_0}\right) = -\ln\left(\frac{1}{2}\right) = \ln 2,$$

$$t_{1/2} = \frac{\ln 2}{k}.$$

20.3 First-order reaction,

$$R = k[A]^\alpha \quad \text{and} \quad \alpha = 1$$

20.4 Reaction rate

$$R = -\frac{d[ZnS]}{dt} = -\frac{1}{2}\frac{d[HCl]}{dt} = \frac{d[H_2S]}{dt} = \frac{d[ZnCl_2]}{dt}.$$

The stoichometric coefficients are 1 for [ZnS], 2 for [HCl], 1 for [H_2S] and 1 for [$ZnCl_2$].

20.5 Experimental method to determine reaction order.
Assume the law of mass action $R = k[A]^\alpha$.
Method of initial rates – measure dependence of initial reaction rate on the concentration,

$$R_0 = k[A_0]^\alpha.$$

This implies $\log R_0 = \log k + \alpha \log[A_0]$.
A plot of $\log R_0$ versus $\log[A_0]$ gives a straight line and the gradient gives the reaction order.

20.6 All the fundamental physical laws apply to biology as well. This may seem obvious now, but was a major step in the early 1900s. Your list could include:

Conservation of mass,
Conservation of charge,
Minimisation of free energy,
Law of mass action,
Conservation of energy,
0th, 1st and 2nd laws of thermodynamics,
Maxwell's equations,
Special relativity,
General relativity,
Quantum field theory (required for the calculation of van der Waals forces) etc.

20.7 Gibbs free energy,

$$F_G = U_{int} + PV - TS \tag{1}$$

Helmholtz free energy,

$$F_H = U_{int} - TS \tag{2}$$

The two expressions are equivalent in some cases, e.g. for an ideal gas $PV = nRT = \text{const} = c$. Therefore

$$F_G - c = U_{\text{int}} - TS.$$

However, only changes in free energies can be measured, so equations (1) and (2) are equivalent in this case.

20.8 a. The reaction rate is

$$R = -\frac{1}{x}\frac{d[A]}{dt} = -\frac{1}{y}\frac{d[B]}{dt} = \frac{d[C]}{dt}.$$

The law of mass action is $R = k[A]^\alpha[B]^\beta$ and therefore $K = \dfrac{[C]^\gamma}{[A]^\alpha[B]^\beta}$.

b. Vary [A], keeping [B] constant and measure the rate R. Plot log R versus log [A] the gradient is α. Vary [B], keeping [B] constant and measure the rate R. Plot log R versus log [B] the gradient is β.

20.9 Zeroth order $\dfrac{da}{dt} = -k$,

First order $\dfrac{da}{dt} = -ka$,

Second order $\dfrac{da}{dt} = -ka^2$.

20.10 This is a surprisingly hard question. The first step is relatively straightforward to write down the rate equations,

$$\frac{da}{dt} = -k_1 a \tag{1}$$

$$\frac{dx}{dt} = k_1 a - k_2 x \tag{2}$$

$$\frac{dp}{dt} = k_2 x \tag{3}$$

Equation (1) is just first-order kinetics and the solution is easy

$$a(t) = a_0 e^{-k_1 t}.$$

The next step requires the integrating factor method of solving first-order ordinary differential equations. If you want to brush up on this area check K.F. Riley, 'Maths methods for physics and engineering', CUP. The required integrating factor is

$$\mu(t) = e^{\int k_2 dt} = e^{k_2 t}.$$

Therefore we have

$$e^{k_2 t}\frac{dx}{dt} + k_2 x e^{k_2 t} = k_1 a_0 e^{-k_1 t} e^{k_2 t},$$

$$x e^{k_2 t} = \int a_0 k_1 e^{t(k_2 - k_1)}\,dt + c = \frac{a_0 k_1}{k_2 - k_1}e^{t(k_2 - k_1)} + c,$$

$$x = \frac{a_0 k_1}{k_2 - k_1} e^{-k_1 t} + c e^{-k_2 t}.$$

The boundary conditions are $x = 0$ when $t = 0$

$$c = -\frac{a_0 k_1}{k_2 - k_1}.$$

The final solution to equation (2) is therefore $x(t) = \dfrac{a_0 k_1}{k_2 - k_1} \left(e^{-k_1 t} - e^{-k_2 t} \right)$ as required.

The solution to the third equation follows easily from conservation of mass, i.e. $p = a_0 - a - x_0$.

Giving $p(t) = a_0 - a_0 e^{-k_1 t} - \dfrac{a_0 k_1}{k_2 - k_1} \left(e^{-k_1 t} - e^{-k_2 t} \right)$ as required.

20.11 Equation (20.69) is $\dfrac{du}{d\tau} + u = \dfrac{1}{1 + \theta}$.

The integrating factor is $e^{\int_0^\tau ds} = e^\tau$,

$$\frac{d}{d\tau}(e^\tau u) = \left(\frac{1}{1 + \theta} \right) e^\tau,$$

$$u = \left(\frac{1}{1 + \theta} \right) (1 + C e^{-\tau}).$$

The boundary conditions are $u = 1$, $v = 0$, $\tau = 0$ and therefore $C = \theta$.

Finally $u = \left(\dfrac{1}{1 + \theta} \right) (1 + \theta e^{-\tau})$ as required.

For v, need to use $u + v = 1$.

$$v = 1 - u = \frac{\theta}{1 + \theta}(1 - e^{-\tau}).$$

Chapter 21

21.1   i. The rate equation is

$$\frac{d[A]}{dt} = k_-[C] - k_+[A][B].$$

ii. If the concentration of B is a constant

$$[A] + [C] = A_0.$$

iii. At equilibrium $\dfrac{d[A]}{dt} = \dfrac{d[B]}{dt} = \dfrac{d[C]}{dt} = 0$,

Therefore $[C]_{eq} = \dfrac{k_+}{k_-}[A]_{eq}[B]_{eq}$.

iv. $[C] = A_0 \dfrac{[B]}{K_{eq} + [B]}$ where $K_{eq} = k_-/k_+$.

Chapter 22

22.1 GCGATATCGCAAA is the template strand.
CGCTATAGCGTTT is the coding strand.
GCGAUAUCGCAAA is the RNA transcript.

Structural difference:

DNA is double helical and semiflexible.
RNA is mostly single stranded and thus much more flexible. Its chemistry is similar to that of DNA, but it has the sugar ribose rather than deoxyribose. One base has uracil rather than thymine with DNA.
RNA might thus be expected to be more chemically active due to its increased accessibility and flexibility.

22.2 Example of a gene construct. You could construct a repressilator in an Ecoli cell with traffic lights fluorescent reporters (i.e. 3 separate fluorescent proteins as reporters).

22.3 There are a huge range of possibilities. There are millions of gene circuits that have been catalogued to date. The flagellar motor in bacteria is controlled by a multioutput feedforward loop (FFL). The temporal order of arginine (peptide) biosynthesis is controlled by a single output module (SIM). Incoherent 1- FFL occurs in the galactose (carbohydrate) system of E-coli. Coherent 1- FFL occurs in the arabinose (carbohydrate) system of E-coli.

22.4 Reaction diffusion equations have been used to model a wide range of biological (and physical) phenomena. They include: morphogenesis (e.g. patterns on animal coats such as leopards, genets, tropic fish, etc.), population biology (epidemics), tumour growth, wound healing, calcium fertlisation waves in frog eggs, and cyclic AMP waves in slime molds.

22.5 $T_{1/2} = \dfrac{K}{2\beta} = \dfrac{1}{10} = 0.1$ s.

22.6 Time shift occurs at $t = 0$.

Before the shift Y reaches a steady state level $Y(t = 0) = Y_{st} = \dfrac{\beta_1}{\alpha}$, (1).

After the shift $\dfrac{dY}{dt} = \beta_2 - \alpha Y$.

General solution of the equation is $Y = C_1 + C_2 e^{-\alpha t}$.

Determine C_1 and C_2 from (1) and Y at long times is $Y = \dfrac{\beta_2}{\alpha}$,

$$Y(t) = \frac{\beta_2}{\alpha} + \left(\frac{\beta_1}{\alpha} - \frac{\beta_2}{\alpha}\right) e^{-\alpha t}.$$

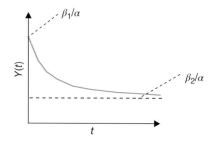

22.7 Need an integrating factor to solve 1st-order differential equation, $\dfrac{dY}{dt} + \alpha Y = \beta$ and the integrating

factor is $e^{\int_0^t \alpha \, ds}$.

This leads to $Y = \dfrac{\beta}{\alpha} + Ce^{-\alpha t}$.

The boundary condition is $Y = 0$ when $t = 0$, so $C = -\dfrac{\beta}{\alpha}$.

The solution is $Y = Y_s(1 - e^{-\alpha t})$ where $Y_s = \dfrac{\beta}{\alpha}$.

22.8 From equation (22.27),

$$0 = D \frac{\partial^2}{\partial x^2} \left(M_0 e^{-x/\lambda} \right) - \alpha M_0 e^{-x/\lambda},$$

$$0 = DM_0 \left(-\frac{1}{\lambda} \right)^2 e^{-x/\lambda} - \alpha M_0 e^{-x/\lambda}.$$

Finally $\lambda = \sqrt{\dfrac{D}{\alpha}}$ as required.

Chapter 23

23.1 $F = QE$

F is the force, Q is the charge and E is the electric field, A is the area. Therefore

$$\frac{F}{A} = \frac{Q}{A} E.$$

From elementary electrostatics of parallel plates,

$$\frac{F}{A} = \frac{\kappa \varepsilon_0 V_{\text{r}}}{b} \frac{V_{\text{r}}}{b} = \frac{5.7 \times 8.854 \times 10^{-12} \times \left(70 \times 10^{-3}\right)^2}{\left(5 \times 10^{-9}\right)^2} = 9892 \, \text{N m}^{-2}$$

where ε_0 is the permittivity of free space, k is the dielectric constant, and V_{r} is the voltage drop across the membrane.

23.2 *Case I*

$$R_1 = \frac{\rho l}{A} = \frac{0.5 \times 0.005}{\pi \left(5 \times 10^{-6}\right)^2} = 3.18 \times 10^7 \, \Omega.$$

Case II

$$R_2 = \frac{\rho l}{A_2} = \frac{0.5 \times 0.005}{\pi \left(500 \times 10^{-6}\right)^2} = 3183 \, \Omega.$$

23.3

$$V_{Na} = \frac{kT}{q} \ln \frac{[Na^+]_o}{[Na^+]_i} = 26 \ln \frac{460}{50} = 57 \, mV,$$

$$V_K = \frac{kT}{q} \ln \frac{[K^+]_o}{[K^+]_i} = 26 \ln \frac{10}{400} = -26 \ln \frac{400}{10} = -77 \, mV,$$

$$V_{Cl} = -\frac{kT}{q} \ln \frac{[Cl^-]_o}{[Cl^-]_i} = 26 \ln \frac{540}{70} = -68 \, mV.$$

23.4 The Nernst potentials are calculated at 23 °C. They are
$V_{Na} = 57$ mV,
$V_K = -77$ mV,
$V_{Cl} = -68$ mV.
The resting potential at 23 °C is therefore

$$V_r = \frac{\sum_i g_i V_i}{\sum_i g_i} = \frac{0 \times 57 + 2.7 \times 10^{-6}(-0.077) + 1 \times 10^{-6}(-0.068)}{2.7 \times 10^{-6} + 1 \times 10^{-6}} = -0.075 \, V.$$

Now the equation for the Nernst potential is $V_C = \frac{kT}{q} \ln \frac{C_0}{C_i}$.

The ratio of the two temperatures is $310/296 = 1.047$.
The new resting potential is therefore $1.047 \times (-0.075) = -0.0785$ V.

23.5 The voltage clamp operates by negative feedback. The membrane potential amplifier measures the membrane voltage and sends an output to the feedback amplifier; this subtracts the membrane voltage from the command voltage (the clamped voltage), which it receives from the signal generator. This signal is amplified and the output is sent into the axon via the current electrode.

The clamp circuit produces a current equal and opposite to the ion channel currents. The intracellular electrode measures the average properties of a large number of ion channels in the cell and can be used to accurately measure the conductance at a constant voltage. The currents that flow are proportional to the number of open ion channels (the conductance). Without the voltage-clamp circuit the functional dependence of the conductance on the voltage cannot be easily measured. It also allows subthreshold voltages to be measured without the runaway creation of an action potential.

Patch clamps extend this technique to measure the conductance of single ion channels (although they function with a single electrode rather than two).

23.6 Below the passive subthreshold voltage the potential propagates passively, i.e. there is an exponential decay as a function of position along the axon.

Above the subthreshold voltage a superexponential response is observed in an all-or-nothing nerve impulse. The reaction eventually becomes so pronounced that the membrane's potential is reversed (it becomes positive).The action potential can propagate wave-like along the length of the axon.

Chapter 24

24.1 There are a huge range of nonlinear phenomena in biology (in practice it is often challenging to find a mechanism that is truly linear!)
  Examples:

Electrical activity in the heart,
Action potential in a neuron,
White blood cell counts from a patient with leukaemia (aperiodic oscillations),
Cheyne–Stokes respiration (oscillatory breathing disease),
Pupil size of the eye.

24.2

$$G_1 = \left(\frac{\partial x}{\partial y}\right) = \frac{1}{2}\left(\frac{y-p}{2}\right)^{-1/2}\frac{1}{2},$$

$$G_2 = \left(\frac{\partial y}{\partial x}\right) = -2x,$$

$$\mathrm{OLG} = G_1 G_2 = -\frac{x}{\sqrt{2}}\frac{1}{(y-p)^{1/2}}.$$

24.3

$$G_1 = \frac{\partial y}{\partial x} = -1,$$

$$G_2 = \frac{\partial x}{\partial y} = \frac{1}{6},$$

$$\mathrm{OLG} = G_1 G_2 = -\frac{1}{6}.$$

24.4 Professional musicians have enlarged areas in the brain for motor skills. Taxi drivers have enlarged hippocampuses (regions that handle the storage and retrieval of memories). It is possible to detect whether people will lie about a subject before they speak (in a statistically significant manner).

24.5 Glial cells:

The old textbook answer was they maintain homeostasis, form myelin and provide support/protection for the brain's neurons; 'a glue for the brain'.
  The modern perspective is that they play a much more active role in signalling, e.g. determining synaptic connections, and are involved in waste removal in sleep/waking cycles.

24.6 a. Problems in electrical activity of waves in the heart.
  Bradycardia – slow pacemaker – slow heart rate and can cause cardiac arrest,
  Tachycardia – fast pacemaker – high heart rate and can cause angina,
  Also observe quasiperiodicity and chaos in cardiac fibrillation.
  b. Many varieties of epilepsy. Some types are associated with ion channel mutations. Seizures have specific patterns of oscillatory electrical activity.

24.7 An array of SQUIDs (super-conducting quantum interference device) or SERFs (spin-exchange relaxation free) are used to measure the minute magnetic fields associated with mental activity.

Chapter 25

25.1 There are magnetic resonance imaging, e.g. BOLD – blood oxygen level dependence, positron emission tomography, optical coherence tomography, electroencephalography and magnetoencephalography (MEG) using SQUIDS.

25.2 Snakes can image in the infrared. So can vampire bats, some beetles (e.g. species that need to lay their eggs in burnt trees), butterflies and insects. Insects can image in the ultraviolet. Many flowers make use of this fact. Most birds have 4 types of cone cells (compared with the 3 in apes) which extends their vision into the ultraviolet.

25.3 Adjacent uniform grey stripes appear to have a gradation in their perceived shading. This is due to a mechanism of contrast enhancement performed by neurons in the eye that are attached to multiple rods.

25.4 Benjamin Libet's most famous experiment demonstrates that unconscious electrical processes in the brain (the readiness potential) precedes and possibly causes volitional acts, which are retrospectively felt to be consciously motivated by the subject. The experiment challenges the notion of 'free will'.

25.5 Eric Kandel (Nobel Prize winner in 2000) demonstrated support for Hebbian learning in Aplysia at the single-cell level. He also made many important contributions to the molecular basis of memory.

25.6 The acoustic frequencies are not directly modulated into the spike trains. An area of the ear measures these high frequencies (the amplitude of oscillations at a particular point on the basilar membrane) and it is the amplitude of these readings that is transferred to the brain.

Chapter 26

26.1 $(1\ 0\ 1\ 0\ 1\ 0) \begin{pmatrix} 1 & 1 & 0 & 0 \\ 0 & 0 & 0 & 0 \\ 1 & 1 & 0 & 0 \\ 0 & 0 & 0 & 0 \\ 1 & 1 & 0 & 0 \\ 0 & 0 & 0 & 0 \end{pmatrix} = (3\ 3\ 0\ 0) \rightarrow (1\ 1\ 0\ 0).$

The first two of the output neurons are active in the recalled memory.

26.2 Memories are not stored in single cells, but in the connections between cells. An estimate gives 1 byte of storage per connection (an underestimate, since the connective strength could adopt a range of values). There are 100 trillion connections in an average human brain and thus an estimate of the storage capacity of a human brain is 100 terabytes! More sophisticated calculations are also possible based on more detailed models of neural networks.

26.3 Have a look at the web.

Index

Page numbers in bold refer to the recommended page number(s).

The Physics of Living Processes: A Mesoscopic Approach, First Edition. Thomas Andrew Waigh.
© 2014 John Wiley & Sons, Ltd. Published 2014 by John Wiley & Sons, Ltd.